Mike Holt's Illustrated Guide

Electrical Exam Preparation

Based on the

2002

National Electrical Code®

Includes:
Electrical Theory
National Electrical Code
NEC Calculations
Over 3,000 Practice Questions
and 25 Practice Exams

Mike Holt

NOTICE TO THE READER

Publisher does not warrant or guarantee any of the products described herein or perform any independent analysis in connection with any of the product information contained herein. Publisher does not assume, and expressly disclaims, any obligation to obtain and include information other than that provided to it by the manufacturer.

The reader is expressly warned to consider and adopt all safety precautions that might be indicated by the activities herein and to avoid all potential hazards. By following the instructions contained herein, the reader willingly assumes all risks in connections with such instructions.

The publisher makes no representation or warranties of any kind, including but not limited to, the warranties of fitness for particular purpose or merchantability, nor are any such representations implied with respect to the material set forth herein, and the publisher takes no responsibility with respect to such material. The publisher shall not be liable for any special, consequential, or exemplary damages resulting, in whole or part, from the readers' use of, or reliance upon, this material.

Cover Design: Paul Wright - wright1156@myacc.net

Design, Layout, and Typesetting: Paul Wright - wright1156@myacc.net
Graphic Illustrations: Mike Culbreath

Revised July 5, 2002

Printed July, 2002.

For more information, contact:

Mike Holt Enterprises, Inc.
7310 West McNab Road
Suite 201
Tamarac, Florida 33321

This logo is a registered trademark of Mike Holt Enterprises, Inc.

Library of Congress Cataloging-in-Publication Data
Holt, Charles Michael. Master electrician's exam preparation: electrical theory, National Electrical Code, NEC calculations: contains over 3,000 practice questions/Michael Holt. Includes index. 1. Electric engineering United States Examinations, questions, etc.2. Electricians Licenses United States.3. Electric engineering Problems, exercises, etc.

ISBN: 0-9710307-0-7

I dedicate this book to the Lord Jesus Christ,
my mentor and teacher

Preface

INTRODUCTION

Passing an important electrical exam is the dream of all those who care about improving themselves and their families. Unfortunately, many don't pass the exam at all, and few pass it the first time. The primary reasons that people fail their exam is because they are not prepared on the technical material and/or they don't know how to take the exam.

Most electrical exams contain questions on electrical theory, basic electrical calculations, the *National Electrical Code*, and important and difficult *National Electrical Code* Calculations. This book contains hundreds of illustrations, examples, over 3,000 practice questions covering all of these subjects and 25 practice exams.

The writing style of this book is intended to be informal and relaxed. Graphics in color are included to make the learning easy and fun.

After you've read each unit, you need to take about ten hours to complete the two hundred unit practice questions, and then complete the *NEC* practice exams. If you have difficulty with a question or section in the book, skip it for now and get back to it later. Don't frustrate yourself! The answer key is intended to be very clear, so be sure to review those questions that you miss.

ERRORS IN BOOK

I have taken great care in researching the *Code* rules in this book, but I'm not perfect. If you feel that I've made an error, please let me know by contacting me directly at Mike@MikeHolt.com, or 1-888-NEC CODE (1-888-632-2633). In addition, visit www.NECcode.com for the most current answer key.

HOW TO USE THIS BOOK

Each unit of this book contains objectives, explanations with graphics, examples, steps for calculations, formulas and practice calculations, and 2002

NEC questions and exams. As you read this book, review the author's comments, graphics, and examples with your 2002 *Code* book.

Journeyman's Exam. If you are using this book to pass a Journeyman's Electrician Exam, then you are only required to complete Units 1 through 9 in its entirety, as well as the *NEC* questions and exams in Units 10 through 12.

Other Electrical Exams (Contractors and Inspectors). Complete all units including all practice questions and all exams

> **Note:** This book contains many cross references to other related Code rules. Please, take the time to review the cross references.

As you progress through this book, you will find some formulas, rules or some comments that you don't understand. Don't get frustrated. Highlight the section in the book that you are having a problem with. Discuss it with your boss, inspector, or co-worker, etc., and maybe they'll have some additional feedback. If necessary, just skip the difficult points and get back to them later. Once you've completed the book, review those highlighted sections again and see if you understand.

> **Note.** Some words are italicized to bring them to your attention. Be sure that you understand the terms before you continue with each unit.

DIFFERENT INTERPRETATIONS

Electricians, contractors, some inspectors, and others love arguing *Code* interpretations and discussing *Code* requirements. As a matter of fact, discussing the *NEC* and its application is a great way to increase your knowledge of the *Code* and its intended use. The best way to discuss requirements with others is by referring to a specific Section in the *Code*, rather than by talking in vague generalities.

ABOUT THE AUTHOR

Mike Holt worked his way up through the electrical trade from apprentice electrician to master electrician and electrical inspector. Mike did not complete high school due to circumstances. Realizing that success depends on one's education, he immediately attained his GED, and ten years later he attended the University of Miami's Graduate School for a Master's in Business Administration (MBA). Today Mike continues to stay current in all aspects of his education. Because of his educational experiences, he understands the needs of his students, and strongly encourages and motivates them to continue their own education.

Mike is nationally recognized as one of America's most knowledgeable electrical trainers. He has touched the lives of many thousands of electricians, inspectors, contractors and engineers. His dynamic and animated teaching style is relaxed, direct and fun. Perhaps Mike's best quality is his ability to motivate his students to become successful.

Mike resides in Central Florida, is the father of seven children, and has many outside interests and activities. In 1988, he became the National Barefoot Waterskiing Champion. He has set five national records and, in 1999, he again captured the National Barefoot Waterskiing Championship. Mike enjoys white-water rafting, racquetball, playing his guitar, and spending time with his family and his new sport – Motocross racing (if he could just stop breaking his bones). His commitment to God has helped him develop a lifestyle that balances God, family, career, and self.

ACKNOWLEDGMENTS

I would like to say thank you to all the people in my life who believed in me, even those who didn't. There are many people who played a role in the development and production of this book. I will start with Mike Culbreath (Master Electrician), who helped me transform my words and visions into lifelike graphics. I could not have produced such a fine book without his help.

Next, Paul Wright of Digital Design Group, for the excellent job of electronic production and typesetting.

To my beautiful wife, Linda, and my seven children: Belynda, Melissa, Autumn, Steven, Michael, Meghan, and Brittney, thank you for loving me.

Also thanks to all those who helped me in the electrical industry, the Electrical Construction and Maintenance magazines for my first "big break," and to Joe McPartland, "my mentor," who was there to help and encourage me. Joe, I'll never forget to help others as you've helped me.

In addition, I would like to thank Joe Salimando, the former publisher of Electrical Contractor magazine, and Dan Walters of the National Electrical Contractors Association (NECA) for my second "big break."

Thanks also go to the following industry leaders: Phil Simmons, Brooke Stauffer, Jeff Sargent, More Ode, Joe Ross, Morris Trimmer, James Stallcup, Dick Lloyd, D.J. Clements, John Calloggero, Tony Selvestri, and Marvin Weiss for helping me become a leader and for being special people in my life.

The final personal thank you goes to Sarina, my long-time friend and office manager. Thank you for covering the office for me while I spend so much time writing books, doing seminars, and producing videos and software. Your love and concern for me has helped me through many difficult times.

I would also like to thank the following individuals who reviewed the manuscript and offered invaluable suggestions and feedback. Their assistance is greatly appreciated.

Mel Amundson
Instructor, Amstar
Warrenton, IN

Timothy A. Bartek
Instructor, Greater Altoona Career & Tech Center
Altoona, PA

Robert Cheek
Instructor, Atlanta Electrical Apprenticeship and Training
Atlanta, GA

Scott Davis
Sr. Electrical Systems Specialist
City of Santa Ana Planning & Building Agency

Dennis J. Downer
IBM Electrical Inspector
Morrisville, VT

Douglas Egan
Electrical Engineer
Mundelein, IL

David Anson Forbes
Project Manager, Davco Electrical Contractors Corp.
Lake Worth, FL

Frank Gren, PE
Director of Electrical Engineering
Cleveland, OH

Bob Huddleston
Eastman Chemical Company
Kingsport, TN

Craig H. Matthews, P.E.
Austin Brockenbrough and Associates, L.L.
Chester, VA

Ted W. Kessner
Electrical Contractor/Instructor
Autryville, NC

Gene Reinbout
Instructor, Local 449
Idaho Falls, ID

Wayne Sargent
Electrical Contractor
Silverton, OR

Erik Spielvogel
Project Manager
South Jordan, Utah

Brooke Stauffer
National Electrical Contractors Association
Bethesda, MD

Mike Thaman
Instructor, Rhodes State College
Lima, OH

James L. Thomas
Instructor at James Sprunt Community College
Kenansville, NC

Daniel Turcotte
Electrician
Richford, Vt

Paul V. Westrom
Electrical Instructor, Dunwoody College of Technology
Minneapolis, MN

Ed Williams
Electrical Consultant
Albuquerque, NM

Gary J. Wood
Director, Continuing Education
Associated Builders & Contractors of San Diego

Finally, I would like to thank the following individuals would worked tirelessly to proofread and edit the final stages of this publication: Toni Culbreath, Amanda Mullins, Barbara Parks and Sarina Snow. Their attention to detail and dedication to this project is greatly appreciated.

THE EMOTIONAL ASPECT OF LEARNING

To learn effectively, you must develop an attitude that learning is a process that will help you grow both personally and professionally. The learning process has an emotional as well as an intellectual component that we must recognize. To understand what affects our learning, consider the following:

Positive Image. Many feel disturbed by the expectations of being treated like children and we often feel threatened with the learning experience.

Uniqueness. Each of us will understand the subject matter from different perspectives and we all have some unique learning problems and needs.

Resistance to Change. People tend to resist change and resist information that appears to threaten their comfort level of knowledge. However, we often support new ideas that support our existing beliefs.

Dependence and Independence. The dependent person is afraid of disapproval and often will not participate in class discussion and will tend to wrestle alone. The independent person spends too much time asserting differences and too little time trying to understand others' views.

Fearful. Most of us feel insecure and afraid of learning, until we understand the process. We fear that our performance will not match the standard set by us or by others.

Egocentric. Our ego tendency is to prove someone is wrong, with a victorious surge of pride. Learning together without a win/lose attitude can be an exhilarating learning experience.

Emotional. It is difficult to discard our cherished ideas in the face of contrary facts when overpowered by the logic of others.

HOW TO GET THE BEST GRADE

Studies have concluded that for students to get their best grades, they must learn to get the most from their natural abilities. It's not how long you study or how high your IQ is, it's what you do and how you study that counts the most. To get your best grade, you must make a decision to do your best and follow as many of the following techniques as possible.

Reality. These instructions are a basic guide to help you get the maximum grade. It is unreasonable to think that all of the instructions can be followed to the letter all of the time. Day-to-day events and unexpected situations must be taken into consideration.

Support. You need encouragement in your studies and you need support from your loved ones and employer. To properly prepare for your exam, you need to study 10 to 15 hours per week for about 3 to 6 months.

Communication with Your Family. Good communication with your family is very important because studying every night and on weekends can cause much tension and stress. Try to get their support, cooperation, and encouragement during this difficult time. Let them know the benefits to the family and what passing the exam means. Be sure to plan some special time with them during this preparation period; don't go overboard and leave them alone too long.

Stress. Stress can really take the wind out of you. It takes practice, but get into the habit of relaxing before you begin your studies. Stretch, do a few sit-ups and push-ups, take a 20-minute walk, or a few slow, deep breaths. Close your eyes for a couple of minutes, and deliberately relax the muscle groups that are associated with tension, such as the shoulders, back, neck, and jaw.

Attitude. Maintaining a positive attitude is important. It helps keep you going and helps keep you from getting discouraged.

Training. Preparing for the exam is the same as training for any event. Get plenty of rest and avoid intoxicating drugs, including alcohol. Stretching or exercising each day for at least 10 minutes helps you get in a better mood. Eat light meals such as pasta, chicken, fish, vegetables, fruit, etc. Try to avoid red meat, butter, sugar, salt and high-fat content foods. They slow you down and make you tired and sleepy.

Eye Care. It is very important to have your eyes checked! Human beings were not designed to do constant seeing less than arm's length away. Our eyes were designed for survival, spotting food and enemies at a distance. Your eyes will be under tremendous stress because of prolonged, near-vision reading, which can result in headaches, fatigue, nausea, squinting, or eyes that burn, ache, water, or tire easily.

Be sure to tell your eye doctor that you are studying to pass an exam (bring this book and the *Code* Book with you) and that you expect to do a tremendous amount of reading and writing. Reading glasses can reduce eye discomfort.

Reducing Eye Strain. Be sure to look up occasionally, away from near tasks to distant objects. Your work area should be three times brighter than the rest of the room. Don't read under a single lamp in a dark room. Try to eliminate glare. Mixing of fluorescent and incandescent lighting can be helpful.

Sit up straight, chest up, shoulders back, so both eyes are an equal distance from what you are viewing.

Getting Organized. Our lives are so busy that simply making time for homework and exam preparation is almost impossible. You can't waste time looking for a pencil or missing paper. Keep everything you need together. Maintain folders, one for notes, one for exams and answer keys, and one for miscellaneous items.

It is very important that you have a private study area available at all times. Keep your materials there. The dining room table is not a good spot.

Time Management. Time management and planning are very important. There simply are not enough hours in the day to get everything done. Make a schedule that allows time for work, rest, study, meals, family, and recreation. Establish a schedule that is consistent from day-to-day.

Have a calendar and immediately plan your exam preparation schedule. Try to follow the same routine each week and try not to become overtired. Learn to pace yourself to accomplish as much as you can without the need for cramming.

Clean Up Your Act. Keep all of your papers neat, clean, and organized. Now is not the time to be sloppy. If you are not neat, now is an excellent time to begin.

Speak Up In Class. If you are in a classroom setting, the most important part of the learning process is class participation. If you don't understand the instructor's point, ask for clarification. Don't try to get attention by asking questions you already know the answer to.

Study With A Friend. Studying with a friend can make learning more enjoyable. You can push and encourage each other. You are more likely to study if someone else is depending on you.

Students who study together perform above average because they try different approaches and explain their solutions to each other. Those who study alone spend most of their time reading and rereading the text and trying the same approach time after time even though it is unsuccessful.

Study Anywhere/Anytime. To make the most of your limited time, always keep a copy of the book(s) with you. Any time you get a minute free, study! Continue to study any chance you get. You can study at the supply house when waiting for your material; you can study during your coffee break, or even while you're at the doctor's office. Become creative!

You need to find your best study time. For some, it could be late at night when the house is quiet. For others, it's the first thing in the morning before things get going.

Set Priorities. Once you begin your study, stop all phone calls, TV shows, radio, snacks, and other interruptions. You can always take care of it later.

HOW TO TAKE THE EXAM

Being prepared for an exam means more than just knowing electrical concepts, the *Code*, and the calculations. Have you felt prepared for an exam, then choke when actually taking it? Many good and knowledgeable people didn't pass their exam because they did not know "how to take an exam."

Taking exams is a learned process that takes practice and involves strategies. The following suggestions are designed to help you learn these methods:

Relax. This is easier said than done, but it is one of the most important factors in passing your exam. Stress and tension cause us to choke or forget. Everyone has had experiences where they get tense and couldn't think straight. The first step is becoming aware of the tension, and the second step is to make a deliberate effort to relax. Make sure you're comfortable; remove clothes if you are hot, or put on a jacket if you are cold.

There are many ways to relax and you have to find a method that works for you. Two of the easiest methods that work very well for many people follow:

Breathing Technique: Take a few slow deep breaths every few minutes. Do not confuse this with hyperventilation, which is abnormally fast breathing.

Single-Muscle Relaxation: When we are tense or stressful, many of us do things like clench our jaw, squint our eyes, or tense our shoulders without even being aware of it. If you find a muscle group that does this, deliberately relax that one group. The rest of the muscles will automatically relax also. Try to repeat this every few minutes, and it will help you stay more relaxed during the exam.

Have the Proper Supplies. First of all, make sure you have everything needed several days before the exam. The night before the exam is not the time to be out buying pencils, calculators, and batteries. The night before the exam, you should have a checklist (prepared in advance) of everything you could possibly need.

The following is a sample checklist to get you started:

- Six sharpened #2H pencils or two mechanical pens with extra #2H leads. The type with the larger lead is better for filling in the answer key.

- Two calculators. Most examining boards require quiet, paperless calculators. Solar calculators are great, but there may not be enough light to operate them.

- Spare batteries. Two sets of extra batteries should be taken. It's very unlikely you'll need them, but…

- Extra glasses if you use them.

- A wrist watch to ensure that you stay on track.

- All your reference materials, even the ones not on the list. Let the proctors tell you which ones are not permitted.

- Bring something to drink. Coffee is excellent.

- Some fruit, nuts, aspirin, etc.

- Know where the exam is going to take place and how long it takes to get there. Arrive at least 30 minutes early.

> *Note.* It is also a good idea to pack a lunch rather than going out. It can give you a little extra time to review the material for the afternoon portion of the exam, and it reduces the chance of coming back late.

Understand the Question. To answer a question correctly, you must first understand the question. One word in a question can totally change the meaning of it. Carefully read every word of every question. Underlining key words in the question will help you focus.

Skip the Difficult Questions. Contrary to popular belief, you do not have to answer one question before going on to the next one. The irony is that the question you get stuck on is one that you'll probably get wrong. This will result in not having enough time to answer the easy questions. You will get all stressed-out worrying that you will not complete the exam on time, and a chain reaction is started.

More people fail their exams this way than for any other reason.

The following strategy should be used to avoid getting into this situation:

- **First Pass:** Answer the questions you know. Give yourself about 30 seconds for each question. If you can't find the answer in your reference book within the 30 seconds, go on to the next question. Chances are that you'll come across the answers while looking up another question. The total time for the first pass should be 25 percent of the exam time.

- **Second Pass:** This pass is done the same as the first pass except that you allow a little more time for each question, about 60 seconds. If you still can't find the answer, go on to the next one. Don't get stuck. Total time for the second pass should be about 30 percent of the exam time.

- **Third Pass:** See how much time is left and subtract 30 minutes. Spend the remaining time equally on each question. If you still haven't answered the question, it's time to make an educated guess. Never leave a question unanswered.

- **Fourth pass:** Use the last 30 minutes of the exam to transfer your answers from the exam booklet to the answer key. Read each question and verify that you selected the correct answer on the test book. Transfer the answers carefully to the answer key. With the remaining time, see if you can find the answer to those questions you guessed at.

Guessing. When time is running out and you still have unanswered questions, GUESS! Never leave a question unanswered.

You can improve your chances of getting a question correct by the process of elimination. When one of the choices is "none of these," or "none of the above," it is usually not the correct answer. This improves your chances from one-out-of-four (25 percent), to one-out-of-three (33 percent). Guess "all of these" or "all of the above," and don't select the high or low number.

How do you pick one of the remaining answers? Some people toss a coin, others will count up how many of the answers were A's, B's, C's, and D's and use the one with the most as the basis for their guess.

CHECKING YOUR WORK

The first thing to check (and you should be watching out for this during the whole exam) is to make sure you mark the answer in the correct spot. People have failed the exam by one-half of a point. When they reviewed their exam, they found they correctly answered several questions on the test booklet, but marked the wrong spot on the exam answer sheet. They knew the answer was "(b) False" but marked in "(d)" in error.

Another thing to be very careful of, is marking the answer for, let's say question 7, in the spot reserved for question 8.

CHANGING ANSWERS

When re-reading the question and checking the answers during the fourth pass, resist the urge to change an answer. In most cases, your first choice is best and if you aren't sure, stick with the first choice. Only change answers if you are sure you made a mistake. Multiple choice exams are graded electronically so be sure to thoroughly erase any answer that you changed. Also erase any stray pencil marks from the answer sheet.

ROUNDING OFF

You should always round your answers to the same number of places as the exam's answers.

Example: If an exam has multiple choice of:

(a) 2.2   (b) 2.1
(c) 2.3   (d) none of these

And your calculation comes out to 2.16, do not choose the answer (d) none of these. The correct answer is (a) 2.2, because the answers in this case are rounded off to the tenth.

Example: It could be rounded to tens, such as:

(a) 50   (b) 60
(c) 70   (d) none of these.

For this group, an answer such as 67 would be (c) 70, while an answer of 63 would be (b) 60. The general rule is to check the question's choice of answers then round off your answer to match it.

SUMMARY

- Make sure everything is ready and packed the night before the exam.
- Don't try to cram the night before the exam – if you don't know it by then, it's too late!
- Have a good breakfast. Get the thermos and energy snacks ready.
- Take all your reference books. Let the proctors tell you what you can't use.
- Know where the exam is to be held and arrive early.

- Bring identification and your confirmation papers from the license board if this is required.
- Review your *NEC* while you wait for your exam to begin.
- Try to stay relaxed.
- Determine the time per question for each pass and don't forget to save 30 minutes for transferring your answers to the answer key.
- Remember – in the first pass answer only the easy questions. In the second pass, spend a little more time per question, but don't get stuck. In the third pass, use the remainder of the time minus 30 minutes. In the fourth pass, check your work and transfer the answers to the answer key.

THINGS TO BE CAREFUL OF

- Don't get stuck on any one question.
- Read each question carefully.
- Be sure you mark the answer in the correct spot on the answer sheet.
- Don't get flustered or extremely tense.

The National Electrical Code

The *NEC* is to be used by experienced persons having an understanding of electrical terms, theory, safety procedures, and trade practices. These individuals include electrical contractors, electrical inspectors, AHJ's (Authorities Having Jurisdiction), electrical engineers, designers, and qualified persons. The *Code* was not written to serve as an instructive or teaching manual [90.1(C)] for untrained individuals.

Learning to use the *NEC* is somewhat like learning to play chess. You first must learn the names for the game pieces, how each piece moves and how the pieces are placed on the board. Once you understand these fundamentals, you're ready to start playing the game, but all you can do is make crude moves because you really don't understand what you're doing. To play the game well, you'll need to study the rules, learn subtle and complicated strategies, and then practice, practice, practice.

Electrical work is similar in some ways, though it's certainly not a game; it must be taken very seriously. Learning the terms, concepts, and basic layout of the *NEC* gives you just enough knowledge to be dangerous.

There are thousands of specific and sometimes unique applications of electrical installations and the *Code* does not have a rule for every one of them. To properly apply the *NEC*, you must understand how each rule affects not only the electrical, but also the safety aspects of the installation. I suggest that you have a copy of the 2002 *NEC* nearby, as I will make specific references to it throughout this text.

NEC TERMS AND CONCEPTS

The *NEC* contains many technical terms, so it is crucial that the user understand their meanings. If you do not understand a term used in a *Code* rule, it will be next to impossible to properly apply the rule.

It's not only the technical words that require close attention, because even the simplest of words can make a big difference in the intent of a rule. The word "or" can imply alternate choices for equipment wiring methods, while "and" can mean an additional requirement.

> **Note:** *Electricians, engineers, and other trade-related professionals use terms and phrases (slang or jargon) that are not used in the NEC. The problem with slang terms is that not everybody understands their meaning and it can lead to additional confusion.*

Understanding the safety-related concepts behind many of the *NEC* rules requires one to have a good foundation of electrical fundamentals. For example:

• How is it possible that a bird can sit on an energized power line safely?

• Why can't a single-circuit conductor be installed within a metal raceway?

• Why must we reduce the ampacity of current-carrying conductors if we install four or more in a raceway?

• Why does one section of the *NEC* require a 40A circuit breaker to protect a 12 AWG conductor, yet in another it allows a 20A circuit breaker?

• Why are bonding jumpers sometimes required for metal raceways containing 480Y/277V circuits, but not for 120/240V circuits?

If you have a good understanding of alternating current and impedance fundamentals, you'll find understanding and properly applying the requirements of the *NEC* much easier.

THE NEC STYLE

Contrary to popular belief, the *NEC* is fairly well-organized, although understanding the *NEC* structure and writing style is extremely important to learn how to use the *Code* effectively. *The National Electrical Code* is organized into 11 components:

1. **Table of Contents**
2. **Chapters (major categories)**
3. **Articles (individual subjects)**
4. **Parts (divisions of an Article)**
5. **Sections, Lists, and Tables (Code rules)**
6. **Exceptions (Code rules)**
7. **Fine Print Notes (explanatory material, not mandatory Code language)**
8. **Definitions (Code rules)**
9. **Marginal Notations, Code changes (|)**
10. **Index**
11. **Annexes**

1. Table of Contents. The Table of Contents (pages 2 through 8) displays the layout of the Chapters, Articles, and Parts as well as their page numbers.

2. Chapters. There are nine chapters, each of which contains chapter-specific articles. The nine chapters fall into four groupings, general requirements, specific requirements, communications systems, and tables.

General Requirements: Chapters 1 through 4
Chapter 1. General
Chapter 2. Wiring and Protection
Chapter 3. Wiring Methods and Materials
Chapter 4. Equipment for General Use
Specific Requirements and Conditions: Chapters 5 through 7
Chapter 5. Special Occupancies
Chapter 6. Special Equipment
Chapter 7. Special Conditions
Chapter 8. Communications Systems (Telephone, Radio/Television, and Cable TV Systems.)
Chapter 9. Tables (Conductor and Raceway Specifications and Properties.)

3. Articles. The *NEC* contains approximately 140 Articles, each of which covers a specific subject, for example:

Article 110. General Requirements
Article 250. Grounding
Article 300. Wiring Methods
Article 430. Motors
Article 500. Hazardous (classified) Locations
Article 680. Swimming Pools
Article 725. Class 1, Class 2, and Class 3 Remote-Control, Signaling, and Power-Limited Circuits
Article 800. Communications Wiring

4. Parts. Many articles are subdivided into Parts. For example, Article 250 is so large that it has been divided into nine parts, such as:

Part I. General
Part II. Circuit and System Grounding
Part III. Grounding Electrode System

> **CAUTION:** *The "Parts" of a Code article are not included in the section numbers. Because of this, we have a tendency to forget what "Part" the Code rule is relating to. For example, Table 110.34 gives the working-space clearances in front of electrical equipment. If we aren't careful,*

we might think this table applies to all electrical installations, but Table 110.34 is contained in Part III, Over 600V, Nominal of Article 110. The rule for working clearance in front of electrical equipment for systems under 600V is in Table 110.26, which is Part II, 600V, Nominal, or less.

5. Sections, Lists, and Tables.

Sections. Each *Code* rule is called a Section, such as Section 225.26. A *Code* section may be broken down into subsections by letters in parentheses, such as (A), and numbers in parentheses may further break down each subsection and the third level to lower-case letters. For example, the rule requiring all receptacles in a dwelling-unit bathroom to be GFCI-protected is contained in 210.8(A)(1).

> **Note:** *Many in the electrical industry incorrectly use the term "Article" when referring to a Code section. For example, they will say "Article 210.8," when they should be saying "Section 210.8."*

From this point on in the book, I will use section references without using the word "Section" to follow the 2002 format. For example, "Section 250.118" will be referred to as just "250.118." Where I use a section number in text as a reference, it may be inside brackets, for example Section 90.2(A)(4) would be [90.2(A)(4)].

Lists. If a rule contains a list of items, then the items are identified as (1), (2), (3), (4), etc. See 250.118. If a list is part of a numeric subsection, then the items are listed as a., b., c., etc. See 250.114(4).

Tables. Many *Code* requirements are contained within Tables, which are lists of *Code* rules placed in a systematic arrangement.

Many times notes are provided with each table; be sure to read them, as they are also part of the *Code*. For example, Note 1 for Table 300.5 explains how to measure the cover when burying cables and raceways, and Note 5 explains what to do if solid rock is encountered.

6. Exceptions. Exceptions are *Code* rules that provide an alternative method to a specific requirement. There are two types of exceptions: mandatory and permissive. When a rule has several exceptions, those exceptions with mandatory requirements are listed before the permissive exceptions.

Mandatory Exception. A mandatory exception uses the words "must" or "must not." The word "must" in an exception means that if you are using the exception, you are required to do it in a particular way. The term "must not" means it is not permitted; for an example, see 230.24(A) Ex. 1.

Permissive Exception. A permissive exception uses such words as "must be permitted," which means that it is acceptable to do it in this way. See 230.24(A) Ex. 2.

7. Fine Print Note (FPN). A Fine Print Note contains explanatory material intended to clarify a rule or give assistance, but it is not a *Code* requirement. FPNs often use the word 'may,' but never "must." See 90.1(B) FPN.

8. Definitions. Definitions of specific words and terms are listed in Article 100, within some articles, and parts of some articles.

Article 100. Definitions are listed in Article 100 and throughout the *NEC*. In general, the definitions listed in Article 100 apply to more than one *Code* article, such as 'Branch Circuit,' which is used in many articles.

Definitions contained in Articles. Definitions at the beginning of an article apply only to that specific article. For example, the definition of "Swimming Pool"' is contained in 680.2 because this term applies only to the requirements contained in Article 680 – Swimming Pools.

Part. Definitions located in a Part of an article apply only to that part of the article. For example, the definition of "motor control circuit" at 430.71 applies only to the rules in Part VI, Motor Control Circuits of Article 430. When a definition is listed in the general definitions of an article, it can be correctly applied to the entire article, but when it is found within a part of an article, it is to be correctly applied only to that part of the article in which it is found.

Section. Definitions located in a Section apply only to that section. For example, the definition of a "hallway" in 210.52(H) only applies to that section.

9. Changes. Changes to the *NEC* are identified in the margins of the 2002 *NEC* by a vertical line (I). For example, see 90.1(D) Relation to International Standards.

Many rules in the 2002 *NEC* were relocated. The place from which the *Code* rule was removed has no identifier and the place where the rule was inserted has no identifier. For example, the rules for receptacles that were located in 210-7 of the 1999 *NEC* were relocated to 406.3, without any identifying marks.

10. Index. The 30-page index in the 2002 *NEC* (pages 681 through 711) is excellent and should be very helpful to locate the rule in question.

11. Annexes. Annexes are not a part of the *NEC*, and are included for informational purposes only. They include the following:

Annex A. Product Safety Standards
Annex B. Conductor Ampacity Under Engineering Supervision
Annex C. Raceway Fill Tables
Annex D. Examples
Annex E. Types of Construction
Annex F. Cross-Reference Tables

HOW TO LOCATE SPECIFIC MATERIAL IN THE CODE

How you go about finding what you are looking for in the *NEC* depends to some degree on your experience with it. Many experienced *Code* users will consult the Table of Contents instead of the Index to locate specific information.

For example, what *Code* rule specifies the maximum number of disconnects permitted for a service? If you're an experienced *Code* user, you'll know that Article 230 applies to "Services" and it contains a Part "VI. Disconnection Means." With this knowledge, you can quickly go to page 2 of the Table of Contents and see that the rule is on page 70-81.

You might wonder why the number 70 precedes all of the page numbers. This is because the NFPA has numbered all of its standards, and the *NEC* is standard number 70. Since I will be referencing only the *NEC*, I will drop the 70 in front of the page numbers.

If you used the Index (subjects listed in alphabetical order) and looked up the term "service disconnect," you will see that there is no such listing. If you tried "disconnecting means," then "services," you would find that the Index specifies that the rule is at 230-VI! Because the *NEC* does not give a page number in the Index, you'll need to use the Table of Contents to get the page number, or flip through the page of Article 230 until you find Part VI, which is on page 81.

Many people complain that the *Code* only confuses them by taking them in circles. As you gain experience in using the *Code* and deepen your understanding of *Code* words, terms, principles and practices, you will find the *Code* much easier to understand. You may even come to appreciate the value of the multiple section references contained in the articles.

CUSTOMIZING YOUR CODE BOOK

One way for you to increase your comfort level with the *Code* book is to customize it to meet your needs. You can do this by highlighting and underlining important *Code* rules, and by attaching convenient page tabs.

Highlighting. As you read this book, be sure you highlight those rules in the *Code* that are important to you. Use yellow for general interest and orange for important rules you want to find quickly. Be sure to highlight terms in the Index and Table of Contents as you use them.

Because of the size of the 2002 *NEC*, I strongly recommend that you highlight in green the parts of larger articles that are important for your application, particularly:

Article 230. Services
Article 250. Grounding
Article 410. Luminaires
Article 430. Motors
Trust me; you'll be glad you did!

Underlining. Underline or circle key words and phrases in the *NEC* with a red pen (not a lead pencil) and use a 6-inch ruler to keep lines straight and neat. This is a very handy way to have important words or terms stand out. A small 6-inch ruler also comes in handy when trying to locate specific information in the many tables in the *Code*.

Tabbing the NEC. Placing tabs on important *Code* articles, sections, and tables will make it very easy to access important *Code* rules. However, too many tabs will defeat the purpose. You can order a set of *Code* tabs designed by Mike Holt online at www.mikeholt.com or by phone at 1-888-NEC CODE (1-888-632-2633).

GLOBAL CHANGES TO THE 2002 NEC

Some changes were made in the *NEC* to facilitate its international adoption.

Format – The format system for the 2002 *NEC* was changed from the 'hyphen' system to the scientific method of notation typically called the 'dot' system. For example, Section 110-26 will now be 110.26. The first letter will be capitalized and the following characters will be lower case. For example: 110.26(A), 110.26(A)(1), 110.26(A)(1)(a).

New Articles – Many new articles were added, for example:

Article 80. Administration and Enforcement
Article 285. Transient Voltage Surge Suppressors (TVSSs)
Article 406. Receptacles, Cord Connectors, and Attachment Plugs (Caps)
Article 647. Sensitive Electronic Equipment
Article 692. Fuel Cells

Metric Units – Units of measurement are now listed with the metric unit first, followed by the foot/pound measurement in parentheses. See 90.9.

Parts – Article parts in the 1999 *NEC* were identified by an upper case letter, such as Part A, Part B, and Part C. The 2002 *NEC* identifies article parts by a Roman numeral, such as Part I, Part II, and Part III.

Sections Relocated – Many *Code* sections were relocated. For example, the requirements for receptacles contained in Articles 210 and 410 were moved to the new Article 406 covering the requirements for receptacles only.

To achieve a uniform numbering system, all wiring method articles had their sections renumbered. For example, the section numbers for all raceways will be as follows:

I General

3XX.1 Scope

3XX.2 Definitions

3XX.3 Other Articles

3XX.6 Listing Requirements

II Installation

3XX.10 Uses Permitted

3XX.12 Uses Not Permitted

3XX.14 Dissimilar Metals

3XX.16 Temperature Limits

3XX.20 Size

3XX.22 Number of Conductors

3XX.24 Bends–How Made

3XX.26 Bends–Number in One Run

3XX.28 Reaming and Threading (Metal)

3XX.28 Trimming (Nonmetallic)

3XX.30 Securing and Supporting

3XX.40 Boxes and Fittings

3XX.42 Coupling and Connectors

3XX.44 Expansion Fittings

3XX.46 Bushings

3XX.48 Joints

3XX.50 Conductor Terms

3XX.56 Splices and Taps

3XX.60 Grounding

AUTHOR'S COMMENT: 3XX represents the actual article number, which could be Articles: 320 – Armored Cable, 334 – Nonmetallic-Sheathed Cable, 348 – Flexible Metal Conduit, or 358 – Electrical Metallic Tubing.

Exceptions – Many exceptions were rewritten into complete sentences and converted into positive text.

New Definitions – New definitions were added to Article 100 as well as other articles. For example, the term "luminaire" replaces "fixture," and "lighting fixture."

While the *NEC* does have a specific article titled definitions, it is not intended to be a dictionary. The official dictionary of the NFPA is the IEEE dictionary. When a technical definition is needed, only an approved technical dictionary such as the IEEE dictionary should be consulted.

Wire Identifiers – The common abbreviation 'No.' is no longer used to describe conductor size throughout the *NEC*. For example, a No. 8 is now identified as 8 AWG.

Article Numbers – The article numbers for all wiring methods were renumbered.

New	Old	
312	373	Cabinets and Cutout Boxes
314	370	Outlet, Device, Pull, and Junction Boxes; Conduit Bodies; Fittings; and Manholes
320	333	Armored Cable – Type AC
324	328	Flat Conductor Cable – Type FCC
330	334	Metal Clad Cable – Type MC
332	330	Mineral Insulated Cable – Type MI
334	336	Nonmetallic-Sheathed Cable – Type NM
336	340	Tray Cable – Type TC
338	338	Service Entrance Cable – Type SE
340	339	Underground Feeder Cable – Type UF
342	345	Intermediate Metal Conduit – IMC
344	346	Rigid Metal Conduit – RMC
348	350	Flexible Metal Conduit – FMC
350	351A	Liquidtight Flexible Metal – LFMC
352	347	Rigid Nonmetallic Conduit – RNC
354	343	Nonmetallic Underground Conduit with Conductors – NUCC
356	351B	Liquidtight Nonmetallic Conduit – LFNC
358	348	Electrical Metallic Tubing – EMT
362	331	Electrical Nonmetallic Tubing – ENT
366	374	Auxiliary Gutters
368	364	Busways
370	365	Cablebus
372	358	Cellular Concrete Raceway
374	356	Cellular Metal Floor Raceway
376	362A	Metal Wireways
378	362B	Nonmetallic Wireways
380	353	Multioutlet Assembly
382	342	Nonmetallic Extensions
384	352C	Strut-Type Channel Raceways
386	352A	Surface Metal Raceways
388	352B	Surface Nonmetallic Raceways
390	354	Underfloor Raceways
392	318	Cable Trays
394	324	Concealed Knob-and-Tube
396	321	Messenger Supported Wiring
398	320	Open Wiring on Insulators
527	305	Temporary Installations

Instructors who would like a review copy of this book
are invited to contact us directly at:

1-877-NEC CODE
Fax: 1-954-720-7944

Mike Holt Online
www.NECcode.com
Sales@mikeholt.com

CHAPTER 1
Electrical Theory and Code Questions

Scope of Chapter 1

Unit 1   Electrician's Math and Basic Electrical Formulas

Unit 2   Electrical Circuits

Unit 3   Understanding Alternating Current

Unit 4   Motors and Transformers

Unit 1

Electrician's Math and Basic Electrical Formulas

OBJECTIVES

After reading this unit, the student should be able to briefly explain the following concepts:

Part A - Electrician's Math
Fractions
Kilo
Knowing your answer
Multiplier
Parentheses
Percent increase
Percentage

Reciprocals
Square root
Squaring
Transposing formulas
Part B - Basic Electrical Formulas
Conductance and resistance
Electric meters
Electrical circuit

Electrical circuit values
Ohm's law
PIE circle formula
Power changes with
  the square of the
  voltage
Power source
Power wheel

After reading this unit, the student should be able to briefly explain the following terms:

Part A - Electrician's Math
Fractions
Kilo
Multiplier
Parentheses
Percentage
Ratio
Reciprocals
Rounding off
Square root
Squaring a number
Transposing
Part B - Basic Electrical Formulas
A – Ampere
Alternating current
Ammeter

Ampere
Armature
Clamp-on ammeters
Conductance
Conductors
Current
Direct current
Directly proportional
E – Electromotive Force
Electric meters
Electromagnetic
Electromagnetic field
Electron pressure
Helically wound
Intensity
Inversely proportional

Megohmmeter
Ohmmeter
Ohms
P – Power
Perpendicular
Polarity
Polarized
Power
Power source
Resistance
Shunt bar
Shunt meter
Solenoid
V – Voltage
Voltmeter
W – Watt

PART A – ELECTRICIAN'S MATH

1–1 FRACTIONS

Fractions represent a part of a number. To change a fraction to a decimal form, divide the numerator (top number of the fraction) by the denominator (bottom number of the fraction).

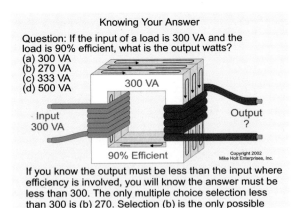

Knowing Your Answer

Question: If the input of a load is 300 VA and the load is 90% efficient, what is the output watts?
(a) 300 VA
(b) 270 VA
(c) 333 VA
(d) 500 VA

300 VA

Input
300 VA

Output
?

90% Efficient

Copyright 2002
Mike Holt Enterprises, Inc.

If you know the output must be less than the input where efficiency is involved, you will know the answer must be less than 300. The only multiple choice selection less than 300 is (b) 270. Selection (b) is the only possible answer and no calculation is necessary.

Figure 1–1
Knowing Your Answer

Converting Percentages to Decimals

Percentage	Drop "%"	Decimal
32.5%	.32.5%	0.325

Move the decimal point two places to the left.
Copyright 2002 Mike Holt Enterprises, Inc.

Figure 1–2
Converting Percentages to Decimals

❏ Fractions to Decimal

Convert the following fractions to a decimal.

$1/6$ = one divided by six = 0.167

$5/4$ = five divided by four = 1.25

$7/2$ = seven divided by two = 3.5

1–2 KILO

The letter "k" is the abbreviation of *kilo,* which represents 1,000 in the metric system of measurement.

❏ Kilo

What is the wattage for an 8 kW rated range?

(a) 8W   (b) 8,000W   (c) 4,000W   (d) none of these

• Answer: (b) 8,000W

Wattage = kW × 1,000. In this case 8 kW × 1,000 = 8,000W.

❏ Kilo

What is the kVA rating of a 300 VA load?

(a) 300 kVA   (b) 3,000 kVA   (c) 30 kVA   (d) 0.3 kVA

• Answer: (d) 0.3 kVA, kVA = VA divided by 1,000

In this case 300 VA/1,000 = 0.3 kVA

Note: The use of "k" is not limited to kW or kVA, it is also used to express wire size, such as 250 kcmil (250,000 cm).

1–3 KNOWING YOUR ANSWER

When working with mathematical calculations you should know if the answer is greater than, or less than, the values given.

❏ Knowing Your Answer

If the input of a load is 300 VA and the load is 90% efficient, what is the output VA?

Note: Because efficiency is always less than 100%, the output is always less than the input. Figure 1–1.

(a) 300 VA   (b) 270 VA   (c) 333 VA   (d) 500 VA

• Answer: (b) 270 VA

Since the use of efficiency in the question implied that the output had to be less than the input, the answer must be less than 300 VA. The only choice that is less than 300 VA is (b) 270 VA.

1–4 MULTIPLIER

Often a number is required to be increased or decreased by a percentage. When a percentage or fraction is used as a multiplier, follow these steps:

Step 1:   Convert the multiplier to a decimal form, then

Step 2:   Multiply the number by the decimal value from Step 1.

❏ Increase by 125 Percent

An overcurrent protection device (breaker or fuse) shall be sized no less than 125% of the continuous load. If the load is 80A, the overcurrent protection device would have to be sized no less than _____.

(a) 80A   (b) 100A   (c) 125A   (d) none of these
    • Answer: (b) 100A

Step 1:   Convert 125% to a decimal: 1.25.

Step 2:   Multiply the load rating by the multiplier: 80A × 1.25 = 100A.

❏ Limit to 80 Percent

The maximum continuous load on an overcurrent protection device is limited to 80% of the device rating. If the device is rated 50A, what is the maximum continuous load?

(a) 80A   (b) 125A   (c) 50A   (d) 40A
    • Answer: (d) 40A

Step 1:   Convert 80% to a decimal: 0.8.

Step 2:   Multiply the overcurrent protection device rating by the multiplier: 50A × 0.8 = 40A.

1–5 PARENTHESES

Whenever numbers are in *(parentheses)*, we must complete the mathematical function within the parentheses before proceeding with the rest of the problem.

❏ Parentheses

What is the voltage drop of two 14 AWG conductors carrying 16A for a distance of 100 ft? Use the following example for the answer:

$$VD = \frac{(2 \text{ wires} \times 12.9 \ \Omega \times 16 \text{ A} \times 100 \text{ ft})}{4{,}110 \text{ circular mils}}$$

(a) 3V   (b) 3.6V   (c) 10.04V   (d) none of these
    • Answer: (c) 10.04V

Step 1:   Calculate the value of the parentheses first: (2 wires × 12. 9Ω × 16A × 100 ft) = 41,280.

Step 2:   Divide the top number by the bottom number: $\frac{41{,}280}{4{,}110} = \frac{10.4}{VD}$

1–6 PERCENTAGES

A *percentage (%)* is a ratio of two numbers. When changing a percent to a decimal or whole number, simply move the decimal point two places to the left. Figure 1–2.

❏ Percentage

32.5% = 0.325   100% = 1.00   125% = 1.25   300% = 3.00

1–7 PERCENT INCREASE

Increasing a number by a specific percentage is accomplished by:

Step 1:   Converting the percentage to a decimal.

Step 2:   Determining the multiplier by adding one to the decimal value from Step 1.

Step 3:   Multiplying the number by the multiplier from Step 2.

❏ **Percent Increase**

Increase the whole number 45 by 35%.

(a) 61   (b) 74   (c) 83   (d) 104

     • Answer: (a) 61

Step 1:   Convert 35% to 0.35.

Step 2:   Multiplier: Add one to the decimal value from Step 1: $1 + 0.35 = 1.35$.

Step 3:   Multiply the number 45 by the multiplier: $45 \times 1.35 = 60.75$.

Step 4:   Round 60.75 up to 61.

1–8 PERCENTAGE RECIPROCALS

A reciprocal is a whole number converted into a fraction, with the number *one* as the numerator (top number). This fraction is then converted to a decimal.

Step 1:   Convert the number to a decimal.

Step 2:   Divide the number into one.

❏ **Reciprocal**

What is the reciprocal of 80%?

(a) 0.80%   (b) 100%   (c) 125%   (d) none of these

     • Answer: (c) 1.25 or 125%

Step 1:   Convert 80% to a decimal: $80\% = 0.8$.

Step 2:   Divide 0.8 into one: $^1/_{0.8} = 1.25$, which is the same as 125%.

❏ **Reciprocal**

A continuous load requires an overcurrent protection device sized no smaller than 125% of the load [210.20(A)]. What is the maximum continuous load permitted on a 100A overcurrent protection device?

(a) 100A   (b) 125A   (c) 80A   (d) none of these

     • Answer: (c) 80A

Step 1:   Convert 125% to a decimal: $125\% = 1.25$.

Step 2:   Divide 1.25 into one: $^1/_{1.25} = 0.8$ or 80%.

         If the overcurrent device is sized no less than 125% of the load, the load is limited to 80% of the overcurrent protection device rating (reciprocal). Therefore, the maximum load is limited to: $100A \times 0.8 = 80A$.

1–9 ROUNDING

Numbers below five are rounded down, while numbers five and above are rounded up.

Note: Rounding to three significant figures should be sufficient for most calculations, such as:

0.1245   = 0.125
1.674   = 1.67
21.94   = 21.9
367.28 = 367

Rounding for Exams

You should always round your answer to the same magnitude as the answers. Do not choose "none of these" in an exam until you have checked all the answers. If after rounding your answer to the exam format and there is no answer, then you should choose "none of these."

❏ **Rounding**

The sum of 12, 17, 28, and 40 is equal to _____.

(a) 80 　　　　　　(b) 90 　　　　　　(c) 100 　　　　　　(d) none of these

• Answer: (c) 100

The answer is actually 97, but there is no 97 as a choice. Do not choose "none of these" in an exam until you have checked how the choices are rounded off. The choices in this case are all rounded off to the nearest ten.

1–10 SQUARING

Squaring a number is multiplying a number by itself, such as: $23^2 = 23 \times 23 = 529$.

❏ **Squaring**

What is the power consumed, in watts, of a 12 AWG conductor that is 200 ft long and has a resistance of 0.4Ω? The current flowing in the circuit is 16A. Formula: $I^2 \times R$

(a) 50W 　　　　　　(b) 150W 　　　　　　(c) 100W 　　　　　　(d) 200W

• Answer: (c) 100W

Step 1:　　$P = I^2 \times R$, $I = 16A$, $R = 0.4Ω$.

Step 2:　　$P = 16A^2 \times 0.4Ω$.

Step 3:　　$P = 102.4W$, answers are rounded to 50's.

❏ **Squaring**

What is the cross-sectional area in square inches of a 1 in. trade-size raceway whose actual internal diameter is 1.049 in.? Formula: Area = $\pi \times r^2$, $\pi = 3.14$, r = radius ($\frac{1}{2}$ the diameter).

(a) 1 sq in. 　　　　　　(b) 0.86 sq in. 　　　　　　(c) 0.34 sq in. 　　　　　　(d) 0.5 sq in.

• Answer: (b) 0.86 sq in.

Step 1:　　Raceway area = $\pi \times r^2 = 3.14 \times (\frac{1}{2} \times 1.049)^2 = 3.14 \times 0.5245^2$.

Step 2:　　Raceway area = $3.14 \times (0.5245 \times 0.5245) = 3.14 \times 0.2751 = 0.86$ sq in.

1–11 SQUARE ROOT

The *square root* of a number is the opposite of squaring a number. For all practical purposes, you must use a calculator with a square root key to determine the square root of a number. For an electrician's exam, the only square root number you need to know is the square root of three ($\sqrt{3}$), which is 1.732. To multiply, divide, add, or subtract a number by a square root value, determine the square root value first, then perform the math function.

Step 1:　　**Enter the number in a calculator.**

Step 2:　　**Press the $\sqrt{\ }$ key of the calculator.**

❏ **Square Root Sample**

What is the $\sqrt{3}$?

(a) 1.55 　　　　　　(b) 1.73 　　　　　　(c) 1.96 　　　　　　(d) none of these

• Answer: (b) 1.73

Step 1:　　Enter the number 3 in a calculator.

Step 2:　　Press the $\sqrt{\ }$ key = 1.732.

❏ **Square Root**

36,000W/(208V $\times \sqrt{3}$) is equal to _____.

(a) 120A   (b) 208A   (c) 360A   (d) 100A

• Answer: (d) 100A

Step 1:   Determine the $\sqrt{3} = 1.732$.

Step 2:   Multiply 208V $\times$ 1.732 = 360V.

Step 3:   36,000W/360V = 100A.

❏ **Square Root**

The phase voltage is equal to $\dfrac{208V}{\sqrt{3}}$ _____.

(a) 120V   (b) 208V   (c) 360V   (d) none of these

• Answer: (a) 120V

Step 1:   Determine the $\sqrt{3} = 1.732$.

Step 2:   Divide 208V by 1.732 = 120V.

PART B – BASIC ELECTRICAL FORMULAS

1–12 ELECTRICAL CIRCUIT

An electric circuit consists of power source, conductors, and load. For current to travel in the circuit, there must be a complete path from one terminal of the power supply, through the conductors and the load, back to the other terminal of the power supply. Figure 1–3.

1–13 ELECTRON FLOW

Inside a dc power source (such as a battery) the electrons travel from the positive terminal to the negative terminal; however, outside of the power source, electrons travel from the negative terminal to the positive terminal. Figure 1–4.

1–14 POWER SOURCE

In any completed circuit, it takes a force to push the electrons through the power source, conductor, and load. The two types of electric current are *direct current (dc)* and *alternating current (ac)*.

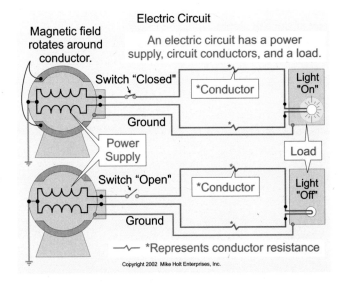

Copyright 2002 Mike Holt Enterprises, Inc.

Figure 1–3

Electric Circuit

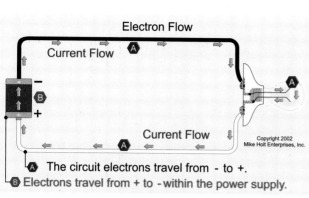

Figure 1–4

Electron Flow

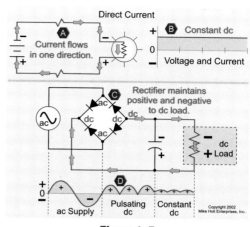

Figure 1–5

Direct Current

Direct Current

The polarity from direct-current power sources never changes; that is, the current flows out of the negative terminal of the power source in the same direction. When the power supply is a *battery,* the polarity and the voltage magnitude remain the same. Figure 1–5.

Alternating Current

Alternating-current power sources produce a voltage and current that has a constant change in polarity and magnitude at a constant frequency. Alternating-current flow is produced by a *generator* or an *alternator.* Figure 1–6.

1–15 CONDUCTANCE AND RESISTANCE

Conductance

Conductance is the property of metal that permits current to flow. The best conductors, in order of their conductivity are: silver, copper, gold, and aluminum. Although silver is a better conductor of electricity than copper, copper is used most widely because it is less expensive. Figure 1–7 Part A.

Resistance

Resistance is the opposite of conductance. It is the property that opposes the flow of electric current. The resistance of a conductor is measured in ohms, according to a standard length of 1,000 ft. Figure 1–7 Part B. This value is listed in the National Electrical Code (NEC), Chapter 9, Table 8, for direct-current (dc) circuits.

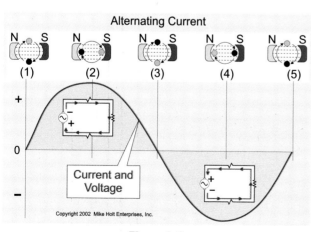

Figure 1–6

Alternating Current

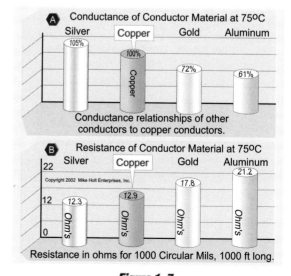

Figure 1–7

Conductance and Resistance

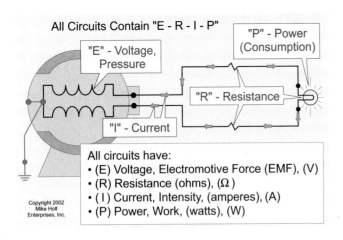

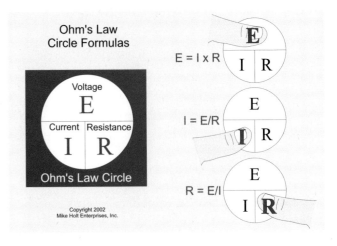

Figure 1–8
Electrical Circuit Values

Figure 1–9
Ohm's Law Circle Formulas

1–16 ELECTRICAL CIRCUIT VALUES

In an electrical circuit, there are four circuit values that must be understood. They are voltage, *resistance, current,* and *power. Figure* 1–8.

Voltage

Electron pressure is called *electromotive force (E or EMF)* and is measured by the unit *volt (V),* abbreviated by the letter *E, EMF* or *V*. Voltage is also a term used to describe the difference of potential between any two points.

Resistance

The friction opposition to the flow of electrons is called *resistance (R),* and the unit of measurement is the *ohm Ω.* Every component of an electric circuit contains resistance, including the power supply.

Current

Free electrons moving in the same direction in a conductor produce an electrical *current* sometimes called *intensity (I).* The rate at which electrons move is measured by the unit called *ampere (A).*

Power

The rate of work that can be produced by the movement of electrons is called *power (P),* and the unit is the *watt (W).*

Note: A 100W lamp consumes 100W of power per hour.

1–17 OHM's LAW I = E/R

Ohm's Law, $I = {}^E\!/_R$, demonstrates the relationship between current, voltage, and resistance in a dc or an ac circuit that supplies only resistive loads. Figure 1–9. Other derived formulas include:

I = E/R   **E = I $\times$ R**   **R = E/I**

Ohm's law states that:

Current is directly proportional to voltage. If the voltage is increased by a given percentage, current increases by that same percentage. If the voltage is decreased by a given percentage, current decreases by the same percentage. Figure 1–10 Part A.

Current is inversely proportional to resistance. An increase in resistance results in a decrease in current. A decrease in resistance results in an increase in current. Figure 1–10 Part B.

Opposition to Current Flow

In a dc circuit, the physical resistance of the conductor opposes the flow of electrons. In an ac circuit, three factors oppose current flow. They are *conductor resistance, inductive reactance*, and *capacitive reactance*. The opposition to current flow, due to a combination of resistance and reactance, is called *impedance.* Impedance is measured in ohms, and is abbreviated with the letter Z. Impedance will be covered later, so for now assume that all circuits have very little or no reactance.

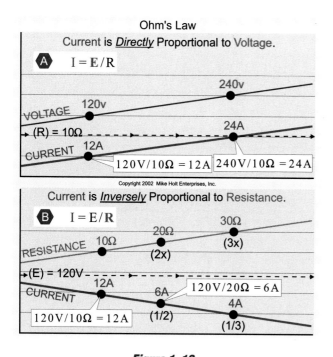

Figure 1–10

Part A – Current Proportional to Voltage

Part B – Current Inversely Proportional to Resistance

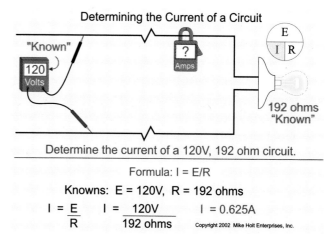

Figure 1–11

Determining the Current

of a Circuit

❏ **Ampere**

A 120V power source supplies a lamp with a resistance of 192 ohms (Ω). What is the current flow of the circuit? Figure 1–11.

   (a) 0.6A   (b) 0.5A   (c) 2.5A   (d) 1.3A

    • Answer: (a) 0.6A

  Step 1:   What is the current (I)?

  Step 2:   What do you know?
             E = 120V, R = 192Ω.

  Step 3:   The formula is: I = E/R.

  Step 4:   I = 120V/192Ω = 0.625A.

❏ **Voltage**

What is the voltage drop of two 12 AWG conductors that supply a 16A load located 50 ft from the power supply? The total resistance of both conductors is 0.2Ω. Figure 1–12.

   (a) 16V   (b) 32V

   (c) 1.6V   (d) 3.2V

    • Answer: (d) 3.2V

  Step 1:   What is voltage drop, (E)?

  Step 2:   What do you know about the conductors?
             I = 16A, R = 0.2Ω

  Step 3:   The formula is: E = I $\times$ R.

  Step 4:   E = 16A $\times$ 0.2Ω = 3.2V.

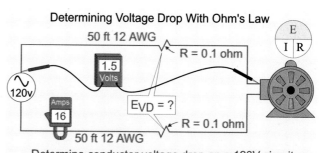

Figure 1–12

Voltage Example

❏ **Resistance**

What is the resistance of the circuit conductors when the conductor voltage drop is 3V and the current flowing through the conductors is 100A? Figure 1–13.

(a) 0.03Ω   (b) 0.2Ω   (c) 3Ω   (d) 30Ω

• Answer: (a) 0.03Ω

Step 1:   What is the resistance, (R)?

Step 2:   What do you know about the conductors? E = 3 VD, I = 100A.

Step 3:   The formula is: R = E/I.

Step 4:   R = 3V/100A = 0.03Ω.

1–18 PIE CIRCLE FORMULA

The PIE circle formula shows the relationship between power, current, and voltage. Figure 1–14.

$$P = E \times I \qquad\qquad I = P/E \qquad\qquad E = P/I$$

❏ **Power**

What is the power loss, in watts, for two conductors that carry 12A and have a voltage drop of 3.6 V? Figure 1–15.

(a) 4.3W   (b) 43W   (c) 432W   (d) none of these

• Answer: (b) 43W

Step 1:   What is the power, (P)?

Step 2:   What do you know? E = 3.6 VD, I = 12A.

Step 3:   The formula is: P = E × I.

Step 4:   The answer is: P = 3.6V × 12A = 43.2W per hour.

❏ **Current**

What is the current flow, in amperes, in the circuit conductors that supply a 7.5 kW heat strip rated 240V when connected to a 240V power supply? Figure 1–16.

(a) 25A   (b) 31A   (c) 39A   (d) none of these

• Answer: (b) 31A

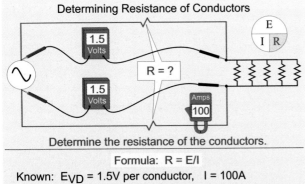

Figure 1–13

Resistance Example

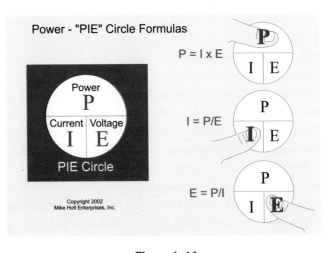

Figure 1–14

Pie Circle Formulas

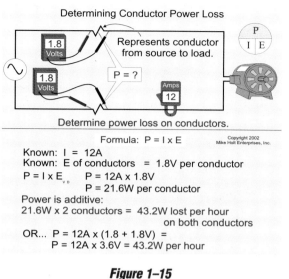

Formula: P = I x E

Known: I = 12A
Known: E of conductors = 1.8V per conductor
P = I x E$_{vd}$ P = 12A x 1.8V
              P = 21.6W per conductor
Power is additive:
21.6W x 2 conductors = 43.2W lost per hour
                           on both conductors
OR... P = 12A x (1.8 + 1.8V) =
        P = 12A x 3.6V = 43.2W per hour

Figure 1–15

Determining Conductor Power Loss

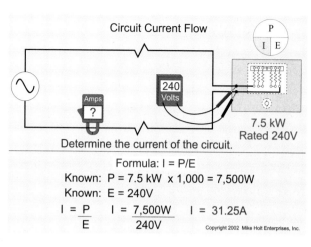

Formula: I = P/E

Known: P = 7.5 kW x 1,000 = 7,500W
Known: E = 240V

$I = \dfrac{P}{E}$ $I = \dfrac{7,500W}{240V}$ I = 31.25A

Copyright 2002 Mike Holt Enterprises, Inc.

Figure 1–16

Circuit Current Flow

Step 1:   What is the question? What is the current, (I)?

Step 2:   What do you know? P = 7,500W, E = 240V.

Step 3:   The formula is: I = P/E.

Step 4:   I = 7,500W/240V = 31.25A.

1–19 FORMULA WHEEL

The formula wheel combines the Ohm's Law and the PIE formulas. The formula wheel is divided up into four sections with three formulas in each section. Figure 1–17.

❑ **Resistance**

What is the resistance of a 75W light bulb that is rated 120V? Figure 1–18.

(a) 100Ω   (b) 192Ω   (c) 225Ω   (d) 417Ω

• Answer: (b) 192Ω

R = E^2/P (Formula 1 of 12), E = 120V rating, P = 75W rating, R = 120 V^2/75W = 192Ω

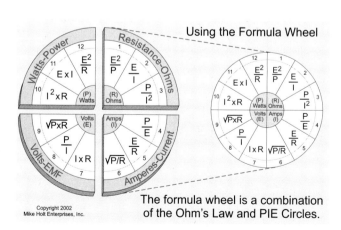

Using the Formula Wheel

The formula wheel is a combination of the Ohm's Law and PIE Circles.

Copyright 2002
Mike Holt Enterprises, Inc.

Figure 1–17

Using the Formula Wheel

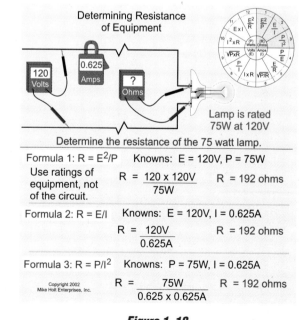

Determining Resistance of Equipment

Determine the resistance of the 75 watt lamp.

Formula 1: R = E^2/P   Knowns: E = 120V, P = 75W
Use ratings of
equipment, not $R = \dfrac{120 \times 120V}{75W}$ R = 192 ohms
of the circuit.

Formula 2: R = E/I   Knowns: E = 120V, I = 0.625A
                      $R = \dfrac{120V}{0.625A}$ R = 192 ohms

Formula 3: R = P/I^2 Knowns: P = 75W, I = 0.625A
                      $R = \dfrac{75W}{0.625 \times 0.625A}$ R = 192 ohms

Copyright 2002
Mike Holt Enterprises, Inc.

Figure 1–18

Determining Resistance of Equipment

❏ Current

What is the current flow of a 10 kW heat strip connected to a 230V (single-phase) power supply? Figure 1–19.

(a) 13A   (b) 26A

(c) 43A   (d) 52A

• Answer: (c) 43A

I = P/E (Formula 4 of 12)

P = 10,000W, E = 230V

I = 10,000W/230V = 43A

Note: Always assume 1Ø (single-phase), unless 3Ø (3-phase) is specified in the question.

❏ Voltage

What is the voltage drop of 200 ft of 12 AWG conductor that carries 16A? Figure 1–20. The resistance of 12 AWG copper conductor is 2Ω per 1,000 ft.

(a) 1.6 VD   (b) 2.9 VD

(c) 3.2 VD   (d) 6.4 VD

• Answer: (d) 6.4 VD

$E = I \times R$ (Formula 7 of 12)

$I = 16A, R = 2\Omega/1,000 = 0.002\Omega$ per ft $\times$ 200 ft $= 0.4\Omega$

$E = 16A \times 0.4\Omega = 6.4$ VD

❏ Power

The total resistance of two 12 AWG copper conductors, 75 ft long, is 0.3Ω (0.15Ω for each conductor). The current of the circuit is 16A. What is the power loss of the conductors in watts per hour? Figure 1–21.

(a) 19W   (b) 77W   (c) 172.8W   (d) none of these

• Answer: (b) 77W

$P = I^2R$ (Formula 10 of 12)

$I = 16A, R = 0.3\Omega$

$P = 16A^2 \times 0.3\Omega = 76.8W$ per hour

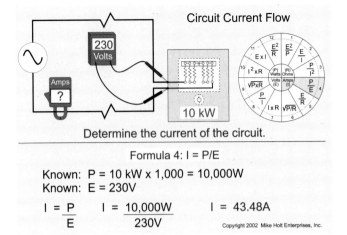

Circuit Current Flow

Determine the current of the circuit.

Formula 4: I = P/E

Known: P = 10 kW x 1,000 = 10,000W
Known: E = 230V

$I = \dfrac{P}{E}$ $I = \dfrac{10,000W}{230V}$ I = 43.48A

Copyright 2002 Mike Holt Enterprises, Inc.

Figure 1–19

Circuit Current Flow

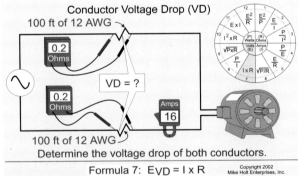

Conductor Voltage Drop (VD)

100 ft of 12 AWG

VD = ?

100 ft of 12 AWG

Determine the voltage drop of both conductors.

Formula 7: E_VD = I x R   Copyright 2002 Mike Holt Enterprises, Inc.

Known: I = 16A
Known: R of one conductor = *0.2 ohm
*0.2 ohm determined by NEC Chapter 9, Table 9 (per ft method)

E_VD = I x R   Or
E_VD = 16A x 0.2 ohms | E_VD = I x R
E_VD = 3.2V on one conductor | R = 0.2Ω x 2 conductors = 0.4Ω
E_VD = 3.2V x 2 conductors | E_VD = 16A x 0.4 ohms
E_VD = 6.4V on both conductors | E_VD = 6.4 volts dropped

Figure 1–20

Conductor Voltage Drop (VD)

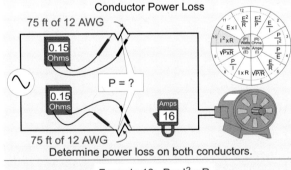

Conductor Power Loss

75 ft of 12 AWG

P = ?

75 ft of 12 AWG

Determine power loss on both conductors.

Formula 10: P = I² x R

Known: I = 16A
Known: R of one conductor = *0.15 ohms
*0.15 ohm determined by NEC Ch 9, Tbl 9 (per ft method)

P = I² x R   Or   P = I² x R
P = (16A x 16A) x 0.15 ohms   P = (16A x 16A) x 0.3 ohm
P = 256A x 0.15 ohm   P = 256A x 0.3 ohm
P = 38.4W on one conductor   P = 76.8W on both conductors
P = 38.4W x 2 conductors
P = 76.8W on both conductors   Copyright 2002 Mike Holt Enterprises, Inc.

Figure 1–21

Conductor Power Loss

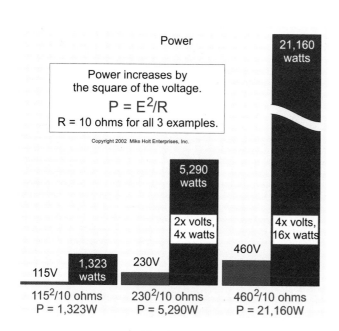

Figure 1–22
Power Changes with the Square of the Voltage

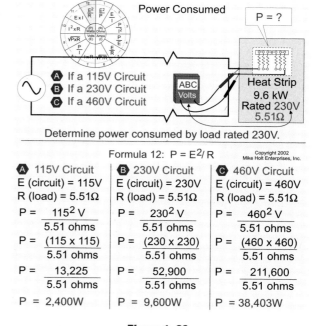

Figure 1–23
Power Changes with the Square of the Voltage

1–20 POWER CHANGES WITH THE SQUARE OF THE VOLTAGE

The power consumed by a resistor is affected by the voltage applied. Power is proportional to the square of the voltage and directly proportional to the resistance. Figure 1–22.

This means that when the voltage goes up, the power goes up by the square of the change and the opposite when the voltage is reduced. For example, if you double the voltage, the power goes up four times (2^2), or when the voltage drops to one-half its level the power goes down to one-quarter the value ($1/2^2$).

P = E²/R

❑ Power Changes with the Square of the Voltage

What is the power consumed by a 9.6 kW heat strip rated 230V connected to 115, 230, and 460V power supplies? Note: The resistance of the heat strip is 5.51Ω. Figure 1–23.

Step 1:   What is the question? What is the power consumed, (P)?

Step 2:   What do you know about the heat strip?
E = 115V, 230V and 460V, R = 5.51Ω

Step 3:   The formula to determine power is, P = E²/R.
Power at 115V = 115V²/5.51Ω, P = 2,400W.
Power at 230V = 230V²/5.51Ω, P = 9,600W (Increase voltage 2× = Power increased 4×).
Power at 460V = 460V²/5.51Ω, P = 38,403W (Increase voltage 4× = Power increased 16×).

1–21 ELECTRIC METERS

Basic electrical meters use a *helically* wound coil of conductor, called a *solenoid,* to produce a strong electromagnetic field to attract an iron bar inside the coil. The iron bar that moves inside the coil is called an *armature.* When a meter has positive (+) and negative (–) shown for the meter leads, the meter is said to be *polarized.* The negative (–) lead must be connected to the negative terminal of the power source, and the positive (+) lead must be connected to the positive terminal.

Ammeter (Not clamp-on type)

An ammeter is a meter that has a *helically* wound coil, and it uses the circuit energy to measure dc. As current flows through the meter's coil, the coil's electromagnetic field draws in the iron bar (armature). The greater the current flow through

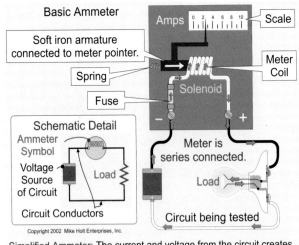

Figure 1–24
Basic Ammeter

Simplified Ammeter: The current and voltage from the circuit creates a magnetic field in the meter coil that attracts the soft iron bar with the current indicator. If the circuit's current was stronger, a larger magnetic field would be created and it would pull the armature in farther.

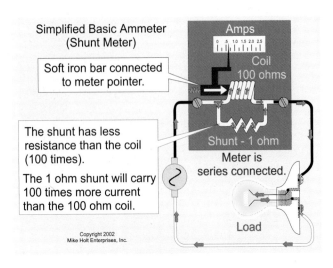

Figure 1–25
Simplified Basic Ammeter (Shunt Meter)

the meter's coil, the greater the electromagnetic field and the further the armature is drawn into the coil. Ammeters are connected in series with the circuit and are used to measure only dc. Figure 1–24.

Ammeters are connected in *series* with the power supply and the load. If the ammeter is accidentally connected in *parallel* to the power supply, the current flow through the meter will be extremely high. The excessive high current through the meter will destroy the meter due to excessive heat.

If the ammeter is not connected to the proper *polarity* when measuring dc, the meter's needle will quickly move in the reverse direction and possibly damage the meter's calibration.

Ammeters that measure currents larger than 10 milliamperes often contain a device called a *shunt,* which is placed in parallel with the meter coil. This permits the current flow to divide between the meter's coil and the shunt bar. The current through the meter coil depends on the resistance of the shunt bar.

❏ **Shunt Ammeter Current**

What is the current flow through the meter if the shunt bar is 1Ω and the coil is 100Ω? Figure 1–25.

(a) The same as the shunt   (b) 1/10 the shunt amperage
(c) 1/100 the shunt amperage   (d) 1/1,000 the shunt amperage

• Answer: (c) 1/100 the shunt amperage

Since the shunt is 100 times less resistant than the meter's coil, the shunt bar will carry 100 times more current than the meter's coil, or the meter's coil will carry 1/100 the shunt amperage.

Clamp-on Ammeter

Clamp-on ammeters are used to measure ac. They are connected *perpendicularly* around the conductor (90 degrees) without breaking the circuit. A clamp-on ammeter indirectly utilizes the circuit energy by induction of the electromagnetic field. The clamp-on ammeter is actually a transformer (sometimes called a *current transformer*). The primary winding is the phase conductor, and the secondary winding is the meter's coil. Figure 1–26.

The electromagnetic field around the phase conductor expands and collapses, which causes electrons to flow in the meter's circuit. As current flows through the meter's coil, the electromagnetic field of the meter draws in the armature. Since the phase conductor serves as the primary (one turn), the current to be measured must be high enough to produce an electromagnetic field that is strong enough to cause the meter to operate.

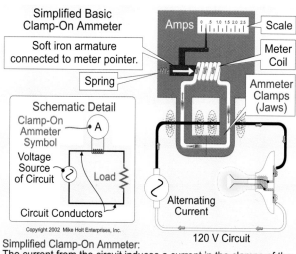

Simplified Basic Clamp-On Ammeter

Soft iron armature connected to meter pointer.

Spring

Schematic Detail
Clamp-On Ammeter Symbol
Voltage Source of Circuit
Load
Circuit Conductors

Copyright 2002 Mike Holt Enterprises, Inc.

Amps
Scale
Meter Coil
Ammeter Clamps (Jaws)

Alternating Current

120 V Circuit

Simplified Clamp-On Ammeter:
The current from the circuit induces a current in the clamps of the ammeter, which creates current in the coil of the meter. This creates a magnetic field in the meter coil that attracts the soft iron bar with the current indicator. The clamps of the ammeter must be perpendicular to the circuit conductor to get an accurate reading.

Figure 1–26

Clamp-on Ammeters

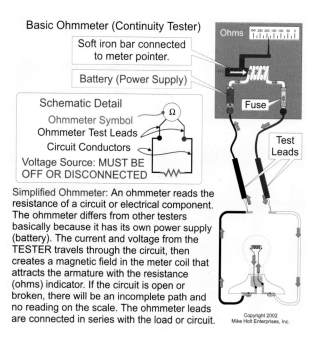

Basic Ohmmeter (Continuity Tester)

Soft iron bar connected to meter pointer.

Battery (Power Supply)

Schematic Detail
Ohmmeter Symbol
Ohmmeter Test Leads
Circuit Conductors
Voltage Source: MUST BE OFF OR DISCONNECTED

Ohms

Fuse

Test Leads

Simplified Ohmmeter: An ohmmeter reads the resistance of a circuit or electrical component. The ohmmeter differs from other testers basically because it has its own power supply (battery). The current and voltage from the TESTER travels through the circuit, then creates a magnetic field in the meter coil that attracts the armature with the resistance (ohms) indicator. If the circuit is open or broken, there will be an incomplete path and no reading on the scale. The ohmmeter leads are connected in series with the load or circuit.

Copyright 2002 Mike Holt Enterprises, Inc.

Figure 1–27

Ohmmeter

Ohmmeter and Megohmmeter

Ohmmeters are used to measure the resistance of a circuit or component and can be used to locate open circuits or shorts. An ohmmeter has an armature, a coil and its own power supply, generally a battery. Ohmmeters are always connected to de-energized circuits and polarity need not be observed. When an ohmmeter is used, current flows through the meter's coil causing an electromagnetic field around the coil and drawing in the armature. The greater the current flow, as a result of lower resistance ($I = {}^E/_R$), the greater the magnetic field and the further the armature is drawn into the coil. A short circuit is indicated by a reading of zero and an open circuit will be indicated by infinity (∞). Figure 1–27.

A *megohmmeter* or *megohmer* is an instrument designed to measure very high resistances such as those found in cable insulation between motor or transformer windings.

Voltmeter

Voltmeters are used to measure both dc and ac voltage. A voltmeter contains a *resistor* in *series* with the coil and utilizes the circuit energy for its operation. The purpose of the resistor in the meter is to reduce the current flow through the meter. As current flows through the meter's helically wound coil, the combined electromagnetic field of the coil draws in the iron bar. The greater the circuit voltage, the greater the current flow through the meter's coil ($I = {}^E/_R$). The greater the current flow through the meter's coil, the greater the electromagnetic field and the further the armature is drawn into the coil. Figure 1-28.

Polarity must be observed when connecting voltmeters to dc circuits. If the meter is not connected to the proper polarity, the meter's needle will quickly move in the reverse direction and damage the meter's calibration. Polarity is not required when connecting voltmeters to ac circuits.

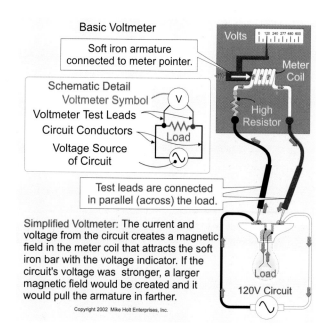

Basic Voltmeter

Soft iron armature connected to meter pointer.

Schematic Detail
Voltmeter Symbol
Voltmeter Test Leads
Circuit Conductors
Voltage Source of Circuit
Load

Volts

Meter Coil
High Resistor

Test leads are connected in parallel (across) the load.

Simplified Voltmeter: The current and voltage from the circuit creates a magnetic field in the meter coil that attracts the soft iron bar with the voltage indicator. If the circuit's voltage was stronger, a larger magnetic field would be created and it would pull the armature in farther.

Load
120V Circuit

Copyright 2002 Mike Holt Enterprises, Inc.

Figure 1–28

Voltmeter

Unit 1 – Electrician's Math and Basic Electrical Formulas

Calculations Questions

Part A – Electrician's Math (• Indicates that 75% or less of exam takers get the question correct)

Note. Always assume copper conductors for all answers, unless aluminum is specified in the question.

1–1 Fractions

1. The decimal equivalent for the fraction one-half is _____.
   (a) 0.5   (b) 5   (c) 2   (d) 0.2

2. The approximate decimal equivalent for the fraction $^4/_{18}$ is _____.
   (a) 4.51   (b) 1.52   (c) 2.53   (d) 0.22

1–2 Kilo

3. A 75W load expressed in terms of kW is _____.
   (a) 75 kW   (b) 7.5 kW   (c) 0.75 kW   (d) 0.075 kW

1–3 Knowing Your Answer

4. • At full load, the output VA of a transformer winding is 100W. What is the input power in watts, if the transformer winding is 90 percent efficient?
   NOTE: Because the efficiency is less than 100 percent, the input power will be greater than the output power.
   (a) 90W   (b) 111W   (c) 100W   (d) 125W

1–4 Multiplier

5. The method of altering the value of a number by multiplying it by another number is called _____.
   (a) percentage   (b) decimal   (c) fraction   (d) multiplication

6. An overcurrent protection device (circuit breaker or fuse) shall be sized no less than 115 percent of the load. If the load is 20A, the overcurrent protection device would have to be sized at no less than _____.
   (a) 20A   (b) 23A   (c) 17A   (d) 30A

7. The overcurrent protection device shall be sized no smaller than 125 percent of the continuous load. If the overcurrent device is rated 90A, what is the maximum continuous load?
   (a) 72A   (b) 70A   (c) 68A   (d) 66A

8. Which of the following statements is/are correct if the ampacity of a 50A wire needs to be adjusted by a multiplier of 0.80?
   (a) The answer will be less than 50A   (b) 80% of 50A
   (c) The math is 50A × 0.8   (d) all the statements

1–5 Parentheses

9. What is the length (distance) of two 14 AWG conductors carrying 16A, with a voltage drop of 10V?

   Formula: $D = \dfrac{(4,100 \text{ cm} \times 10V)}{(2 \text{ wires} \times 12.9A \times 16\Omega)}$

   (a) 50 ft   (b) 75 ft   (c) 100 ft   (d) 150 ft

10. What is the line current of a 3Ø, 18 kW, 208V load? Formula: $I = \dfrac{W}{E \times \sqrt{3}}$
     (a) 25A   (b) 50A
     (c) 100A   (d) 150A

1–6 Percentages

11. When changing the value of a number from a percentage to a decimal or whole number, move the decimal point (from the percent value) _____ place(s) to the left.
     (a) one   (b) two   (c) three   (d) four

12. The decimal equivalent for 75 percent is _____.
     (a) 0.075   (b) 0.75   (c) 7.5   (d) 75

13. The decimal equivalent for 225 percent is _____.
     (a) 225   (b) 22.5   (c) 2.25   (d) 0.225

14. The decimal equivalent for 300 percent is _____.
     (a) 0.03   (b) 0.3   (c) 3   (d) 30.0

1-7 Percent Increase

15. The feeder demand load for an 8 kW load increased by 20 percent is _____.
     (a) 8 kW   (b) 9.6 kW   (c) 6.4 kW   (d) 10 kW

1–8 Percentage Reciprocals

16. What is the reciprocal of 125 percent?
     (a) 0.8   (b) 100%   (c) 125%   (d) none of these

17. A continuous load requires an overcurrent protection device sized no smaller than 125 percent of the load. What is the maximum continuous load permitted on a 100A overcurrent protection device?
     (a) 140A   (b) 150A   (c) 80A   (d) 110A

1–9 Rounding Off

18. • The sum of 5, 7, 8 and 9 is approximately _____.
     (a) 20   (b) 25   (c) 30   (d) 35

1–10 Squaring

19. What is the power consumed in watts of a 12 AWG conductor that is 100 ft long, has a resistance (R) 0.2Ω and the current (I) in the circuit is 16A? Formula: Power = I^2R
     (a) 25W   (b) 50W   (c) 75W   (d) 10W

20. • What is the approximate cross-section area in sq in. of a raceway that has a diameter of 2 in. (radius of 1 in.)? Formula: $\pi = 3.14 \times r^2$, r = $^1/_2$ diameter
     (a) 1.04 sq in.   (b) 2.13 sq in.   (c) 3.14 sq in.   (d) 4.31 sq in.

21. The numeric equivalent of 4^2 is _____.
     (a) 2   (b) 8   (c) 16   (d) 32

22. The numeric equivalent of 12^2 is _____.
     (a) 3.46   (b) 24   (c) 144   (d)1,728

1–11 Square Root

23. What is the approximate square root of 1,000 ($\sqrt{1,000}$)?
     (a) 3   (b) 32   (c) 100   (d) 500

24. The square root of 3 ($\sqrt{3}$) is _____.
    (a) 1.732   (b) 9   (c) 729   (d) 1.5

Part B – Basic Electrical Formulas

1–12 Electrical Circuits

25. An electric circuit consists of the _____.
    (a) power source   (b) conductors   (c) load   (d) all of these

1–13 Electron Flow

26. Inside a dc power source, electrons travel from the positive terminal to the negative terminal.
    (a) True   (b) False

1–14 Power Source

27. The polarity of a(n) _____current power source never changes. One terminal is always negative, the other is always positive and current will flow in one constant direction.
    (a) static   (b) direct   (c) alternating   (d) all of these

28. _____current power sources produce a voltage that has a constant change in polarity and magnitude in one direction exactly the same as it does in the other.
    (a) Static   (b) Direct   (c) Alternating   (d) all of these

1–15 Conductance and Resistance

29. Conductance is a property that permits current to flow. The best conductors in order of conductivity are _____.
    (a) gold, silver, copper, aluminum   (b) gold, aluminum, silver, copper
    (c) gold, copper, silver, aluminum   (d) silver, copper, gold, aluminum

1–16 Electrical Circuit Values

30. The _____ is the pressure required to force one ampere of current (electrons) to flow through a one ohm resistor.
    (a) ohm   (b) watt   (c) volt   (d) ampere

31. All conductors have resistance that opposes the flow of electrons. Some materials have more resistance than others. _____ has the lowest resistance and _____ is more resistant than copper.
    (a) Silver, gold   (b) Gold, aluminum (c) Gold, silver   (d) None of these

32. Resistance is represented by the letter R and is expressed in _____.
    (a) volts   (b) impedance   (c) capacitance   (d) ohms

33. The opposition to the flow of current can be thought of as restricting the flow of electrons in the circuit. Every component of an electric circuit contains resistance, except the power supplies such as the generator or transformer.
    (a) True   (b) False

34. In electrical systems, the volume of electrons that move through a conductor in one second is called the _____ of the circuit.
    (a) intensity (I - current)   (b) voltage (V - volt)
    (c) power (W - watts)   (d) resistance (ohms - resistance)

35. The rate of work that can be produced by the movement of electrons is called _____.
    (a) voltage   (b) current   (c) power   (d) none of these

1–17 Ohm's Law

36. Ohm's Law formula demonstrates that current is _____ proportional to the voltage and _____ proportional to the resistance.
    (a) indirectly, inversely   (b) inversely, directly   (c) inversely, indirectly   (d) directly, inversely

37.   In an ac circuit, which of the following factors affect the total opposition to current flow?
      (a) Resistance.   (b) Capacitive reactance.
      (c) Inductive reactance.   (d) all of these

38.   The combined opposition to current flow in an ac circuit is called _____ and it is often represented by the letter Z.
      (a) resistance   (b) capacitance   (c) induction   (d) impedance

39.   • What is the voltage drop of two 12 AWG conductors supplying a 16A load located 100 ft from the power supply?
      Formula: $E_{VD} = I \times R$, I = 16A, R = 0.4Ω (200 ft of 12 AWG copper wire)
      (a) 6.4V   (b) 12.8V   (c) 1.6V   (d) 3.2V

40.   What is the resistance of the circuit conductors when the conductor voltage drop is 7.2V and the current flow is 50A?
      (a) 0.14Ω   (b) 0.3Ω   (c) 3Ω   (d) 14Ω

1–18 Pie Circle Formula

41.   What is the power loss in watts for a conductor that carries 24A and has a voltage drop of 7.2V?
      (a) 173W   (b) 350W   (c) 700W   (d) 2,400W

42.   What is the current flow of a 10 kW heat strip rated 230V, 1Ø?
      (a) 35A   (b) 38A   (c) 43A   (d) 60A

1–19 Formula Wheel

43.   • The formulas listed in the formula wheel apply to _____.
      (a) dc circuits   (b) ac circuits with unity power factor
      (c) a and b   (d) none of these

44.   When working any circuit formula, the key to getting the correct answer is following these four simple steps:
      Step 1: Know what the question is asking.
      Step 2: Determine what you know about the circuit or resistors.
      Step 3: Select the formula.
      Step 4: Work out the formula calculation.
      (a) True   (b) False

45.   The total resistance of two 12 AWG copper conductors 150 ft long is 0.6Ω and the current of the circuit is 16A. What is the power loss of the conductors in watts per hour?
      (a) 74W   (b) 154W   (c) 197W   (d) 214W

46.   • What is the power loss in a conductor, in watts, for a 120V circuit that has a voltage drop of 3.6V and carries a current flow of 12A?
      (a) 43W   (b) 86W   (c) 172W   (d) 722W

47.   • Approximately what does it cost per year (24 hours per day, 365 days per year at 10 cents per kWh) for the power loss of two 12 AWG conductors, each 150 ft long (0.3Ω), if the circuit carries 12A and runs continuously?
      (a) $38   (b) $56   (c) $72   (d) $91

1–20 Power Changes with the Square of the Voltage

48.   What is the energy consumed by a 10 kW heat strip rated 230V that is connected to a 115V circuit?
      (a) 10 kW   (b) 2.5 kW   (c) 5 kW   (d) 20 kW

1–21 Electric Meters

49.   A noninductive _____ is connected in series with the load.
      (a) watthour meter   (b) voltmeter   (c) power meter   (d) ammeter

50.   • A noninductive ammeter is used to measure _____. It is connected in series with the circuit.
      (a) dc   (b) power   (c) voltage   (d) all of these

51. Inductive (clamp-on) ammeters have one coil _____ around the circuit conductor.
     (a) in series   (b) in parallel
     (c) in series-parallel   (d) at right angles (perpendicular)

52. An ohmmeter has _____ connected in series with the resistor. As current flows through the meter coil, the magnetic field around the coil acts on (draws in) the soft iron bar (armature). The greater the current flow through the circuit, the greater the magnetic field and the further the iron bar armature is drawn into the magnetic coil.
     (a) a coil and resistor   (b) a coil and power supply
     (c) two coils   (d) none of these

☆ Challenge Questions

1–2 Kilo

53. One kVA is equal to _____.
     (a) 100 VA   (b) 1,000V   (c) 1,000W   (d) 1,000 VA

1–15 Conductance and Resistance

54. • _____ is not an insulator.
     (a) Bakelite   (b) Oil   (c) Air   (d) Salt water

1–16 Electrical Circuit Values

55. • _____ is not the force that moves electrons.
     (a) EMF   (b) Voltage   (c) Potential   (d) Current

56. Conductor resistance varies with _____.
     (a) material   (b) voltage   (c) current   (d) power

1–17 Ohm's Law (I = $^E/_R$)

57. • If the contact resistance of a connection increases and the current of the circuit (load) remains the same, the voltage dropped across the connection _____.
     (a) will increase   (b) will decrease   (c) will remain the same   (d) cannot be determined

58. • To double the current of a circuit when the voltage remains constant, the R (resistance) must be _____.
     (a) doubled   (b) reduced by half (c) increased   (d) none of these

59. • An ohmmeter is being used to test a relay coil. The equipment instructions indicate that the resistance of the coil should be between 30 and 33Ω. The ohmmeter indicates that the actual resistance is less than 22Ω. This reading would most likely indicate _____.
     (a) the coil is okay (b) an open coil   (c) a shorted coil (d) a meter problem

1–19 Formula Wheel

60. • To calculate the energy consumed in watts by a resistive appliance, you need to know the _____ of the circuit.
     (a) voltage and current   (b) current and resistance
     (c) voltage and resistance   (d) any of these pairs of variables

61. • The number of watts of heat given off by a resistor is expressed by the formula $I^2 \times R$. If 10V is applied to a 5Ω resistor, then _____ of heat will be given off.
     (a) 500W   (b) 250W   (c) 50W   (d) 20W

62. • Power loss in a circuit because of heat can be determined by the formula _____.
     (a) $P = R \times I$ (b) $P = I \times R$ (c) $P = I^2 \times R$ (d) none of these

63. • The energy consumed (watts) by a 5Ω resistor will be _____ than the energy consumed by a 10Ω resistor assuming the current in both circuits is the same.
     (a) more   (b) less

64. When a lamp that is rated 500W at 115V is connected to a 120V power supply, the current of the circuit will be _____. Tip: The power does not remain the same when voltage is changed and you need to determine the resistance of the load first.
     (a) 3.8A   (b) 4.5A   (c) 2.7A   (d) 5.5A

1–20 Power Changes with the Square of the Voltage

65. A 120V rated toaster will produce _____ heat when supplied by 115V.
     (a) more   (b) less   (c) the same   (d) none of these

66. • When a 144Ω resistive 100W load operates at a voltage that is 10 percent higher than its nameplate rating, the appliance will consume approximately _____.
     (a) 50W   (b) 75W   (c) 121W   (d) 150W

67. • A 1,500W resistive heater is rated 230V and it is connected to a 208V supply. The power consumed for this load at 208V will be approximately _____. Tip: The power does not remain the same when voltage is changed and you need to determine the resistance of the load first.
     (a) 1,625W   (b) 1,750W   (c) 1,850W   (d) 1,225W

68. • The total resistance of a circuit is 12Ω. The load has a resistance of 10Ω and the wire has a resistance of 2Ω. If the current of the circuit is 3A, then the power consumed by the circuit conductors is _____.
     (a) 28W   (b) 18W   (c) 90W   (d) 75W

1–21 Electric Meters

69. • The best instrument for detecting an electric current is a(n) _____.
     (a) ohmmeter   (b) voltmeter   (c) ammeter   (d) wattmeter

70. When measuring the resistance of a circuit with an ohmmeter, polarity (+ or -) must be observed.
     (a) True   (b) False

71. • When the test leads of an ohmmeter are shorted together, the meter will read _____ on the scale.
     (a) zero or near zero ohms   (b) 1,000
     (c) infinity   (d) any of these

72. • A short circuit is indicated by a reading of _____ when tested with an ohmmeter.
     (a) zero or near zero ohms   (b) megohms
     (c) infinity   (d) R

73. Voltmeters can be used to measure _____.
     (a) grounded circuits only   (b) voltage potential
     (c) ac voltages only   (d) dc voltages only

74. Voltmeters shall be connected in _____ with the circuit component being tested.
     (a) series   (b) parallel   (c) series-parallel (d) multiwire

75. To measure the voltage across a load, you would connect a(n) _____.
     (a) voltmeter across the load   (b) ammeter across the load
     (c) voltmeter in series with the load   (d) ammeter in series with the load

76. A voltmeter is connected in _____ to the load.
     (a) series   (b) parallel   (c) series-parallel (d) none of these

77. • In the course of normal operation, the least effective instrument for determining that a generator may overheat because it is overloaded is a(n) _____.
     (a) ammeter   (b) voltmeter   (c) wattmeter   (d) none of these

78. • A direct-current voltmeter (not a digital meter) can be used to measure _____.
     (a) power   (b) frequency   (c) polarity   (d) power factor

79. • Polarity must be observed when connecting an analog voltmeter to _____ current circuit.
     (a) an alternating   (b) a direct
     (c) any   (d) polarity doesn't matter

80. The minimum number of wattmeters necessary to measure the power in the load of a balanced 3Ø, 4-wire system is _____.
     (a) 1   (b) 2   (c) 3   (d) 4

NEC Questions – Articles 90-110

Article 90 Introduction

81. The *NEC* is _____.
     (a) intended to be a design manual
     (b) meant to be used as an instruction guide for untrained persons
     (c) for the practical safeguarding of persons and property
     (d) published by the Bureau of Standards

82. Compliance with the provisions of the *Code* will result in _____.
     (a) good electrical service
     (b) an efficient electrical system
     (c) an electrical system essentially free from hazard
     (d) all of these

83. • The *Code* contains provisions considered necessary for safety regardless of _____.
     (a) efficient use
     (b) convenience
     (c) good service or future expansion of electrical use
     (d) all of these

84. Hazards often occur because of _____.
     (a) overloading of wiring systems by methods or usage not in conformity with this *Code*
     (b) initial wiring not providing for increases in the use of electricity
     (c) a and b
     (d) none of these

85. The following systems shall be installed in accordance with the *NEC*:
     (a) signaling   (b) communications   (c) power and lighting   (d) all of these

86. • The *Code* applies to the installation of _____.
     (a) electrical conductors and equipment within or on public and private buildings
     (b) outside conductors and equipment on the premises
     (c) optical fiber cable
     (d) all of these

87. The *Code* does not cover installations in ships, watercraft, railway rolling stock, aircraft, or automotive vehicles.
     (a) True   (b) False

88. The *Code* covers underground installations in mines and self-propelled mobile surface mining machinery and its attendant electrical trailing cable.
     (a) True   (b) False

89. Installation of communications equipment that is under the exclusive control of communications utilities, and located outdoors or in building spaces used exclusively for such installations _____ covered by the *Code*.
     (a) are   (b) are sometimes   (c) are not   (d) might be

90. The *Code* covers all of the following electrical installations except _____.
    (a) floating buildings
    (b) in or on private and public buildings
    (c) industrial substations
    (d) electrical generation installations on property owned or leased by the electric utility company

91. Service laterals installed by an electrical contractor shall be installed in accordance with the *NEC*.
    (a) True   (b) False

92. Chapters 1 through 4 of the *NEC* apply _____.
    (a) generally to all electrical installations
    (b) to special installations and conditions
    (c) to special equipment and material
    (d) all of these

93. Communications wiring such as telephone, antenna and CATV wiring within a building is not required to comply with the installation requirements of Chapters 1 through 7, except where it is specifically referenced therein.
    (a) True   (b) False

94. The requirements in "Annexes" shall be complied with.
    (a) True   (b) False

95. The authority having jurisdiction for enforcement of the *Code* has the responsibility _____.
    (a) for making interpretations of the rules of the *Code*
    (b) for deciding upon the approval of equipment and materials
    (c) for waiving specific requirements in the *Code* and allowing alternate methods and material if safety is maintained
    (d) all of these

96. The authority having jurisdiction is not required to enforce any requirements of Chapter 7 (Signaling Circuits) or Chapter 8 (Communications Circuits), because this is not within the scope of enforcement.
    (a) True   (b) False

97. A *Code* rule may be waived or alternative methods of installation approved that may be contrary to the *NEC*, if the AHJ gives verbal or written consent.
    (a) True   (b) False

98. In the event the *Code* requires new products, constructions, or materials that are not yet available at the time a new edition is adopted, the _____ may permit the use of the products, constructions, or materials that comply with the most recent previous edition of this *Code* adopted by the jurisdiction.
    (a) architect   (b) master electrician   (c) authority having jurisdiction   (d) supply house

99. Explanatory material, such as references to other standards, references to related sections of this *Code*, or information related to a *Code* rule, is included in the form of Fine Print Notes (FPNs).
    (a) True   (b) False

100. Equipment that is listed by a qualified electrical testing laboratory is not required to have the factory-installed _____ wiring inspected at the time of installation except to detect alterations or damage.
    (a) external   (b) associated   (c) internal   (d) all of these

101. Compliance with either the metric or the inch-pound unit of measurement system is permitted.
    (a) True   (b) False

Chapter 1 General

Article 100 Definitions

102.   Admitting close approach, not guarded by locked doors, elevation, or other effective means, is commonly referred to as _____.

(a) accessible (equipment)   (b) accessible (wiring methods)
(c) accessible, readily   (d) all of these

103.   Capable of being reached quickly for operation, renewal, or inspections without resorting to portable ladders and such is known as _____.
(a) accessible (equipment)   (b) accessible (wiring methods)
(c) accessible, readily   (d) all of these

104.   Capable of being removed or exposed without damaging the building structure or finish, or not permanently closed in by the structure or finish of the building defines _____.
(a) accessible (equipment)   (b) accessible (wiring methods)
(c) accessible, readily   (d) all of these

105.   A junction box located above a suspended ceiling having removable panels is considered to be _____.
(a) concealed   (b) accessible   (c) readily accessible   (d) recessed

106.   Acceptable to the authority having jurisdiction means _____.
(a) identified   (b) listed   (c) approved   (d) labeled

107.   • A generic term for a group of nonflammable synthetic chlorinated hydrocarbons used as electrical insulating media is _____.

(a) oil   (b) girasol   (c) askarel   (d) phenol

108.   • A device that, by insertion in a receptacle, establishes a connection between the conductors of the attached flexible cord and the conductors connected permanently to the receptacle is called a(n) _____.
(a) attachment plug   (b) plug cap   (c) plug   (d) any of these

109.   Where no statutory requirement exists, the AHJ could be a property owner or his/her agent, such as an architect or engineer.
(a) True   (b) False

110.   According to the *Code*, automatic is self-acting, operating by its own mechanism when actuated by some impersonal influence, such as _____.
(a) change in current strength   (b) temperature
(c) mechanical configuration   (d) all of these

111.   The connection between the grounded conductor and the equipment grounding conductor at the service is accomplished by installing a(n) _____ jumper.
(a) main bonding   (b) bonding   (c) equipment bonding   (d) circuit bonding

112.   The conductors between the final overcurrent protection device and the outlet(s) are known as the _____ conductors.
(a) feeder   (b) branch-circuit   (c) home run   (d) none of these

113.   A(n) _____ branch circuit supplies energy to one or more outlets to which appliances are to be connected and that has no permanently connected luminaires (lighting fixtures) that are not a part of an appliance.
(a) general purpose   (b) multiwire   (c) individual   (d) appliance

114.   A branch circuit that supplies only one utilization equipment is a(n) _____ branch circuit.
(a) individual   (b) general-purpose   (c) isolated   (d) special-purpose

115. • For a circuit to be considered a multiwire branch circuit, it must have _____.
      (a) two or more ungrounded conductors with a voltage potential between them
      (b) a grounded conductor having equal voltage potential between it and each ungrounded conductor of the circuit
      (c) a grounded conductor connected to the grounded (neutral) terminal of the system
      (d) all of these

116. The *Code* defines a(n) _____ as a structure that stands alone or that is cut off from adjoining structures by firewalls, with all openings therein protected by approved fire doors.
      (a) unit   (b) apartment   (c) building   (d) utility

117. A circuit breaker is a device designed to _____ a circuit by nonautomatic means and to open the circuit automatically on a predetermined overcurrent without damage to itself when properly applied within its rating.
      (a) blow   (b) disconnect   (c) connect   (d) open and close

118. _____ is a qualifying term indicating that there is a purposely introduced delay in the tripping action of the circuit breaker, which delay decreases as the magnitude of the current increases.
      (a) Adverse-time   (b) Inverse-time   (c) Time delay   (d) Timed unit

119. NM cable is considered _____ if rendered inaccessible by the structure or finish of the building.
      (a) inaccessible   (b) concealed   (c) hidden   (d) enclosed

120. • A conductor encased within material of composition or thickness that is not recognized by this *Code* as electrical insulation is considered _____.
      (a) noninsulating   (b) bare   (c) covered   (d) protected

121. A separate portion of a conduit or tubing system that provides access through a removable cover(s) to the interior of the system at a junction of two or more sections of the system or at a terminal point of the system, is defined as a(n) _____.
      (a) junction box   (b) accessible raceway   (c) conduit body   (d) pressure connector

122. A solderless pressure connector is a device that _____ between two or more conductors or between one or more conductors and a terminal by means of mechanical pressure and without the use of solder.
      (a) provides access   (b) protects the wiring
      (c) is never needed   (d) establishes a connection

123. A load is considered to be continuous if the maximum current is expected to continue for _____ or more.
      (a) $\frac{1}{2}$ hour   (b) 1 hour   (c) 2 hours   (d) 3 hours

124. A device or group of devices that serves to govern in some predetermined manner the electric power delivered to the apparatus to which it is connected is a _____.
      (a) relay   (b) breaker   (c) transformer   (d) controller

125. • The _____ of any system is the ratio of the maximum demand of a system, or part of a system, to the total connected load of a system under consideration.
      (a) load   (b) demand factor   (c) minimum load   (d) computed factor

126. A unit of an electrical system that is intended to carry but not utilize electric energy is a(n) _____.
      (a) raceway   (b) fitting   (c) device   (d) enclosure

127. Which of the following does the *Code* recognize as a device?
      (a) Switch.   (b) Light bulb.   (c) Transformer.   (d) Motor.

128. A _____ is a device, group of devices, or other means by which the conductors of a circuit can be disconnected from their source of supply.
      (a) feeder   (b) enclosure
      (c) disconnecting means   (d) conductor interrupter

129. Constructed so that dust will not enter the enclosing case under specified test conditions is known as _____.
      (a) dusttight   (b) dustproof   (c) dust rated   (d) all of these

130.   Continuous duty is defined as _____.
        (a) when the load is expected to continue for three hours or more
        (b) operation at a substantially constant load for an indefinitely long time
        (c) operation at loads and for intervals of time, both of which may be subject to wide variations
        (d) operation at which the load may be subject to maximum current for six hours or more

131.   Varying duty is defined as _____.
        (a) intermittent operation in which the load conditions are regularly recurrent
        (b) operation at a substantially constant load for an indefinite length of time
        (c) operation for alternate intervals of load and rest, or load, no load, and rest
        (d) operation at loads and for intervals of time, both of which may be subject to wide variations

132.   Surrounded by a case, housing, fence, or wall(s) that prevents persons from accidentally contacting energized parts is called _____.
        (a) guarded   (b) covered   (c) protection   (d) an enclosure

133.   An apparatus enclosed in a case that is capable of withstanding an explosion of a specified gas or vapor that may occur within it, and of preventing the ignition of a specified gas or vapor surrounding the enclosure by sparks, flashes, or explosion of the gas or vapor within, and that operates at such an external temperature that a surrounding flammable atmosphere will not be ignited is defined as a(n) _____.
        (a) overcurrent protection device   (b) thermal apparatus
        (c) explosionproof apparatus   (d) bomb casing

134.   When the term exposed, as it relates to live parts, is used by the *Code*, it refers to _____.
        (a) capable of being inadvertently touched or approached nearer than a safe distance by a person
        (b) parts that are not suitably guarded, isolated, or insulated
        (c) wiring on, or attached to, the surface or behind panels designed to allow access
        (d) a and b

135.   The *NEC* term to define wiring methods that are not concealed is _____.
        (a) open   (b) uncovered   (c) exposed   (d) bare

136.   The *Code* defines a _____ as: "all circuit conductors between the service equipment, the source of a separately derived system, or other power supply source and the final branch-circuit overcurrent device."
        (a) feeder   (b) branch circuit   (c) service   (d) all of these

137.   A _____ is a building or portion of a building in which one or more self-propelled vehicles can be kept for use, sale, storage, rental, repair, exhibition, or demonstration purposes.
        (a) garage   (b) residential garage   (c) service garage   (d) commercial garage

138.   Connected to earth or to some conducting body that serves in place of the earth is called _____.
        (a) grounding   (b) bonded   (c) grounded   (d) all of these

139.   A system or circuit conductor that is intentionally grounded is a(n) _____.
        (a) grounding conductor   (b) unidentified conductor
        (c) grounded conductor   (d) none of the above

140.   _____ is defined as intentionally connected to earth through a ground connection or connection of sufficiently low impedance and having sufficient current-carrying capacity to prevent the buildup of voltages that may result in undue hazards to connected equipment or to persons.
        (a) Effectively grounded   (b) A proper wiring system
        (c) A lighting rod   (d) A grounded conductor

141.   A "Class A" GFCI protection device is designed to de-energize the circuit when the ground-fault current is approximately _____.
        (a) 4 mA   (b) 5 mA   (c) 6 mA   (d) any of these

142. A device intended for the protection of personnel, that functions to de-energize a circuit within an established period of time when a current-to-ground exceeds some predetermined value less than that required to operate the overcurrent protection device of the supply circuit, is a(n) _____.
      (a) dual-element fuse   (b) inverse-time breaker
      (c) ground-fault circuit interrupter   (d) safety switch

143. A system intended to provide protection of equipment from damaging line-to-ground fault currents by operating to cause a disconnecting means to open all ungrounded conductors of the faulted circuit is defined as _____.
      (a) ground-fault protection of equipment   (b) guarded
      (c) personal protection   (d) automatic protection

144. • In a grounded system, the conductor that connects the circuit-grounded conductor at the service to the grounding electrode is called the _____ conductor.
      (a) main grounding   (b) common main
      (c) equipment grounding   (d) grounding electrode

145. A hoistway is any _____ in which an elevator or dumbwaiter is designed to operate.
      (a) hatchway or well hole   (b) vertical opening or space
      (c) shaftway   (d) all of these

146. Recognized as suitable for the specific purpose, function, use, environment, and application is the definition of _____.
      (a) labeled   (b) identified (as applied to equipment)
      (c) listed   (d) approved

147. The highest current at rated voltage that a device is intended to interrupt under standard test conditions is the _____.
      (a) interrupting rating   (b) manufacturer's rating
      (c) interrupting capacity   (d) GFCI rating

148. _____ means that an object is not readily accessible to persons unless special means for access are used.
      (a) Isolated   (b) Secluded   (c) Protected   (d) Locked

149. Equipment or materials to which a symbol or other identifying mark of an organization that is acceptable to the authority having jurisdiction has been attached is known as _____.
      (a) listed   (b) labeled   (c) approved   (d) rated

150. An outlet intended for the direct connection of a lampholder, a luminaire, or a pendant cord terminating in a lampholder is a(n) _____.
      (a) outlet   (b) receptacle outlet (c) lighting outlet (d) general-purpose outlet

151. The environment of a wiring method under the eaves of a house having a roofed open porch would be considered a _____ location.
      (a) dry   (b) damp   (c) wet   (d) moist

152. A _____ location may be temporarily subject to dampness and wetness.
      (a) dry   (b) damp   (c) moist   (d) wet

153. Conduit installed underground or encased in concrete slabs that are in direct contact with the earth shall be considered a _____ location.
      (a) dry   (b) damp   (c) wet   (d) moist

154. The term "luminaire" replaces the terms "fixture" and "lighting fixture" throughout the 2002 *NEC*.
      (a) True   (b) False

155. A _____ circuit, would be a circuit, other than field wiring, in which any arc or thermal effect produced under intended operating conditions of the equipment is not capable, under specified test conditions, of igniting the flammable gas-air, vapor-air, or dust-air mixture.
      (a) nonconductive   (b) branch   (c) nonincendive   (d) closed

156.   A(n) _____ is a point on the wiring system at which current is taken to supply utilization equipment.
        (a) box   (b) receptacle   (c) outlet   (d) device

157.   Any current in excess of the rated current of equipment or the ampacity of a conductor is called _____.
        (a) trip current   (b) faulted   (c) overcurrent   (d) shorted

158.   An overload may be caused by a short circuit or ground fault.
        (a) True   (b) False

159.   A single panel or group of panel units designed for assembly in the form of a single panel is called a _____.
        (a) switchboard   (b) disconnect   (c) panelboard   (d) switch

160.   The *Code* defines a(n) _____ as one familiar with the construction and operation of the electrical equipment and installations, and who has received safety training on the hazards involved.
        (a) inspector   (b) master electrician
        (c) journeyman electrician   (d) qualified person

161.   Something constructed, protected, or treated so as to prevent rain from interfering with the successful operation of the apparatus under specified test conditions is defined as _____.
        (a) raintight   (b) waterproof   (c) weathertight   (d) rainproof

162.   A raintight enclosure is constructed or protected so that exposure to a beating rain will not result in the entrance of water under specified test conditions.
        (a) True   (b) False

163.   A contact device installed at an outlet for the connection of an attachment plug is known as a(n) _____.
        (a) attachment point   (b) tap   (c) receptacle   (d) wall plug

164.   A single receptacle is a single contact device with no other contact device on the same _____.
        (a) circuit   (b) yoke   (c) run   (d) equipment

165.   When one electrical circuit controls another circuit through a relay, that first circuit is called a _____.
        (a) control circuit   (b) remote-control circuit (c) signal circuit   (d) controller

166.   Equipment enclosed in a case or cabinet that is provided with a means of sealing or locking so that live parts cannot be made accessible without opening the enclosure is said to be _____.
        (a) guarded   (b) protected   (c) sealable   (d) lockable

167.   A(n) _____ system is a premises wiring system whose power is derived from a battery, from a solar photovoltaic system, or from a generator, transformer, or converter windings, and that has no direct electrical connection, including a solidly connected grounded circuit conductor, to supply conductors originating in another system.
        (a) separately derived   (b) classified
        (c) direct   (d) emergency

168.   Service conductors only originate from the service point and terminate at the service equipment (disconnect).
        (a) True   (b) False

169.   Overhead-service conductors from the last pole or other aerial support to and including the splices, if any, are called _____ conductors.
        (a) service-entrance   (b) service-drop
        (c) service   (d) overhead service

170.   The service conductors between the terminals of the service equipment and a point usually outside the building, clear of building walls, where joined by tap or splice to the service drop is called _____ service entrance conductors.
        (a) underground   (b) complete   (c) overhead   (d) grounded

171. The _____ is the necessary equipment, usually consisting of a circuit breaker(s) or switch(es) and fuse(s) and their accessories, connected to the load end of service conductors to a building or other structure, or an otherwise designated area, and intended to constitute the main control and cutoff of the supply.
(a) service equipment   (b) service
(c) service disconnect   (d) service overcurrent protection device

172. • Underground service conductors between the street main and the first point of connection to the service entrance are known as the _____.
(a) utility service   (b) service lateral   (c) service drop   (d) main service conductors

173. The _____ is the point of connection between the facilities of the serving utility and the premises wiring.
(a) service entrance   (b) service point
(c) overcurrent protection   (d) beginning of the wiring system

174. A signaling circuit is any electric circuit that energizes signaling equipment.
(a) True   (b) False

175. The combination of all components and subsystems that convert solar energy into electrical energy is called a _____ system.
(a) solar   (b) solar voltaic
(c) separately derived source   (d) solar photovoltaic

176. A _____ switch is a manually operated device used in conjunction with a transfer switch to provide a means of directly connecting load conductors to a power source, and of disconnecting the transfer switch.
(a) transfer   (b) motor-circuit
(c) general-use snap   (d) bypass isolation

177. A form of general-use switch constructed so that it can be installed in device boxes or on box covers, or otherwise used in conjunction with wiring systems recognized by this *Code* is called a _____ switch.
(a) transfer   (b) motor-circuit   (c) general-use snap   (d) bypass isolation

178. An isolating switch is one that is _____.
(a) not readily accessible to persons unless special means for access is used
(b) capable of interrupting the maximum operating overload current of a motor
(c) intended for use in general distribution and branch circuits
(d) intended for isolating an electrical circuit from the source of power

179. A large single panel, frame, or assembly of panels on which switches, overcurrent and other protective devices, buses, and instruments are mounted is a _____. They are generally accessible from the rear as well as from the front and are not intended to be installed in cabinets.
(a) switchboard   (b) panel box   (c) switch box   (d) panelboard

180. A thermal protector may consist of one or more heat-sensing elements integral with the motor or motor-compressor and an external control device.
(a) True   (b) False

181. Utilization equipment is equipment that utilizes electricity for _____.
(a) chemicals   (b) heating   (c) lighting   (d) any of these

182. The voltage of a circuit is defined by the *Code* as the _____ root-mean-square (effective) difference of potential between any two conductors of the circuit.
(a) lowest   (b) greatest   (c) average   (d) nominal

183. A value assigned to a circuit or system for the purpose of conveniently designating its voltage class such as 120/240V is called _____ voltage.
(a) root-mean-square
(b) circuit
(c) nominal
(d) source

184. An enclosure or device constructed so that moisture will not enter the enclosure or device under specific test conditions is called _____.
(a) watertight
(b) moistureproof
(c) waterproof
(d) rainproof

185. A(n) _____ enclosure is so constructed or protected that exposure to the weather will not interfere with successful operation.
(a) weatherproof
(b) weathertight
(c) weather-resistant
(d) all weather

Article 110 Requirements for Electrical Installations

186. In determining equipment to be installed, considerations such as the following should be evaluated:
(a) Mechanical strength.
(b) Cost.
(c) Arcing effects.
(d) a and c

187. To be *Code* compliant, listed or labeled equipment shall be installed and used in accordance with any instructions included in the _____.
(a) catalog
(b) product
(c) listing or labeling
(d) all of these

188. When referencing the *NEC*, conductors shall be _____ unless otherwise provided in the *Code*.
(a) bare
(b) stranded
(c) copper
(d) aluminum

189. Conductor sizes are expressed in American Wire Gage (AWG) or in _____.
(a) inches
(b) circular mils
(c) square inches
(d) AWG

190. All wiring shall be installed so that the completed system will be free from _____, other than required or permitted in Article 250.
(a) short circuits
(b) grounds
(c) a and b
(d) none of these

191. Only wiring methods recognized as _____ are included in the *Code*.
(a) identified
(b) efficient
(c) suitable
(d) cost-effective

192. Equipment intended to break current at other than fault levels shall have an interrupting rating at system voltage sufficient for the current that must be interrupted.
(a) True
(b) False

193. Circuit-protective devices are used to clear a fault without the occurrence of extensive damage to the electrical components of the circuit. This fault shall be assumed to be either between two or more of the _____ or between any circuit conductor and the grounding conductor or enclosing metal raceway.
(a) bonding jumpers
(b) grounding jumpers
(c) wiring harnesses
(d) circuit conductors

194. The _____ of the circuit shall be so selected and coordinated as to permit the circuit protective devices to clear a fault without extensive damage to the electrical components of the circuit.
(a) overcurrent protective devices
(b) total circuit impedance
(c) component short-circuit current ratings
(d) all of these

195. Unless identified for use in the operating environment, no conductors or equipment shall be _____ having a deteriorating effect on the conductors or equipment.
(a) located in damp or wet locations
(b) exposed to fumes, vapors, or gases
(c) exposed to liquids or excessive temperatures
(d) all of these

196. • Equipment approved for use in dry locations only shall be protected against permanent damage from the weather during _____.
(a) design   (b) building construction   (c) inspection   (d) none of these

197. Some spray cleaning and lubricating compounds contain chemicals that cause severe deteriorating reactions with plastics.
(a) True   (b) False

198. The *NEC* requires that electrical work be installed _____.
(a) in a neat and workmanlike manner
(b) under the supervision of a qualified person
(c) completed before being inspected
(d) all of these

199. Conductors shall be _____ to provide ready and safe access in underground and subsurface enclosures into which persons enter for installation and maintenance.
(a) bundled   (b) tied together   (c) color-coded   (d) racked

200. The *Code* prohibits damage to the internal parts of electrical equipment by foreign material such as paint, plaster, cleaners, etc. Precautions shall be taken to provide protection from the detrimental effects of paint, plaster, cleaners, etc. on internal parts such as _____.
(a) busbars   (b) wiring terminals   (c) insulators   (d) all of these

Unit 1 NEC Exam – NEC Code Order 90.1 – 110.12

1. The *Code* contains provisions considered necessary for safety regardless of _____.
   (a) efficient use
   (b) convenience
   (c) good service or future expansion of electrical use
   (d) all of these

2. Hazards often occur because of _____.
   (a) overloading of wiring systems by methods or usage not in conformity with this *Code*
   (b) initial wiring not providing for increases in the use of electricity
   (c) a and b
   (d) none of these

3. The *Code* applies to the installation of _____.
   (a) electrical conductors and equipment within or on public and private buildings
   (b) outside conductors and equipment on the premises
   (c) optical fiber cable
   (d) all of these

4. The *Code* covers all of the following electrical installations except _____.
   (a) floating buildings
   (b) in or on private and public buildings
   (c) industrial substations
   (d) electrical generation installations on property owned or leased by the electric utility company

5. Equipment that is listed by a qualified electrical testing laboratory is not required to have the factory-installed _____ wiring inspected at the time of installation except to detect alterations or damage.
   (a) external   (b) associated   (c) internal   (d) all of these

6. Admitting close approach, not guarded by locked doors, elevation or other effective means, is commonly referred to as _____.
   (a) accessible (equipment)   (b) accessible (wiring methods)
   (c) accessible, readily   (d) all of these

7. A generic term for a group of nonflammable synthetic chlorinated hydrocarbons used as electrical insulating media is _____.
   (a) oil   (b) girasol   (c) askarel   (d) phenol

8. A device that, by insertion in a receptacle, establishes a connection between the conductors of the attached flexible cord and the conductors connected permanently to the receptacle is called a(n) _____.
   (a) attachment plug   (b) plug cap
   (c) plug   (d) any of these

9. The connection between the grounded conductor and the equipment grounding conductor at the service is accomplished by installing a(n) _____ jumper.
   (a) main bonding   (b) bonding
   (c) equipment bonding   (d) circuit bonding

10. For a circuit to be considered a multiwire branch circuit, it must have _____.
    (a) two or more ungrounded conductors with a voltage potential between them
    (b) a grounded conductor having equal voltage potential between it and each ungrounded conductor of the circuit
    (c) a grounded conductor connected to the grounded (neutral) terminal of the system
    (d) all of these

11. A conductor encased within material of composition or thickness that is not recognized by this *Code* as electrical insulation is considered _____.
    (a) noninsulating   (b) bare   (c) covered   (d) protected

12. The _____ of any system is the ratio of the maximum demand of a system, or part of a system, to the total connected load of a system under consideration.
    (a) load
    (b) demand factor
    (c) minimum load
    (d) computed factor

13. Surrounded by a case, housing, fence, or wall(s) that prevents persons from accidentally contacting energized parts is called _____.
    (a) guarded   (b) covered   (c) protection   (d) an enclosure

14. When the term exposed, as it relates to live parts, is used by the *Code* it refers to _____.
    (a) capable of being inadvertently touched or approached nearer than a safe distance by a person
    (b) parts that are not suitably guarded, isolated, or insulated
    (c) wiring on or attached to the surface or behind panels designed to allow access
    (d) a and b

15. In a grounded system, the conductor that connects the circuit grounded conductor at the service to the grounding electrode is called the _____ conductor.
    (a) main grounding
    (b) common main
    (c) equipment grounding
    (d) grounding electrode

16. An overload may be caused by a short circuit or ground fault.
    (a) True   (b) False

17. When one electrical circuit controls another circuit through a relay, that first circuit is called a _____.
    (a) control circuit
    (b) remote-control circuit
    (c) signal circuit
    (d) controller

18. The _____ is the necessary equipment, usually consisting of a circuit breaker(s) or switch(es) and fuse(s) and their accessories, connected to the load end of service conductors to a building or other structure, or an otherwise designated area, and intended to constitute the main control and cutoff of the supply.
    (a) service equipment
    (b) service
    (c) service disconnect
    (d) service overcurrent protection device

19. Underground service conductors between the street main and the first point of connection to the service entrance are known as the _____.
    (a) utility service
    (b) service lateral
    (c) service drop
    (d) main service conductors

20. In determining equipment to be installed, considerations such as the following should be evaluated:
    (a) mechanical strength
    (b) cost
    (c) arcing effects
    (d) a and c

21. To be *Code* compliant, listed or labeled equipment shall be installed and used in accordance with any instructions included in the _____.
    (a) catalog
    (b) product
    (c) listing or labeling
    (d) all of these

22. Only wiring methods recognized as _____ are included in the *Code*.
    (a) identified   (b) efficient   (c) suitable   (d) cost-effective

23. Equipment approved for use in dry locations only shall be protected against permanent damage from the weather during _____.
    (a) design
    (b) building construction
    (c) inspection
    (d) none of these

24.   The *NEC* requires that electrical work be installed _____.
      (a) in a neat and workmanlike manner
      (b) under the supervision of a qualified person
      (c) completed before being inspected
      (d) all of these

25.   The *Code* prohibits damage to the internal parts of electrical equipment by foreign material such as paint, plaster, cleaners, etc. Precautions must be taken to provide protection from the detrimental effects of paint, plaster, cleaners, etc. on internal parts such as _____.
      (a) busbars   (b) wiring terminals   (c) insulators   (d) all of these

Unit 2

Electrical Circuits

OBJECTIVES

After reading this unit, the student should be able to briefly explain the following concepts:

Series circuit calculations

Parallel circuit calculations

Series-parallel circuit

Calculations

Multiwire circuit calculations

Neutral current calculations

Dangers of multiwire circuits

After reading this unit, the student should be able to briefly explain the following terms:

Amp-hour

Equal resistor method

Kirchoff's law

Multiwire circuits

Neutral conductor

Nonlinear leads

Parallel circuits

Phases

Pigtailing

Product of the sum method

Reciprocal method

Series circuit

Series-parallel circuit

Unbalanced current

Ungrounded conductors
(hot wires)

PART A – SERIES CIRCUITS

INTRODUCTION TO SERIES CIRCUITS

A series circuit is a circuit in which the current leaves the voltage source and flows through every electrical device with the same intensity before it returns to the voltage source. If any part of a series circuit is opened, the current stops flowing in the entire circuit. Figure 2–1.

For most practical purposes, series (closed-loop) circuits are not used for building wiring, but they are important for the operation of many control and signal circuits. Figure 2–2. Motor control circuit stop switches are generally wired in series with the starter's coil and the line conductors. Dual-rated motors, such as 460/230V, have their windings connected in series when supplied by the higher voltage, and in parallel when supplied by the lower voltage.

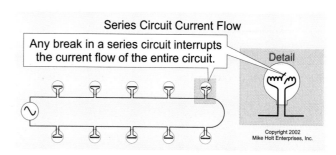

Series Circuit Current Flow

Any break in a series circuit interrupts the current flow of the entire circuit.

Detail

Copyright 2002
Mike Holt Enterprises, Inc.

Figure 2–1
Series Circuit Current Flow

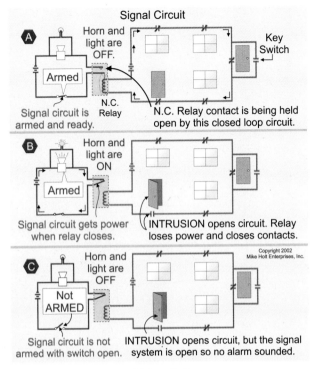

Figure 2–2

Signal Circuit

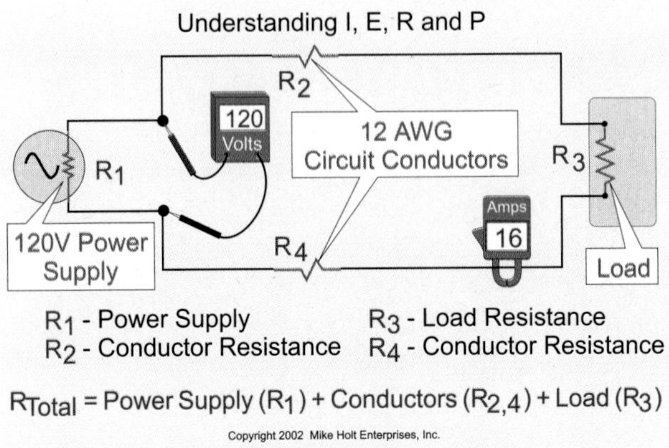

Figure 2–3

Understanding I, E, R, and P

2–1 UNDERSTANDING SERIES CALCULATIONS

It is important to understand the relationship between resistance, current, voltage, and power of series circuits. Figure 2–3.

Calculating Resistance Total

In a series circuit, the total resistance of the circuit is equal to the sum of all the series resistors' resistance, according to the formula:

$$R_T = R_1 + R_2 + R_3 + R_4$$

❏ **Total Resistance**

What is the total resistance of the loads in Figure 2–4?

(a) 2.5Ω   (b) 5.5Ω

(c) 7.5Ω   (d) 10Ω

• Answer: (c) 7.5Ω

R_1 – Power Supply	0.05Ω
R_2 – Conductor 1	0.15Ω
R_3 – Appliance	7.15Ω
R_4 – Conductor 2	0.15Ω
Total Resistance:	7.50Ω

Calculating Voltage Drop

The result of current flowing through a resistor is a reduction in voltage across the resistor, which is called voltage drop. In a closed-loop circuit, the sum of the voltage drops of all the loads is equal to the voltage source. Figure 2–5. This is known as Kirchoff's First Law of Voltage. The voltage drop (VD) of each resistor can be determined by the formula:

$$E_{VD} = I \times R$$

I = Current of the circuit

R = Resistance of the resistor

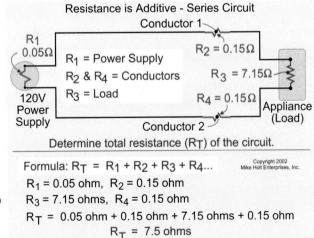

Determine total resistance (R_T) of the circuit.

Formula: $R_T = R_1 + R_2 + R_3 + R_4...$

$R_1 = 0.05$ ohm, $R_2 = 0.15$ ohm

$R_3 = 7.15$ ohms, $R_4 = 0.15$ ohm

$R_T = 0.05$ ohm + 0.15 ohm + 7.15 ohms + 0.15 ohm

$R_T = 7.5$ ohms

Figure 2–4

Resistance is Additive

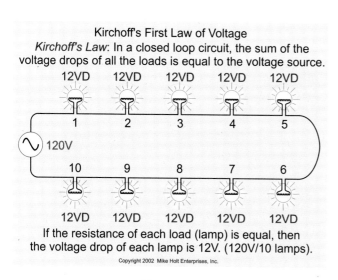

Kirchoff's First Law of Voltage

Kirchoff's Law: In a closed loop circuit, the sum of the voltage drops of all the loads is equal to the voltage source.

If the resistance of each load (lamp) is equal, then the voltage drop of each lamp is 12V. (120V/10 lamps).

Copyright 2002 Mike Holt Enterprises, Inc.

Figure 2–5
Voltage Drop Distribution

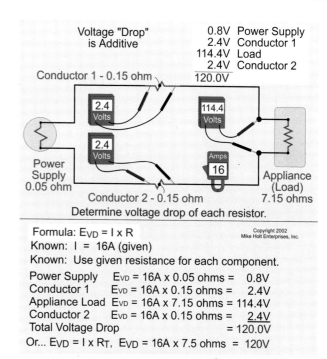

Determine voltage drop of each resistor.

Formula: $E_{VD} = I \times R$ Copyright 2002 Mike Holt Enterprises, Inc.
Known: I = 16A (given)
Known: Use given resistance for each component.

Power Supply E_{VD} = 16A x 0.05 ohms =   0.8V
Conductor 1 E_{VD} = 16A x 0.15 ohms =   2.4V
Appliance Load E_{VD} = 16A x 7.15 ohms = 114.4V
Conductor 2 E_{VD} = 16A x 0.15 ohms =   2.4V
Total Voltage Drop   = 120.0V
Or... $E_{VD} = I \times R_T$, E_{VD} = 16A x 7.5 ohms = 120V

Figure 2–6
Series – Voltage Drop Example

❏ **Voltage Drop**

What is the voltage drop across each resistor in Figure 2–6 using the formula $VD = I \times R$?

• Answer: $VD = I \times R$

$VD = I \times R$

VD_{R1} – Power Supply	$16A \times 0.05\Omega$ =	0.8V
VD_{R2} – Conductor 1	$16A \times 0.15\Omega$ =	2.4V
VD_{R3} – Appliance	$16A \times 7.15\Omega$ =	114.4V
VD_{R4} – Conductor 2	$16A \times 0.15\Omega$ =	2.4V
Total Voltage Drop	$16A \times 7.50\Omega$ =	120.0V

Note: Due to rounding, the sum of the voltage drops may be slightly different than the voltage source.

Voltage of Series Connected Power Supplies

When *power supplies* are connected in series, the voltage of each power supply will add together (sum), providing that all the polarities are connected properly.

❏ **Series Connected Power Supplies**

What is the total voltage output of four 1.5V batteries connected in series? Figure 2–7.

(a) 1.5V   (b) 3.0V
(c) 4.5V   (d) 6.0V
• Answer: (d) 6V

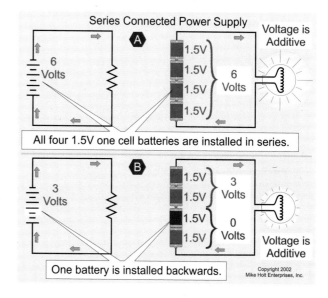

Series Connected Power Supply

All four 1.5V one cell batteries are installed in series.

One battery is installed backwards.

Copyright 2002 Mike Holt Enterprises, Inc.

Figure 2–7
Series Connected Power Supply

Current of Resistor

In a series circuit, the current throughout the circuit is constant and does not change. The current through each resistor of the circuit can be determined by the formula:

$$I = \frac{E}{R}$$

E = Voltage drop of the load or circuit

R = Resistance of the load or circuit

Note: If the resistance is not given, you can determine the resistance of a resistor (if you know the nameplate voltage and power rating of the load) by the formula: $R = E^2/P$, E^2 = Nameplate voltage rating (squared), P = Nameplate power rating.

❏ Current of Circuit

What is the current flow through the series circuit in Figure 2–8? Note: In a series circuit, current remains the same.

(a) 4A (b) 8A
(c) 12A (d) 16A

• Answer: (d) 16A

$I = \frac{E}{R}$
I_{R1} – Power Supply   0.8 VD/0.05Ω = 16A
I_{R2} – Conductor 1   2.4 VD /0.15Ω = 16A
I_{R3} – Appliance   114.4 VD/7.15Ω = 16A
I_{R4} – Conductor 2   2.4 VD/0.15Ω = 16A
Circuit Current   120V/7.5Ω = 16A

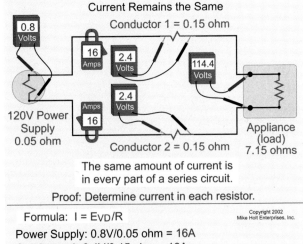

Current Remains the Same
Conductor 1 = 0.15 ohm

0.8 Volts
16 Amps
2.4 Volts
114.4 Volts
2.4 Volts
Amps
16
120V Power Supply
0.05 ohm
Conductor 2 = 0.15 ohm
Appliance (load)
7.15 ohms

The same amount of current is in every part of a series circuit.

Proof: Determine current in each resistor.

Copyright 2002
Mike Holt Enterprises, Inc.

Formula: I = E_{VD}/R
Power Supply: 0.8V/0.05 ohm = 16A
Conductor 1: 2.4V/0.15 ohm = 16A
Appliance Load: 114.4V/7.15 ohms = 16A
Conductor 2: 2.4V/0.15 ohm = 16A

Or... use total resistance, I = E_T/R_T
E = 120V, R_T = 0.05 + 0.15 + 7.15 + 0.15 = 7.5 ohms
I = 120V/7.5 ohms = 16A

Figure 2–8
Current Remains the Same

Power of Resistor or Circuit

The power consumed in a series circuit is equal to the sum of the power of all of the resistors in the series circuit. The resistor with the highest resistance will consume the most power, and the resistor with the smallest resistance will consume the least power. You can calculate the power consumed (watts) of each resistor or of the circuit by the formula:

$$P = I^2R, \quad I^2 = \text{Current of the circuit (squared)}, \quad R = \text{Resistance of circuit or resistor}$$

❏ Power

What is the power consumed of each resistor in Figure 2–9?

• Answer: P = I^2R

$P = I^2R$
P_{R1} – Power Supply $16A^2 \times 0.05Ω$ = 12.8W
P_{R2} – Conductor 1 $16A^2 \times 0.15Ω$ = 38.4W
P_{R3} – Appliance $16A^2 \times 7.15Ω$ = 1,830W
P_{R4} – Conductor 2 $16A^2 \times 0.15Ω$ = 38W

2–2 SERIES CIRCUIT SUMMARY

Figure 2–10, Series Circuit Summary.

Note 1:   The total resistance of a series circuit is equal to the sum of the resistors in the circuit.

Note 2:   Current is constant.

Note 3:   The sum of the voltage drop from all resistors equals the voltage source.

Note 4:   The sum of the power from all resistors equals the total power of the circuit.

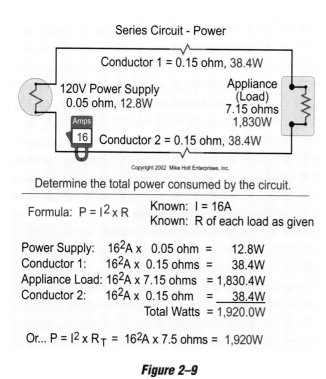

Series Circuit - Power

Conductor 1 = 0.15 ohm, 38.4W

120V Power Supply
0.05 ohm, 12.8W

Appliance
(Load)
7.15 ohms
1,830W

Amps
16

Conductor 2 = 0.15 ohm, 38.4W

Copyright 2002 Mike Holt Enterprises, Inc.

Determine the total power consumed by the circuit.

Formula: $P = I^2 \times R$

Known: I = 16A
Known: R of each load as given

Power Supply: $16^2A \times 0.05$ ohm =   12.8W
Conductor 1: $16^2A \times 0.15$ ohms =   38.4W
Appliance Load: $16^2A \times 7.15$ ohms = 1,830.4W
Conductor 2: $16^2A \times 0.15$ ohm =   38.4W
                              Total Watts = 1,920.0W

Or... $P = I^2 \times R_T = 16^2A \times 7.5$ ohms = 1,920W

Figure 2–9

Series Circuit – Power is Additive

Series Circuit - Summary

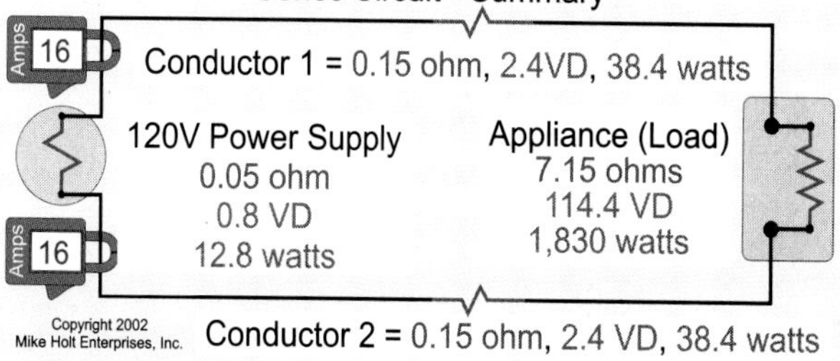

Amps
16

Conductor 1 = 0.15 ohm, 2.4VD, 38.4 watts

120V Power Supply
0.05 ohm
0.8 VD
12.8 watts

Appliance (Load)
7.15 ohms
114.4 VD
1,830 watts

Amps
16

Copyright 2002
Mike Holt Enterprises, Inc.

Conductor 2 = 0.15 ohm, 2.4 VD, 38.4 watts

Series Circuit Summary:

1: The total resistance (RT) is equal to the sum of all the resistors of the circuit. $R_T = R_1 + R_2 + R_3$...
   RT = 0.05Ω + 0.15Ω + 7.15Ω + 0.15Ω = 7.5 ohms

2: Current is constant. Current is the same in every part of the circuit. $I = E_T/R_T = 120V/7.5$ ohms = 16A

3: The sum of the voltage drop of all resistors must equal the voltage source. Voltage Source (VS) = $VD_1 + VD_2 + VD_3$...
   VS = 0.8V + 2.4V + 114.4V + 2.4V = 120V

4: The sum of the power of all resistors equals the total power of the circuit. $P_T = P_1 + P_2 + P_3$...
   P_T = 12.8W + 38.4W + 1,830W + 38.4W = 1,920W

Figure 2–10

Series Circuit Summary

PART B – PARALLEL CIRCUITS

INTRODUCTION TO PARALLEL CIRCUITS

A parallel circuit is a circuit in which current leaves the voltage source, branches through different parts of the circuit, and then returns to the voltage source. Figure 2–11. Parallel is a term used to describe a method of connecting electrical components so that the current can flow through two or more different branches of the circuit.

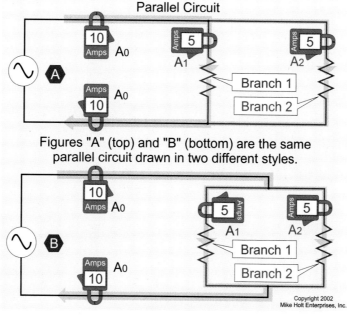

Parallel Circuit

Figures "A" (top) and "B" (bottom) are the same parallel circuit drawn in two different styles.

Copyright 2002
Mike Holt Enterprises, Inc.

Figure 2–11

Parallel Circuit

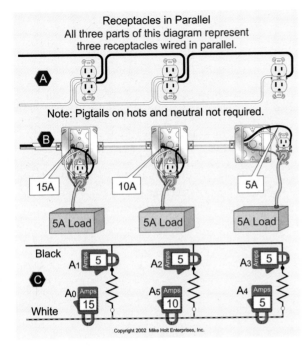

Receptacles in Parallel
All three parts of this diagram represent three receptacles wired in parallel.

Note: Pigtails on hots and neutral not required.

15A   10A   5A

5A Load   5A Load   5A Load

Black

White

Copyright 2002 Mike Holt Enterprises, Inc.

Figure 2–12
Receptacles In Parallel

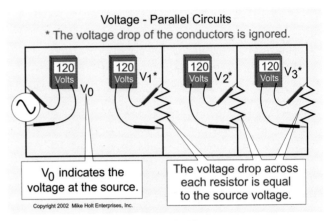

Voltage - Parallel Circuits
* The voltage drop of the conductors is ignored.

V_0 indicates the voltage at the source.

The voltage drop across each resistor is equal to the source voltage.

Copyright 2002 Mike Holt Enterprises, Inc.

Figure 2–13
Voltage in Parallel Circuits

2–3 PRACTICAL USES OF PARALLEL CIRCUITS

Parallel circuits are commonly used for building wiring. Figure 2–12. In addition, parallel (open-loop) circuits are often used for fire alarm systems and the internal wiring of many types of electrical equipment, such as motors and transformers. Dual-rated motors, such as 460/230V, have their windings connected in parallel when supplied by the lower voltage, and in series when supplied by the higher voltage.

2–4 UNDERSTANDING PARALLEL CALCULATIONS

It is important to understand the relationship between voltage, current, power, and resistance of a parallel circuit.

Voltage of Each Branch

The voltage drop across loads connected in parallel is equal to the voltage that supplies each parallel branch. Figure 2–13.

Power Supplies Connected in Parallel

When power supplies are connected in parallel, voltage remains the same but the current, or amp-hour, capacity is increased. When connecting batteries in parallel, always connect batteries of the same voltage with the proper polarity.

❏ Parallel Connected Power Supplies

If two 12V batteries are connected in parallel, what is the output voltage? Figure 2–14.

(a) 3V   (b) 6V
(c) 12V   (d) 24V

• Answer: (c) 12V. The voltage remains the same when power supplies are connected in parallel.

Note: Two 12V batteries connected in parallel will result in an output voltage of 12V, but the amp-hour capacity will be increased, resulting in longer service.

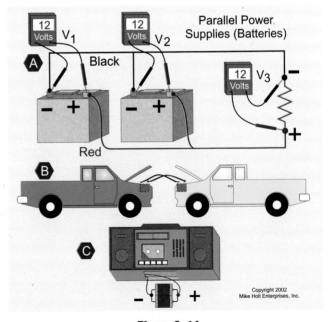

Parallel Power Supplies (Batteries)

Black

Red

Copyright 2002
Mike Holt Enterprises, Inc.

Figure 2–14
Parallel Connected Power Supplies

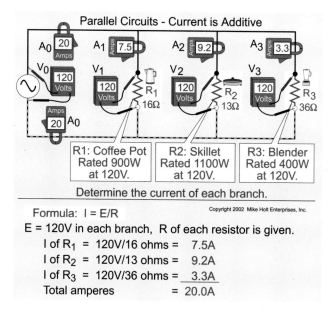

R1: Coffee Pot Rated 900W at 120V.
R2: Skillet Rated 1100W at 120V.
R3: Blender Rated 400W at 120V.

Determine the current of each branch.

Formula: I = E/R

Copyright 2002 Mike Holt Enterprises, Inc.

E = 120V in each branch, R of each resistor is given.

$$I \text{ of } R_1 = 120V/16 \text{ ohms} = 7.5A$$
$$I \text{ of } R_2 = 120V/13 \text{ ohms} = 9.2A$$
$$I \text{ of } R_3 = 120V/36 \text{ ohms} = \underline{3.3A}$$
$$\text{Total amperes} = 20.0A$$

Figure 2–15
Parallel Circuits – Current is Additive

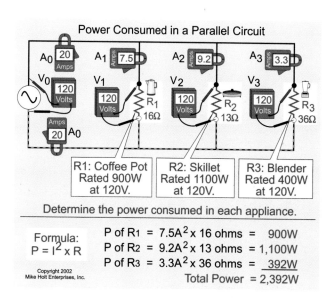

R1: Coffee Pot Rated 900W at 120V.
R2: Skillet Rated 1100W at 120V.
R3: Blender Rated 400W at 120V.

Determine the power consumed in each appliance.

Formula:
$P = I^2 \times R$

Copyright 2002 Mike Holt Enterprises, Inc.

$$P \text{ of } R_1 = 7.5A^2 \times 16 \text{ ohms} = 900W$$
$$P \text{ of } R_2 = 9.2A^2 \times 13 \text{ ohms} = 1,100W$$
$$P \text{ of } R_3 = 3.3A^2 \times 36 \text{ ohms} = \underline{392W}$$
$$\text{Total Power} = 2,392W$$

Figure 2–16
Power Consumed in a Parallel Circuit

Current Through Each Branch

The current from the power supply is equal to the sum of the branch-circuit currents. The current in each branch depends on the branch voltage and branch resistance and can be calculated by the formula:

$$I = {}^{E}/_{R}$$

E = Voltage of Branch

R = Resistance of Branch

❏ Current Through Each Branch

What is the current of each appliance in Figure 2–15?

• Answer: $I = {}^{E}/_{R}$

$I = {}^{E}/_{R}$

I_{R1} – Coffee Pot	120V/16Ω	=	7.50A
I_{R2} – Skillet	120V/13Ω	=	9.20A
I_{R3} – Blender	120V/36Ω	=	3.30A
Total Current (7.5A + 9.17A + 3.33A)			20.00A

Power Consumed of Each Branch

The total power consumed by any circuit is equal to the sum of the branch powers. Each branch power depends on the branch current and resistance. The power can be found by the formula:

$$P = I^2R$$

I = Current of Each Branch

R = Resistance of Each Branch

❏ Power of Each Branch

What is the power consumed by each appliance in Figure 2–16?

• Answer: $P = I^2R$

$P = I^2R$

I_{R1} – Coffee Pot	$7.5A^2 \times 16Ω$	=	900W
I_{R2} – Skillet	$9.2A^2 \times 13Ω$	=	1,093W
I_{R3} – Blender	$3.3A^2 \times 36Ω$	=	399W
Total Power (900W + 1,093W + 399W)			2,392W

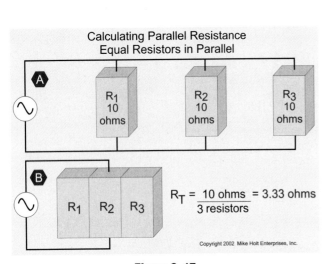

Figure 2–17
Calculating Parallel Resistance

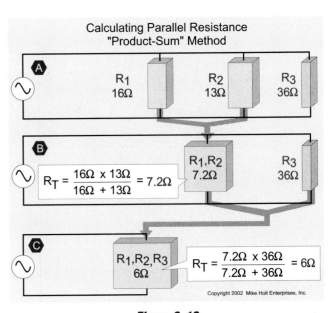

Figure 2–18
Resistance Product Over the Sum Method

2–5 PARALLEL CIRCUIT RESISTANCE CALCULATIONS

In a parallel circuit, the total circuit resistance is always less than the smallest resistor and can be determined by one of three methods:

Equal Resistor Method

The equal resistor method can be used when all the resistors of the parallel circuit have the same resistance. Simply divide the resistance of one resistor by the number of resistors in parallel.

$$R_T = R/N$$

R = Resistance of One Resistor, N = Number of Resistors

❏ Equal Resistors Method

The total resistance of three 10Ω resistors is _____. Figure 2–17.

(a) 10Ω (b) 20Ω (c) 30Ω (d) none of these

• Answer: (d) none of these

$$R_T = \frac{\text{Resistance of One Resistor}}{\text{Number of Resistors}} = \frac{10\Omega}{3 \text{ resistors}} = 3.33\Omega$$

The Product-Over-the-Sum Method

The product-over-the-sum method can be used to calculate the resistance of two resistors.

$$R_T = \frac{R_1 \times R_2 \text{ (Product)}}{R_1 + R_2 \text{ (Sum)}}$$

The term *product* means the answer of numbers that are multiplied together. The term *sum* is the answer to numbers that are added together. The product-over-the-sum method can be used for more than two resistors, but only two can be calculated at a time.

❏ Product-Over-the-Sum Method

What is the total resistance of a 16Ω coffee pot and a 13Ω skillet connected in parallel? Figure 2–18.

(a) 16Ω (b) 13.09Ω (c) 29.09Ω (d) 7.2Ω

• Answer: (d) 7.2Ω

The total resistance of a parallel circuit is always less than the smallest resistor (13Ω).

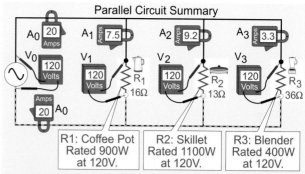

Parallel Circuit Summary

R1: Coffee Pot Rated 900W at 120V.
R2: Skillet Rated 1100W at 120V.
R3: Blender Rated 400W at 120V.

Copyright 2002 Mike Holt Enterprises, Inc.

Parallel Circuit Summary:
1: Total resistance is always less than smallest resistor.
2: Total current is equal to the sum of the currents of the individual branches. $A_0 = A_1 + A_2 + A_3...$
$A_0 = 7.5A + 9.2A + 3.3A = 20A$
3: Voltage is constant.
4: Total power is equal to the sum of the power of the individual branches. $P_T = P_1 + P_2 + P_3...$
$P_T = 900W + 1100W + 400W = 2,400W$

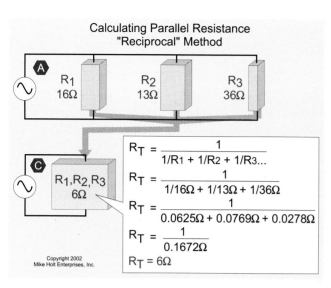

Calculating Parallel Resistance "Reciprocal" Method

$$R_T = \frac{1}{1/R_1 + 1/R_2 + 1/R_3...}$$

$$R_T = \frac{1}{1/16\Omega + 1/13\Omega + 1/36\Omega}$$

$$R_T = \frac{1}{0.0625\Omega + 0.0769\Omega + 0.0278\Omega}$$

$$R_T = \frac{1}{0.1672\Omega}$$

$$R_T = 6\Omega$$

Copyright 2002
Mike Holt Enterprises, Inc.

Figure 2–19
Calculating Parallel Resistance "Reciprocal" Method

Figure 2–20
Parallel Circuit Summary

$$R_T = \frac{R_1 \times R_2}{R_1 + R_2} = \frac{16\Omega \times 13\Omega}{16\Omega + 13\Omega} = 7.2\Omega$$

Reciprocal Method

The advantage of the reciprocal method is that this formula can be used for an unlimited number of parallel resistors.

$$R_T = \frac{1}{{}^1/_{R1} + {}^1/_{R2} + {}^1/_{R3} ...}$$

❏ **Reciprocal Method**

What is the resistance total of a 16Ω, 13Ω and 36Ω resistor connected in parallel? Figure 2–19.

(a) 13Ω   (b) 16Ω   (c) 36Ω   (d) 6Ω

• Answer: (d) 6Ω

$$R_T = \frac{1}{{}^1/_{16}\,\Omega + {}^1/_{13}\,\Omega + {}^1/_{36}\,\Omega} = \frac{1}{0.0625\Omega + 0.0769\Omega + 0.0278\Omega}$$

$$R_T = \frac{1}{0.1672\,\Omega} = 6\Omega$$

2–6 PARALLEL CIRCUIT SUMMARY

Note 1:   The total resistance of a parallel circuit is always less than the smallest resistor. Figure 2–20.

Note 2:   Current total of a parallel circuit is equal to the sum of the currents of the individual branches.

Note 3:   Voltage is constant.

Note 4:   Power total is equal to the sum of the power in all the individual branches.

PART C – SERIES–PARALLEL AND MULTIWIRE BRANCH CIRCUITS

INTRODUCTION TO SERIES-PARALLEL CIRCUITS

A series-parallel circuit is a circuit that contains some resistors in series and some resistors in parallel to each other. Figure 2–21.

2–7 REVIEW OF SERIES AND PARALLEL CIRCUITS

For a better understanding of series-parallel circuits, let's review the rules for series and parallel circuits. That portion of the circuit that contains resistors in series must comply with the rules of series circuits, and that portion of the circuit that is connected in parallel must comply with the rules of parallel circuits. Figure 2–22.

Series Circuit Rules:

Note 1:   Resistance is additive.

Note 2:   Current is constant.

Note 3:   Voltage is additive.

Note 4:   Power is additive.

Parallel Circuit Rules:

Note 1:   Resistance is less than the smallest resistor.

Note 2:   Current is additive.

Note 3:   Voltage is constant.

Note 4:   Power is additive.

2–8 SERIES-PARALLEL CIRCUIT RESISTANCE CALCULATIONS

When determining the resistance total of a series-parallel circuit, it is best to redraw the circuit so you can see the series components and the parallel branches. Determine the resistance of the series components first, or the parallel components depending on the circuit, then determine the resistance total of all the branches. Keep breaking the circuit down until you have determined the total effective resistance of the circuit as one resistor.

❏ **Series-Parallel Circuit Resistance**

What is the resistance total of the circuit shown in Figure 2–23, Part A?

(a) 2Ω   (b) 5Ω   (c) 7Ω   (d) 10Ω

• Answer: (d) 10Ω

Series - Parallel Circuits

Figure 2–21
Series – Parallel Circuits

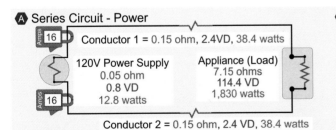

A Series Circuit - Power

1: The total resistance (RT) is equal to the sum of all the resistors of the circuit.
$R_T = R_1 + R_2 + R_3...$, $R_T = 0.05\Omega + 0.15\Omega + 7.15\Omega + 0.15\Omega = 7.5$ ohms
2: Current is constant. Current is the same in every part of the circuit.
$I = E_T/R_T = 120V/7.5$ ohms = 16A
3: The sum of the voltage drop of all resistors must equal the voltage source.
Voltage Source (VS) = $VD_1 + VD_2 + VD_3...$
VS = 0.8V + 2.4V + 114.4V + 2.4V = 120V
4: The sum of the power of all resistors equals the total power of the circuit.
$P_T = P_1 + P_2 + P_3...$, $P_T = 12.8W + 38.4W + 1,830W + 38.4W = 1,920W$

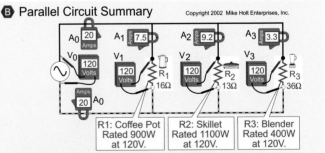

B Parallel Circuit Summary   Copyright 2002 Mike Holt Enterprises, Inc.

R1: Coffee Pot Rated 900W at 120V.
R2: Skillet Rated 1100W at 120V.
R3: Blender Rated 400W at 120V.

1: Total resistance is always less than the smallest resistor.
2: Total current is equal to the sum of the currents of the individual branches.
$A_0 = A_1 + A_2 + A_3...$, $A_0 = 7.5A + 9.2A + 3.3A = 20A$
3: Voltage is constant.
4: Total power is equal to the sum of the power of the individual branches.
$P_T = P_1 + P_2 + P_3...$, $P_T = 900W + 1100W + 400W = 2,400W$

Figure 2–22
Series and Parallel Circuit Summary

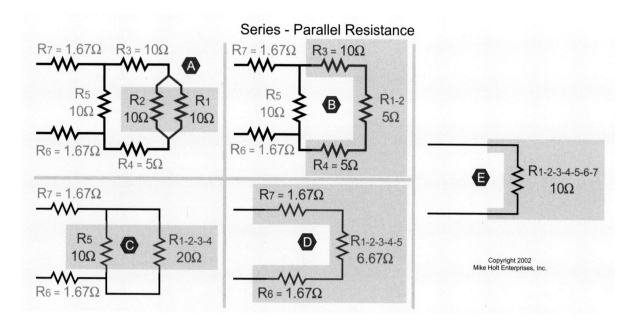

Figure 2–23
Series-Parallel Resistance

Step 1:   Determine the equal parallel resistors: R_1 (10Ω) and R_2 (10Ω) Figure 2–23, Part A:
$R_T = {}^R/_N$, R = Resistance of one resistor, N = Number of resistors.
$R_T = 10\Omega/2$ resistors, $R_T = 5\Omega$

Step 2:   Redraw the circuit – Figure 2–23, Part B.
Determine the series resistance of $R_{1,2}$ (5Ω), plus R_3 (10Ω), plus R_4 (5Ω).
Series resistance total is equal to $R_{1,2} + R_3 + R_4$.
$R_T = 5\Omega + 10\Omega + 5\Omega$, $R_T = 20\Omega$

Step 3:   Redraw the circuit – Figure 2–23, Part C.
Determine the parallel resistance of R_5, plus the resistance of $R_{1,2,3,4}$. Remember the resistance total of a parallel circuit is always less than the smallest parallel branch (10Ω). Since we have only two parallel branches, the resistance can be determined by the product of the sum method.

$$R_T = \frac{R_5 \times R_{1,2,3,4}\ (\text{Product})}{R_1 + R_2\ (\text{Sum})} = \frac{(10\Omega \times 20\Omega)}{(10\Omega + 20\Omega)} = \frac{200\Omega}{30\Omega} = 6.67\Omega$$

Step 4:   Redraw the circuit – Figure 2–23, Part D.
Determine the series resistance total of R_7, plus R_6, plus $R_{1,2,3,4,5}$
Series resistance total is equal to $R_7 + R_6 + R_{1,2,3,4,5}$
$R_T = 1.67\Omega + 1.67\Omega + 6.67\Omega$, $R_T = 10\Omega$

PART D – MULTIWIRE BRANCH CIRCUITS

INTRODUCTION TO MULTIWIRE BRANCH CIRCUITS

A multiwire branch circuit consists of two or more ungrounded conductors (hot or phase conductors) having a potential difference between them and an equal difference of potential between each hot wire and grounded (neutral) conductor. Figure 2–24.

Note: The *NEC* contains specific requirements on multiwire branch circuits. See Article 100 definition of multiwire branch circuit, 210.4 branch circuit requirements, and 300.13(B) requirements for pigtailing neutral conductors.

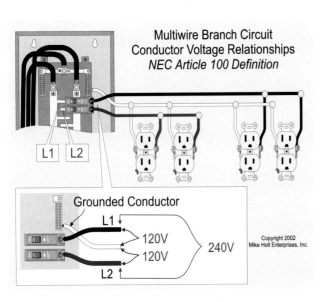

Figure 2–24
Multiwire Branch Circuit

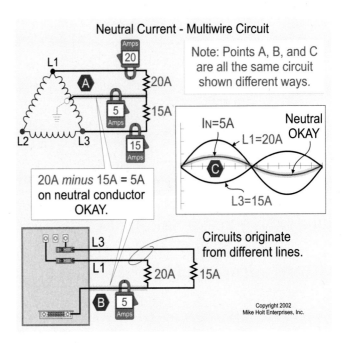

Figure 2–25
Neutral Current in Multiwire Branch Circuit

2–9 NEUTRAL CURRENT CALCULATIONS

When current flows in the *neutral* conductor of a multiwire branch circuit, the current is called *unbalanced current*. This current can be determined according to the following:

120/240V, 3-Wire Circuits

In a 120/240V, 3-wire circuit consisting of two hot wires and a grounded (neutral) conductor, the third wire will carry no current when the circuit is balanced. However, the neutral (or grounded) conductor will carry unbalanced current when the circuit is not balanced. The neutral current can be calculated as:

$$I_{Neutral} = I_{Line\ 1} - I_{Line\ 2}$$

$I_{Line\ 1}$ = Line 1 Neutral Current

$I_{Line\ 2}$ = Line 2 Neutral Current

❏ 120/240V, 3-Wire Neutral Current

What is the neutral current if current is: Line 1 = 20A and Line 2 = 15A? Figure 2–25.

(a) 0A   (b) 5A   (c) 10A   (d) 35A

• Answer: (b) 5A

$I_{Neutral} = I_{Line\ 1} - I_{Line\ 2}$

$I_{Neutral}$ = 20A less 15A

$I_{Neutral}$ = 5A

Wye 3-Wire 1Ø Neutral Current

A 3-wire, 208Y/120 or 480Y/277V, 1Ø circuit consisting of two phases and a neutral always carries unbalanced current. The current on the grounded (neutral) conductor is determined by the formula:

$$I_{Neutral} = \sqrt{(I_{Line\ 1}{}^2 + I_{Line\ 2}{}^2) - (I_{Line\ 1} \times I_{Line\ 2})}$$

❏ 3-Wire Wye Circuit Neutral Current

What is the neutral current for a 20A, 1Ø, 3-wire circuit (two hots and a neutral)? Power is supplied from a 208Y/120V feeder. Figure 2–26.

(a) 40A   (b) 20A   (c) 60A   (d) 0A

• Answer: (b) 20A

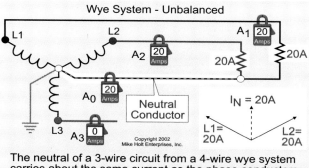

Wye System - Unbalanced

The neutral of a 3-wire circuit from a 4-wire wye system carries about the same current as the phase conductors.

$I_N = \sqrt{(L1^2 + L2^2 + L3^2) - [(L1 \times L2) + (L2 \times L3) + (L1 \times L3)]}$

$I_N = \sqrt{(20^2 + 20^2 + 0^2) - [(20 \times 20) + (20 \times 0) + (20 \times 0)]}$

$I_N = \sqrt{(400 + 400 + 0) - (400 + 0 + 0)}$

$I_N = \sqrt{800 - 400}$ $I_N = \sqrt{400}$ $I_N = 20A$

Figure 2–26
Neutral Current – Unbalanced Wye Circuit

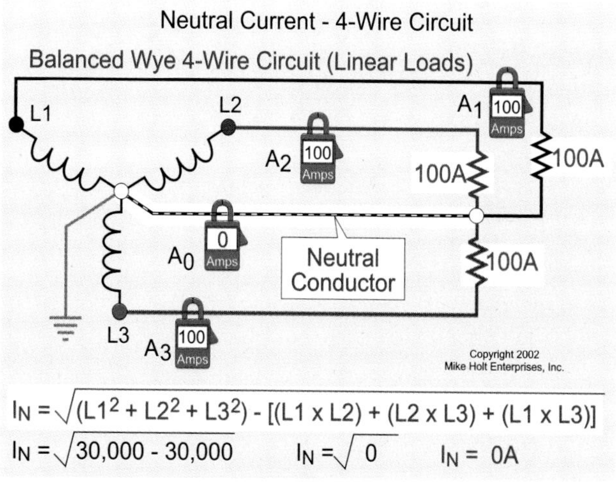

Neutral Current - 4-Wire Circuit

Balanced Wye 4-Wire Circuit (Linear Loads)

$I_N = \sqrt{(L1^2 + L2^2 + L3^2) - [(L1 \times L2) + (L2 \times L3) + (L1 \times L3)]}$

$I_N = \sqrt{30,000 - 30,000}$ $I_N = \sqrt{0}$ $I_N = 0A$

Figure 2–27
Neutral Current – Balanced Wye 4-Wire Circuit

$I_{Neutral} = \sqrt{(I_{Line\ 1}^2 + I_{Line\ 2}^2) - (I_{Line\ 1} \times I_{Line\ 2})}$

$I_{Neutral} = \sqrt{(20A^2 + 20A^2) - (20A \times 20A)} = \sqrt{400}$

$I_{Neutral} = 20A$

Wye 4-wire, 3Ø Circuit Neutral Current

A 4-wire, 3Ø wye, 208Y/120 or 480Y/277V circuit will carry no current when the circuit is balanced, but will carry unbalanced current when the circuit is not balanced.

$I_{Neutral} = \sqrt{(I_{Line\ 1}^2 + I_{Line\ 2}^2 + I_{Line\ 3}^2) - [(I_{Line\ 1} \times I_{Line\ 2}) + (I_{Line\ 2} \times I_{Line\ 3}) + (I_{Line\ 1} \times I_{Line\ 3})]}$

$I_{Line\ 1}$ = Neutral Current Line 1

$I_{Line\ 2}$ = Neutral Current Line 2

$I_{Line\ 3}$ = Neutral Current Line 3

❏ Balanced Circuits

What is the neutral current for a 3Ø, 4-wire, 208Y/120V feeder where L1 = 100A, L2 = 100A, and L3 = 100A? Figure 2–27.

(a) 50A   (b) 100A   (c) 125A   (d) 0A

• Answer: (d) 0A

$I_{Neutral} = \sqrt{(I_{Line\ 1}^2 + I_{Line\ 2}^2 + I_{Line\ 3}^2) - [(I_{Line\ 1} \times I_{Line\ 2}) + (I_{Line\ 2} \times I_{Line\ 3}) + (I_{Line\ 1} \times I_{Line\ 3})]}$

$I_{Neutral} = \sqrt{(100A^2 + 100A^2 + 100A^2) - [(100A \times 100A) + (100A \times 100A) + (100A \times 100A)]} = \sqrt{0}$

$I_{Neutral} = 0A$

❏ Unbalanced Circuits

What is the neutral current for a 3Ø, 4-wire, 208Y/120V feeder where L_1 = 100A, L_2 = 100A, and L_3 = 50A? Figure 2–28.

(a) 50A   (b) 100A   (c) 125A   (d) 0A

• Answer: (a) 50A

$I_{Neutral} = \sqrt{(I_{Line\ 1}^2 + I_{Line\ 2}^2 + I_{Line\ 3}^2) - [(I_{Line\ 1} \times I_{Line\ 2}) + (I_{Line\ 2} \times I_{Line\ 3)} + (I_{Line\ 1} \times I_{Line\ 3})]}$

$I_{Neutral} = \sqrt{(100A^2 + 100A^2 + 50A^2) - (100A \times 100A) + (100A \times 50A) + (100A \times 50A)]} = \sqrt{2,500}$

$I_{Neutral} = 50A$

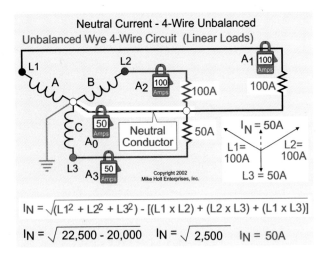

$$I_N = \sqrt{(L1^2 + L2^2 + L3^2) - [(L1 \times L2) + (L2 \times L3) + (L1 \times L3)]}$$

$$I_N = \sqrt{22,500 - 20,000} \qquad I_N = \sqrt{2,500} \qquad I_N = 50A$$

Figure 2-28
Neutral Current – 4-Wire Wye – Unbalanced

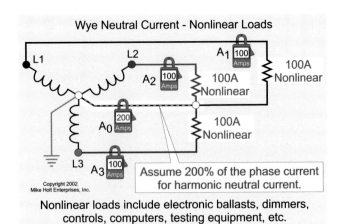

Assume 200% of the phase current for harmonic neutral current.

Nonlinear loads include electronic ballasts, dimmers, controls, computers, testing equipment, etc.

Figure 2-29
Neutral Current – Nonlinear Loads

Nonlinear Load Neutral Current

The neutral conductor of a balanced wye, 4-wire, 3Ø circuit will carry current when supplying power to nonlinear loads, such as computers, copy machines, laser printers, fluorescent lighting, etc. The current can be as much as two times the phase current depending on the harmonic content of the loads.

❑ Nonlinear Loads

What is the neutral current for a 3Ø, 4-wire, 208Y/120V feeder supplying power to a nonlinear load where L1 = 100A, L2 = 100A, and L3 = 100A? Assume that the harmonic content results in neutral current equal to 200% of the phase current. Figure 2–29.

(a) 80A   (b) 100A   (c) 125A   (d) 200A

• Answer: (d) 200A

2–10 DANGERS OF MULTIWIRE BRANCH CIRCUIT

Improper wiring, or mishandling of multiwire branch circuits, can cause excessive neutral current (overload) or destruction of electrical equipment because of overvoltage if the neutral conductor is opened.

Overloading the Neutral

If the ungrounded conductors (hot wires) of a multiwire branch circuit are connected to different phases, the current on the grounded (neutral) conductor will cancel. If the ungrounded conductors are not connected to different phases, the current from each phase will add on the grounded (neutral) conductor. This can result in an overload of the grounded (neutral) conductor. Figure 2–30.

Note: Overloading of the grounded (neutral) conductor will cause the insulation to look discolored due to excessive heat. Now you know why the neutral wires sometimes look burned.

Overvoltage

If the neutral conductor of a multiwire branch circuit is opened, the multiwire branch circuit changes from a parallel circuit into a series circuit. Instead of two 120V circuits, there is now one 240V circuit which can result in fires and the destruction of the electric equipment because of overvoltage. Figure 2–31.

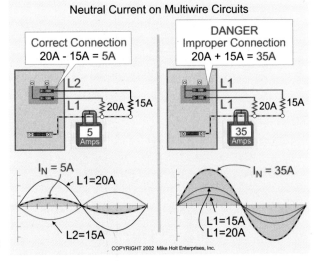

Figure 2-30
Neutral Current on Multiwire Circuits

To determine the operating voltage of each load in an open multiwire branch circuit, use the following steps:

Step 1:   Determine the resistance of each appliance, $R = E^2/P$.
E = Appliance voltage nameplate rating.
P = Appliance power nameplate rating.

Step 2:   Determine the circuit resistance, $R_T = R_1 + R_2$.

Step 3:   Determine the current of the circuit, $I = E_S/R_T$.
E_S = Voltage Source.
R_T = Resistance total, from Step 2.

Step 4:   Determine the voltage for each appliance, $E = I_T \times R$.
I_T = Current of the circuit.
R = Resistance of each resistor.

Step 5:   Determine the power consumed by each appliance, $P = E^2/R$.
E = Voltage the appliance operates at (squared).
R = Resistance of the appliance.

❑ **At what voltage does each of the loads operate if the neutral is opened? Figure 2–31.**

Step 1:   Determine the resistance of each appliance, $R = E^2/P$.
Hair dryer rated 1,200W at 120V.
$R = 120 V^2/1,200W = 12\Omega$
Television rated 600W at 120V.
$R = 120 V^2/600W = 24\Omega$

Step 2:   Determine the circuit resistance, $R_T = R_1 + R_2$.
$R_T = 12\Omega + 24\Omega = 36\Omega$

Step 3:   Determine the current of the circuit, $I = E_S/R_T$.
E_S = Voltage Source, R_T = Resistance Total
$I = \dfrac{240V}{36\Omega} = 6.67A$

Step 4:   Determine the voltage for each appliance, $E = I_T \times R$.
I_T = Current of the circuit, R = Resistance of each resistor
Hair dryer: $6.67A \times 12\Omega = 80V$
Television: $6.67A \times 24\Omega = 160V$

Step 5:   Determine the power consumed by each appliance, $P = E^2/R$.
E^2 = Voltage the appliance operates at (squared)
R = Resistance of the appliance
Hair Dryer: $P = 80V^2/12\Omega = 533$ W
Television: $P = 160V^2/24\Omega = 1,067W$

Note: The 600W, 120V rated TV operates at 160V and consumes 1,067W. You can kiss this TV good-bye because of the dangers associated with multiwire branch circuits. Don't use them for sensitive or expensive equipment such as computers, stereos, etc.

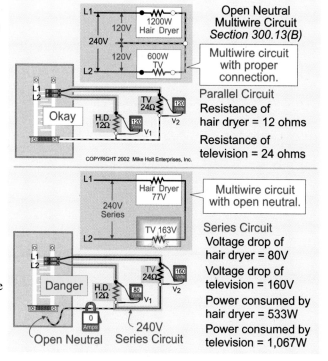

Figure 2-31
Neutral Current on Multiwire Circuits

Unit 2 – Electrical Circuits Summary Questions

Calculations Questions

Part A – Series Circuits

1. A closed-loop circuit is a circuit in which a specific amount of current leaves the voltage source and flows through every electrical device in a single path before it returns to the voltage source.
   (a) True   (b) False

2. Closed-loop circuits are typically used for signal and control circuits.
   (a) True   (b) False

2–1 Understanding Series Calculations

3. Resistance opposes the flow of electrons. In a series circuit, the total resistance of the circuit is equal to the sum of the resistance of all the resistors.
   (a) True   (b) False

4. The opposition to current flow results in _____.
   (a) current   (b) voltage   (c) voltage drop   (d) none of these

5. In a series circuit, the current is _____ through the transformer winding, the conductors, and the appliance.
   (a) proportional   (b) distributed   (c) additive   (d) constant

6. • When power supplies are connected in series, the voltage remains the same provided all the polarities are connected properly.
   (a) True   (b) False

7. The power consumed in a series circuit is equal to the power of the largest resistor in the series circuit.
   (a) True   (b) False

Part B – Parallel Circuits

Introduction to Parallel Circuits

8. A _____ circuit is a circuit in which current leaves the voltage source, branches through different parts of the circuit in different magnitudes, and then branches back to the voltage source.
   (a) series   (b) parallel   (c) series-parallel   (d) multiwire

2–4 Understanding Parallel Calculations

9. • The power supply provides the pressure needed to move the electrons, however, the _____ oppose(s) the current flow.
   (a) power supply   (b) conductors   (c) appliances   (d) all of these

10. When power supplies are connected in parallel, the amp-hour capacity remains the same.
    (a) True   (b) False

11. The total current of a parallel circuit is equal to the sum of the branch currents. The current in each branch can be calculated by the formula: $I = {}^E/_R$.
    (a) True   (b) False

12. When current flows through a resistor, power is consumed. The power consumption of each branch in a parallel circuit can be determined by the formula: P = I² × R. The total power consumed in a parallel circuit is equal to the largest branch power.
(a) True (b) False

2–5 Parallel Circuit Resistance Calculations

13. The basic method(s) of calculating total resistance of a parallel circuit is/are _____.
(a) equal resistor method (b) reciprocal method
(c) product over the sum method (d) all of these

14. The total resistance of three 6Ω resistors in parallel is _____.
(a) 6Ω (b) 12Ω (c) 18Ω (d) none of these

15. The circuit resistance of a 600W coffee pot and a 1,000W skillet is _____ when connected to a 120V parallel circuit.
(a) 24Ω (b) 14.4Ω (c) 38.4Ω (d) 9Ω

16. The resistance total of 20Ω, 20Ω, and 10Ω resistors in parallel is _____.
(a) 5Ω (b) 20Ω (c) 30Ω (d) 50Ω

2–6 Parallel Circuit Summary

17. • Which of the following statements is/are true about parallel circuits?
(1) The total resistance of a parallel circuit is less than the smallest resistor of the circuit.
(2) Total circuit current is equal to the sum of the branch currents.
(3) The power of all resistors is equal to the sum of the branch powers.
(a) 1 and 2 (b) 1, 2 and 3
(c) 2 and 3 (d) 1 and 3

Part C – Series–Parallel and Multiwire Branch Circuits

Introduction to Series-Parallel Circuits

18. A _____ is a circuit that contains some resistors in series and some resistors in parallel to each other.
(a) parallel circuit (b) series circuit (c) series-parallel circuit (d) none of these

Part D – Multiwire Branch Circuits

Introduction to Multiwire Branch Circuits

19. A multiwire branch circuit has two or more ungrounded (hot) conductors having a potential difference between them and an equal difference of potential between each ungrounded (hot) conductor and the grounded (neutral) conductor.
(a) True (b) False

2–9 Neutral Current Calculations

20. • The current on the grounded (neutral) conductor of a 2-wire circuit will be _____ of the current on the ungrounded (hot) conductor.
(a) 50% (b) 70% (c) 80% (d) 100%

Three-Wire Circuits

21. • A balanced 120/240V, 1Ø, 3-wire circuit is properly connected to Line 1 and Line 2. The current flowing through the grounded (neutral) conductor will be equal to _____ of the current flowing through the ungrounded conductor..
(a) 0% (b) 60% (c) 80% (d) 90%

22. A 3-wire, 120/240V circuit carries 10A of unbalanced neutral current if:
    Line 1 = 20A and Line 2 = 10A.
    (a) True   (b) False

23. • What is the neutral current for a 20A, 3-wire, 208Y/120V circuit?
    (a) 0A   (b) 10A   (c) 20A   (d) 40A

Four-Wire Circuits

24. The neutral of a 3Ø, 4-wire, 208Y/120 or 480Y/277V circuit carries the unbalanced current when the circuit is balanced.
    (a) True   (b) False

25. What is the neutral current for a 3Ø, 4-wire, 208Y/120V circuit where L1 = 20A, L2 = 20A and L3 = 20A?
    (a) 0A   (b) 10A   (c) 20A   (d) 40A

26. • The grounded (neutral) conductor of a balanced, 4-wire, 3Ø wye circuit carries no current when supplying power to balanced nonlinear loads.
    (a) True   (b) False

27. • A 4-wire, 3Ø multiwire branch circuit has 20A on each 120V phase. The neutral conductor could carry as much as _____ if it supplies nonlinear loads.
    (a) 0A   (b) 10A   (c) 15A   (d) 40A

2–10 Dangers of Multiwire Circuit

28. Improper wiring or mishandling of multiwire branch circuits can cause _____ connected to the circuit.
    (a) overloading of the ungrounded (hot) conductors
    (b) overloading of the grounded (neutral) conductors
    (c) destruction of equipment because of overvoltage
    (d) b and c

29. • Because of the dangers associated with an open neutral (grounded conductor), the continuity of the _____ conductor cannot be dependent on the connection to the receptacle.
    (a) ungrounded   (b) grounded   (c) a and b   (d) none of these

☆ Challenge Questions

Part A – Series Circuits

30. • A series circuit contains two resistors, one rated 4Ω and the other rated 8Ω. If the total voltage drop across both resistors equals 12V, then the current that passes through either resistor would be _____.
    (a) 1A   (b) 2A   (c) 4A   (d) 8A

31. • A series circuit has four 40Ω resistors and the power supply is 120V. The voltage drop of each resistor would be

    _____.
    (a) one-quarter of the source voltage   (b) 30V
    (c) the same across each resistor   (d) all of these

32. • The power consumed in a series circuit is _____.
    (a) the sum of the power consumed of each load
    (b) determined by the formula $P_T = I^2 \times R_T$
    (c) determined by the formula $P_T = E \times I$
    (d) all of these

33. • The reading on voltmeter 2 (V₂) is _____ (See Figure 2-32).

(a) 5V   (b) 7V
(c) 10V   (d) 6V

34. The voltmeter connected across the switch would read _____ (See Figure 2-33).

(a) 3V   (b) 12V
(c) 6V   (d) 18V

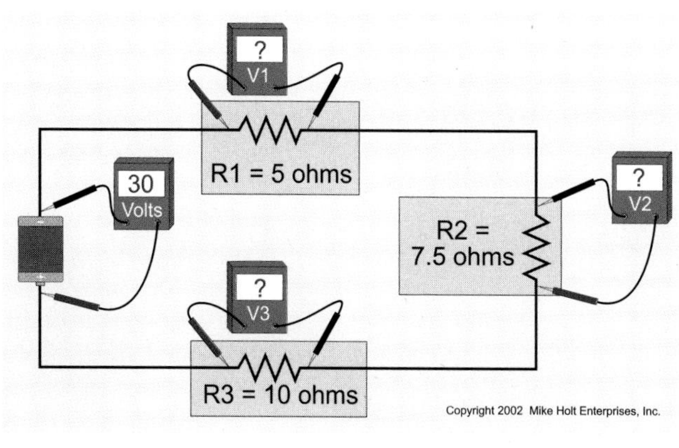

Part B – Parallel Circuits

35. • In general, when multiple light bulbs are wired in a single luminaire, they are connected in _____ to each other.

(a) series   (b) series-parallel
(c) parallel   (d) order of wattage

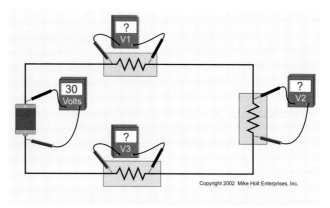

Figure 2-32

36. • A 1Ø, dual-rated, 120/240V motor will have its winding connected in _____ when supplied by 120V.

(a) series   (b) parallel   (c) series-parallel   (d) parallel-series

37. • The voltmeters shown in Figure 2-34 are connected _____ each of the loads.

(a) in series to   (b) across   (c) in parallel to   (d) b and c

38. • If the supply voltage is 100V, the total energy consumed for four 10Ω resistors will be more if all resistors are connected _____.

(a) in series   (b) series-parallel   (c) in parallel   (d) any of these

Circuit Resistance

39. A parallel circuit has three resistors. One resistor is rated 2Ω, one is 3Ω, and the other is 7Ω. The total resistance of the parallel circuit is _____. Remember the total resistance of any parallel circuit is always less than the smallest resistor.

(a) 12Ω   (b) 1Ω   (c) 42Ω   (d) 1.35Ω

Figure 2–35 applies to the next three questions.

40. • The total current of the circuit can be measured by ammeter _____ (See Figure 2-35).

(a) A1   (b) A2   (c) A3   (d) none of these

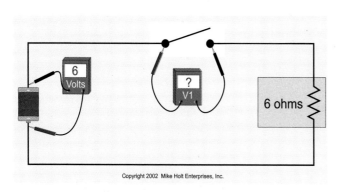

Figure 2-33

Figure 2-34

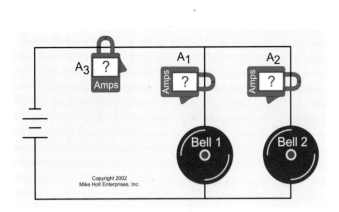

Figure 2-35

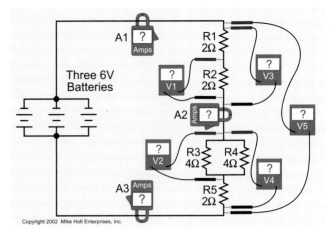

Figure 2-36

41. • If Bell 2 consumes 12W of power when supplied by two 12V batteries (connected in series), the resistance of this bell is _____ (See Figure 2-35).
     (a) 9.6Ω   (b) 44Ω   (c) 576Ω   (d) 48Ω

42. Determine the total circuit resistance of the parallel circuit based on the following facts (See Figure 2-35):
     1. The current on ammeter 1 reads 0.75A.
     2. The voltage of the circuit is 30V.
     3. Bell 2 has a resistance of 48Ω.
     Tip: Total resistance of a parallel circuit is always less than the smallest resistor.
     (a) 22Ω   (b) 48Ω   (c) 1,920Ω   (d) 60Ω

Part C – Series-Parallel Circuits

Figure 2–36 applies to the next three questions:

43. • The total current of this circuit can be read on _____ (See Figure 2-36).
     (a) Ammeter 1   (b) Ammeter 2   (c) Ammeter 3   (d) all of these

44. • The reading of V_2 is _____ (See Figure 2-36).
     (a) 1.5V   (b) 4V   (c) 4V   (d) 8V

45. • The reading of V_4 is _____ (See Figure 2-36).
     (a) 1.5V   (b) 3V   (c) 5V   (d) 8V

Figure 2–37 applies to the next two questions:

46. Resistor R_1 has a resistance of 5Ω and resistors R_2, R_3, and R_4 have a resistance of 15Ω each. The total resistance of this series-parallel circuit is _____ (See Figure 2-37).
     (a) 50Ω   (b) 35Ω   (c) 25Ω   (d) 10Ω

47. • What is the voltage drop across R_1, if R_1 is equal to 5Ω and total resistance of R_2, R_3, and R_4 is 5Ω (See Figure 2-37)?
     (a) 60V   (b) 33V   (c) 40V   (d) 120V

Part D – Multiwire Branch Circuits

48. • If the neutral of the circuit in the diagram is opened, the circuit becomes one series circuit of 240V. Under this condition, the current of the circuit is _____. Tip: Determine the total resistance (See Figure 2-38).
     (a) 0.67A   (b) 0.58A   (c) 2.25A   (d) 0.25A

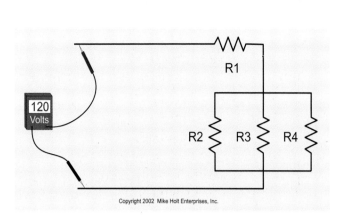

Figure 2-37

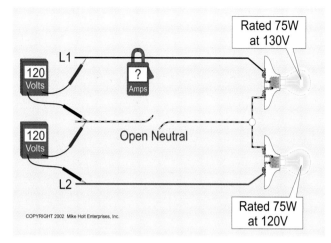

Figure 2-38

NEC Questions - Articles 110-220

Article 110 Requirements for Electrical Installations

49. Electrical equipment that depends on the _____ principles for cooling of exposed surfaces shall be installed so that airflow over such surfaces is not prevented by walls or by adjacent installed equipment.
(a) introduced
(b) natural circulation of air and convection
(c) artificial cooling and circulation
(d) magnetic induction

50. Many terminations and equipment are marked with _____.
(a) an etching tool   (b) a removable label
(c) a tightening torque   (d) the manufacturer's initials

51. Connection of conductors to terminal parts shall ensure a thoroughly good connection without damaging the conductors and shall be made by means of _____.
(a) solder lugs   (b) pressure connectors
(c) splices to flexible leads   (d) any of these

52. Connection by means of wire-binding screws, studs, or nuts having upturned lugs or the equivalent shall be permitted for _____ or smaller conductors.
(a) 10 AWG   (b) 8 AWG   (c) 6 AWG   (d) none of these

53. Soldered splices shall first be spliced or joined so as to be mechanically and electrically secure without solder and then be soldered.
(a) True   (b) False

54. The temperature rating associated with the ampacity of a _____ shall be so selected and coordinated so as to not exceed the lowest temperature rating of any connected termination, conductor, or device.
(a) terminal   (b) conductor   (c) device   (d) all of these

55. • What size THHN conductor is required for a 50A circuit if the equipment is listed and identified for use with a 75°C conductor? Tip: Table 310.16 lists conductor ampacities.
(a) 10 AWG   (b) 8 AWG   (c) 6 AWG   (d) all of these

56. Separately installed pressure connectors shall be used with conductors at the _____ not exceeding the ampacity at the listed and identified temperature rating of the connector.
(a) voltages   (b) temperatures   (c) listings   (d) ampacities

57. The *NEC* requires series-rated installations to be field-marked to indicate the maximum level of fault-current for which the system has been installed.
(a) True   (b) False

58. Sufficient access and _____ shall be provided and maintained about all electrical equipment to permit ready and safe operation and maintenance of such equipment.
(a) ventilation   (b) cleanliness   (c) circulation   (d) working space

59. Enclosures housing electrical apparatus that are controlled by lock and key shall be considered _____ to qualified persons.
(a) readily accessible   (b) accessible   (c) available   (d) none of these

60. Working-space distances for enclosed live parts shall be measured from the _____ of equipment or apparatus, if such are enclosed.
(a) enclosure   (b) opening   (c) a or b   (d) none of these

61. A minimum of _____ of working clearance is required to live parts operating at 300V, nominal to ground, where there are exposed live parts on one side and no live or grounded parts on the other side.
(a) 2 ft   (b) 3 ft   (c) 4 ft   (d) 6 ft

62. The minimum working clearance on a circuit that is 120 volts-to-ground, with exposed live parts on one side and no live or grounded parts on the other side of the working space, is _____.
(a) 1 ft   (b) 3 ft   (c) 4 ft   (d) 6 ft

63. • The required working clearance for access to live parts operating at 300V, nominal-to-ground, where there are exposed live parts on one side and grounded parts on the other side, is _____ according to Table 110.26(A).
(a) 3 ft   (b) $3^1/_2$ ft   (c) 4 ft   (d) $4^1/_2$ ft

64. Concrete, brick or tile walls shall be considered as _____, as it applies to working-space requirements.
(a) inconsequential   (b) in the way   (c) grounded   (d) none of these

65. The dimension of working clearance for access to live parts operating at 300V, nominal-to-ground, where there are exposed live parts on both sides of the workspace (not guarded) with the operator between, is _____ according to Table 110.26(A).
(a) 3 ft   (b) $3^1/_2$ ft   (c) 4 ft   (d) $4^1/_2$ ft

66. The working space in front of the electric equipment shall not be less than _____ wide or less than the width of the equipment, whichever is greater.
(a) 15 in.   (b) 30 in   (c) 40 in   (d) 60 in

67. Equipment such as raceways, cables, wireways, cabinets, panels, etc. can be located above or below other electrical equipment when the associated equipment does not extend more than _____ from the front of the electrical equipment.
(a) 3 in   (b) 6 in   (c) 12 in   (d) 30 in

68. When normally enclosed live parts are exposed for inspection or servicing, the working space, if in a passageway or general open space, shall be suitably _____.
(a) accessible   (b) guarded   (c) open   (d) enclosed

69. Working space shall not be used for _____.
(a) storage   (b) raceways   (c) lighting   (d) accessibility

70. For equipment rated 1,200A or more and over 6 ft wide that contains overcurrent devices, switching devices or control devices, there shall be one entrance to the required working space not less than 24 in. wide and $6^1/_2$ ft high at each end of the working space. Where the depth of the working space is twice that required by 110.26(A)(1), _____ entrance(s) shall be permitted.
(a) one   (b) two   (c) three   (d) none of these

71. The minimum headroom of working spaces about motor control centers shall be _____.
     (a) 3 ft   (b) 5 ft   (c) 6 ft   (d) 6 ft 6 in.

72. The minimum headroom for working spaces about service equipment, switchboards, panelboards or motor control centers shall be 6 ft 6 in., except for service equipment or panelboards in existing dwelling units that do not exceed 200A.
     (a) True   (b) False

73. The dedicated equipment space for electrical equipment that is required for panelboards is measured from the floor to a height of _____ above the equipment, or to the structural ceiling, whichever is lower.
     (a) 3 ft   (b) 6 ft   (c) 12 ft   (d) 30 ft

74. • Ventilated heating or cooling equipment (including ducts) that service the electrical room or space cannot be installed in the dedicated space above a panelboard or switchboard.
     (a) True   (b) False

75. The dedicated space above a panelboard extends to the structural ceiling which can be a suspended ceiling.
     (a) True   (b) False

76. Unless specified otherwise, live parts of electrical equipment operating at _____ or more shall be guarded.
     (a) 12V   (b) 15V   (c) 50V   (d) 24V

77. To guard live parts over 50V but less than 600V, the equipment can be _____.
     (a) isolated in a room accessible to qualified persons only
     (b) located on a balcony
     (c) elevated 8 ft or more above the floor
     (d) any of these

78. Live parts of electrical equipment operating at _____ or more shall be guarded against accidental contact by approved enclosures or by suitable permanent, substantial partitions or screens arranged so that only qualified persons have access to the space within reach of the live parts.
     (a) 20V   (b) 30V   (c) 50V   (d) 100V

79. In locations where electrical equipment is likely to be exposed to _____, enclosures or guards shall be so arranged and of such strength as to prevent such damage.
     (a) air circulation   (b) physical damage   (c) magnetic fields   (d) weather

80. Entrances to rooms and other guarded locations containing exposed live parts shall be marked with conspicuous _____ forbidding unqualified persons to enter.
     (a) warning signs   (b) alarms   (c) a and b   (d) neither a nor b

81. Electrical installations over 600V located in _____, where access is controlled by lock and key or other approved means, shall be considered to be accessible to qualified persons only.
     (a) a room or closet
     (b) a vault
     (c) an area surrounded by a wall, screen or fence
     (d) any of these

82. Openings in ventilated dry-type _____ or similar openings in other equipment over 600V shall be designed so that foreign objects inserted through these openings will be deflected from energized parts.
     (a) lampholders   (b) motors   (c) fuseholders   (d) transformers

83. On circuits over 600V, nominal, where energized live parts are exposed, the minimum clear workspace shall not be less than _____ high.
     (a) 3 ft   (b) 5 ft   (c) 6 ft 3 in.   (d) 6 ft 6 in.

84. • For switchboards and control panels, operating at over 600V, nominal, and exceeding 6 ft in width, there shall be one entrance at each end of the equipment. _____ entrance(s) is (are) required for the working space if the depth of the working space is twice that required by 110.34(A).
(a) One   (b) Two   (c) Three   (d) Four

85. At least one entrance, not less than 24 in. wide and 6 ft, 6 in. high, shall be provided to give access to the working space about electrical equipment that operates at over 600V, nominal. For switchboards and control panels that exceed 6 ft in width, there shall be one entrance at each end of such equipment, except where the working space is twice that required in 110.34(A).
(a) True   (b) False

86. _____ shall be provided to give safe access to the working space around equipment over 600V installed on platforms, balconies, mezzanine floors, or in attic or roof rooms or spaces.
(a) Ladders   (b) Platforms or ladders
(c) Permanent ladders or stairways   (d) Openings

87. The minimum depth of clear working space in front of electrical equipment for 5,000V, nominal-to-ground is _____ when there are exposed live parts on both sides of the workspace with the operator between.
(a) 4 ft   (b) 5 ft   (c) 6 ft   (d) 9 ft

88. When switches cutouts, or other equipment operating at 600V, nominal, or less are installed in a room or enclosure where there are exposed live parts or exposed wiring operating at over 600V, nominal, the high-voltage equipment shall be effectively separated from the space occupied by _____ by a suitable partition, fence or screen.
(a) the access area   (b) the low-voltage equipment
(c) unauthorized persons   (d) motor-control equipment

89. Where switches, cutouts or similar equipment operating at 600V, nominal, or less are installed in a room where they are exposed to energized parts at over 600V, nominal, the high-voltage equipment shall be effectively separated from the space occupied by the low-voltage equipment by a suitable _____.
(a) partition   (b) fence   (c) screen   (d) any of these

90. Switches or other equipment operating at 600V, nominal, or less and serving only equipment within a high-voltage vault, room or enclosure shall be permitted to be installed in the _____ enclosure, room, or vault if accessible to qualified persons only.
(a) restricted   (b) medium-voltage   (c) sealed   (d) high-voltage

91. • Warning signs for over 600V shall read: Warning - High Voltage - Keep Out.
(a) True   (b) False

92. Illumination shall be provided for all working spaces about electrical equipment over 600V. The lighting outlets shall be arranged so that persons changing lamps or making repairs on the lighting system are not endangered by _____ or other equipment.
(a) live parts   (b) rotating parts   (c) bright lamps   (d) panelboards

93. Unguarded live parts operating at 30,000V located above a working space shall be elevated at least _____ above the working space.
(a) 24 ft   (b) 18 ft   (c) 12 ft   (d) $9^{1}/_{2}$ ft

Chapter 2 Wiring and Protection

Article 200 Use and Identification of Grounded Conductors

94. Article 200 contains the requirements for _____.
(a) identification of terminals
(b) grounded conductors in premises wiring systems
(c) identification of grounded conductors
(d) all of these

95. In Article 200, connected so as to be capable of carrying current (as distinguished from connection through electromagnetic induction) defines the term _____.
(a) effectively grounded
(b) electrically connected
(c) a grounded system
(d) none of these

96. Premises wiring shall not be electrically connected to a supply system unless the supply system contains, for any grounded conductor of the interior system, a corresponding conductor that is grounded.
(a) True
(b) False

97. An insulated grounded conductor of _____ or smaller shall be identified by a continuous white or gray outer finish, or by three continuous white stripes on other than green insulation along its entire length.
(a) 3 AWG
(b) 4 AWG
(c) 6 AWG
(d) 8 AWG

98. Application of distinctive marking at the terminals during the process of installation shall identify the grounded conductors of _____ metal-sheathed cable.
(a) armored
(b) mineral-insulated
(c) copper
(d) aluminum

99 Grounded (neutral) conductors _____ and larger shall be identified by a continuous white or gray outer finish along their entire lengths, or by distinctive white markings such as tape, paint or other effective means at their terminations.
(a) 10 AWG
(b) 8 AWG
(c) 6 AWG
(d) 4 AWG

100. A cable containing an insulated conductor with a white outer finish can be used for 3-way or 4-way switch loops, if it is permanently reidentified by painting or other effective means at its termination, and at each location where the conductor is visible and accessible.
(a) True
(b) False

101. Receptacles, polarized attachment plugs and cord connectors for plugs and polarized plugs must have the terminal intended for connection to the grounded conductor identified. Identification shall be by a metal or metal coating that is substantially _____ in color or by the word white or the letter W located adjacent to the identified terminal.
(a) green
(b) white
(c) gray
(d) b or c

102. No _____ shall be attached to any terminal or lead so as to reverse designated polarity.
(a) grounded conductor
(b) grounding conductor
(c) ungrounded conductor
(d) grounding connector

Article 210 Branch Circuits

103. 208Y/120V or 480Y/227V, 3Ø, 4-wire, wye systems used to supply nonlinear loads such as personal computers, energy-efficient electronic ballasts, electronic dimming, etc., cause distortion of the phase and neutral currents producing high, unwanted and potentially hazardous harmonic neutral currents. The *Code* cautions us that the system design should allow for the possibility of high harmonic neutral currents.
(a) True
(b) False

104. • Multiwire branch circuits shall _____.
(a) supply only line-to-neutral loads
(b) provide a means to disconnect simultaneously all ungrounded conductors if the multiwire branch circuit supplies more than one device or equipment on the same yoke for commercial installations
(c) have their conductors originate from different panelboards
(d) none of these

105. When more than one nominal voltage system exists in a building, each ungrounded system conductor shall be identified by phase and system. The means of identification shall be permanently posted at each branch-circuit panelboard.
(a) True
(b) False

106. Where more than one nominal voltage system exists in a building, each ungrounded conductor of a multiwire branch circuit, where accessible, shall be identified by phase and system. The identification can be _____ permanently posted at each branch-circuit panelboard.
(a) color-coding   (b) phase tape   (c) tagging   (d) any of these

107. Where more than one nominal voltage system exists in a building, each _____ conductor of a multiwire branch circuit, where accessible, shall be identified by phase and system.
(a) grounded   (b) ungrounded   (c) grounding   (d) all of these

108. In dwelling units, the voltage between conductors shall not exceed 120V, nominal, between conductors that supply the terminals of _____.
(a) luminaries
(b) cord-and-plug-connected loads of 1,440 VA, nominal, or less
(c) more than $1/4$-hp
(d) a and b

109. A branch-circuit voltage that exceeds 277 volts-to-ground and does not exceed 600V between conductors is used to wire the auxiliary equipment of electrical discharge lamps mounted on poles. The minimum height of these luminaires shall not be less than _____.
(a) 31 ft   (b) 15 ft   (c) 18 ft   (d) 22 ft

110. All ungrounded (hot) conductors terminating on multiple receptacles on the same yoke must have a means to be disconnected simultaneously in _____ occupancies.
(a) dwelling unit   (b) commercial   (c) industrial   (d) all of these

111. Feeders supplying 15 and 20A receptacle branch circuits shall be permitted to be protected by a ground-fault circuit interrupter in lieu of the provisions for such interrupters as specified in 210.8 and Article 527.
(a) True   (b) False

112. All 125V, 1Ø, 15 and 20A receptacles installed in bathrooms of _____ shall have ground-fault circuit interrupter (GFCI) protection for personnel.
(a) guest rooms in hotels/motels
(b) dwelling units
(c) office buildings
(d) all of these

113. GFCI protection for personnel is required for all 125V, 1Ø, 15 and 20A receptacles installed in a dwelling unit _____.
(a) attic   (b) garage   (c) laundry   (d) all of these

114. GFCI protection is required for all 125V, 1Ø, 15 and 20A receptacles in accessory buildings that have a floor located at or below grade level not intended as _____ and limited to storage areas.
(a) habitable   (b) finished   (c) a or b   (d) none of these

115. • A _____ receptacle without GFCI protection is permitted to be located in a dwelling unit garage for one appliance if located within the dedicated space for the appliance.
(a) multioutlet   (b) duplex   (c) single   (d) none of these

116. GFCI protection for personnel is required for fixed electric snow-melting or deicing equipment receptacles that are not readily accessible and are supplied by a dedicated branch circuit.
(a) True   (b) False

117. All 125V, 1Ø, 15 and 20A receptacles installed in crawl spaces at or below grade level and in _____ of dwelling units shall have GFCI protection for personnel.
(a) unfinished attics
(b) finished attics
(c) unfinished basements
(d) finished basements

118. Dwelling unit GFCI protection is required for all 125V, 15 and 20A receptacles installed to serve the countertop surfaces in residential kitchens.
(a) True   (b) False

119. All 125V, 1Ø, 15 and 20A receptacles installed in dwelling unit boathouses shall be GFCI-protected.
(a) True   (b) False

120. GFCI protection is required for all 125V, 1Ø, 15 and 20A receptacles installed _____ of commercial, industrial and all other nondwelling occupancies.
(a) in storage rooms   (b) in equipment rooms   (c) in warehouses   (d) in bathrooms

121. GFCI protection is required for all 125V, 15 and 20A receptacles installed on rooftops, including those for fixed electric snow-melting or deicing equipment.
(a) True   (b) False

122 All 125V, 1Ø, 15 and 20A receptacles on(in) _____ of commercial occupancies shall be GFCI-protected.
(a) bathrooms   (b) rooftops   (c) kitchens   (d) all of these

123. Where the load is computed on a volt-amperes or square foot basis, the wiring system up to and including the branch-circuit _____ shall be provided to serve not less than the calculated load.
(a) wiring   (b) protection   (c) panelboard(s)   (d) all of these

124. There shall be a minimum of one _____ branch circuit for the laundry outlet(s) in a dwelling unit.
(a) 15A   (b) 20A   (c) 30A   (d) b and c

125. A dedicated 20A circuit is permitted to supply power to a dwelling unit bathroom for receptacle outlet(s) and other equipment within the same bathroom.
(a) True   (b) False

126. All branch circuits that supply 125V, 1Ø, 15 and 20A outlets installed in dwelling unit bedrooms shall be protected by a(n) _____ listed to provide protection of the entire branch circuit.
(a) AFCI   (b) GFCI   (c) a and b   (d) none of these

127. Branch-circuit conductors that supply a continuous load or any combination of continuous and non-continuous loads shall have an ampacity of not less than 125 percent of the continuous load, plus 100 percent of the noncontinuous load.
(a) True   (b) False

128. The recommended maximum total voltage drop on both the feeder and branch-circuit conductors is _____ percent.
(a) 3   (b) 2   (c) 5   (d) 3.6

129. The grounded conductor of a 3-wire branch circuit supplying a household electric range shall be permitted to be smaller than the ungrounded conductors when the maximum demand of an 8.75 kW range has been computed according to Column C of Table 220.19. However, the ampacity of the neutral conductor shall not be less than _____ percent of the branch-circuit rating and not be smaller than _____ AWG.
(a) 50, 6   (b) 70, 6   (c) 50, 10   (d) 70, 10

130. Where a branch circuit supplies continuous loads or any combination of continuous and noncontinuous loads, the rating of the overcurrent device shall not be less than the noncontinuous load plus 125 percent of the continuous load.
(a) True   (b) False

131. • A single receptacle installed on an individual branch circuit shall be rated at least _____ percent of the rating of the circuit.
(a) 50   (b) 60   (c) 90   (d) 100

132. When connected to a branch circuit supplying _____ or more receptacles or outlets, a receptacle shall not supply a total cord-and-plug-connected load in excess of the maximum specified in Table 210.21(B)(2).
(a) two   (b) three   (c) four   (d) five

133. What is the maximum cord-and-plug-connected load permitted on a 15A multioutlet receptacle that is supplied by a 20A circuit?
      (a) 12A   (b) 16A   (c) 20A   (d) 24A

134. • If a 20A branch circuit supplies multiple 125V receptacles, the receptacles must have an ampere rating of no less than _____.
      (a) 10A   (b) 15A   (c) 20A   (d) 30A

135. It shall be permitted to base the _____ rating of a range receptacle on a single range demand load specified in Table 220.19.
      (a) circuit   (b) voltage   (c) ampere   (d) resistance

136. The total rating of utilization equipment fastened in place shall not exceed _____ percent of the branch-circuit ampere rating where the circuit supplies receptacles for cord-and-plug-connected equipment.
      (a) 50   (b) 75   (c) 100   (d) 125

137. Multioutlet circuits rated 15 or 20A can supply fixed appliances (utilization equipment fastened in place) as long as the fixed appliances do not exceed _____ percent of the circuit rating.
      (a) 125   (b) 100   (c) 75   (d) 50

138. _____ in dwelling units shall supply only loads within that dwelling unit or loads associated only with that dwelling unit.
      (a) Service-entrance conductors   (b) Ground-fault protection
      (c) Branch circuits   (d) none of these

139. A permanently installed cord connector on a cord pendant shall be considered a receptacle outlet.
      (a) True   (b) False

140. No point along the floor line in any wall space of a dwelling unit may be more than _____ from an outlet.
      (a) 12 ft   (b) 10 ft   (c) 8 ft   (d) 6 ft

141. In dwelling units, when determining the spacing of general-use receptacles, _____ on exterior walls are not considered wall space.
      (a) fixed panels   (b) fixed glass   (c) sliding panels   (d) all of these

142. In a dwelling unit, each wall space of _____ or wider requires a receptacle.
      (a) 2 ft   (b) 3 ft   (c) 4 ft   (d) 5 ft

143. Receptacle outlets shall, insofar as practicable, be spaced equal distances apart in a dwelling unit. Receptacle outlets in floors shall not be counted as part of the required number of receptacle outlets unless they are located within _____ of the wall.
      (a) 6 in.   (b) 12 in.   (c) 18 in.   (d) close to the wall

144. In dwelling units, outdoor receptacles can be connected to the 20A small-appliance branch circuit.
      (a) True   (b) False

145. The small-appliance branch circuits can supply the _____ as well as the kitchen.
      (a) dining room   (b) refrigerator   (c) breakfast room   (d) all of these

146. Two 20A small-appliance branch circuits can supply more than one kitchen in a dwelling.
      (a) True   (b) False

147. A receptacle outlet shall be installed at each wall counter space that is 12 in. or wider so that no point along the wall line is more than _____, measured horizontally, from a receptacle outlet in that space.
      (a) 10 in.   (b) 12 in.   (c) 16 in.   (d) 24 in.

148. A receptacle outlet shall be installed in dwelling units for every kitchen and dining area countertop space _____, and no point along the wall line shall be more than 2 ft measured horizontally, from a receptacle outlet in that space.
      (a) wider than 10 in.   (b) wider than 3 ft   (c) 18 in. or wider   (d) 12 in. and wider

149.   One receptacle outlet shall be installed at each island or peninsular countertop having a long dimension of _____ in. or greater and a short dimension of _____ in. or greater.
        (a) 12, 24   (b) 24, 12   (c) 24, 48   (d) 48, 24

150.   For the purpose of determining the placement of receptacles in a dwelling unit kitchen, a(n) _____ countertop is measured from the connecting edge.
        (a) island   (b) usable   (c) peninsular   (d) cooking

151.   When breaks occur in dwelling unit kitchen-countertop spaces for ranges, refrigerators, sinks, etc., each countertop surface is considered a separate counter space for determining receptacle placement.
        (a) True   (b) False

152.   Kitchen and dining room countertop receptacle outlets in dwelling units shall be installed above the countertop surface, and not more than ___ above the countertop.
        (a) 12 in.   (b) 20 in.   (c) 24 in.   (d) none of these

153.   Island or peninsular countertop receptacle outlets can be installed below the countertop surface in dwelling units when necessary for the physically impaired or if no wall space is available above the countertop.
        (a) True   (b) False

154.   The maximum distance the required receptacle for a dwelling unit countertop surface can be above a dwelling unit kitchen counter surface is _____
        (a) 10 in.   (b) 12 in.   (c) 18 in.   (d) 20 in.

155.   In dwelling units, at least one wall receptacle outlet shall be installed in bathrooms within _____ of the outside edge of each basin. The receptacle outlet shall be located on a wall or partition that is adjacent to the basin or basin countertop.
        (a) 12 in.   (b) 18 in.   (c) 24 in.   (d) 36 in.

156.   A one-family or two-family dwelling unit requires a minimum of _____ GFCI receptacle(s) to be installed outdoors.
        (a) zero   (b) one   (c) two   (d) three

157.   A receptacle outlet for the laundry is not required in a dwelling unit in a multifamily building when laundry facilities are provided on the premises that are available to all building occupants.
        (a) True   (b) False

158.   For a one-family dwelling, at least one receptacle outlet is required in each _____.
        (a) basement   (b) attached garage
        (c) detached garage with electric power   (d) all of these

159.   Where a portion of the dwelling unit basement is finished into one or more habitable rooms, each separate unfinished portion shall have a receptacle outlet installed.
        (a) True   (b) False

160.   Hallways in dwelling units that are _____ long or longer require a receptacle outlet.
        (a) 12 ft   (b) 10 ft   (c) 8 ft   (d) 15 ft

161.   Receptacles installed behind a bed in the guest rooms in hotels and motels shall be located so as to prevent the bed from contacting an attachment plug, or the receptacle shall be provided with a suitable guard.
        (a) True   (b) False

162.   At least one receptacle outlet shall be installed directly above a show-window for each ____, or major fraction thereof, of show-window area measured horizontally at its maximum width.
        (a) 10 ft   (b) 12 ft   (c) 18 ft   (d) 24 ft

163.   A 125V, 1Ø, 15 or 20A rated receptacle outlet shall be installed at an accessible location for the servicing of heating, air-conditioning and refrigeration equipment. The receptacle shall be located on the same level and within _____ of the heating, air-conditioning and refrigeration equipment.
        (a) 10 ft   (b) 15 ft   (c) 20 ft   (d) 25 ft

164. A 125V, 1Ø, 15 or 20A receptacle outlet shall be located within 25 ft of heating, air-conditioning, and refrigeration equipment for _____ occupancies.
      (a) dwelling   (b) commercial   (c) industrial   (d) all of these

165. In a dwelling unit, at least _____ wall switch-controlled lighting outlet(s) shall be installed in every dwelling unit habitable room and bathroom.
      (a) one   (b) three   (c) six   (d) none of these

166. Which rooms in a dwelling unit must have a switch-controlled lighting outlet?
      (a) Every habitable room.   (b) Bathrooms.
      (c) Hallways and stairways.   (d) all of these

167. In _____ rooms other than kitchens and bathrooms of dwelling units, one or more receptacles controlled by a wall switch shall be permitted in lieu of lighting outlets.
      (a) habitable   (b) finished   (c) all   (d) a and b

168. In a dwelling unit, illumination from a lighting outlet shall be provided at the exterior side of each outdoor entrance or exit that has grade-level access.
      (a) True   (b) False

169. When considering lighting outlets in dwelling units, a vehicle door in a garage is considered as an outdoor entrance.
      (a) True   (b) False

170. Where a lighting outlet(s) is installed for interior stairways, there shall be a wall switch at each floor landing that includes an entryway where the stairway between floor levels has four risers or more.
      (a) True   (b) False

171. In a dwelling unit, illumination on the exterior side of outdoor entrances or exits that have grade-level access can be controlled by _____.
      (a) home automation devices   (b) motion sensors
      (c) photocells   (d) any of these

172. In a dwelling unit, at least one lighting outlet _____ located at the point of entry to the attic, underfloor space, utility room, and basement shall be installed where these spaces are used for storage or contain equipment requiring servicing.
      (a) that is unswitched and   (b) containing a switch
      (c) controlled by a wall switch   (d) b or c

173. For other than dwelling units, a wall-switch lighting outlet is required near equipment requiring servicing in attics or underfloor spaces, and the switch shall be located at the point of entrance to the attic or underfloor space.
      (a) True   (b) False

Article 215 Feeders

174. The feeder conductor ampacity shall not be less than that of the service-entrance conductors where the feeder conductors carry the total load supplied by service-entrance conductors with an ampacity of _____ or less.
      (a) 100A   (b) 60A   (c) 55A   (d) 30A

175. Dwelling unit or mobile home feeder conductors not over 400A can be sized according to 310.15(B)(6).
      (a) True   (b) False

176. Where installed in a metal raceway, all feeder conductors using a common grounded conductor shall be _____.
      (a) insulated for 600V   (b) enclosed within the same raceway
      (c) shielded   (d) none of these

177. If required by the authority having jurisdiction, a diagram showing feeder details shall be provided _____ of the feeders.
      (a) after the installation   (b) prior to the installation
      (c) before the final inspection   (d) diagrams are not required

178.   When a feeder supplies _____ in which equipment grounding conductors are required, the feeder shall include or provide a grounding means to which the equipment grounding conductors of the branch circuits shall be connected.
        (a) equipment disconnecting means   (b) electrical systems
        (c) branch circuits   (d) electric-discharge lighting equipment

179.   • Ground-fault protection is required for the feeder disconnect if _____.
        (a) the feeder is rated 1,000A or more
        (b) it is a solidly-grounded wye system
        (c) more than 150 volts-to-ground, but not exceeding 600V phase-to-phase
        (d) all of these

180.   Ground-fault protection of equipment shall not be required if ground-fault protection of equipment is provided on the _____ side of the feeder.
        (a) load   (b) supply   (c) service   (d) none of these

Article 220 Branch-Circuit, Feeder and Service Calculations

181.   When computations result in a fraction of an ampere that is less than_____, such fractions shall be permitted to be dropped.
        (a) 0.49   (b) 0.50   (c) 0.51   (d) none of these

182.   The 3 VA per-square-foot general lighting load for dwelling units shall not include _____.
        (a) open porches
        (b) garages
        (c) unused or unfinished spaces not adaptable for future use
        (d) all of these

183.   • When determining the load for recessed luminaires for branch circuits, the load shall be based on the _____.
        (a) wattage rating of the luminaire   (b) VA rating of the equipment and lamps
        (c) wattage rating of the lamps   (d) none of these

184.   Where fixed multioutlet assemblies used in other than dwelling units or the guest rooms of hotels or motels are employed, each _____ or fraction thereof of each separate and continuous length of multioutlet assembly shall be considered as one outlet of not less than 180 VA capacity where appliances are unlikely to be used simultaneously.
        (a) 5 ft   (b) 5¹/₂ ft   (c) 6 ft   (d) 6¹/₂ ft

185.   A single piece of equipment consisting of a multiple receptacle comprised of _____ or more receptacles shall be computed at not less than 90 VA per receptacle.
        (a) 1   (b) 2   (c) 3   (d) 4

186.   • The demand factors of Table 220.11 shall apply to the computed load of feeders to areas in hospitals, hotels, and motels where the entire lighting is likely to be used at one time, as in operating rooms, ballrooms, or dining rooms.
        (a) True   (b) False

187.   The minimum feeder load for show-window lighting will be _____ per-linear-foot.
        (a) 400 VA   (b) 200 VA   (c) 300 VA   (d) 180 VA

188.   For other than dwelling units, the feeder and service load calculation for track lighting is to be determined at 150 VA for every _____ of track installed.
        (a) 4 ft   (b) 6 ft   (c) 2 ft   (d) none of these

189.   • Receptacle loads for nondwelling units, computed at not more than 180 VA per outlet in accordance with 220.3(B)(9), shall be permitted to be _____.
        (a) added to the lighting loads and made subject to the demand factors of Table 220.11
        (b) made subject to the demand factors of Table 220.13
        (c) made subject to the lighting demand loads of Table 220.3(B)
        (d) a or b

190. Loads that are computed for dwelling unit small-appliance branch circuits can be included with the _____ load and subject to the demand factors permitted in Table 220.11 for the general lighting load.
     (a) general lighting   (b) feeder   (c) appliance   (d) receptacle

191. When sizing a feeder, the appliance loads in dwelling units can apply a demand factor of 75 percent of nameplate ratings for _____ or more appliances fastened in place on the same feeder.
     (a) two   (b) three   (c) four   (d) five

192. Using standard load calculations, the feeder demand factor for five household clothes dryers is _____ percent.
     (a) 70   (b) 85   (c) 50   (d) 100

193. To determine the feeder demand load for ten 3 kW household cooking appliances, use _____ of Table 220.19.
     (a) Column A   (b) Column B   (c) Column C   (d) none of these

194. The feeder demand load for four 6 kW cooktops is _____ kW.
     (a) 17   (b) 4   (c) 12   (d) 24

195. The demand factors of Table 220.20 apply to space heating, ventilating or air-conditioning equipment.
     (a) True   (b) False

196. Where it is unlikely that two or more loads will be in use simultaneously, it shall be permissible to use only the _____ loads on at any given time in computing the total load to a feeder.
     (a) smaller of the   (b) larger of the
     (c) difference between the   (d) none of these

197. • The maximum unbalanced feeder load for household electric ranges, wall-mounted ovens, counter-mounted cooking units and electric dryers shall be considered as _____ percent of the load on the ungrounded conductors as determined in accordance with Table 220.19 for ranges and Table 220.18 for dryers.
     (a) 50   (b) 70   (c) 85   (d) 115

198. There shall be no reduction in the size of the grounded conductor on _____ type lighting loads.
     (a) dwelling unit   (b) hospital   (c) nonlinear   (d) motel

199. If a dwelling unit is served by 1Ø, 3-wire, 120/240V service-entrance or feeder conductors with an ampacity of _____ or greater, it shall be permissible to compute the feeder and service loads in accordance with 220.30 instead of the method specified in Part II of Article 220.
     (a) 100   (b) 125   (c) 150   (d) 175

200. Feeder and service-entrance conductors with demand loads determined by the use of 220.30 shall be permitted to have the _____ load determined by 220.22.
     (a) feeder   (b) circuit   (c) neutral   (d) none of these

Unit 2 NEC Exam – NEC Code Order 110.14 – 220.22

1. What size THHN conductor is required for a 50A circuit if the equipment is listed and identified for use with a 75°C conductor? Tip: Table 310.16 lists conductor ampacities.
    (a) 10 AWG   (b) 8 AWG   (c) 6 AWG   (d) all of these

2. Working-space distances for enclosed live parts shall be measured from the _____ of equipment or apparatus if such are enclosed.
    (a) enclosure   (b) opening   (c) a or b   (d) none of these

3. The required working clearance for access to live parts operating at 300V, nominal-to-ground, where there are exposed live parts on one side and grounded parts on the other side, is _____, according to Table 110.26(A).
    (a) 3 ft   (b) 3$^1/_2$ ft   (c) 4 ft   (d) 4$^1/_2$ ft

4. Working space shall not be used for _____.
    (a) storage   (b) raceways   (c) lighting   (d) accessibility

5. Ventilated heating or cooling equipment (including ducts) that service the electrical room or space cannot be installed in the dedicated space above a panelboard or switchboard.
    (a) True   (b) False

6. On circuits over 600V nominal, where energized live parts are exposed, the minimum clear workspace shall not be less than _____ high.
    (a) 3 ft   (b) 5 ft   (c) 6 ft 3 in.   (d) 6 ft 6 in.

7. For switchboards and control panels, operating at over 600V, nominal, and exceeding 6 ft in width, there shall be one entrance at each end of the equipment. _____ entrance(s) is (are) required for the working space if the depth of the working space is twice that required by 110.34(A).
    (a) One   (b) Two   (c) Three   (d) Four

8. The minimum depth of clear working space in front of electrical equipment for 5,000V, nominal-to-ground is _____ when there are exposed live parts on both sides of the workspace with the operator between.
    (a) 4 ft   (b) 5 ft   (c) 6 ft   (d) 9 ft

9. Switches or other equipment operating at 600V, nominal or less and serving only equipment within a high-voltage vault, room, or enclosure shall be permitted to be installed in the _____ enclosure, room, or vault if accessible to qualified persons only.
    (a) restricted   (b) medium-voltage (c) sealed   (d) high-voltage

10. Warning signs for over 600V shall read: Warning - High Voltage - Keep Out.
    (a) True   (b) False

11. Multiwire branch circuits shall _____.
    (a) supply only line-to-neutral loads
    (b) provide a means to disconnect simultaneously all ungrounded conductors if the multiwire branch circuit supplies more than one device or equipment on the same yoke for commercial installations
    (c) have their conductors originate from different panelboards
    (d) none of these

12. A _____ receptacle without GFCI protection is permitted to be located in a dwelling unit garage for one appliance if located within the dedicated space for the appliance.
    (a) multioutlet   (b) duplex   (c) single   (d) none of these

13. A single receptacle installed on an individual branch circuit must be rated at least _____ percent of the rating of the circuit.
    (a) 50   (b) 60   (c) 90   (d) 100

14. If a 20A branch circuit supplies multiple 125V receptacles, the receptacles must have an ampere rating of no less than _____.
    (a) 10A   (b) 15A   (c) 20A   (d) 30A

15. It shall be permitted to base the _____ rating of a range receptacle on a single range demand load specified in Table 220.19.
(a) circuit   (b) voltage   (c) ampere   (d) resistance

16. In dwelling units, outdoor receptacles can be connected to the 20A small-appliance branch circuit.
(a) True   (b) False

17. For the purpose of determining the placement of receptacles in a dwelling unit kitchen, a(n) _____ countertop is measured from the connecting edge.
(a) island   (b) usable   (c) peninsular   (d) cooking

18. Which rooms in a dwelling unit must have a switch-controlled lighting outlet?
(a) Every habitable room   (b) Bathrooms
(c) Hallways and stairways   (d) all of these

19. Ground-fault protection is required for the feeder disconnect if _____.
(a) the feeder is rated 1,000A or more
(b) it is a solidly grounded wye system
(c) more than 150 volts-to-ground, but not exceeding 600V phase-to-phase
(d) all of these

20. The 3 VA per-square-foot general lighting load for dwelling units shall not include _____.
(a) open porches
(b) garages
(c) unused or unfinished spaces not adaptable for future use
(d) all of these

21. When determining the load for recessed luminaires for branch circuits, the load shall be based on the _____.
(a) wattage rating of the luminaire
(b) VA rating of the equipment and lamps
(c) wattage rating of the lamps
(d) none of these

22. The demand factors of Table 220.11 shall apply to the computed load of feeders to areas in hospitals, hotels and motels where the entire lighting is likely to be used at one time, as in operating rooms, ballrooms or dining rooms.
(a) True   (b) False

23. Receptacle loads for nondwelling units, computed at not more than 180 VA per outlet in accordance with 220.3(B)(9), shall be permitted to be _____.
(a) added to the lighting loads and made subject to the demand factors of Table 220.11
(b) made subject to the demand factors of Table 220.13
(c) made subject to the lighting demand loads of Table 220.3(B)
(d) a or b

24. To determine the feeder demand load for ten 3 kW household cooking appliances, use _____ of Table 220.19.
(a) Column A   (b) Column B   (c) Column C   (d) none of these

25. The maximum unbalanced feeder load for household electric ranges, wall-mounted ovens, counter-mounted cooking units and electric dryers shall be considered as _____ percent of the load on the ungrounded conductors as determined in accordance with Table 220.19 for ranges and Table 220.18 for dryers.
(a) 50   (b) 70   (c) 85   (d) 115

Unit 2 NEC Exam – Random Order 90.3 – 220.3

1. A 125V, 1Ø, 15 or 20A rated receptacle outlet shall be installed at an accessible location for the servicing of heating, air-conditioning and refrigeration equipment. The receptacle shall be located on the same level and within _____ of the heating, air-conditioning and refrigeration equipment.
    (a) 10 ft   (b) 15 ft   (c) 20 ft   (d) 25 ft

2. A receptacle outlet must be installed in dwelling units for every kitchen and dining area countertop space _____ and no point along the wall line shall be more than 2 ft measured horizontally from a receptacle outlet in that space.
    (a) wider than 10 in.   (b) wider than 3 ft
    (c) 18 in. or wider   (d) 12 in. and wider

3. Dwelling unit GFCI protection is required for all 125V, 15 and 20A receptacles installed to serve the countertop surfaces in residential kitchens.
    (a) True   (b) False

4. A _____ location may be temporarily subject to dampness and wetness.
    (a) dry   (b) damp   (c) moist   (d) wet

5. A junction box located 35 ft above the ground is considered to be _____.
    (a) concealed   (b) accessible
    (c) readily accessible   (d) recessed

6. A permanently installed cord connector on a cord pendant shall be considered a receptacle outlet.
    (a) True   (b) False

7. A raintight enclosure is constructed or protected so that exposure to a beating rain will not result in the entrance of water under specified test conditions.
    (a) True   (b) False

8. All 125V, 1Ø, 15 and 20A receptacles installed in bathrooms of _____ shall have ground-fault circuit interrupter (GFCI) protection for personnel.
    (a) guest rooms in hotels/motels
    (b) dwelling units
    (c) office buildings
    (d) all of these

9. Application of distinctive marking at the terminals during the process of installation shall identify the grounded conductors of _____ metal-sheathed cable.
    (a) armored   (b) mineral-insulated
    (c) copper   (d) aluminum

10. Chapters 1 through 4 of the *NEC* apply _____.
    (a) generally to all electrical installations
    (b) to special installations and conditions
    (c) to special equipment and material
    (d) all of these

11. Dwelling unit or mobile home feeder conductors not over 400A can be sized according to 310.15(B)(6).
    (a) True   (b) False

12. Entrances to rooms and other guarded locations containing exposed live parts shall be marked with conspicuous _____ forbidding unqualified persons to enter.
    (a) warning signs   (b) alarms
    (c) a and b   (d) neither a nor b

13. Ground-fault protection of equipment shall not be required if ground-fault protection of equipment is provided on the _____ side of the feeder.
    (a) load   (b) supply   (c) service   (d) none of these

14. Illumination shall be provided for all working spaces about electrical equipment over 600V. The lighting outlets shall be arranged so that persons changing lamps or making repairs on the lighting system are not endangered by _____ or other equipment.
   (a) live parts
   (b) rotating parts
   (c) bright lamps
   (d) panelboards

15. In dwelling units, when determining the spacing of general-use receptacles, _____ on exterior walls are not considered wall space.
   (a) fixed panels
   (b) fixed glass
   (c) sliding panels
   (d) all of these

16. Many terminations and equipment are marked with _____.
   (a) an etching tool
   (b) a removable label
   (c) a tightening torque
   (d) the manufacturer's initials

17. Multioutlet circuits rated 15 or 20A can supply fixed appliances (utilization equipment fastened in place) as long as the fixed appliances do not exceed _____ percent of the circuit rating.
   (a) 125
   (b) 100
   (c) 75
   (d) 50

18. The dedicated space above a panelboard extends to the structural ceiling, which can be a suspended ceiling.
   (a) True
   (b) False

19. The feeder demand load for four 6 kW cooktops is _____ kW.
   (a) 17
   (b) 4
   (c) 12
   (d) 24

20. The recommended maximum total voltage drop on both the feeder and branch-circuit conductors is _____ percent.
   (a) 3
   (b) 2
   (c) 5
   (d) 3.6

21. When a feeder supplies _____ in which equipment grounding conductors are required, the feeder shall include or provide a grounding means to which the equipment grounding conductors of the branch circuits shall be connected.
   (a) equipment disconnecting means
   (b) electrical systems
   (c) branch circuits
   (d) electric-discharge lighting equipment

22. Where fixed multioutlet assemblies used in other than dwelling units or the guest rooms of hotels or motels are employed, each _____ or fraction thereof of each separate and continuous length of multioutlet assembly shall be considered as one outlet of not less than 180 VA capacity where appliances are unlikely to be used simultaneously.
   (a) 5 ft
   (b) $5^1/_2$ ft
   (c) 6 ft
   (d) $6^1/_2$ ft

23. Where more than one nominal voltage system exists in a building, each _____ conductor of a multiwire branch circuit, where accessible, shall be identified by phase and system.
   (a) grounded
   (b) ungrounded
   (c) grounding
   (d) all of these

24. _____ is a qualifying term indicating that there is a purposely introduced delay in the tripping action of the circuit breaker, which delay decreases as the magnitude of the current increases.
   (a) Adverse-time
   (b) Inverse-time
   (c) Time delay
   (d) Timed unit

25. _____ shall be provided to give safe access to the working space around equipment over 600V installed on platforms, balconies, mezzanine floors, or in attic or roof rooms or spaces.
   (a) Ladders
   (b) Platforms or ladders
   (c) Permanent ladders or stairways
   (d) Openings

26. A _____ circuit is a circuit in which any arc or thermal effect produced under intended operating conditions of the equipment is not capable, under specified test conditions, of igniting the flammable gas-air, vapor-air, or dust-air mixture.
   (a) nonconductive
   (b) branch
   (c) nonincendive
   (d) closed

27. A 125V, 1Ø, 15 or 20A receptacle outlet must be located within 25 ft of heating, air-conditioning and refrigeration equipment for _____ occupancies.
    (a) dwelling   (b) commercial   (c) industrial   (d) all of these

28. A cable containing an insulated conductor with a white outer finish can be used for 3-way or 4-way switch loops if the white conductor is not used for the switch leg, and if it is permanently reidentified by painting or other effective means at its termination and at each location where the conductor is visible and accessible.
    (a) True   (b) False

29. A "Class A" GFCI protection device is designed to de-energize the circuit when the ground-fault current is approximately _____.
    (a) 4 mA   (b) 5 mA   (c) 6 mA   (d) any of these

30. A *Code* rule may be waived or alternative methods of installation approved that may be contrary to the *NEC*, if the AHJ gives verbal or written consent.
    (a) True   (b) False

31. A dedicated 20A circuit is permitted to supply power to a dwelling unit bathroom for receptacle outlet(s) and other equipment within the same bathroom.
    (a) True   (b) False

32. A hoistway is any _____ in which an elevator or dumbwaiter is designed to operate.
    (a) hatchway or well hole   (b) vertical opening or space
    (c) shaftway   (d) all of these

33. A single piece of equipment consisting of a multiple receptacle comprised of _____ or more receptacles shall be computed at not less than 90 VA per receptacle.
    (a) 1   (b) 2   (c) 3   (d) 4

34. A solderless pressure connector is a device that _____ between two or more conductors or between one or more conductors and a terminal by means of mechanical pressure and without the use of solder.
    (a) provides access   (b) protects the wiring
    (c) is never needed   (d) establishes a connection

35. A(n) _____ is a point on the wiring system at which current is taken to supply utilization equipment.
    (a) box   (b) receptacle   (c) outlet   (d) device

36. All 125V, 1Ø, 15 and 20A receptacles in/on _____ of commercial occupancies must be GFCI-protected.
    (a) bathrooms   (b) rooftops   (c) kitchens   (d) all of these

37. All 125V, 1Ø, 15 and 20A receptacles installed in dwelling unit boathouses shall be GFCI-protected.
    (a) True   (b) False

38. All branch circuits that supply 125V, 1Ø, 15 and 20A outlets installed in dwelling unit bedrooms shall be protected by a(n) _____ listed to provide protection of the entire branch circuit.
    (a) AFCI   (b) GFCI   (c) a and b   (d) none of these

39. All ungrounded (hot) conductors terminating on multiple receptacles on the same yoke must have a means to be disconnected simultaneously in _____ occupancies.
    (a) dwelling unit   (b) commercial
    (c) industrial   (d) all of these

40. An isolating switch is one that is _____.
    (a) not readily accessible to persons unless special means for access are used
    (b) capable of interrupting the maximum operating overload current of a motor
    (c) intended for use in general distribution and branch circuits
    (d) intended for isolating an electrical circuit from the source of power

41. An outlet intended for the direct connection of a lampholder, a luminaire or a pendant cord terminating in a lampholder is a(n) _____.
    (a) outlet
    (b) receptacle outlet
    (c) lighting outlet
    (d) general-purpose outlet

42. At least one entrance, not less than 24 in. wide and 6 ft, 6 in. high, shall be provided to give access to the working space about electrical equipment that operates at over 600V, nominal. For switchboards and control panels that exceed 6 ft in width, there shall be one entrance at each end of such equipment, except where the working space is twice that required in 110.34(A).
    (a) True
    (b) False

43. Branch-circuit conductors that supply a continuous load or any combination of continuous and noncontinuous loads shall have an ampacity of not less than 125 percent of the continuous load, plus 100 percent of the noncontinuous load.
    (a) True
    (b) False

44. Capable of being removed or exposed without damaging the building structure or finish, or not permanently closed in by the structure or finish of the building defines _____.
    (a) accessible (equipment)
    (b) accessible (wiring methods)
    (c) accessible, readily
    (d) all of these

45. Communications wiring such as telephone, antenna and CATV wiring within a building is not required to comply with the installation requirements of Chapters 1 through 7, except where it is specifically referenced therein.
    (a) True
    (b) False

46. Compliance with either the metric or the inch-pound unit of measurement system is permitted.
    (a) True
    (b) False

47. Concrete, brick or tile walls shall be considered as _____ as it applies to working-space requirements.
    (a) inconsequential
    (b) in the way
    (c) grounded
    (d) none of these

48. Conductor sizes are expressed in American Wire Gage (AWG) or in _____.
    (a) inches
    (b) circular mils
    (c) square inches
    (d) AWG

49. Conduit installed underground or encased in concrete slabs that are in direct contact with the earth shall be considered a _____ location.
    (a) dry
    (b) damp
    (c) wet
    (d) moist

50. Connected to earth or to some conducting body that serves in place of the earth is called _____.
    (a) grounding
    (b) bonded
    (c) grounded
    (d) all of these

Unit 3

Understanding Alternating Current

OBJECTIVES

After reading this unit, the student should be able to briefly explain the following concepts:

Part A – Alternating-Current Fundamentals

Alternating current
Armature turning frequency
Current flow
Generator
Magnetic cores
Phase differences in degrees
Phase – in and out
Values of ac
Waveform

Part B - Induction and Capacitance Charge, discharging and testing of capacitors

Conductor impedance
Conductor shape
Induced voltage and applied current
Magnetic cores
Use of capacitors

Part C - Power Factor and Efficiency

Apparent power (volt-ampere)
Efficiency
Power factor
True power (watts)

After reading this unit, the student should be able to briefly explain the following terms:

Part A – Alternating-Current Fundamentals

Ampere-turns
Armature speed
Coil
Conductor cross-sectional area
Effective (RMS)
Effective to peak
Electromagnetic field
Frequency
Impedance
Induced voltage
Magnetic field
Magnetic flux lines
Peak
Peak to effective
Phase relationship

RMS
Root-mean-square
Self-inductance
Skin effect
Waveform

Part B and C - Induction and Capacitance

Back-EMF
Capacitance
Capacitive reactance
Capacitor
Coil
Counter-electromotive force
Eddy currents
Farads
Frequency
Henrys
Impedance
Induced voltage
Induction

Inductive reactance
Phase relationship
Self-inductance
Skin effect

Part D - Power Factor and Efficiency

Apparent power
Efficiency
Input watts
Output watts
Power factor
True power
Volt-ampere
Wattmeter

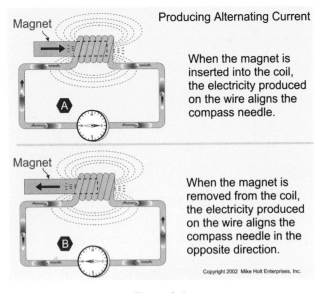

Producing Alternating Current

When the magnet is inserted into the coil, the electricity produced on the wire aligns the compass needle.

When the magnet is removed from the coil, the electricity produced on the wire aligns the compass needle in the opposite direction.

Copyright 2002 Mike Holt Enterprises, Inc.

Figure 3-1
Producing Alternating Current

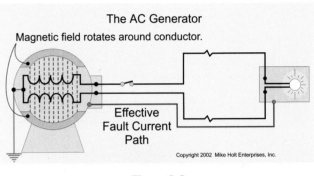

The AC Generator
Magnetic field rotates around conductor.

Effective Fault Current Path

Copyright 2002 Mike Holt Enterprises, Inc.

Figure 3-2
The AC Generator

PART A – ALTERNATING-CURRENT FUNDAMENTALS

3-1 CURRENT FLOW

For current to flow in a circuit, the circuit must be a closed loop and the power supply must push the electrons through the completed circuit. The transfer of electrical energy can be accomplished by electrons flowing in one constant direction (dc) or by electrons flowing in one direction and then reversing in the other direction (ac).

3-2 ALTERNATING CURRENT

Alternating current is generally used instead of dc for electrical systems and building wiring. This is because ac can easily have voltage variations (transformers). In addition, ac can be transmitted inexpensively at high voltage over long distances resulting in reduced voltage drop of the power distribution system as well as smaller power distribution wire and equipment. Alternating current can be used for certain applications for which dc is not suitable.

Alternating current is produced when electrons in a conductor are forced to move because there is a moving magnetic field. The lines of force from the magnetic field cause the electrons in the wire to flow in a specific direction. When the lines of force of the magnetic field move in the opposite direction, the electrons in the wire are forced to flow in the opposite direction. Figure 3–1. Electrons will flow only when there is relative motion between the conductors and the magnetic field.

3-3 ALTERNATING-CURRENT GENERATOR

An ac generator consists of many loops of wire that rotate between the flux lines of a magnetic field. Each conductor loop travels through the magnetic lines of force in opposite directions, causing the electrons within the conductor to move in a specific direction. Figure 3–2. The force on the electrons caused by the magnetic flux lines is called voltage, or electromotive force (EMF), and is abbreviated as V or E.

The magnitude of the electromotive force is dependent on the number of turns (wraps) of wire, the strength of the magnetic field and the speed at which the coil rotates. The rotating conductor loop mounted on a shaft is called a rotor or armature. Slip, or collector rings, and carbon brushes are used to connect the output voltage from the generator to an external circuit.

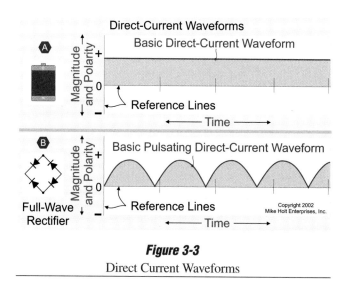

Figure 3-3

Direct Current Waveforms

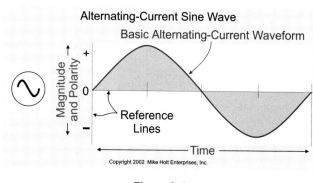

Figure 3-4

Alternating Current Sine Wave

3–4 WAVEFORM

A waveform is a pictorial view of the shape and magnitude of the level and direction of current or voltage over a period of time. The waveform represents the magnitude (level) and direction (polarity) of the current or voltage over time in relationship with the generator's armature.

Direct Current Waveform

The polarity of the voltage of dc is positive and the current flows through the circuit in the same direction at all times. In general, dc voltage and current remain the same magnitude, particularly when supplied from batteries or capacitors. Figure 3–3.

Alternating Current Waveform

The waveform for ac displays the level (magnitude) and direction (polarity) of the current and voltage for every instant of time for one full revolution of the armature. When the waveform of an ac circuit is symmetrical, with positive polarity above and negative below the zero reference level, the waveform is called a sine wave. Figure 3–4.

3–5 ARMATURE TURNING FREQUENCY

Frequency is a term used to indicate the number of times a generator's armature turns one full revolution (360 degrees) in one second. This is expressed as hertz, or cycles per second, and is abbreviated as Hz or cycles per second (cps). In the Unites States, frequency is not a problem for most electrical work because it remains constant at 60 hertz. Figure 3–5.

Armature Speed

A current or voltage that is produced when the generator's armature makes one cycle (360 degrees) in 1/60th of a second is called 60 Hz. This is because the armature travels at a speed of one cycle every 1/60th of a second. A 60 Hz circuit takes 1/60th of a second for the armature to complete one full cycle (360 degrees), and it will take 1/120th of a second to complete $1/2$ cycle (180 degrees). Figure 3–6.

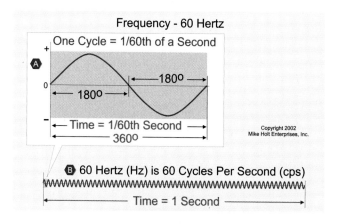

Figure 3-5

Frequency – 60 Hertz

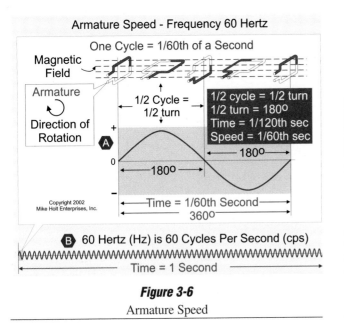

Figure 3-6

Armature Speed

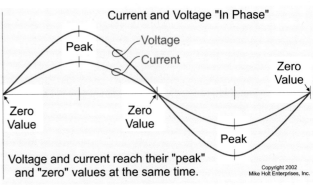

Figure 3-7

Current and Voltage "In-Phase"

3–6 PHASE – IN AND OUT

Phase is a term used to indicate the time relationship between two waveforms, such as voltage-to-current or voltage-to-voltage. *In-phase* means that two waveforms are in step with each other; that is, they cross the horizontal zero axis at the same time. Figure 3–7. The terms lead and lag are used to describe the relative position, in degrees, between two waveforms. The waveform that is ahead *leads* and the waveform behind *lags*. Figure 3–8.

3–7 PHASE DIFFERENCES IN DEGREES

Phase differences are expressed in *degrees,* with one full revolution of the armature equal to 360 degrees. Single-phase, 120/240V power is out-of-phase by 180 degrees, and 3Ø power is out-of-phase by 120 degrees. Figure 3–9.

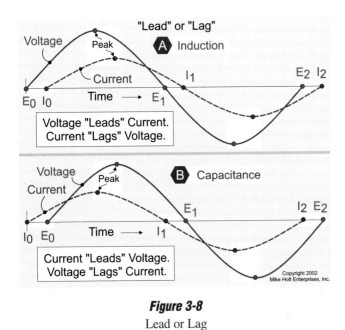

Figure 3-8

Lead or Lag

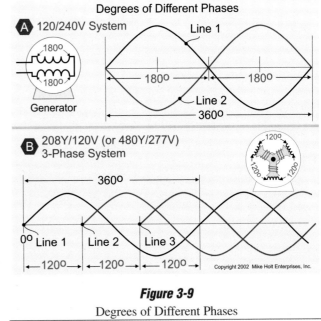

Figure 3-9

Degrees of Different Phases

3–8 VALUES OF ALTERNATING CURRENT

There are many different types of values in ac; the most important are peak (maximum) and effective. In ac systems, effective is the value that will cause the same amount of heat to be produced, as in a dc circuit containing only resistance. Another term for effective is RMS, which is root-mean-square. In the field, whenever you measure voltage, you are always measuring effective voltage. Newer types of clamp-on ammeters have the ability to measure peak as well as effective (RMS) current. Figure 3–10.

Note: The values of dc generally remain constant and are equal to the effective values of ac.

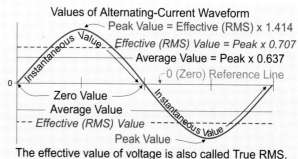

The effective value of voltage is also called True RMS. This is what most voltmeters read. The values shown in this figure apply to voltage waveforms as well as current waveforms.

Copyright 2002 Mike Holt Enterprises, Inc.

Figure 3-10

Values of Alternating-Current Waveform

❏ Current

Twelve amperes effective ac will have the same heating effect as _____ dc.

   (a) 7A   (b) 9A   (c) 11A   (d) 12A

    • Answer: (d) 12A

❏ Voltage

One hundred and twenty volts effective ac will produce the same heating effect as _____ dc.

   (a) 70V   (b) 90V   (c) 120V   (d) 170V

    • Answer: (c) 120V

Effective-to-Peak

To convert effective-to-peak, the following formulas should be helpful:

Peak = Effective (RMS) × 1.414, or

$$\text{Peak} = \frac{\text{Effective (RMS)}}{0.707}$$

❏ Effective-to-Peak

What is the peak voltage of a 120V effective circuit?

   (a) 120V   (b) 170V   (c) 190V   (d) 208V

    • Answer: (b) 170V

    Peak Volts = Effective (RMS)V × 1.414

    Peak Volts = 120V × 1.414 = 170V

Peak-to-Effective

Peak to effective can be determined by the formula:

Effective (RMS) = Peak × 0.707

❏ Peak-to-Effective

What is the effective current of 60A peak?

   (a) 42A   (b) 60A   (c) 72A   (d) none of these

    • Answer: (a) 42A

    Effective Current = Peak Current × 0.707

    Effective Current = 60A × 0.707 = 42A

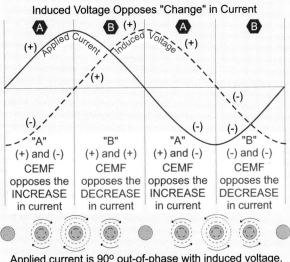

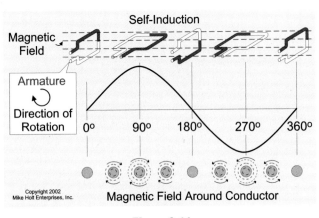

Figure 3-11

Self-Induction

Figure 3-12

Induced Voltage

PART B – INDUCTION

INDUCTION INTRODUCTION

When a magnetic field moves through a conductor, it causes the electrons in the conductor to move. The movement of electrons caused by electromagnetism is called *induction*. In an ac circuit, the current movement of the electrons increases and decreases with the rotation of the *armature* through the magnetic field. As the current flowing through the conductor increases or decreases, it causes an expanding and collapsing electromagnetic field within the conductor. Figure 3–11. This varying electromagnetic field within the conductor causes the electrons in the conductor to move at 90 degrees to the flowing electrons within the conductor. This movement of electrons because of the conductor's electromagnetic field is called self-induction. Self-induction (induced voltage) is commonly called *counter-electromotive force (CEMF)*, or back-EMF.

3–9 INDUCED VOLTAGE AND APPLIED CURRENT

The induced voltage in the conductor because of the conductor's counter-electromotive force (CEMF) is always 90 degrees out-of-phase with the applied current. Figure 3–12. When the current in the conductor increases, the induced voltage tries to prevent the current from increasing by storing some of the electrical energy in the magnetic field around the conductor. When the current in the conductor decreases, the induced voltage attempts to keep the current from decreasing by releasing the energy from the magnetic field back into the conductor. The opposition to any change in current because of induced voltage is called inductive reactance, which is measured in ohms, is abbreviated as X_L and is known as *Lenz's Law*. Inductive reactance changes proportionally with frequency and can be found by the formula:

$X_L = 2\pi fL$

$\pi = 3.14$

f = frequency

L = Inductance in henrys

❏ **Inductive Reactance**

What is the inductive reactance of a conductor in a transformer that has an inductance of 0.001 henrys? Calculate at 60 Hz, 180 Hz, and 300 Hz. Figure 3–13.

• Answers:

$X_L = 2\pi fL$

X_L at 60 Hz = $2 \times 3.14 \times 60$ Hz $\times 0.001$ henrys = 0.377Ω

X_L at 180 Hz = $2 \times 3.14 \times 180$ Hz $\times 0.001$ henrys = 1.13Ω

X_L at 300 Hz = $2 \times 3.14 \times 300$ Hz $\times 0.001$ henrys = 1.884Ω

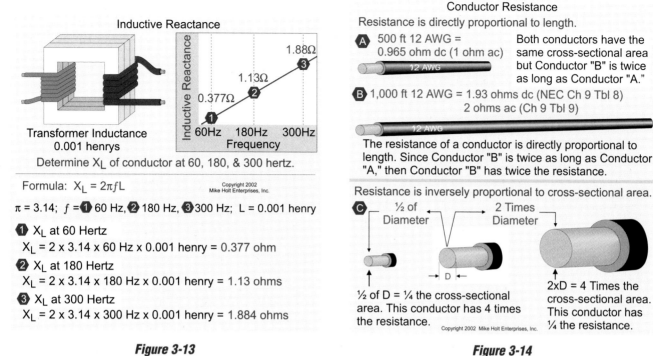

Inductive Reactance

Transformer Inductance
0.001 henrys
Determine X_L of conductor at 60, 180, & 300 hertz.

Formula: $X_L = 2\pi f L$ Copyright 2002
Mike Holt Enterprises, Inc.

$\pi = 3.14$; $f =$ ❶ 60 Hz, ❷ 180 Hz, ❸ 300 Hz; L = 0.001 henry

❶ X_L at 60 Hertz
$X_L = 2 \times 3.14 \times 60\ Hz \times 0.001\ henry = 0.377\ ohm$

❷ X_L at 180 Hertz
$X_L = 2 \times 3.14 \times 180\ Hz \times 0.001\ henry = 1.13\ ohms$

❸ X_L at 300 Hertz
$X_L = 2 \times 3.14 \times 300\ Hz \times 0.001\ henry = 1.884\ ohms$

Figure 3-13
Inductive Reactance

Conductor Resistance
Resistance is directly proportional to length.

Ⓐ 500 ft 12 AWG = 0.965 ohm dc (1 ohm ac)   Both conductors have the same cross-sectional area but Conductor "B" is twice as long as Conductor "A."

Ⓑ 1,000 ft 12 AWG = 1.93 ohms dc (NEC Ch 9 Tbl 8)
2 ohms ac (Ch 9 Tbl 9)

The resistance of a conductor is directly proportional to length. Since Conductor "B" is twice as long as Conductor "A," then Conductor "B" has twice the resistance.

Resistance is inversely proportional to cross-sectional area.

Ⓒ ½ of Diameter   2 Times Diameter

½ of D = ¼ the cross-sectional area. This conductor has 4 times the resistance.   Copyright 2002 Mike Holt Enterprises, Inc.

2xD = 4 Times the cross-sectional area. This conductor has ¼ the resistance.

Figure 3-14
Conductor Resistance

3–10 CONDUCTOR IMPEDANCE

The resistance of a conductor is a physical property of the conductors that oppose the flow of electrons. This resistance is directly proportional to the conductor length and is inversely proportional to the conductor cross-sectional area. This means that the longer the wire, the greater the conductor resistance; the smaller the cross-sectional area of the wire, the greater the conductor resistance. Figure 3–14.

The opposition to ac flow due to inductive reactance and conductor resistance is called impedance (Z). Impedance is only present in ac circuits and is measured in ohms. This opposition to ac (Z) is greater than the resistance (R) of a dc circuit because of *counter-electromotive force, eddy currents* and *skin effect*.

Eddy Currents and Skin Effect

Eddy currents are small independent currents that are produced as a result of the expanding and collapsing magnetic field of ac flow. Figure 3–15. The expanding and collapsing magnetic field of an ac circuit also induces a counter-electromotive force in the conductors that repel the flowing electrons toward the surface of the conductor. Figure 3–16. The counter-electromotive force causes the circuit current to flow near the surface of the conductor rather than at the conductor's center. The flow of electrons near the surface is known as skin effect. Eddy currents and skin effect decrease the effective conductor cross-sectional area for current flow, which results in an increase in the conductor's impedance which is measured in ohms.

3–11 INDUCTION AND CONDUCTOR SHAPE

The shape into which the conductor is configured affects the amount of *self-induced voltage* in the conductor. If a conductor is coiled into adjacent loops *(winding)*, the electromagnetic field *(flux lines)* of each conductor loop adds together to create a stronger overall magnetic field. Figure 3–17. The amount of self-induced voltage created within a conductor is directly proportional to the current flow, the length of the conductor and the frequency at which the magnetic fields cut through the conductors. Figure 3–18.

3–12 INDUCTION AND MAGNETIC CORES

Self-induction in a *coil* of conductors is affected by the winding current magnitude, the core material, the number of conductor loops *(turns)* and the spacing of the winding. Figure 3–19. An *Iron core* within the conductor winding permits an easy path for the electromagnetic flux lines, which produce a greater counter-electromotive force than air core windings. In addition, conductor coils are measured in *ampere-turns*; that is, the current in amperse times the number of conductor loops or turns.

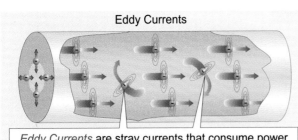

Eddy Currents are stray currents that consume power and oppose current flow. They are produced by the expanding and collapsing magnetic field of alternating-current circuits.

Copyright 2002 Mike Holt Enterprises, Inc.

Figure 3-15
Eddy Currents

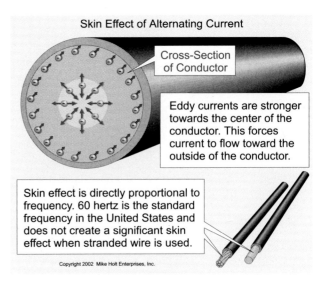

Skin Effect of Alternating Current

Cross-Section of Conductor

Eddy currents are stronger towards the center of the conductor. This forces current to flow toward the outside of the conductor.

Skin effect is directly proportional to frequency. 60 hertz is the standard frequency in the United States and does not create a significant skin effect when stranded wire is used.

Copyright 2002 Mike Holt Enterprises, Inc.

Figure 3-16
Skin Effect of Alternating Current

Ampere-Turns = Coil Ampere × Number of Coil Turns

❑ **Ampere-Turns**

A magnetic coil has 1,000 turns and carries five amperes. What is the ampere-turns rating of this coil? Figure 3–20.

(a) 5,000 Ampere-Turns   (b) 7,500 Ampere-Turns   (c) 10,000 Ampere-Turns   (d) none of these

• Answer: (a) 5,000 A-turns

Ampere-Turns = Amperes × Number of Coil Turns

Ampere-Turns = 5A × 1,000 turns = 5,000

PART C – CAPACITANCE

CAPACITANCE INTRODUCTION

Capacitance is the property of an electric circuit that enables the storage of electric energy by means of an electrostatic field, such as the way a spring stores mechanical energy. Figure 3–21. Capacitance exists whenever an insulating material separates two objects that have a difference of potential. A capacitor is two metal foils separated by insulating material (wax paper) rolled together. Devices that intentionally introduce capacitance into circuits are called capacitors or condensers.

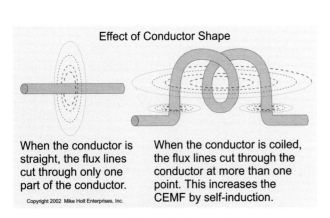

Effect of Conductor Shape

When the conductor is straight, the flux lines cut through only one part of the conductor.

When the conductor is coiled, the flux lines cut through the conductor at more than one point. This increases the CEMF by self-induction.

Copyright 2002 Mike Holt Enterprises, Inc.

Figure 3-17
Effect of Conductor Shape

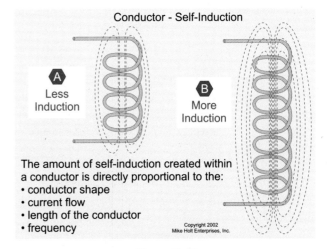

Conductor - Self-Induction

A
Less Induction

B
More Induction

The amount of self-induction created within a conductor is directly proportional to the:
• conductor shape
• current flow
• length of the conductor
• frequency

Copyright 2002 Mike Holt Enterprises, Inc.

Figure 3-18
Conductor – Self-Induction

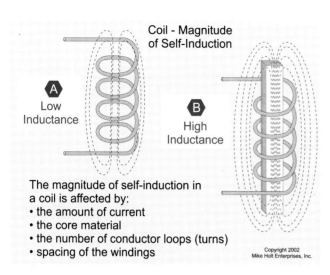

Figure 3-19

Coil – Magnitude of Self-Induction

The magnitude of self-induction in a coil is affected by:
• the amount of current
• the core material
• the number of conductor loops (turns)
• spacing of the windings

Copyright 2002
Mike Holt Enterprises, Inc.

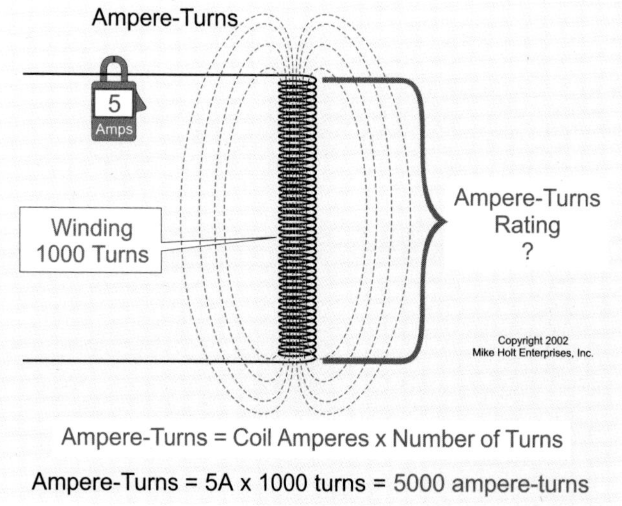

Figure 3-20

Ampere Turns

Ampere-Turns = Coil Amperes x Number of Turns

Ampere-Turns = 5A x 1000 turns = 5000 ampere-turns

Copyright 2002
Mike Holt Enterprises, Inc.

3–13 CHARGE, TESTING AND DISCHARGING

When a capacitor has a potential difference between the plates, it is said to be *charged.* One plate has an excess of free electrons (–) and the other plate has a lack of electrons (+). Because of the insulation *(dielectric)* between the capacitor foils, electrons cannot flow from one plate to the other. However, there are electric lines of force between the plates, and this force is called the *electric field.* Figure 3–22.

Factors that determine capacitance are; the surface area of the plates, the distance between the plates, and the dielectric insulating material between the plates. Capacitance is measured in *farads,* but the opposition offered to the flow of current by a capacitor is called *capacitive reactance*, which is measured in *ohms.* Capacitive reactance is abbreviated X_C, is inversely proportional to frequency and can be calculated by the equation:

$$X_C = \frac{1}{2\pi fC}$$

π = 3.14, f = frequency, C = capacitance measure in farads

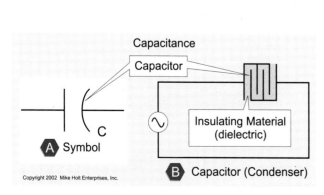

Figure 3-21

Capacitance

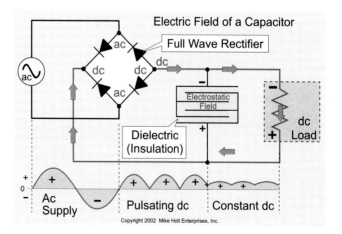

Figure 3-22

Electric Field of a Capacitor

❑ Capacitive Reactance

What is the capacitive reactance of a 0.000001 farad capacitor supplied by 60 Hz, 180 Hz and 300 Hz? Figure 3–23.

Answer: $X_C = \dfrac{1}{2\pi fC}$.

$$X_C \text{ at } 60 \text{ Hz} = \frac{1}{2 \times 3.14 \times 60 \text{ Hz} \times 0.000001 \text{ farads}} = 2,654\Omega$$

$$X_C \text{ at } 180 \text{ Hz} = \frac{1}{2 \times 3.14 \times 180 \text{ Hz} \times 0.000001 \text{ farads}} = 885\Omega$$

$$X_C \text{ at } 300 \text{ Hz} = \frac{1}{2 \times 3.14 \times 300 \text{ Hz} \times 0.000001 \text{ farads}} = 530.46\Omega$$

Capacitor Short and Discharge

The more the capacitor is charged, the stronger the electric field. If the capacitor is overcharged, the electrons from the negative plate could be pulled through the dielectric insulation to the positive plate resulting in a short of the capacitor. To discharge a capacitor, all that is required is a conducting path between the capacitor plates.

Testing a Capacitor

If a capacitor is connected in series with a dc power supply and a test lamp, the lamp will be continuously illuminated if the capacitor is shorted, but will remain dark if the capacitor is good. Figure 3–24.

3–14 USE OF CAPACITORS

Capacitors (condensers) are used to prevent arcing across the *contacts* of low ampere switches and dc power supplies for electronic loads such as computers. Figure 3–25. In addition, capacitors cause the current to lead the voltage by as much as 90 degrees and can be used to correct poor power factor due to inductive reactive loads such as motors. Poor power factor, due to induction, causes current to lag voltage by as much as 90 degrees. Correction of the power factor to *unity* (100%) is known as *resonance* and occurs when inductive reactance is equal to capacitive reactance in a series circuit $X_L = X_C$.

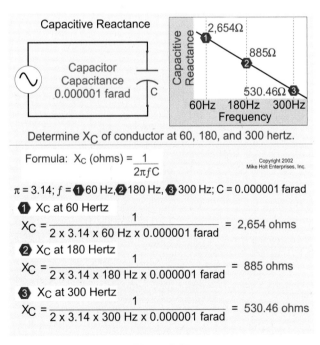

Figure 3-23

Capacitive Reactance

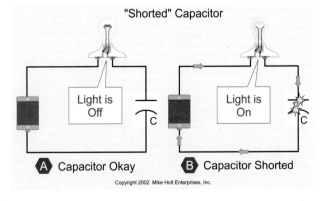

Figure 3-24

Shorted Capacitor

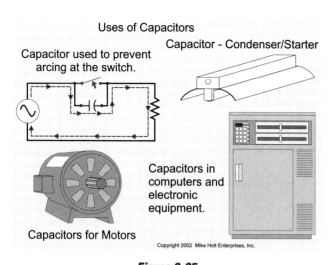

Figure 3-25
Uses of Capacitors

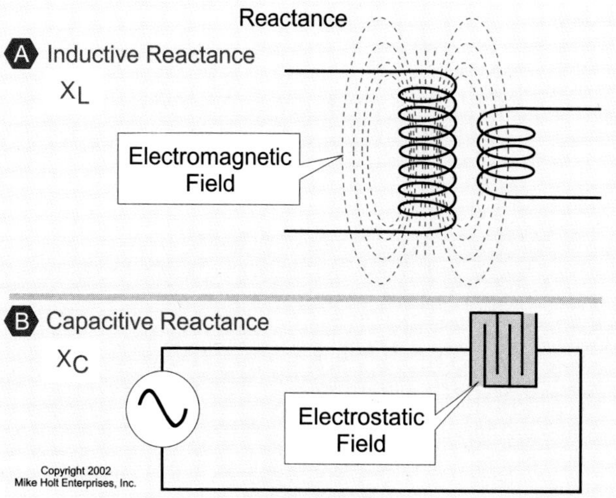

Figure 3-26
Reactance

PART D – POWER FACTOR AND EFFICIENCY

POWER FACTOR INTRODUCTION

Alternating-current circuits develop inductive and capacitive reactance, which causes some power to be stored temporarily in the electromagnetic field of the inductor or in the electrostatic field of the capacitor. Figure 3–26. Because of the temporary storage of energy, the voltage and current of inductive and capacitive circuits are not in phase. This out-of-phase relationship between voltage and current is called *power factor*. Figure 3–27. Power factor affects the calculation of power in ac circuits, but not in dc circuits.

3–15 APPARENT POWER (Volt-Ampere)

In ac circuits, the value from multiplying volts times amperes equals *apparent power* and is commonly referred to as *volt-ampere*, (VA), or *kilovolt-ampere* (kVA). Apparent power (VA) is equal to or greater than *true power* (watts) depending on the *power factor*. When sizing circuits or equipment, always size the circuit equipment according to the apparent power (volt-ampere), not the true power (watt).

Apparent Power (1Ø) = Volts × Amperes

Apparent Power (3Ø) = Volts × Amperes × √3

Apparent Power = True Power/Power Factor

❏ **Apparent Power**

What is the apparent power of a 16A load operating at 120V? Figure 3–28.

(a) 2,400 VA   (b) 1,920 VA   (c) 1,632 VA   (d) none of these

• Answer: (b) 1,920 VA

❏ **Apparent Power**

What is the apparent power of a 250W luminaire that has a power factor of 80%? Figure 3–29.

(a) 200 VA   (b) 250 VA   (c) 313 VA   (d) 375 VA

• Answer: (c) 313 VA

Apparent Power = True Power/Power Factor

Apparent Power = 250W/0.8 PF

Apparent Power = 313 VA

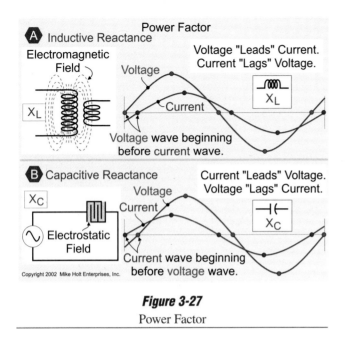

Figure 3-27

Power Factor

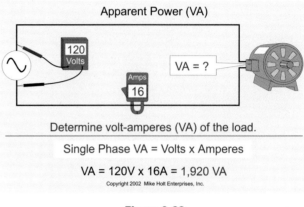

Figure 3-28

Apparent Power (VA)

3–16 POWER FACTOR

Power Factor is the ratio of active power (watts) to apparent power (VA) expressed as a percent that does not exceed 100%. Power Factor is a form of measurement of how far the voltage and current are out-of-phase with each other. In an ac circuit that supplies power to resistive loads such as incandescent lighting, heating elements, etc., the circuit voltage and current will be *in-phase*. The term "in-phase" means that the voltage and current reach their zero and peak values at the same time resulting in a power factor of 100 percent or *unity*. Unity can occur if the circuit only supplies resistive loads or capacitive reactance (X_C) is equal to inductive reactance(X_L).

The formulas for determining power factor are:

Power Factor = True Power (W)/Apparent Power (VA)

Power Factor = Watts/Volt-Ampere

Power Factor = Resistance (R)/Impedance (Z)

The formulas for current:

Amperes (1Ø) = Watts/(Volts Line-to-Line × Power Factor)

Amperes (3Ø) = Watts/(Volts Line-to-Line × 1.732 × Power Factor)

❏ **Power Factor**

What is the power factor of a fluorescent lighting luminaire that produces 160W of light and has a ballast rated for 200 VA? Figure 3–30.

(a) 80%   (b) 90%   (c) 100%   (d) none of these

• Answer: (a) 80%

$$\text{Power Factor} = \frac{W}{VA} = \frac{160W}{200\ VA} = 0.80 \text{ or } 80\%$$

❏ **Current**

What is the current flow for three 5 kW, 230V, 1Ø, loads that have a power factor of 90%?

(a) 72A   (b) 90A   (c) 100A   (d) none of these

• Answer: (a) 72A

I = Watts/(Volts × Power Factor) = 15,000W/(230V × 0.9) = 72A

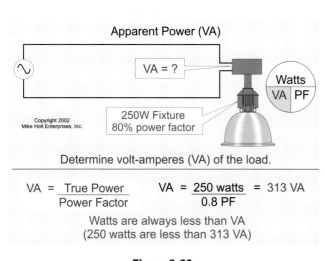

Determine volt-amperes (VA) of the load.

$$VA = \frac{True\ Power}{Power\ Factor} \qquad VA = \frac{250\ watts}{0.8\ PF} = 313\ VA$$

Watts are always less than VA
(250 watts are less than 313 VA)

Figure 3-29
Apparent Power (VA)

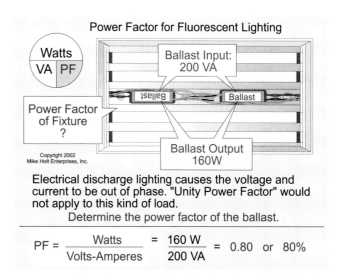

Electrical discharge lighting causes the voltage and current to be out of phase. "Unity Power Factor" would not apply to this kind of load.

Determine the power factor of the ballast.

$$PF = \frac{Watts}{Volts\text{-}Amperes} = \frac{160\ W}{200\ VA} = 0.80\ or\ 80\%$$

Figure 3-30
Power Factor for Fluorescent Lighting

3–17 TRUE POWER (Watts)

ALTERNATING CURRENT

True power is the energy consumed for work, expressed by the unit called the watt. Power is measured by a wattmeter, which is connected series-parallel [series (measures amperes) and parallel (measures volts)] to the circuit conductors. The true power of a circuit that contains inductive or capacitance reactance can be calculated by the use of one of the following formulas:

True Power (1Ø) = Volts × Amperes × Power Factor

True Power (3Ø) = Volts × Amperes × Power Factor × $\sqrt{3}$

Note: True power for a dc circuit is calculated as Volts × Amperes.

❑ **True Power**

What is the true power of a 16A load operating at 120V with a power factor of 85%? Figure 3–31.

(a) 2,400W   (b) 1,920W   (c) 1,632W   (d) none of these

   • Answer: (c) 1,632W

True Power = Volts × Amperes × Power Factor

True Power = (120V × 16A) × 0.85 = 1,632W per hour

Note: True power is always equal to or less than apparent power.

DIRECT CURRENT

In dc or ac circuits at unity power factor:

True Power = Volts × Amperes

❑ **True Power**

What is true power of a 240V, 30A, (1Ø) resistive load that has unity power factor? Figure 3–32.

(a) 4,300W   (b) 7,200W   (c) 7,200 VA   (d) none of these

   • Answer: (b) 7,200W

True Power = Volts × Amperes

True Power = 240V × 30A = 7,200W

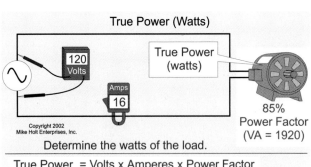

True Power (Watts)

True Power
(watts)

120 Volts

Amps 16

85%
Power Factor
(VA = 1920)

Copyright 2002
Mike Holt Enterprises, Inc.

Determine the watts of the load.

True Power = Volts x Amperes x Power Factor

Volts = 120V, Amperes = 16A, Power Factor = 85% or 0.85

True Power = (120V x 16A) x 0.85 = 1,632W

Note: True power is always less than
VA when power factor is involved.

Figure 3-31
True Power (Watts)

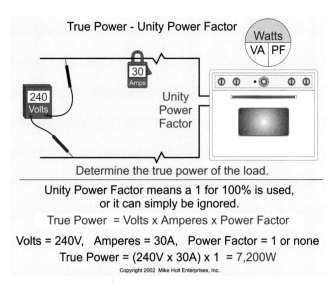

True Power - Unity Power Factor

Watts
VA | PF

30 Amps

240 Volts

Unity
Power
Factor

Determine the true power of the load.

Unity Power Factor means a 1 for 100% is used,
or it can simply be ignored.

True Power = Volts x Amperes x Power Factor

Volts = 240V, Amperes = 30A, Power Factor = 1 or none
True Power = (240V x 30A) x 1 = 7,200W

Copyright 2002 Mike Holt Enterprises, Inc.

Figure 3-32
True Power – Unity Power Factor

3–18 EFFICIENCY

Efficiency has nothing to do with power factor. Efficiency is the ratio of output power to input power, where power factor is the ratio of true power (watts) to apparent power (Volt Ampere). Energy that is not used for its intended purpose is called power loss. Power losses can be caused by conductor resistance, friction, mechanical loading, etc. Power losses of equipment are expressed by the term efficiency, which is the ratio of the input power to the output power. Figure 3–33. The formulas for efficiency are:

> **Efficiency = Output watts/Input watts**
>
> **Input watts = Output watts/Efficiency**
>
> **Output watts = Input watts × Efficiency**

❏ **Efficiency**

If the input of a load is 800W and the output is 640W, what is the efficiency of the equipment? Figure 3–34.
(a) 60%   (b) 70%   (c) 80%   (d) 100%

   • Answer: (c) 80%

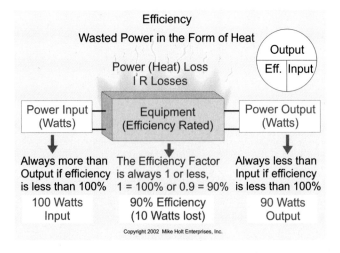

Efficiency
Wasted Power in the Form of Heat

Power (Heat) Loss
I R Losses

Output
Eff. | Input

Power Input
(Watts)

Equipment
(Efficiency Rated)

Power Output
(Watts)

Always more than
Output if efficiency
is less than 100%

The Efficiency Factor
is always 1 or less,
1 = 100% or 0.9 = 90%

Always less than
Input if efficiency
is less than 100%

100 Watts
Input

90% Efficiency
(10 Watts lost)

90 Watts
Output

Copyright 2002 Mike Holt Enterprises, Inc.

Figure 3-33
Efficiency

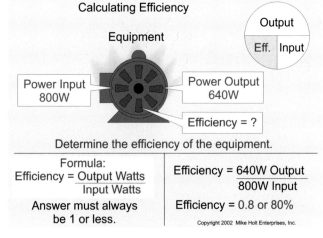

Calculating Efficiency

Output
Eff. | Input

Equipment

Power Input
800W

Power Output
640W

Efficiency = ?

Determine the efficiency of the equipment.

Formula:
Efficiency = Output Watts / Input Watts

Answer must always
be 1 or less.

Efficiency = 640W Output / 800W Input

Efficiency = 0.8 or 80%

Copyright 2002 Mike Holt Enterprises, Inc.

Figure 3-34
Calculating Efficiency

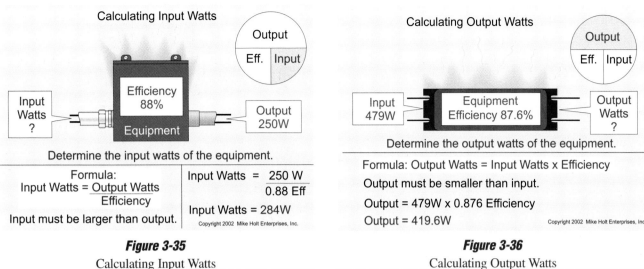

Figure 3-35

Calculating Input Watts

Figure 3-36

Calculating Output Watts

$$\text{Efficiency} = \frac{\text{Output}}{\text{Input}}$$

$$\text{Efficiency} = \frac{640 \text{ W}}{800 \text{ W}} = 0.8 \text{ or } 80\%$$

Note. Efficiency is always less than 100%.

❏ **Input**

If the output of a load is 250W and the equipment is 88% efficient, what are the input watts? Figure 3–35.

(a) 200W   (b) 250W   (c) 285W   (d) 325W

• Answer: (c) 285W

$$\text{Input} = \frac{\text{Output}}{\text{Efficiency}}$$

$$\text{Input} = \frac{250W}{0.88 \text{ efficiency}}$$

Input = 284W

Note. Input is always greater than the output.

❏ **Output**

If a load is 87.6% efficient, for every 479W of input there will be _____ watts of output. Figure 3–36.

(a) 440   (b) 480   (c) 390   (d) 420

• Answer: (d) 420W

Output = Input × Efficiency

Output = 479 × 0.876,

Output = 419.6W

Note: Output is always less than input.

Unit 3 – Understanding Alternating Current Summary Questions

Calculations Questions

Part A – Alternating-Current Fundamentals

3–1 Current Flow

1. The effects of electron movement are the same regardless of the direction of the current flow.
    (a) True   (b) False

3–2 Alternating Current

2. AC is primarily used because it can be transmitted inexpensively and because it can be used for certain applications for which dc is not suitable.
    (a) True   (b) False

3. Faraday's experiments revealed that when a magnetic field moves through a coil of wire, the lines of force of the magnetic field makes the electrons in the wire flow in a specific direction. When the magnetic field moves in the opposite direction, electrons in the wire flow continue to flow in the same direction.
    (a) True   (b) False

3–3 Alternating-Current Generator

4. A simple ac generator consists of a rotating loop of wire between the lines of force of opposite poles of a magnet. The magnitude of the voltage is dependent on the _____.
    (a) number of turns of wire
    (b) strength of the magnetic field
    (c) speed at which the coil rotates
    (d) all of these

3–4 Waveform

5. • A waveform is used to display the level and direction of current and voltage. The waveform for _____ circuits displays the level and direction of the current and voltage for every instant of time for one full revolution of the armature.
    (a) dc   (b) ac   (c) a and b   (d) none of these

6. For ac circuits, the waveform is called the _____.
    (a) frequency   (b) cycle   (c) degree   (d) none of these

3–5 Armature Turns Frequency

7. The number of times the armature turns in one second is called the frequency. Frequency is expressed as _____ or cycles per second.
    (a) degrees   (b) sine wave   (c) phase   (d) hertz

3–6 Phase – In and Out

8. When two waveforms are in step with each other, they are said to be in-phase. In a purely resistive ac circuit, the current and voltage are in-phase. This means that they both reach their zero and _____values at the same time.
    (a) peak   (b) effective   (c) average   (d) none of these

3–7 Phase Difference in Degrees

9. Phase differences are expressed in _____.
    (a) sine   (b) phase   (c) hertz   (d) degrees

10. The terms _____ and _____ are used to describe the relative positions in time of two waveforms (voltage or current).
    (a) hertz, phase   (b) frequency, phase   (c) sine, degrees   (d) lead, lag

3–8 Values of Alternating Current

11. • _____is the value of the voltage or current at any one particular moment of time.
    (a) Peak   (b) Root-mean-square   (c) Effective   (d) Instantaneous

12. _____is the maximum value that ac voltage reaches.
    (a) Peak   (b) Root-mean-square   (c) Instantaneous   (d) none of these

13. • "Effective" is the ac voltage or current value that produces the same amount of heat in a resistor that would be produced by the same amount of dc voltage or current. _____ is the same as effective.
    (a) Peak   (b) Root-mean-square   (c) Instantaneous   (d) none of these

Part B – Induction

14. The movement of electrons because of electromagnetism is called _____.
    (a) flux lines   (b) voltage   (c) induction   (d) magnetic field

15. The induction of voltage in a conductor because of expanding and collapsing magnetic fields is known as _____.
    (a) flux lines   (b) power
    (c) self-induced voltage   (d) a magnetic field

16. Change in current produces a magnetic field through the conductor, which produces an induced voltage that always opposes the change in current. The induced voltage that opposes the change in current is called _____.
    (a) CEMF
    (b) counter-electromotive force
    (c) back-EMF
    (d) all of these

3–9 Induced Voltage and Applied Current

17. When the conductor current increases, the direction of the induced voltage (CEMF) in the conductor is opposite to the direction of the conductor current and tries to prevent the current from increasing.
    (a) True   (b) False

18. When the current in the conductor decreases, the direction of the induced voltage in the conductor attempts to keep the current from decreasing by releasing the energy from the magnetic field back into the conductor.
    (a) True   (b) False

3-10 Conductor Impedance

19. In dc circuits, the only property that affects current and voltage flow is _____, which is a physical property of the conductors that oppose current flow.
    (a) voltage   (b) CEMF   (c) back-EMF   (d) none of these

20. Conductor resistance is directly proportional to the conductor length and inversely proportional to conductor cross-sectional area.
    (a) True   (b) False

21. Alternating current produces a CEMF that is set up inside the conductor which increases the effective resistance of the conductor because of eddy currents and skin effect.
    (a) True   (b) False

22. The opposition to current flow in a conductor because of resistance and induction is called _____.
    (a) resistance   (b) capacitance   (c) induction   (d) impedance

23. • Eddy currents are small independent currents that are produced as a result of dc. Eddy currents flow erratically through a conductor, consume power and increase the effective conductor resistance by opposing the current flow.
    (a) True   (b) False

24. The expanding and collapsing magnetic field of the conductors caused by the current flow in a conductor induces a voltage in the conductors, which repels the flowing electrons toward the surface of the conductor. This has the effect of decreasing the effective conductor cross-sectional area which causes an increase in the conductor impedance. The flow of electrons near the surface is known as _____.
    (a) eddy currents   (b) induced voltage   (c) impedance   (d) skin effect

3-11 Induction and Conductor Shape

25. The amount of the self-induced voltage created within the conductor is directly proportional to the current flow, the length of the conductor, and the frequency at which the magnetic fields cut through the conductors.
    (a) True   (b) False

3-12 Induction and Magnetic Cores

26. • Self-inductance (CEMF) in a coil is affected by the _____.
    (a) winding length and shape
    (b) core material
    (c) frequency
    (d) all of these

Part C – Capacitance

Capacitance Introduction

27. _____is the property of an electric circuit that enables it to store electric energy by means of an electrostatic field and to release this energy at a later time.
    (a) Capacitance   (b) Induction   (c) Self-induction   (d) none of these

3–13 Charge, Testing and Discharging

28. When a capacitor has a potential difference between the plates, it is said to be _____. One plate has an excess of free electrons and the other plate has fewer electrons.
     (a) induced   (b) charged   (c) discharged   (d) shorted

29. If the capacitor is overcharged, the electrons from the negative plate may be pulled through the insulation to the positive plate. The capacitor is said to have _____.
     (a) charged   (b) discharged   (c) induced   (d) shorted

30. To discharge a capacitor, all that is required is a(n) _____ path between the capacitor plates.
     (a) conducting   (b) insulating   (c) isolating   (d) none of these

3–14 Uses of Capacitors

31. What helps prevent arcing across the contacts of electric switches?
     (a) Springs   (b) A condenser   (c) An inductor   (d) A resistor

32. The current in a purely capacitive circuit _____.
     (a) leads the applied voltage by 90 degrees
     (b) lags the applied voltage by 90 degrees
     (c) leads the applied voltage by 180 degrees
     (d) lags the applied voltage by 180 degrees

33. Circuits containing inductive or capacitive reactance temporarily store power in the electromagnetic field of induction and the electrostatic field of capacitors.
     (a) True   (b) False

Part D – Power Factor and Efficiency

3–15 Apparent Power

34. If you measure voltage and current in an inductive or capacitive circuit and then multiply them together, you would obtain the circuit's _____.
     (a) true power   (b) power factor   (c) apparent power   (d) power loss

35. Apparent power is equal to or greater than true power depending on the power factor.
     (a) True   (b) False

36. When sizing circuits or equipment, always size the circuit components and transformers according to the apparent power, not the true power.
     (a) True   (b) False

3–16 Power Factor

37. Power factor is a measurement of how far the voltage and current are out-of-phase with each other. Power factor is the ratio between true power (resistive load) and apparent power (reactive load). Power factor can be expressed by the formula: _____.
     (a) $PF = {}^P/_E$ (b) $PF = {}^R/_Z$ (c) $PF = I^2R$ (d) $PF = {}^Z/_R$

38. • When current and voltage are in-phase, the power factor is _____.
     (a) 100%   (b) unity   (c) 90 degrees   (d) a and b

39. In an ac circuit that supplies power to resistive loads such as incandescent lighting, heating elements, etc., the circuit voltage and current are said to be _____, resulting in a power factor of unity.
(a) out-of-phase
(b) leading by 90 degrees
(c) 90 degrees out-of-phase
(d) none of these

40. True power is the energy consumed for work expressed by the term watts. To determine the true power of a circuit that contains inductive or capacitance reactance, we must multiply the volts times the current times the _____.
(a) efficiency   (b) sine wave   (c) power factor   (d) none of these

41. In dc circuits, the voltage and current are constant and the true power is simply volts times amperes.
(a) True   (b) False

42. What does it cost per year (9 cents per kWh) for ten 150W recessed luminaires to operate if they are on 6 hours per day?
(a) $153   (b) $296   (c) $235   (d) $180

43. A 2 × 4 recessed luminaire contains four 34W lamps and the ballast is rated 1.5A at 120V. What is the power factor of the ballast assuming 100 percent efficiency?
(a) 55%   (b) 65%   (c) 70%   (d) 75%

44. • What is the apparent power of a 20A load operating at 120V with a power factor of 85 percent?
(a) 2,400W   (b) 1,920W   (c) 1,632W   (d) 2,400 VA

45. • What is the true power of a 20A load operating at 120V with a power factor of 85 percent?
(a) 2,400W   (b) 1,920W   (c) 1,632W   (d) none of these

46. • Since power factor cannot be greater than 100 percent, true power is equal to or less than the apparent power. Because of power factor, the VA of the load is greater than the watts, which results in less loads per circuit, more circuits and larger kVA transformers.
(a) True   (b) False

47. • What is the true power of a 10A circuit operating at 120V with unity (100 percent) power factor?
(a) 1,200 VA   (b) 2,400 VA   (c) 1,200W   (d) 2,400W

3–17 True Power (Watts)

48. What size transformer is required for a 240V, 125A, 1Ø load?
(a) 3 kVA   (b) 30 kVA   (c) 12.5 kVA   (d) 15 kVA

49. What size transformer is required for a 30 kW load that has a power factor of 85 percent?
(a) 12.5 kVA   (b) 35 kVA   (c) 7.5 kVA   (d) 15 kVA

50. • How many 20A, 120V circuits are required for forty-two 300W recessed luminaires (noncontinuous load)? Tip: Only so many luminaires are permitted on a circuit.
(a) 3 circuits   (b) 4 circuits   (c) 5 circuits   (d) 6 circuits

3–18 Efficiency

51. • How many 20A, 120V circuits are required for forty-two 300W recessed luminaires (noncontinuous load) with a power factor of 85 percent?
     (a) 5 circuits   (b) 6 circuits   (c) 7 circuits   (d) 8 circuits

52. Efficiency is the ratio of the output to input power.
     (a) True   (b) False

53. If the output is 1,320W and the input is 1,800W, what is the efficiency of the equipment?
     (a) 62%   (b) 73%   (c) 0%   (d) 100%

54. If the output is 160W and the equipment is 88 percent efficient, what are the input amperes if the voltage is 120V?
     (a) 0.75A   (b) 1.5A   (c) 2.275A   (d) 3.25A

☆ Challenge Questions

Part A – Alternating-Current Fundamentals

3–2 Alternating Current

55. A transformer winding that is 97 percent efficient produces _____ output for every 1 kW input.
     (a) 970W   (b) 1,000W   (c) 1,030W   (d) 1,300W

56. The primary reason(s) for high-voltage transmission lines is (are) _____.
     (a) reduced voltage drop   (b) smaller wire
     (c) smaller equipment   (d) all of these

57. One of the advantages of a higher-voltage system as compared to a lower-voltage system (for the same wattage loads) is _____.
     (a) reduced voltage drop   (b) reduced power use
     (c) large currents   (d) lower electrical pressure

58. The advantage of ac over dc is that ac provides for _____.
     (a) better speed control   (b) ease of voltage variation
     (c) lower resistance at high currents   (d) none of these

3–4 Waveforms

59. A waveform represents _____.
     (a) the magnitude and direction of current or voltage
     (b) how current or voltage can vary with time
     (c) how output voltage can vary with the generator armature
     (d) all of these

3–5 Armature Turning Frequency

60. Frequency of an ac waveform is the number of times the current or voltage goes through 360 degrees in _____.
     (a) $1/10$ second   (b) 5 seconds   (c) 1 second   (d) 60 seconds

61. How much time does it take for 60 Hz ac to travel through 180 degrees?
     (a) $1/40$ second   (b) $1/40$ second   (c) $1/180$ second   (d) none of these

3–8 Values of Alternating Current

62. • The heating effects of 10A of ac as compared with 10A of dc is _____.
    (a) the same   (b) less   (c) greater   (d) none of these

63. If the maximum value of an ac system is 50A, the RMS value would be approximately _____.
    (a) 25A   (b) 30A   (c) 35A   (d) 40A

64. • The maximum value of 120V dc is equal to the maximum value of an equivalent ac.
    (a) True   (b) False

65. • You are getting a 120V reading on your voltmeter. This is an indication of the _____ value of the voltage source.
    (a) average   (b) peak   (c) effective   (d) instantaneous

Part B – Induction

3–9 Voltage and Applied Current

66. _____Law states that a change in current produces a counter-electromotive force whose direction is such that it opposes the change in current.
    (a) Kirchoff's Second   (b) Kirchoff's First
    (c) Lenz's   (d) Hertz's

67. Inductive reactance is abbreviated as _____.
    (a) I^2R (b) L_X (c) X_L (d) Z

68. Inductive reactance changes proportionally with frequency.
    (a) True   (b) False

69. Inductive reactance is measured in _____.
    (a) farads   (b) watts   (c) ohms   (d) coulombs

70. • If the frequency is constant, the inductive reactance of a circuit will _____.
    (a) remain constant regardless of the current and voltage changes
    (b) vary directly with the voltage
    (c) vary directly with the current
    (d) not affect the impedance

3–10 Conductor Impedance

71. The total opposition to current flow in an ac circuit is expressed in ohms and is called _____.
    (a) impedance   (b) conductance   (c) reluctance   (d) none of these

72. Impedance is present in _____ type circuit(s).
    (a) resistance   (b) direct current   (c) alternating current   (d) none of these

73. Conductor resistance to ac flow is _____ the resistance to dc.
    (a) higher than   (b) lower than   (c) the same as   (d) none of these

Part C – Capacitance

3–13 Charge, Testing and Discharging

74. If a test lamp is placed in series with a capacitor with a dc voltage source and the lamp is continuously illuminated, it is an indication that the capacitor is _____.
    (a) fully charged   (b) shorted   (c) fully discharged   (d) open-circuit

75. The insulating material between the surface plates of a capacitor is called the _____.
    (a) inhibitor   (b) electrolyte   (c) dielectric   (d) regulator

76. Capacitors are measured in _____.
    (a) watts   (b) volts   (c) farads   (d) henrys

77. Capacitive reactance is measured in _____.
    (a) ohms   (b) volts   (c) watts   (d) henrys

3–14 Uses of Capacitors

78. • In a circuit that has only capacitive reactance (X_C), the voltage and current are said to be out-of-phase to each other because the voltage _____.
    (a) leads the current by 90 degrees   (b) lags the current by 90 degrees
    (c) leads the current   (d) none of these

79. Resonance occurs when _____.
    (a) only resistance occurs in the system   (b) the power factor is equal to 0
    (c) $X_L = X_C$ are in a series circuit   (d) when R = Z

Part D – Power Factor and Efficiency

3–15 Apparent Power

80. If you multiply the voltage times the current in an inductive or capacitive circuit, the answer you obtain will be the _____ of the circuit.
    (a) watts   (b) true power   (c) apparent power   (d) all of these

81. The apparent power of a 19.2A, 120V load is _____.
    (a) 2.304 kVA   (b) 2.3 kVA   (c) 2.3 VA   (d) 230 kVA

3–16 Power Factor

82. Power factor in an ac circuit will be unity (100 percent), if the circuit contains only _____.
    (a) induction motors   (b) transformers   (c) reactance coils   (d) resistive loads

83. • Three 8 kW electric-discharge lighting bank circuits, which have a 92 percent power factor, are connected to a 230V, 3Ø source. The current flow of these lights is _____.
    (a) 37A   (b) 50A   (c) 65A   (d) 75A

3–17 True Power

84. • A wattmeter is connected in _____ in the circuit.
    (a) series   (b) parallel   (c) series-parallel   (d) none of these

85. • True power is always voltage times current for _____.
    (a) all ac circuits   (b) dc circuits
    (c) ac circuits at unity power factor   (d) b and c

86. Power consumed (watts) in either a 1Ø ac or dc system is always equal to _____.

    (a) $E \times I$ (b) $E \times R$ (c) $I^2 \times R$ (d) $\dfrac{E}{(I \times R)}$

87. • The power consumed on a 208V, 76A, 3Ø circuit that has a power factor of 89 percent is _____.
    (a) 27,379W   (b) 35,808W   (c) 24,367W   (d) 12,456W

88. • The true power of a 1Ø, 2.1 kVA load with a power factor of 91 percent is _____.
    (a) 2.1 kW   (b) 1.91 kW   (c) 1.75 kW   (d) 1.911 kW

3–18 Efficiency

89. • Motor efficiency can be determined by which of the following formulas?
    (a) Horsepower $\times$ 746   (b) Horsepower $\times$ 746/VA Input
    (c) Horsepower $\times$ 746/W Input   (d) Horsepower $\times$ 746/kVA Input

90. The efficiency ratio of a 4,000 VA transformer winding with a secondary VA of 3,600 VA is _____.
    (a) 80%   (b) 70%   (c) 90%   (d) 110%

NEC Questions – Articles 220-240

Article 220 Branch-Circuit, Feeder, and Service Calculations

91. Under the optional method for calculating a single-family dwelling, general loads beyond the initial 10 kW are to be assessed at a _____ percent demand factor.
    (a) 40   (b) 50   (c) 60   (d) 75

92. A demand factor of _____ percent is applicable for a multifamily dwelling with ten units if the optional calculation method is used.
    (a) 75   (b) 60   (c) 50   (d) 43

93. The connected load to which the demand factors of Table 220.32 apply shall include 3 VA per _____ for general lighting and general-use receptacles.
    (a) inch   (b) foot   (c) square inch   (d) square foot

94. The connected load to which the demand factors of Table 220.32 apply shall include the _____ rating of all appliances that are fastened in place, permanently connected or located to be on a specific circuit; such as ranges, wall-mounted ovens, counter-mounted cooking units, clothes dryers, water heaters and space heaters.
    (a) calculated   (b) nameplate   (c) circuit   (d) overcurrent protection

95. In order to use the optional method for calculating a service to a school, the school shall be equipped with _____.
    (a) cooking facilities   (b) electric space heating   (c) air-conditioning   (d) b or c

96. Feeder conductors for new restaurants shall not be required to be of _____ ampacity than the service-entrance conductors.
    (a) greater   (b) lesser   (c) equal   (d) none of these

97. • Service-entrance or feeder conductors whose demand load is determined by the optional calculation, as permitted in 220.36, shall not be permitted to have the neutral load determined by 220.22.
    (a) True   (b) False

98. When a farm dwelling has electric heat and the farm operation has electric grain-drying systems, Part _____ of Article 220 shall not be used to compute the dwelling load where the dwelling and farm load are supplied by a common service.
    (a) I   (b) II   (c) III   (d) none of these

99.   For each farm building or load supplied by _____ or more branch circuits, the load for feeders, service-entrance conductors and service equipment shall be computed in accordance with demand factors not less than indicated in Table 220.40.
      (a) one   (b) two   (c) three   (d) four

Article 225 Outside Branch Circuits and Feeders

100. Open individual conductors shall not be smaller than _____ AWG copper for spans up to 50 ft in length and _____ AWG copper for a longer span, unless supported by a messenger wire.
      (a) 10, 8   (b) 6, 8   (c) 6, 6   (d) 8, 8

101. The minimum point of attachment of overhead conductors to a building shall in no case be less than _____ above finished grade.
      (a) 8 ft   (b) 10 ft   (c) 12 ft   (d) 15 ft

102. Overhead conductors shall have a minimum of _____ vertical clearance from final grade over residential property and driveways, as well as those commercial areas not subject to truck traffic where the voltage is limited to 300 volts-to-ground.
      (a) 10 ft   (b) 12 ft   (c) 15 ft   (d) 18 ft

103. The minimum clearance for overhead conductors not exceeding 600V that pass over commercial areas subject to truck traffic is _____.
      (a) 10 ft   (b) 12 ft   (c) 15 ft   (d) 18 ft

104. Overhead conductors installed over roofs shall have a vertical clearance of _____ above the roof surface.
      (a) 8 ft   (b) 12 ft   (c) 15 ft   (d) 3 ft

105. If a set of 120/240V overhead conductors terminates at a through-the-roof raceway or approved support, with less than 6 ft of these conductors passing over the roof overhang, the minimum clearance above the roof for these conductors is _____.
      (a) 12 in.   (b) 18 in.   (c) 2 ft   (d) 5 ft

106. The requirement for maintaining a 3 ft vertical clearance from the edge of the roof shall not apply to the final conductor span where the conductors are attached to _____.
      (a) a building pole   (b) the side of a building   (c) an antenna   (d) the base of a building

107. Overhead conductors to a building shall maintain a vertical clearance of final spans above, or within _____ measured horizontally from the platforms, projections or surfaces from which they might be reached.
      (a) 3 ft   (b) 6 ft   (c) 8 ft   (d) 10 ft

108. _____ shall not be installed beneath openings through which materials may be moved, such as openings in farm and commercial buildings, and shall not be installed where they will obstruct entrance to these building openings.
      (a) Overcurrent protection devices
      (b) Overhead branch-circuit and feeder conductors
      (c) Grounding conductors
      (d) Wiring systems

109. A building or structure shall be supplied by a maximum of _____ feeder(s) or branch circuit(s).
      (a) one   (b) two   (c) three   (d) as many as desired

110. A single building or other structure sufficiently large enough to require two or more supplies is permitted by _____.
      (a) architects   (b) special permission   (c) written authorization   (d) master electricians

111. The building disconnecting means shall be installed at a(n) _____ location.
      (a) accessible   (b) readily accessible   (c) outdoor   (d) indoor

112. The disconnecting means is not required to be located at the building or structure where documented safe switching procedures are established and maintained and where the installation is monitored by _____ persons.
      (a) maintenance   (b) management   (c) service   (d) qualified

113. • There shall be no more than _____ disconnects installed for each supply.
   (a) two   (b) four   (c) six   (d) none of these

114. The two to six disconnects as permitted in 225.33 shall be _____. Each disconnect shall be marked to indicate the load served.
   (a) the same size   (b) grouped
   (c) in the same enclosure   (d) none of these

115. The one or more additional disconnecting means for fire pumps or for emergency, legally required standby or optional standby systems as permitted by 225.30, shall be installed sufficiently remote from the one to six disconnecting means for normal supply to minimize the possibility of _____ interruption of supply.
   (a) accidental   (b) intermittent   (c) simultaneous   (d) prolonged

116. In a multiple-occupancy building, each occupant shall have access to his or her own _____.
   (a) disconnecting means   (b) building drops
   (c) building-entrance assembly   (d) lateral conductors

117. In a multiple-occupancy building where electrical maintenance is provided by the building management under continuous building management supervision, the building disconnecting means supplying more than one occupancy can be accessible to authorized _____ only.
   (a) inspectors   (b) tenants
   (c) management personnel   (d) none of these

118. When the building disconnecting means is a power-operated switch or circuit breaker, it shall be able to be opened by hand in the event of a _____.
   (a) ground fault   (b) short circuit   (c) power surge   (d) power-supply failure

119. • The building or structure disconnecting means shall plainly indicate whether it is in the _____ position.
   (a) open or closed   (b) correct   (c) up or down   (d) none of these

120. • In order for a single branch circuit to supply only limited loads, the building disconnecting means shall have a rating of not less than _____.
   (a) 15A   (b) 20A   (c) 25A   (d) 30A

121. For installations consisting of not more than two 2-wire branch circuits, the building disconnecting means shall have a rating of not less than _____.
   (a) 15A   (b) 20A   (c) 25A   (d) 30A

Article 230 Services

122. A building or structure shall be supplied by a maximum of _____ service(s).
   (a) one   (b) two   (c) three   (d) as many as desired

123. A single building or other structure sufficiently large enough to require two or more services is permitted by _____.
   (a) architects   (b) special permission
   (c) written authorization   (d) master electricians

124. Service conductors supplying a building or other structure shall not _____ of another building or other structure.
   (a) be installed on the exterior walls   (b) pass through the interior
   (c) a and b   (d) none of these

125. • Conductors other than service conductors shall not be installed in the same _____.
   (a) service raceway   (b) service cable   (c) enclosure   (d) a or b

126. Service conductors installed as unjacketed multiconductor cable shall have a minimum clearance of _____ from windows that are designed to be opened, doors, porches, stairs, fire escapes and similar equipment.
   (a) 3 ft   (b) 4 ft   (c) 6 ft   (d) 10 ft

127. Overhead-service conductors to a building shall maintain a vertical clearance of final spans above, or within, _____ measured horizontally from the platforms, projections or surfaces from which they might be reached.
      (a) 3 ft   (b) 6 ft   (c) 8 ft   (d) 10 ft

128. _____ shall not be installed beneath openings through which materials may be moved, such as openings in farm and commercial buildings, and shall not be installed where they will obstruct entrance to these building openings.
      (a) Overcurrent protection devices   (b) Overhead-service conductors
      (c) Grounding conductors   (d) Wiring systems

129. Overhead service conductors can be supported to hardwood trees.
      (a) True   (b) False

130. Service-drop conductors shall have _____.
      (a) sufficient ampacity to carry the current for the load
      (b) adequate mechanical strength
      (c) a or b
      (d) a and b

131. The minimum size service-drop conductor permitted by the *Code* is _____ AWG copper or _____ AWG aluminum or copper-clad aluminum.
      (a) 8, 6   (b) 6, 8   (c) 6, 6   (d) 8, 8

132. Service drops installed over roofs shall have a vertical clearance of _____ above the roof surface.
      (a) 8 ft   (b) 12 ft   (c) 15 ft   (d) 3 ft

133. If a set of 120/240V service conductors terminates at a through-the-roof raceway or approved support, with less than 6 ft of these conductors passing over the roof overhang, the minimum clearance above the roof for these service conductors is _____.
      (a) 12 in.   (b) 18 in.   (c) 2 ft   (d) 5 ft

134. The requirement for maintaining a 3 ft vertical clearance from the edge of the roof shall not apply to the final conductor span where the service drop is attached to _____.
      (a) a service pole   (b) the side of a building   (c) an antenna   (d) the base of a building

135. Service-drop conductors shall have a minimum of _____ vertical clearance from final grade over residential property and driveways, as well as those commercial areas not subject to truck traffic where the voltage is limited to 300 volts-to-ground.
      (a) 10 ft   (b) 12 ft   (c) 15 ft   (d) 18 ft

136. The minimum clearance for service drops not exceeding 600V that pass over commercial areas subject to truck traffic is _____.
      (a) 10 ft   (b) 12 ft   (c) 15 ft   (d) 18 ft

137. • Overhead service-drop conductors shall have a horizontal clearance of _____ from a pool.
      (a) 6 ft   (b) 10 ft   (c) 8 ft   (d) 4 ft

138. The minimum point of attachment of the service-drop conductors to a building shall in no case be less than _____ above finished grade.
      (a) 8 ft   (b) 10 ft   (c) 12 ft   (d) 15 ft

139. Where raceway-type service masts are used, all raceway fittings shall be _____ for use with service masts.
      (a) identified   (b) approved   (c) heavy-duty   (d) listed

140. Service-lateral conductors are required to be insulated except for the grounded conductor when it is _____.
      (a) bare copper used in a raceway
      (b) bare copper and part of a cable assembly that is identified for underground use
      (c) copper-clad aluminum with individual insulation
      (d) a and b

141. Service lateral conductors must have _____.
(a) adequate mechanical strength
(b) sufficient ampacity for the loads computed
(c) a and b
(d) none of these

142. Underground-copper service conductors shall not be smaller than _____ AWG copper.
(a) 3   (b) 4   (c) 6   (d) 8

143. To supply power to a 1Ø water heater using underground conductors, the size of the underground service-lateral conductors shall not be smaller than _____.
(a) 4 AWG copper   (b) 8 AWG aluminum   (c) 12 AWG copper   (d) none of these

144. When two to six service disconnecting means in separate enclosures are grouped at one location and supply separate loads from one service drop or lateral, _____ set(s) of service-entrance conductors shall be permitted to supply each or several such service equipment enclosures.
(a) one   (b) two   (c) three   (d) four

145. Service-entrance conductors entering or on the exterior of buildings or other structures shall be insulated.
(a) True   (b) False

146. Service conductors shall be sized no less than _____ percent of the continuous load, plus 100 percent of the noncontinuous load.
(a) 100   (b) 115   (c) 125   (d) 150

147. Wiring methods permitted for service conductors include _____.
(a) rigid metal conduit   (b) electrical metallic tubing
(c) rigid nonmetallic conduit   (d) all of these

148. • Service-entrance conductors shall not be spliced or tapped.
(a) True   (b) False

149. Service-entrance conductors can be spliced or tapped by clamped or bolted connections at any time as long as _____.
(a) the free ends of conductors are covered with an insulation that is equivalent to that of the conductors or with an insulating device identified for the purpose
(b) wire connectors or other splicing means installed on conductors that are buried in the earth are listed for direct burial
(c) no splice is made in a raceway
(d) all of these

150. Service cables, where subject to physical damage, shall be protected.
(a) True   (b) False

151. Service cables that are subject to physical damage shall be protected by _____.
(a) rigid metal conduit   (b) intermediate metal conduit
(c) schedule 80 rigid nonmetallic conduit   (d) any of these

152. Individual open conductors and cables other than service-entrance cables shall not be installed within _____ of grade level or where exposed to physical damage.
(a) 8 ft   (b) 10 ft   (c) 12 ft   (d) 15 ft

153. Service cables mounted in contact with a building shall be supported at intervals not exceeding _____.
(a) 4 ft   (b) 3 ft   (c) 30 in.   (d) 24 in.

154. Where exposed to _____, service conductors shall be mounted on insulators or on insulating supports attached to racks, brackets or other approved means. Where not exposed to the weather, the conductors shall be mounted on glass or porcelain knobs.
(a) a corrosive environment   (b) the weather
(c) the general public   (d) any inspector

155. Where individual open conductors are not exposed to the weather, the conductors shall be mounted on _____ knobs.
(a) door   (b) insulated metal   (c) glass or porcelain   (d) none of these

156. When individual open conductors enter a building or other structure through tubes, _____ shall be formed on the conductors before they enter the tubes.
(a) drop loops   (b) knots   (c) drip loops   (d) none of these

157. When individual open conductors enter a building or other structure, they shall enter through roof bushings or through the wall in an upward slant through individual, noncombustible, nonabsorbent insulating _____.
(a) tubes   (b) raceways   (c) chases   (d) standoffs

158. Service heads for service raceways shall be _____.
(a) raintight   (b) weatherproof   (c) rainproof   (d) watertight

159. Service cables shall be equipped with a raintight _____.
(a) raceway   (b) service head   (c) cover   (d) all of these

160. Type SE cable shall be permitted to be formed in a _____ and taped with self-sealing weather-resistant thermoplastic.
(a) loop   (b) circle   (c) gooseneck   (d) none of these

161. Service heads shall be located _____.
(a) above the point of attachment   (b) below the point of attachment
(c) even with the point of attachment   (d) none of these

162. To prevent water from entering service equipment, service-entrance conductors must _____.
(a) be connected to service-drop conductors below the level of the service head
(b) have drip loops formed on the service-entrance conductors
(c) a or b
(d) a and b

163. Service-drop conductors and service-entrance conductors shall be arranged so that _____ will not enter the service raceway or equipment.
(a) dust   (b) vapor   (c) water   (d) none of these

164. On a 3Ø, 4-wire, delta-connected service where the midpoint of one phase winding is grounded, the service conductor having the higher-phase voltage-to-ground shall be durably and permanently marked by an outer finish that is _____ in color, or by other effective means, at each termination or junction point.
(a) orange   (b) red   (c) blue   (d) any of these

165. The service disconnecting means shall be installed at a(n) _____ location.
(a) dry   (b) readily accessible   (c) outdoor   (d) indoor

166. Service disconnecting means shall not be installed in bathrooms.
(a) True   (b) False

167. Each service disconnect shall be permanently marked to identify it as a service disconnecting means.
(a) True   (b) False

168. Each service disconnect shall be permanently _____ to identify it as part of the service disconnecting means.
(a) identified   (b) positioned   (c) marked   (d) none of these

169. Each service disconnecting means shall be suitable for _____.
(a) hazardous locations   (b) wet locations   (c) dry locations   (d) the prevailing conditions

170. There shall be no more than _____ disconnects installed for each service or for each set of service-entrance conductors as permitted in 230.2 and 230.40.
(a) two   (b) four   (c) six   (d) none of these

171. Disconnecting means used solely for power monitoring equipment, or the control circuit of the ground-fault protection system or power-operable service disconnecting means, installed as part of the listed equipment, shall not be considered a service disconnecting means.
(a) True   (b) False

172. • When the service contains two to six service disconnecting means, they shall be _____.
(a) the same size   (b) grouped at one location
(c) in the same enclosure   (d) none of these

173. The additional service disconnecting means for fire pumps or for emergency, legally required standby, or optional standby services permitted by 230.2, shall be installed sufficiently remote from the one to six service disconnecting means for normal service to minimize the possibility of _____ interruption of supply.
(a) accidental   (b) intermittent   (c) simultaneous   (d) prolonged

174. In a multiple-occupancy building, each occupant shall have access to the occupant's _____.
(a) service disconnecting means   (b) service drops
(c) distribution transformer   (d) lateral conductors

175. In a multiple-occupancy building where electric service and electrical maintenance are provided by the building management under continuous building management supervision, the service disconnecting means supplying more than one occupancy can be accessible to authorized _____ only.
(a) inspectors   (b) tenants
(c) management personnel   (d) none of these

176. When the service disconnecting means is a power-operated switch or circuit breaker, it shall be able to be opened by hand in the event of a _____.
(a) ground fault   (b) short circuit   (c) power surge   (d) power-supply failure

177. • The service disconnecting means shall plainly indicate whether it is in the _____ position.
(a) open or closed   (b) tripped   (c) up or down   (d) correct

178. • In order for a single branch circuit to supply only limited loads, the service disconnecting means shall have a rating of not less than _____.
(a) 15A   (b) 20A   (c) 25A   (d) 30A

179. For installations consisting of not more than two 2-wire branch circuits, the service disconnecting means shall have a rating of not less than _____.
(a) 15A   (b) 20A   (c) 25A   (d) 30A

180. When the service disconnecting means consists of more than one switch or circuit breaker, the combined ratings of all the switches or circuit breakers used _____ than the rating required by 230.79.
(a) can be more   (b) shall not be less
(c) must be more   (d) none of these

181. The service conductors shall be connected to the service disconnecting means by _____ or other approved means.
(a) pressure connectors   (b) clamps   (c) solder   (d) a or b

182. _____ for power-operable service disconnects can be connected on the supply side of the service disconnecting means, if suitable overcurrent protection and disconnecting means are provided.
(a) Control circuits   (b) Distribution panels
(c) Grounding conductors   (d) none of these

183. Each _____ service conductor shall have overload protection.
(a) overhead   (b) underground   (c) ungrounded   (d) none of these

184. In a service, overcurrent protection devices shall never be placed in the grounded service conductor with the exception of a circuit breaker that simultaneously opens all conductors of the circuit.
(a) True   (b) False

185. Where the service overcurrent devices are locked or sealed or are not readily accessible to the _____, branch-circuit overcurrent devices shall be installed on the load side, shall be mounted in a readily accessible location and shall be of lower ampere rating than the service overcurrent device.
(a) inspector   (b) electrician   (c) occupant   (d) all of these

186. Where necessary to prevent tampering, an automatic overcurrent protection device that protects service conductors supplying only a specific load, such as a water heater, shall be permitted to be _____ where located so as to be accessible.
(a) locked   (b) sealed   (c) a or b   (d) none of these

187. Circuits used only for the operation of fire alarms, other protective signaling systems or the supply to fire pump equipment shall be permitted to be connected on the _____ of the service overcurrent protection device where separately provided with overcurrent protection.
(a) base   (b) load side   (c) supply side   (d) top

188. As applicable in 230.95, the rating of the service disconnect is considered to be the rating of the largest _____ that can be installed or the highest continuous-current trip setting for which the actual overcurrent protection device installed in a circuit breaker is rated or can be adjusted.
(a) fuse   (b) circuit   (c) wire   (d) all of these

189. Ground-fault protection of equipment shall be provided for solidly grounded wye electrical services of more than 150 volts-to-ground, but not exceeding 600V phase-to-phase for each service disconnecting means rated _____ or more.
(a) 1,000A   (b) 1,500A   (c) 2,000A   (d) 2,500A

190. The maximum setting of the ground-fault protection in a service disconnecting means shall be _____.
(a) 800A   (b) 1,000A   (c) 1,200A   (d) 2,000A

191. The ground-fault protection system shall be _____ when first installed on site.
(a) inspected   (b) identified   (c) turned on   (d) performance tested

192. Ground-fault protection that functions to open the service disconnect _____ protect(s) service conductors or the service disconnecting means.
(a) will   (b) will not   (c) adequately   (d) totally

193. Ground-fault protection at service equipment may make it necessary to review the overall wiring system for proper selective overcurrent protection _____.
(a) rating   (b) coordination   (c) devices   (d) none of these

194. Where ground-fault protection is provided for the _____ disconnect and interconnection is made with another supply system by a transfer device, means or devices may be needed to ensure proper ground-fault sensing by the ground-fault protection equipment.
(a) circuit   (b) service
(c) switch and fuse combination   (d) b and c

195. For services exceeding 600V, nominal, the isolating switch shall be accessible to _____.
(a) all occupants   (b) qualified personnel only
(c) a height of 8 feet   (d) any of these

Article 240 Overcurrent Protection

196. Overcurrent protection for conductors and equipment is designed to _____ the circuit if the current reaches a value that will cause an excessive or dangerous temperature in conductors or conductor insulation.
(a) open   (b) close   (c) monitor   (d) record

197. A device that, when interrupting currents in its current-limiting range, will reduce the current flowing in the faulted circuit to a magnitude substantially less than that obtainable in the same circuit if the device were replaced with a solid conductor having comparable impedance, is defined as a(n) _____ protective device.
    (a) short-circuit
    (b) overload
    (c) ground-fault
    (d) current-limiting overcurrent

198. Conductor overload protection shall not be required where the interruption of the _____ would create a hazard, such as in a material-handling magnet circuit or fire-pump circuit. However, short-circuit protection is required.
    (a) circuit
    (b) line
    (c) phase
    (d) system

199. Unless specifically permitted in 240.4(E) through 240.4(G), the overcurrent protection shall not exceed _____ after any correction factors for ambient temperature and the number of conductors have been applied.
    (a) 15A for 14 AWG copper
    (b) 20A for 12 AWG copper
    (c) 30A for 10 AWG copper
    (d) all of these

200. Flexible cords approved for use with listed appliances or portable lamps shall be permitted to be on a branch circuit protected by a 20A circuit breaker if they are _____.
    (a) not more than 6 ft in length
    (b) 20 AWG and larger
    (c) tinsel cord or 18 AWG cord and larger
    (d) 16 AWG and larger

Unit 3 NEC Exam – NEC Code Order 220.36 – 240.6

1. Service-entrance or feeder conductors whose demand load is determined by the optional calculation, as permitted in 220.36, shall not be permitted to have the neutral load determined by 220.22.
   (a) True   (b) False

2. When a farm dwelling has electric heat and the farm operation has electric grain-drying systems, Part _____ of Article 220 shall not be used to compute the dwelling load where the dwelling and farm load are supplied by a common service.
   (a) I   (b) II   (c) III   (d) none of these

3. For each farm building or load supplied by _____ or more branch circuits, the load for feeders, service-entrance conductors, and service equipment shall be computed in accordance with demand factors not less than indicated in Table 220.40.
   (a) one   (b) two   (c) three   (d) four

4. The minimum clearance for overhead conductors not exceeding 600V that pass over commercial areas subject to truck traffic is _____.
   (a) 10 ft   (b) 12 ft   (c) 15 ft   (d) 18 ft

5. The two to six disconnects as permitted in 225.33 shall be _____. Each disconnect shall be marked to indicate the load served.
   (a) the same size   (b) grouped
   (c) in the same enclosure   (d) none of these

6. The one or more additional disconnecting means for fire pumps or for emergency, legally required standby or optional standby systems as permitted by 225.30, shall be installed sufficiently remote from the one to six disconnecting means for normal supply to minimize the possibility of _____ interruption of supply.
   (a) accidental   (b) intermittent   (c) simultaneous   (d) prolonged

7. When the building disconnecting means is a power-operated switch or circuit breaker, it shall be able to be opened by hand in the event of a _____.
   (a) ground fault   (b) short circuit
   (c) power surge   (d) power-supply failure

8. The building or structure disconnecting means shall plainly indicate whether it is in the _____ position.
   (a) open or closed   (b) correct
   (c) up or down   (d) none of these

9. In order for a single branch circuit to supply only limited loads, the building disconnecting means shall have a rating of not less than _____.
   (a) 15A   (b) 20A   (c) 25A   (d) 30A

10. Conductors other than service conductors shall not be installed in the same _____.
    (a) service raceway   (b) service cable
    (c) enclosure   (d) a or b

11. The minimum clearance for service drops not exceeding 600V that pass over commercial areas subject to truck traffic is _____.
    (a) 10 ft   (b) 12 ft   (c) 15 ft   (d) 18 ft

12. Overhead service-drop conductors shall have a horizontal clearance of _____ from a pool.
    (a) 6 ft   (b) 10 ft   (c) 8 ft   (d) 4 ft

13. To supply power to a 1Ø water heater using underground conductors, the size of the underground service-lateral conductors shall not be smaller than _____.
    (a) 4 AWG copper   (b) 8 AWG aluminum
    (c) 12 AWG copper   (d) none of these

14. Service-entrance conductors shall not be spliced or tapped.
    (a) True   (b) False

15. When individual open conductors enter a building or other structure, they shall enter through roof bushings or through the wall in an upward slant through individual, noncombustible, nonabsorbent insulating _____.
    (a) tubes   (b) raceways   (c) chases   (d) standoffs

16. On a 3Ø, 4-wire, delta-connected service where the midpoint of one phase winding is grounded, the service conductor having the higher-phase voltage-to-ground shall be durably and permanently marked by an outer finish that is _____ in color, or by other effective means, at each termination or junction point.
    (a) orange   (b) red   (c) blue   (d) any of these

17. When the service contains two to six service disconnecting means, they shall be _____.
    (a) the same size   (b) grouped at one location
    (c) in the same enclosure   (d) none of these

18. The additional service disconnecting means for fire pumps or for emergency, legally required standby, or optional standby services permitted by 230.2, shall be installed sufficiently remote from the one to six service disconnecting means for normal service to minimize the possibility of _____ interruption of supply.
    (a) accidental   (b) intermittent   (c) simultaneous   (d) prolonged

19. When the service disconnecting means is a power-operated switch or circuit breaker, it shall be able to be opened by hand in the event of a _____.
    (a) ground fault   (b) short circuit
    (c) power surge   (d) power-supply failure

20. The service disconnecting means shall plainly indicate whether it is in the _____ position.
    (a) open or closed   (b) tripped
    (c) up or down   (d) correct

21. In order for a single branch circuit to supply only limited loads, the service disconnecting means shall have a rating of not less than _____.
    (a) 15A   (b) 20A   (c) 25A   (d) 30A

22. Where necessary to prevent tampering, an automatic overcurrent protection device that protects service conductors supplying only a specific load, such as a water heater, shall be permitted to be _____ where located so as to be accessible.
    (a) locked   (b) sealed   (c) a or b   (d) none of these

23. The maximum setting of the ground-fault protection in a service disconnecting means shall be _____.
    (a) 800A   (b) 1,000A   (c) 1,200A   (d) 2,000A

24. Where ground-fault protection is provided for the _____ disconnect and interconnection is made with another supply system by a transfer device, means or devices may be needed to ensure proper ground-fault sensing by the ground-fault protection equipment.
    (a) circuit
    (b) service
    (c) switch and fuse combination
    (d) b and c

25. Which of the following is not a standard size for fuses or inverse-time circuit breakers?
    (a) 45   (b) 70   (c) 75   (d) 80

Unit 3 NEC Exam – Random Order 90.2 – 240.5

1. Ground-fault protection that functions to open the service disconnect _____ protect(s) service conductors or the service disconnecting means.
   (a) will   (b) will not   (c) adequately   (d) totally

2. Individual open conductors and cables other than service-entrance cables shall not be installed within _____ of grade level or where exposed to physical damage.
   (a) 8 ft   (b) 10 ft   (c) 12 ft   (d) 15 ft

3. Overcurrent protection for conductors and equipment is designed to _____ the circuit if the current reaches a value that will cause an excessive or dangerous temperature in conductors or conductor insulation.
   (a) open   (b) close   (c) monitor   (d) record

4. Overhead conductors shall have a minimum of _____ vertical clearance from final grade over residential property and driveways, as well as those commercial areas not subject to truck traffic where the voltage is limited to 300 volts-to-ground.
   (a) 10 ft   (b) 12 ft   (c) 15 ft   (d) 18 ft

5. Service cables shall be equipped with a raintight _____.
   (a) raceway   (b) service head   (c) cover   (d) all of these

6. Service lateral conductors must have _____.
   (a) adequate mechanical strength
   (b) sufficient ampacity for the loads computed
   (c) a and b
   (d) none of these

7. A demand factor of _____ percent is applicable for a multifamily dwelling with ten units, if the optional calculation method is used.
   (a) 75   (b) 60   (c) 50   (d) 43

8. A device that, when interrupting currents in its current-limiting range, will reduce the current flowing in the faulted circuit to a magnitude substantially less than that obtainable in the same circuit if the device were replaced with a solid conductor having comparable impedance, is defined as a _____ protective device.
   (a) short-circuit   (b) overload
   (c) ground-fault   (d) current-limiting overcurrent

9. Service disconnecting means shall not be installed in bathrooms.
   (a) True   (b) False

10. A remote-control device with a shunt-trip push button that is used to open the service disconnecting means can be considered the service disconnecting means if it is located in accordance with 230.70(A)(1).
    (a) True   (b) False

11. A single building or other structure sufficiently large enough to require two or more services is permitted by _____.
    (a) architects   (b) special permission
    (c) written authorization   (d) master electricians

12. A single building or other structure sufficiently large enough to require two or more supplies is permitted by _____.
    (a) architects   (b) special permission
    (c) written authorization   (d) master electricians

13. Connection of conductors to terminal parts shall ensure a thoroughly good connection without damaging the conductors and shall be made by means of _____.
    (a) solder lugs   (b) pressure connectors
    (c) splices to flexible leads   (d) any of these

14. Each _____ service conductor shall have overload protection.
    (a) overhead   (b) underground   (c) ungrounded   (d) none of these

15. Each service disconnect shall be permanently _____ to identify it as part of the service disconnecting means.
(a) identified   (b) positioned   (c) marked   (d) none of these

16. Electrical installations over 600V located in _____, where access is controlled by lock and key or other approved means, shall be considered to be accessible to qualified persons only.
(a) a room or closet
(b) a vault
(c) an area surrounded by a wall, screen, or fence
(d) any of these

17. Enclosures housing electrical apparatus that are controlled by lock and key shall be considered _____ to qualified persons.
(a) readily accessible   (b) accessible
(c) available   (d) none of these

18. Equipment such as raceways, cables, wireways, cabinets, panels, etc. can be located above or below other electrical equipment when the associated equipment does not extend more than _____ from the front of the electrical equipment.
(a) 3 in.   (b) 6 in.   (c) 12 in.   (d) 30 in.

19. Explanatory material, such as references to other standards, references to related sections of this *Code*, or information related to a *Code* rule, is included in this *Code* in the form of Fine Print Notes (FPNs).
(a) True   (b) False

20. Feeder and service-entrance conductors with demand load determined by the use of Table 220.30 shall be permitted to have the _____ load determined by 220.22.
(a) feeder   (b) circuit   (c) neutral   (d) none of these

21. Flexible cords approved for use with listed appliances or portable lamps shall be protected by a 20A circuit breaker if they are _____.
(a) not more than 6 ft in length
(b) 20 AWG and larger
(c) tinsel cord or 18 AWG cord and larger
(d) 16 AWG and larger

22. For a one-family dwelling, at least one receptacle outlet is required in each _____.
(a) basement
(b) attached garage
(c) detached garage with electric power
(d) all of these

23. For equipment rated 1,200A or more and over 6 ft wide that contains overcurrent devices, switching devices, or control devices, there shall be one entrance to the required working space not less than 24 in. wide and $6\frac{1}{2}$ ft high at each end of the working space. Where the depth of the working space is twice that required by 110.26(A)(1), _____ entrance(s) shall be permitted.
(a) one   (b) two   (c) three   (d) none of these

24. For other than dwelling units, the feeder and service load calculation for track lighting is to be determined at 150 VA for every _____ of track installed.
(a) 4 ft   (b) 6 ft   (c) 2 ft   (d) none of these

25. GFCI protection for personnel is required for fixed electric snow-melting or deicing equipment receptacles that are not readily accessible and are supplied by a dedicated branch circuit.
(a) True   (b) False

26. GFCI protection is required for all 125V, 1Ø, 15 and 20A receptacles installed on rooftops, including those for fixed electric snow-melting or deicing equipment.
(a) True   (b) False

27. GFCI protection is required for all 125V, 1Ø, 15 and 20A receptacles in accessory buildings that have a floor located at or below grade level not intended as _____ and limited to storage areas.
    (a) habitable   (b) finished   (c) a or b   (d) none of these

28. Grounded (neutral) conductors _____ AWG and larger must be identified by a continuous white or gray outer finish along their entire lengths, or by distinctive white markings such as tape, paint, or other effective means at their terminations.
    (a) 10   (b) 8   (c) 6   (d) 4

29. In _____ rooms other than kitchens and bathrooms of dwelling units, one or more receptacles controlled by a wall switch shall be permitted in lieu of lighting outlets.
    (a) habitable   (b) finished   (c) all   (d) a and b

30. In a dwelling unit, at least _____ wall switch-controlled lighting outlet(s) shall be installed in every dwelling unit habitable room and bathroom.
    (a) one   (b) three   (c) six   (d) none of these

31. In a dwelling unit, at least one lighting outlet _____ located at the point of entry to the attic, underfloor space, utility room, and basement shall be installed where these spaces are used for storage or contain equipment requiring servicing.
    (a) that is unswitched and   (b) containing a switch
    (c) controlled by a wall switch   (d) b or c

32. In a dwelling unit, illumination from a lighting outlet shall be provided at the exterior side of each outdoor entrance or exit that has grade-level access.
    (a) True   (b) False

33. In a dwelling unit, illumination on the exterior side of outdoor entrances or exits that have grade-level access can be controlled by _____.
    (a) home automation devices   (b) motion sensors
    (c) photocells   (d) any of these

34. In a service, overcurrent protection devices shall never be placed in the grounded service conductor with the exception of a circuit breaker that simultaneously opens all conductors of the circuit.
    (a) True   (b) False

35. In Article 200, connected so as to be capable of carrying current (as distinguished from connection through electromagnetic induction) defines the term _____.
    (a) effectively grounded   (b) electrically connected
    (c) a grounded system   (d) none of these

36. In dwelling units, at least one wall receptacle outlet shall be installed in bathrooms within _____ of the outside edge of each basin. The receptacle outlet shall be located on a wall or partition that is adjacent to the basin or basin countertop.
    (a) 12 in.   (b) 18 in.   (c) 24 in.   (d) 36 in.

37. Island or peninsular countertop receptacle outlets can be installed below the countertop surface in dwelling units when necessary for the physically impaired or if no wall space is available above the countertop.
    (a) True   (b) False

38. Kitchen and dining room-countertop receptacle outlets in dwelling units must be installed above the countertop surface, and not more than ___ above the countertop.
    (a) 12 in.   (b) 20 in.   (c) 24 in.   (d) none of these

39. No point along the floor line in any wall space of a dwelling unit may be more than _____ from an outlet.
    (a) 12 ft   (b) 10 ft   (c) 8 ft   (d) 6 ft

40. Overhead conductors to a building shall maintain a vertical clearance of final spans above, or within, _____ measured horizontally from the platforms, projections or surfaces from which they might be reached.
    (a) 3 ft   (b) 6 ft   (c) 8 ft   (d) 10 ft

41. Overhead service conductors can be supported to hardwood trees.
    (a) True   (b) False

42. Overhead-service conductors to a building shall maintain a vertical clearance of final spans above, or within _____ measured horizontally from the platforms, projections or surfaces from which they might be reached.
    (a) 3 ft   (b) 6 ft   (c) 8 ft   (d) 10 ft

43. Receptacle outlets shall, insofar as practicable, be spaced equal distances apart in a dwelling unit. Receptacle outlets in floors shall not be counted as part of the required number of receptacle outlets unless they are located within _____ of the wall.
    (a) 6 in.   (b) 12 in.   (c) 18 in.   (d) close to the wall

44. Receptacles installed behind a bed in the guest rooms in hotels and motels shall be located so as to prevent the bed from contacting an attachment plug, or the receptacle shall be provided with a suitable guard.
    (a) True   (b) False

45. Service cables that are subject to physical damage must be protected by _____.
    (a) Rigid metal conduit
    (b) Intermediate metal conduit
    (c) Schedule 80 rigid nonmetallic conduit
    (d) any of these

46. Service cables, where subject to physical damage, shall be protected.
    (a) True   (b) False

47. Service conductors only originate from the service point and terminate at the service equipment (disconnect).
    (a) True   (b) False

48. Service conductors shall be sized no less than _____ percent of the continuous load, plus 100 percent of the noncontinuous load.
    (a) 100   (b) 115   (c) 125   (d) 150

49. Service laterals installed by an electrical contractor must be installed in accordance with the *NEC*.
    (a) True   (b) False

50. Service-drop conductors shall have _____.
    (a) sufficient ampacity to carry the current for the load
    (b) adequate mechanical strength
    (c) a or b
    (d) a and b

Unit 4

Motors and Transformers

OBJECTIVES

After reading this unit, the student should be able to briefly explain the following concepts:

Part A - Motors	Nameplate ampere	Transformer primary vs. secondary
Alternating-current motors	Reversing dc motors	Transformer current
Dual voltage motors	Reversing ac motors	Transformer kVA rating
Horsepower/watts	Volt A calculations	Transformer power losses
Motor speed control	**Part B - Transformers**	Transformer turns ratio

After reading this unit, the student should be able to briefly explain the following terms:

Part A - Motors	Watts rating	Flux leakage loss
Armature winding	**Part B - Transformers**	Hysteresis losses
Commutator	Auto transformers	Kilo volt A
Dual voltage	Circuit impedance	Line current
Field winding	Conductor losses	Ratio
Horsepower ratings	Core losses	Self-excited
Magnetic field	Counter-electromotive force	Step-down transformer
Motor FLC	Current transformers (CT)	Step-up transformers
Nameplate ampere	Eddy current losses	
Torque	Excitation current	

PART A – MOTORS

MOTOR INTRODUCTION

The *electric motor* operates on the principle of the attracting and repelling forces of magnetic fields. One magnetic field (permanent or electromagnetic) is *stationary* and the other magnetic field (called the *armature* or *rotor*) rotates between the poles of the stationary magnet. Figure 4–1. The turning or repelling forces between the magnetic fields is called *torque*. Torque is dependent on the strength of the stationary magnetic field, the strength of the magnetic field of the armature and the physical construction of the motor.

The repelling force of like magnetic polarities, and the attraction force of unlike polarities, causes the armature to rotate in the electric motor. For a dc motor, a device called a *commutator* is placed on the end of the conductor loop or armature. The purpose of the commutator is to maintain the proper polarity of the loop, so as to keep the armature or rotor turning.

The *armature winding* of a motor carries a starting of current when voltage is first applied to the motor windings. As the armature turns, it cuts the lines of force of the *field winding* resulting in an increase in counter-electromotive force, (CEMF) Figure 4–2. The increased counter-electromotive force results in an increase in inductive reactance, which increases the circuit impedance. The increased circuit impedance causes a decrease in the motor armature running current. Figure 4–3.

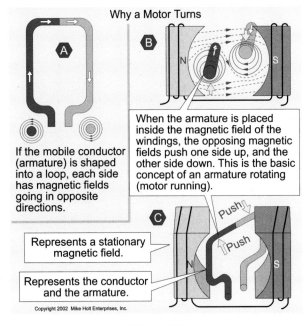

Why a Motor Turns

A If the mobile conductor (armature) is shaped into a loop, each side has magnetic fields going in opposite directions.

B When the armature is placed inside the magnetic field of the windings, the opposing magnetic fields push one side up, and the other side down. This is the basic concept of an armature rotating (motor running).

Represents a stationary magnetic field.

Represents the conductor and the armature.

C Push   Push

Copyright 2002 Mike Holt Enterprises, Inc.

Figure 4-1
Why a Motor Turns

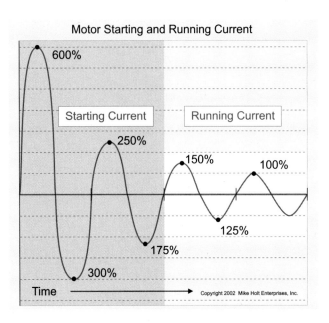

Motor Starting and Running Current

600%

Starting Current   Running Current

250%

150%   100%

125%

175%

300%

Time

Copyright 2002 Mike Holt Enterprises, Inc.

Figure 4-2
Motor Starting and Running Current

4–1 MOTOR SPEED CONTROL

One of the advantages of a *dc motor* over an ac motor is the motor's ability to maintain a constant speed. But series dc motors are susceptible to runaway (increased rotational speed) when not connected to a load.

If the speed of a dc motor is increased, the armature winding is cut by the magnetic field at an increasing rate, resulting in an increase of the armature's counter-electromotive force. The increased CEMF in the armature acts to cut down on the armature current, resulting in the motor slowing down. Placing a load on a dc motor causes the motor to slow down, which reduces the rate at which the armature winding is cut by the magnetic field flux lines. A reduction of the armature flux lines results in a decrease in the armature's CEMF and an increase in motor speed.

4–2 REVERSING A DIRECT-CURRENT MOTOR

To reverse a dc motor, you must reverse either the *magnetic field* of the field winding or the magnetic field of the armature. This is accomplished by reversing either the field or armature current flow. Figure 4–4. Because most dc motors have the field and armature winding connected to the same dc power supply, reversing the polarity of the power supply changes both the field and armature simultaneously. To reverse the rotation of a dc motor, you must reverse either the field or armature leads, but not both.

4–3 ALTERNATING-CURRENT MOTORS

Fractional horsepower motors that can operate on either alternating or dc are called universal motors. A motor that will not operate on dc is called an induction motor.

The induction motor is the purest form of an ac motor, with no physical connection between its rotating member *(rotor)* and stationary member *(stator)*. Two common types of ac induction motors are synchronous and wound-rotor motors.

Synchronous Motors

The synchronous motor's rotor is locked in step with the rotating stator field. Synchronous motors maintain their speed with a high degree of accuracy and are used for electric clocks and other timing devices.

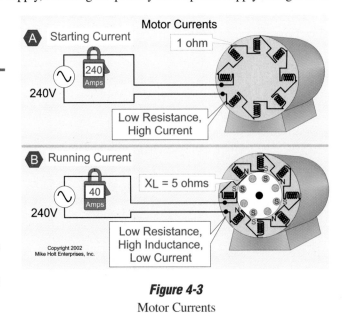

Motor Currents

A Starting Current   1 ohm

240 Amps

240V

Low Resistance, High Current

B Running Current   XL = 5 ohms

40 Amps

240V

Low Resistance, High Inductance, Low Current

Copyright 2002 Mike Holt Enterprises, Inc.

Figure 4-3
Motor Currents

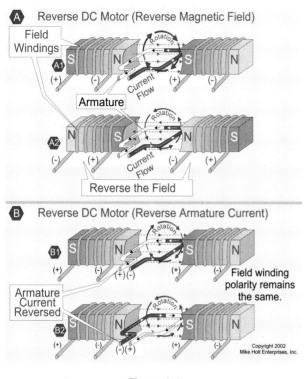

A Reverse DC Motor (Reverse Magnetic Field)

B Reverse DC Motor (Reverse Armature Current)

Figure 4-4
Reverse DC Motor

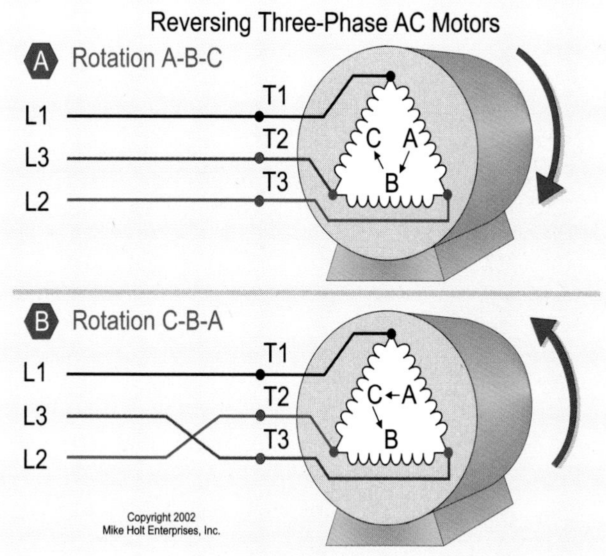

Reversing Three-Phase AC Motors

A Rotation A-B-C

B Rotation C-B-A

Figure 4-5
Reversing Three-Phase AC Motors

Wound-Rotor Motors

Because of their high starting torque design, wound-rotor motors are used only in special applications. They only operate on 3-phase alternating-current circuits.

4–4 REVERSING ALTERNATING-CURRENT MOTORS

Three-phase ac motors can be reversed by reversing any two of the three line conductors that supply the motor. Figure 4–5.

4–5 MOTOR VOLT-AMPERE CALCULATIONS

Dual voltage motors are made with two field windings, each rated for the lower voltage marked on the nameplate of the motor. When the motor is wired for the lower voltage, the field windings are connected in parallel; when wired for the higher voltage, the motor windings are connected in series. Figure 4–6.

Motor Input VA

Regardless of the voltage connection, the *power* consumed by a motor is the same at either voltage. To determine the *motor input* apparent power (VA), use the following formulas:

Motor VA (1Ø) = Volts × Amperes

Motor VA (3Ø) = Volts × Amperes × $\sqrt{3}$

❏ Motor VA Single-Phase

What is the motor VA of a 115/230V, 1Ø, 1-hp motor that has a current rating of 16A at 115V and 8A at 230V? Figure 4–7.

(a) 1,450 VA   (b) 1,600 VA   (c) 1,840 VA   (d) 1,920 VA

• Answer: (c) 1,840 VA

Motor VA = Volts × Amperes

Volts = 115 or 230, Amperes = 16 or 8A

Motor VA = 115V × 16A or 230V × 8A = 1,840 VA

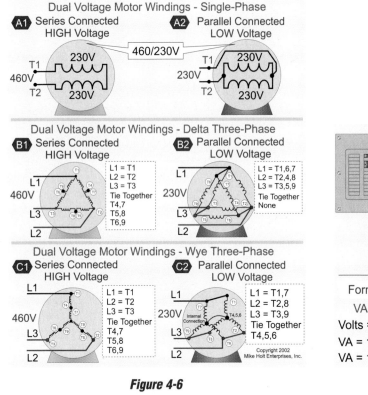

Figure 4-6
Dual Voltage Motor Windings – Single-Phase

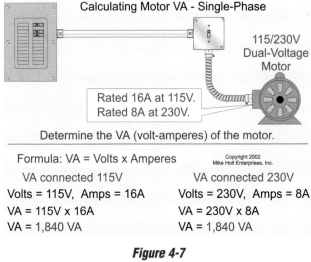

Calculating Motor VA - Single-Phase

115/230V
Dual-Voltage
Motor

Rated 16A at 115V.
Rated 8A at 230V.

Determine the VA (volt-amperes) of the motor.

Formula: VA = Volts x Amperes

Copyright 2002
Mike Holt Enterprises, Inc.

VA connected 115V
Volts = 115V, Amps = 16A
VA = 115V x 16A
VA = 1,840 VA

VA connected 230V
Volts = 230V, Amps = 8A
VA = 230V x 8A
VA = 1,840 VA

Figure 4-7
Calculating Motor VA – Single-Phase

❑ **Motor VA 3Ø**

What is the motor VA of a 230/460V, 3Ø, 30-hp motor that has a current rating of 40A at 460V? Figure 4–8.

(a) 41,450 VA   (b) 31,600 VA   (c) 21,840 VA   (d) 31,869 VA

• Answer: (d) 31,869 VA

Motor VA = Volts $\times$ Amperes $\times \sqrt{3}$

Volts = 460V, Amperes = 40A, $\sqrt{3}$ = 1.732

Motor VA = 460V $\times$ 40A $\times$ 1.732 = 31,869 VA

4–6 MOTOR HORSEPOWER/WATTS

The mechanical work (output) of a motor is rated in horsepower and can be converted to electrical energy as 746W per horsepower.

Horsepower = Output Watts/746W

Motor Output Watts = Horsepower $\times$ 746W

❑ **Motor Horsepower**

What size horsepower motor is required to produce 15 kW output? Figure 4–9.

(a) 5-hp   (b) 10-hp   (c) 20-hp   (d) 30-hp

• Answer: (c) 20-hp

$$\text{Horsepower} = \frac{\text{Output Watts}}{746W} = \frac{15,000W}{746W} = 20\text{-hp}$$

❑ **Motor Output Watts**

What is the output watt rating of a 10-hp motor? Figure 4–10.

(a) 3 kW   (b) 2.2 kVA   (c) 3.2 kW   (d) 7.5 kW

• Answer: (d) 7.5 kW

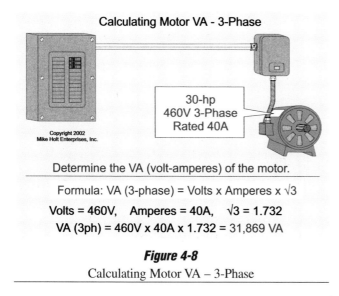

Calculating Motor VA - 3-Phase

30-hp
460V 3-Phase
Rated 40A

Copyright 2002
Mike Holt Enterprises, Inc.

Determine the VA (volt-amperes) of the motor.

Formula: VA (3-phase) = Volts x Amperes x √3

Volts = 460V,   Amperes = 40A,   √3 = 1.732

VA (3ph) = 460V x 40A x 1.732 = 31,869 VA

Figure 4-8

Calculating Motor VA – 3-Phase

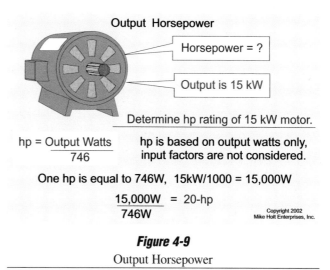

Output Horsepower

Horsepower = ?

Output is 15 kW

Determine hp rating of 15 kW motor.

$$hp = \frac{\text{Output Watts}}{746}$$

hp is based on output watts only, input factors are not considered.

One hp is equal to 746W, 15kW/1000 = 15,000W

$$\frac{15,000W}{746W} = 20\text{-hp}$$

Copyright 2002
Mike Holt Enterprises, Inc.

Figure 4-9

Output Horsepower

Output Watts = 10-hp × 746W = 7,460W

$$\text{Output kW} = \frac{7,460W}{1,000} = 7.46 \text{ kW}$$

4–7 MOTOR NAMEPLATE AMPERE

The motor nameplate indicates the motor operating voltage and current. The actual current drawn by the motor depends on how much the motor is loaded. It is important not to overload a motor above its rated horsepower because the current of the motor will increase to a point that the motor winding will be destroyed from excess heat. The nameplate current can be calculated by the following formulas:

$$\text{Motor Nameplate Ampere 1Ø} = \frac{\text{Motor Horsepower} \times 746W}{(\text{Volts} \times \text{Efficiency} \times \text{Power Factor})}$$

$$\text{Motor Nameplate Ampere 3Ø} = \frac{\text{Motor Horsepower} \times 746W}{(\text{Volts} \times \sqrt{3} \times \text{Efficiency} \times \text{Power Factor})}$$

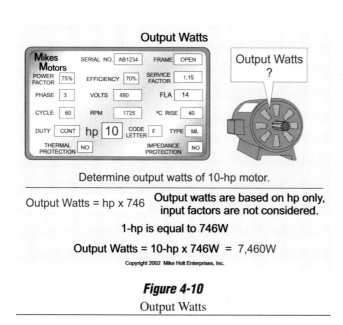

Output Watts

Output Watts ?

Determine output watts of 10-hp motor.

Output Watts = hp x 746   Output watts are based on hp only, input factors are not considered.

1-hp is equal to 746W

Output Watts = 10-hp x 746W = 7,460W

Copyright 2002 Mike Holt Enterprises, Inc.

Figure 4-10

Output Watts

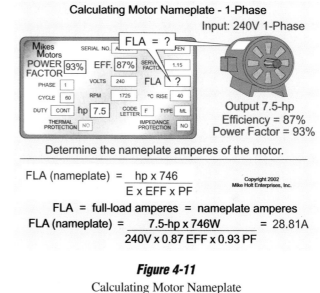

Calculating Motor Nameplate - 1-Phase

Input: 240V 1-Phase

FLA = ?

FLA ?

Output 7.5-hp
Efficiency = 87%
Power Factor = 93%

Determine the nameplate amperes of the motor.

$$\text{FLA (nameplate)} = \frac{\text{hp x 746}}{\text{E x EFF x PF}}$$

Copyright 2002
Mike Holt Enterprises, Inc.

FLA = full-load amperes = nameplate amperes

$$\text{FLA (nameplate)} = \frac{7.5\text{-hp x 746W}}{240V \times 0.87 \text{ EFF} \times 0.93 \text{ PF}} = 28.81A$$

Figure 4-11

Calculating Motor Nameplate

❏ **Nameplate Amperes – 1Ø**

What are the nameplate amperes for a 240V, 1Ø, 7.5-hp motor? The efficiency is 87% and the power factor is 93%. Figure 4–11.

    (a) 19A   (b) 24A   (c) 19A   (d) 29A

    • Answer: (d) 29A

$$\text{Nameplate amperes} = \frac{\text{Horsepower} \times 746\text{W}}{(\text{Volts} \times \text{Efficiency} \times \text{Power Factor})}$$

$$\text{Nameplate amperes} = \frac{7.5\text{ hp} \times 746\text{W}}{(240\text{ V} \times 0.87\text{ Efficiency} \times 0.93\text{ Power Factor})} = 28.81\text{A}$$

❏ **Nameplate Amperes – 3Ø**

What are the nameplate amperes for a 208V, 3Ø, 40-hp motor? The efficiency is 80% and the power factor is 90%. Figure 4–12.

    (a) 85A   (b) 95A   (c) 105A   (d) 115A

    • Answer: (d) 115A

$$\text{Nameplate amperes} = \frac{\text{Horsepower} \times 746\text{W}}{(\text{Volts} \times \sqrt{3} \times \text{Efficiency} \times \text{Power Factor})}$$

$$\text{Nameplate} = \frac{40\text{ Horsepower} \times 746\text{W}}{(208\text{ V} \times 1.732 \times 0.8\text{ Efficiency} \times 0.9\text{ Power Factor})} = 29{,}840/259 = 115\text{A}$$

PART B – TRANSFORMER BASICS

TRANSFORMER INTRODUCTION

A *transformer* is a stationary device used to raise or lower voltage. Transformers have the ability to transfer electrical energy from one circuit to another by *mutual induction* between two conductor coils. Mutual induction occurs between two conductor coils *(windings)* when electromagnetic lines of force within one winding induces a voltage into a second winding. Figure 4–13.

Current transformers use the circuit conductors as the primary winding and step the current down for metering. Often, the *ratio* of the current transformer is 1,000 to 1. This means that if there are 400A on the phase conductors, the current transformer steps the current down to 0.4A for the meter.

The *magnetic coupling (magnetomotive-force, MMF)* between the primary and secondary winding can be increased by increasing the winding *ampere-turns*. Ampere-turns can be increased by increasing the number of coils and/or the current through each coil. When the current in the core of a transformer is raised to a point where there is high *flux density*, additional increases in current will produce few additional flux lines. The transformer iron core is said to be *saturated.*

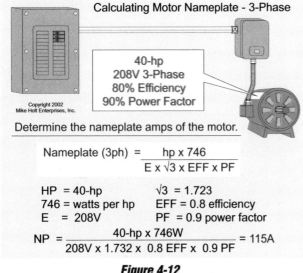

Figure 4-12

Calculating Motor Nameplate – 3-Phase

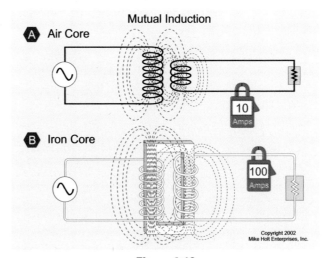

Figure 4-13

Mutual Induction

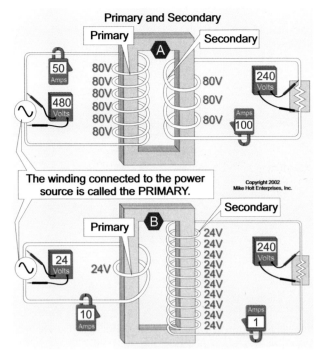

Figure 4-14

Transformer Primary and Secondary

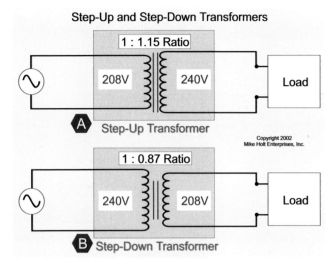

Figure 4-15

Step-Up and Step-Down Trransformers

4–8 TRANSFORMER PRIMARY AND SECONDARY

The transformer *winding* connected to the source is called the *primary winding*. The transformer winding connected to the load is called the *secondary winding*.

4–9 TRANSFORMER SECONDARY AND PRIMARY VOLTAGE

Voltage induced in the secondary winding of a transformer is equal to the sum of the voltages induced in each *loop* of the secondary winding. The voltage induced in the secondary of a transformer depends on the number of secondary conductor turns cut by the primary magnetic flux lines. The greater the number of secondary conductor loops, the greater the secondary voltage. Figure 4–14.

Step-Up and Step-Down Transformers

The secondary winding of a *step-down transformer* has fewer turns than the primary winding, resulting in lower secondary voltage. The secondary winding of a *step-up transformer* has more turns than the primary winding, resulting in higher secondary voltage. Figure 4–15.

4–10 AUTOTRANSFORMERS

Autotransformers are transformers that use a common winding for both the primary and the secondary. The disadvantage of an autotransformer is the lack of isolation between the primary and secondary conductors, but they are often used because they are less expensive. Figure 4–16.

4–11 TRANSFORMER POWER LOSSES

When current flows through the winding of a transformer, power is dissipated in the form of heat. This loss is referred to as conductor I^2R loss. In addition, losses include flux leakage, core loss from eddy currents and hysteresis heating losses. Figure 4–17.

Conductor Resistance Loss

Transformer windings are made of many turns of wire. The resistance of the conductor is directly proportional to the length of the conductor and inversely proportional to the cross-sectional area of the conductor. The more turns there are, the longer the conductor is and the greater the conductor resistance. Conductor losses can be determined by the formula: $P = I^2R$.

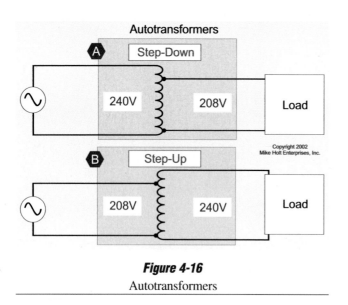

Figure 4-16
Autotransformers

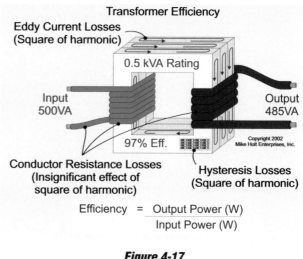

Figure 4-17
Transformer Efficiency

Flux Leakage Loss

The *flux leakage loss* represents the electromagnetic flux lines between the primary and secondary windings that are not used to convert electrical energy from the primary to the secondary. They represent wasted energy.

Core Losses

Iron is the only metal used for transformer cores because it offers low resistance to magnetic flux lines *(low reluctance)*. Iron cores permit more flux lines between the primary and secondary windings, thereby increasing the magnetic coupling between the primary and secondary windings. However, alternating-current circuits produce electromagnetic fields within the windings that induce a circulating current in the iron core. These circulating currents *(eddy currents)* flow within the iron core producing power losses that cannot be transferred to the secondary winding. Transformer iron cores are laminated to have a small cross-sectional area to reduce the eddy currents and their associated losses.

Hysteresis Losses

Each time the primary magnetic field expands and collapses, the transformer's iron-core molecules realign themselves to the changing polarity of the electromagnetic field. The energy required to realign the iron-core molecules to the changing electromagnetic field is called hysteresis losses.

Heating by the Square of the Frequency

Hysteresis and eddy current losses are affected by the square of the ac frequency. For this reason, care must be taken when iron-core transformers are used in applications involving high frequencies and nonlinear loads. If the transformer operates at the third harmonic (180 Hz), the losses will be the square of the multiple of the fundamental frequency. The third harmonic frequency (180 Hz) is three times the fundamental frequency (60 Hz).

❏ Transformer Losses by Square of Frequency

What is the effect on a 60 Hz rated transformer of third harmonic loads (180 Hz)?

(a) heating increases three times   (b) heating increases six times

(c) heating increases nine times   (d) no significant heating

   • Answer: (c) heating increases nine times (3^2)

4–12 TRANSFORMER TURNS RATIO

The relationship of the primary winding voltage to the secondary winding voltage is the same as the relation between the number of primary turns as compared to the number of secondary turns. This relationship is called turns ratio or voltage ratio.

❏ Delta Winding Turns Ratio

What is the turns ratio of a Delta/Delta transformer? The primary winding is 480V and the secondary winding voltage is 240V. Figure 4–18.

(a) 4:1   (b) 1:4   (c) 2:1   (d) 1:2

   • Answer: (c) 2:1

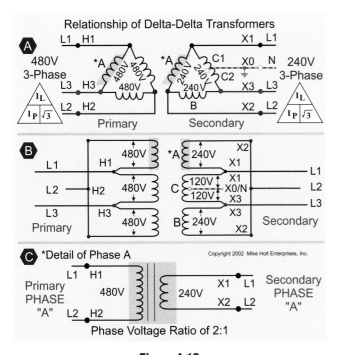

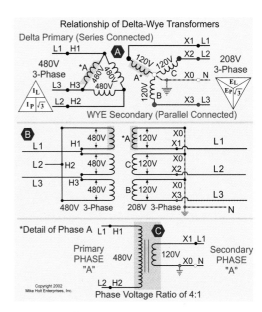

Figure 4-18
Relationships of Delta-Delta Transformers

Figure 4-19
Relationships of Delta-Wye Transformers

The primary winding voltage is 480 and the secondary winding voltage is 240.

This results in a ratio of 480:240 or 2:1.

❏ **Wye Winding Ratio**

What is the turns ratio of a Delta-Wye transformer? The primary winding is 480V and the secondary winding is 120V. Figure 4–19.

(a) 4:1   (b) 1:4   (c) 2.3:1   (d) 1:2.3

• Answer: (a) 4:1

The primary winding voltage is 480 and the secondary winding voltage is 120.

This results in a voltage turn ratio of 480:120 or 4:1.

4–13 TRANSFORMER kVA RATING

Transformers are rated in Kilovolt Ampere, abbreviated as kVA.

4–14 TRANSFORMER CURRENT

Whenever the number of primary turns is greater than the number of secondary turns, the secondary voltage is less than the primary voltage. This results in secondary current being greater than the primary current because the power remains the same, but the voltage changes. Since the secondary voltage of most transformers is less than the primary voltage, the secondary conductors carry more current than the primary. Figure 4–20. Primary and secondary line current can be calculated by:

Current (1Ø) = Volt-Amperes/Volts

Current (3Ø) = Volt-Amperes/(Volts × 1.732)

❏ **Transformer Current – 1Ø**

What is the primary and secondary line current for a 25 kVA, 1Ø transformer, rated 480V primary and 240V secondary? Figure 4–21.

(a) 52A/104A   (b) 104A/52A   (c) 104A/208A   (d) 208A/104A

• Answer: (a) 52A/104A

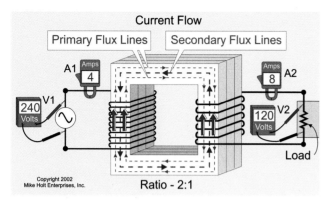

Figure 4-20
Transformer Current Flow

Primary current = VA/E

Primary current = $\dfrac{25{,}000 \text{ VA}}{480\text{V}}$

Primary current = 52A

Secondary current = VA/E

Secondary current = $\dfrac{25{,}000 \text{ VA}}{240\text{V}}$

Secondary current = 104A

❏ **Transformer Current – 3Ø**

What is the primary and secondary line current for a 37.5 kVA 3Ø transformer, rated 480V primary and 208V secondary? Figure 4–22.

(a) 45A/104A   (b) 104A/40A
(c) 208A/140A   (d) 140A/120A

• Answer: (a) 45A/104A

Primary current = $\dfrac{\text{VA}}{\text{E} \times \sqrt{3}}$

Primary current = $\dfrac{37{,}500 \text{ VA}}{(480\text{V} \times 1.732)}$

Primary current = 45A

Secondary current = $\dfrac{\text{VA}}{\text{E} \times \sqrt{3}}$

Secondary current = $\dfrac{37{,}500 \text{ VA}}{(208\text{V} \times 1.732)}$

Secondary current = 104A

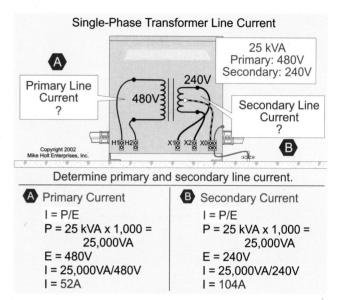

Single-Phase Transformer Line Current

Determine primary and secondary line current.

Ⓐ Primary Current	Ⓑ Secondary Current
I = P/E	I = P/E
P = 25 kVA x 1,000 = 25,000VA	P = 25 kVA x 1,000 = 25,000VA
E = 480V	E = 240V
I = 25,000VA/480V	I = 25,000VA/240V
I = 52A	I = 104A

Figure 4-21
Transformer Line Current

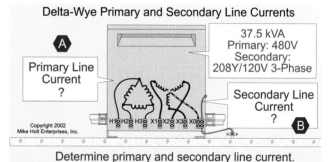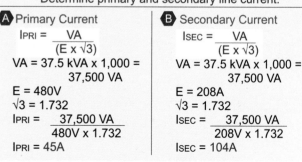

Delta-Wye Primary and Secondary Line Currents

Determine primary and secondary line current.

Ⓐ Primary Current	Ⓑ Secondary Current
$I_{PRI} = \dfrac{VA}{(E \times \sqrt{3})}$	$I_{SEC} = \dfrac{VA}{(E \times \sqrt{3})}$
VA = 37.5 kVA x 1,000 = 37,500 VA	VA = 37.5 kVA x 1,000 = 37,500 VA
E = 480V	E = 208A
$\sqrt{3}$ = 1.732	$\sqrt{3}$ = 1.732
$I_{PRI} = \dfrac{37{,}500 \text{ VA}}{480\text{V} \times 1.732}$	$I_{SEC} = \dfrac{37{,}500 \text{ VA}}{208\text{V} \times 1.732}$
I_{PRI} = 45A	I_{SEC} = 104A

Figure 4-22
Delta-Wye Transformers Primary and
Secondary Line Currents

Unit 4 – Motors and Transformers Summary Questions

Calculations Questions

Part A – Motors

Motor Introduction

1. When voltage is applied to a motor's armature, short-circuit current will flow and the armature will begin to turn. As the armature starts turning, it cuts the lines of force of the field winding resulting in induced CEMF in the armature conductor which reduces the short-circuit current.
   (a) True   (b) False

2. An electric motor works because of the effect a _____ has against a wire carrying an electric current.
   (a) magnetic field   (b) commutator   (c) voltage source   (d) none of these

3. The rotating part of a dc motor or generator is called the _____.
   (a) shaft   (b) rotor   (c) capacitor   (d) field

4. For a dc motor, a device called a _____ is placed on the end of the conductor loop (armature). The polarity of the loop is maintained with the proper magnetic field to keep the opposing magnetic fields pushing each other in the same direction. This, in turn, keeps the loop turning.
   (a) coil   (b) resistor   (c) commutator   (d) none of these

4–1 Motor Speed Control

5. • One of the great advantages of a dc motor is the motor's ability to maintain a constant speed. If the speed of a dc motor is increased, the armature will cut through the field winding magnetic flux at an increasing rate, resulting in a lower CEMF that acts to cut down on the increased armature current, which slows the motor back down.
   (a) True   (b) False

6. • Placing a load on a dc motor causes the motor to slow down which increases the rate at which the field flux lines are cut by the armature. As a result, the armature CEMF increases, resulting in an increase in the applied armature voltage and current. The increase in current results in an increase in motor speed.
   (a) True   (b) False

4–2 Reversing a Direct-Current Motor

7. To reverse a dc motor, you must reverse the direction of the _____.
   (a) field current   (b) armature current   (c) a or b   (d) a and b

4–3 Alternating-Current Motors

8. In a(n) _____ motor the rotor is actually locked in step with the rotating stator field and is dragged along at the synchronous speed of the rotating magnetic field. _____ motors maintain their speed with a high degree of accuracy and are used for electric clocks and other timing devices.
   (a) ac   (b) universal   (c) wound rotor   (d) synchronous

9. _____ motors are used only as special applications because of their high starting torque design and only operate on 3Ø ac power.
   (a) AC   (b) Universal   (c) Wound rotor   (d) Synchronous

10. _____ motors are fractional horsepower motors that operate equally well on ac and dc and are used for vacuum cleaners, electric drills, mixers and light household appliances.
     (a) AC   (b) Universal   (c) Wound rotor   (d) Synchronous

4–4 Reversing Alternating-Current Motors

11. • Three-phase ac motors can be reversed by changing the wiring from ABC phase configuration to _____.
     (a) BCA   (b) CAB   (c) CBA   (d) ABC

4–5 Motor Volt-Ampere Calculations

12. Dual-voltage 277/480V motors are made with two field windings, each rated at 277V. The field windings are connected in parallel for _____V operation and in series for _____V operation.
     (a) 277, 480   (b) 480, 277   (c) 277, 277   (d) 480, 480

4–6 Motor Horsepower/Watts

13. What size motor is required to produce a 30 kW output?
     (a) 20-hp   (b) 30-hp   (c) 40-hp   (d) 50-hp

14. What is the approximate output work in kW for a 15-hp motor?
     (a) 11 kW   (b) 15 kW   (c) 22 kW   (d) 31 kW

15. What is the approximate output work in kW for a 5-hp, 480V, 3Ø motor, efficiency 75 percent and power factor 70 percent?
     (a) 3.7 kW   (b) 7.5 kVA   (c) 7.5 kW   (d) 8.2 kW

4–7 Motor Nameplate Ampere

16. • For practical purposes, you will not need to calculate motor nameplate current, but you should understand how it is calculated. Check the book on this one.
     (a) True   (b) False

17. What are the nameplate amperes for a 5-hp, 240V, 1Ø motor, efficiency 90 percent and power factor 80 percent?
     (a) 19.3A   (b) 21.6A   (c) 28.2A   (d) 31.1A

18. What are the nameplate amperes of a 20-hp 208V, 3Ø motor, efficiency 80 percent and power factor 90 percent?
     (a) 50A   (b) 58A   (c) 65A   (d) 80A

Part B – Transformer Basics

Transformer Introduction

19. A _____ is a stationary device used to raise or lower voltage and has the ability to transfer electrical energy from one circuit to another, with no physical connection between the two.
     (a) capacitor   (b) motor   (c) relay   (d) transformer

20. Transformers operate on the principle of _____.
     (a) magnetoelectricity   (b) turboelectric effect
     (c) thermocouple   (d) mutual induction

4–8 Transformer Primary and Secondary

21. The transformer winding that is connected to the source is called the _____ winding and the transformer winding that is connected to the load is called the _____. Transformers are reversible; and, here either winding can be used as the primary or secondary.
(a) secondary, primary   (b) primary, secondary
(c) depends on the wiring   (d) none of these

22. Voltage induced in the secondary winding of a transformer is dependent on the number of secondary turns as compared to the number of primary turns.
(a) True   (b) False

23. The secondary winding of a step-down transformer has _____ turns than the primary, resulting in a _____ secondary voltage as compared to the primary.
(a) less, higher   (b) more, lower   (c) less, lower   (d) more, higher

24. The secondary winding of a step-up transformer has _____ turns than the primary, resulting in a _____ secondary voltage as compared to the primary.
(a) less, higher   (b) more, lower   (c) less, lower   (d) more, higher

4–10 Autotransformers

25. Autotransformers use the same winding for both the primary and secondary. The disadvantage of an autotransformer is the lack of _____ between the primary and secondary conductors.
(a) power   (b) voltage   (c) isolation   (d) grounding

4–11 Transformer Power Losses

26. The most common causes of power losses for transformer windings are _____.
(a) conductor resistance   (b) eddy currents
(c) hysteresis   (d) all of these

27. The leakage of the electromagnetic flux lines between the primary and secondary windings represents wasted energy.
(a) True   (b) False

28. • The expanding and collapsing electromagnetic field from the transformer winding induces a voltage in the steel core of the transformer. The induced voltage causes _____ to flow within the core, which removes energy from the transformer winding and represents wasted power.
(a) eddy currents   (b) flux   (c) inductive   (d) hysteresis

29. Eddy currents can be reduced by dividing the core into many flat sections or laminations. Because the laminations have a _____ cross-sectional area, the resistance offered to the eddy currents is greatly increased.
(a) round   (b) porous   (c) large   (d) small

30. As current flows through a transformer winding, the iron core is temporarily magnetized by the electromagnetic field created by the ac. Each time the primary magnetic field expands and collapses, the core molecules realign themselves to the changing polarity of the electromagnetic field. The energy required to realign the core molecules to the changing electromagnetic field is called the _____ loss of the core.
(a) eddy current   (b) flux   (c) inductive   (d) hysteresis

4–12 Transformer Turns Ratio

31. The relationship of the primary winding voltage to the secondary winding voltage is the same as the relationship between the numbers of conductor turns on the primary as compared to the secondary. This relationship is called _____.
(a) turns ratio   (b) efficiency   (c) power factor   (d) none of these

32.   • The primary phase voltage is 240 and the secondary phase is 480. This results in a turns ratio of _____.
      (a) 1:2   (b) 2:1   (c) 4:1   (d) 1:4

4–13 Transformer kVA Rating

33.   Transformer windings are rated in _____.
      (a) volt-amperes   (b) kW   (c) watts   (d) kVA

4–14 Transformer Current

34.   • The current flow in the secondary transformer winding creates an electromagnetic field that opposes the primary electromagnetic field resulting in less primary CEMF. The primary current automatically increases in direct proportion to the secondary current.
      (a) True   (b) False

35.   • The transformer winding with the _____ number of turns will have the lower current and the winding with the _____ number of turns will have the higher current.
      (a) lesser, greater (b) most, most   (c) least, least   (d) greater, lesser

☆ Challenge Questions

Part A – Motors

4–1 Motor Speed Control

36.   A(n) _____ type of electric motor tends to run away if it is not always connected to its load.
      (a) dc series   (b) dc shunt   (c) ac induction   (d) ac synchronous

37.   • A(n) _____ motor has a wide speed range.
      (a) ac   (b) dc   (c) synchronous   (d) induction

4–2 Reversing A Direct-Current Motor

38.   • If the two line (supply) leads of a dc series motor are reversed, the motor will _____.
      (a) not run   (b) run backwards
      (c) run the same as before   (d) become a generator

39.   • To reverse a dc series motor, we may simply reverse the supply (power) leads.
      (a) True   (b) False

4–3 Alternating-Current Motors

40.   The _____ induction motor is used only in special applications and is always operated on 3Ø ac power.
      (a) compound   (b) synchronous   (c) split phase   (d) wound rotor

41.   • The rotating part of a dc motor or generator is called the _____.
      (a) shaft   (b) rotor   (c) armature   (d) b or c

4–5 Motor Volt-Ampere Calculations

42.   The input VA of a 5-hp (15.2A), 230V, 3Ø motor is closest to _____.
      (a) 7,500 VA   (b) 6,100 VA   (c) 5,300 VA   (d) 4,600 VA

43.   The input VA of a 1-hp (16A), 115V, 1Ø motor is _____.
      (a) 2,960 VA   (b) 1,840 VA   (c) 3,190 VA   (d) 1,650 VA

Part B – Transformer Basics

4–11 Transformer Power Losses

44.   • When the current in the steel core of a transformer has risen to a point where high flux density has been reached and additional increases in current produce few additional flux lines, the metal core is said to be _____.
   (a) maximum   (b) saturated   (c) full   (d) none of these

45.   • Magnetomotive-force (MMF) can be increased by increasing the _____.
   (a) number of ampere-turns
   (b) current in the coils
   (c) number of coils
   (d) all of these

4–12 Transformer Turns Ratio

46.   • A transformer winding has a voltage turns ratio of 2:1. The current flowing through the secondary winding will be _____ the current flowing through the primary winding.
   (a) higher than   (b) lower than   (c) the same as   (d) none of these

4–13 Transformer kVA Rating

47.   What is the primary kVA rating for a transformer winding that is 100 percent efficient if the 5A load on the secondary operates at 12V?
   (a) 600 kVA   (b) 30 kVA   (c) 6 kVA   (d) 0.06 kVA

4–14 Transformer Current

48.   • The primary winding of a transformer has 100 turns and the secondary winding has 10 turns. What is the current flowing though of the primary winding if the current flowing through the secondary winding is 5A?
   (a) 25A   (b) 10A   (c) 5A   (d) 0.5A

49.   • Which winding of a current transformer (CT) carries more current?
   Tip: An induction clamp-on ammeter (CT) operates on the principle where the meter acts as the secondary winding.
   (a) Primary   (b) Secondary
   (c) Interwinding   (d) Tertiary

50.   If a transformer primary winding has 900 turns and the secondary winding has 90 turns, which winding of the transformer has a larger conductor?
   (a) Primary   (b) Secondary
   (c) Interwinding   (d) None of these

51. If the primary transformer winding operates at 480V and the secondary at 240V, which winding has a larger conductor?
(a) Tertiary   (b) Secondary   (c) Primary   (d) Windings are equal

52. If the transformer winding voltage turns ratio is 5:1 and the current flowing through the secondary winding is 10A, the current flowing through primary winding will be _____ if the secondary current is 10A.
(a) 25A   (b) 10A   (c) 2A   (d) cannot be determined

53. The transformer primary winding operates at 240V, the secondary winding operates at 12V and the secondary load is two 100W lamps. What is the secondary current flow if the transformer winding is 80 percent efficient?
(a) 1A   (b) 17A   (c) 28A   (d) none of these

54. The transformer primary winding operates at 240V, the secondary at 120V and the load is 1,500W. If the transformer is 92 percent efficient, what is the current flowing through the primary winding of this transformer?
(a) 6.8A   (b) 8.6A   (c) 9.9A   (d) 7.8A

55. • The approximate primary kVA of a transformer that supplies a 208V, 100A, 3Ø load is _____. (See Figure 4-23).
(a) 72.5 kVA   (b) 42 kVA   (c) 30 kVA   (d) 21 kVA

56. • The primary current of this transformer is _____. (See Figure 4-23).
(a) 90A   (b) 50A   (c) 25A   (d) 12A

57. • The secondary voltage of this transformer is _____. (See Figure 4-24).
(a) 6V   (b) 12V   (c) 24V   (d) 30V

58. The primary current of the transformer is _____. (See Figure 4-25).
(a) 0.416A   (b) 4.38A   (c) 3.56A   (d) 41.6A

59. The primary load in watts for a transformer (95 percent efficient) that supplies a 500W load is _____. (See Figure 4-25).
(a) 526W   (b) 400W   (c) 475W   (d) 550W

Transformer Miscellaneous

60. The primary winding of a transformer operates at 240V, the secondary at 12V and the load consists of two 100W lamps. What is the load on the secondary winding if the transformer winding is 92 percent efficient?
(a) 200 VA
(b) the same as the primary VA
(c) cannot be calculated
(d) none of these

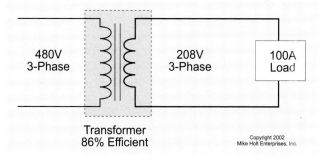

Transformer
86% Efficient

Figure 4-23

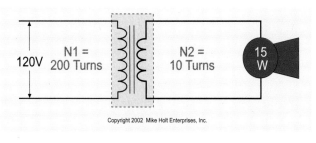

Figure 4-24

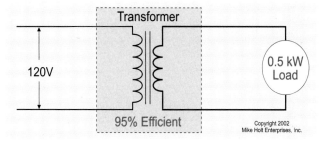

Figure 4-25

61. • The secondary winding of a transformer operates at 24V and it supplies a load that has a rating of 5A. If the transformer is 90 percent efficient, what is the secondary VA load rating?
    (a) 120 VA   (b) The same as the primary VA
    (c) Cannot be calculated   (d) a and b

62. The primary transformer winding operates at 240V, the secondary at 12V and the load is two 100W lamps. What is the primary VA load if the transformer is 92 percent efficient?
    (a) 185 VA   (b) 217 VA   (c) 0.217 VA   (d) cannot be calculated

63. The input power for a load will be approximately _____ if the load operates at an efficiency of 65 percent.
    (a) 11 kW   (b) 20 kW   (c) 31 kW   (d) 33 kW

NEC Questions - Articles 240-250

Article 240 Overcurrent Protection

64. Which of the following is not a standard size for fuses or inverse-time circuit breakers?
    (a) 45   (b) 70   (c) 75   (d) 80

65. The standard ampere ratings for fuses and inverse-time circuit breakers are listed in 240.6. Additional standard ratings for fuses shall be _____.
    (a) 1A   (b) 6A   (c) 601A   (d) all of these

66. When can breakers or fuses be used in parallel?
    (a) When they are factory assembled in parallel
    (b) When they are listed as a unit
    (c) a and b
    (d) a or b

67. Which of the following statements about supplementary overcurrent protection is correct?
    (a) Shall not be used in luminaires.
    (b) May be used as a substitute for a branch-circuit overcurrent protection device.
    (c) May be used to protect internal circuits of equipment.
    (d) Shall be readily accessible.

68. When an orderly shutdown is required to minimize hazard(s) to personnel and equipment, a system of coordination based on two conditions shall be permitted. Those two conditions are (1) _____ short-circuit protection, and (2) _____ indication based on monitoring systems or devices.
    (a) uncoordinated, overcurrent
    (b) coordinated, overcurrent
    (c) coordinated, overload
    (d) none of these

69. Ground-fault protection of equipment shall be provided in accordance with the provisions of 230.95 for solidly grounded wye electrical systems of more than 150 volts-to-ground, but not exceeding 600V phase-to-phase for each individual device used as a building or structure main disconnecting means rated _____, or more.
    (a) 1,000A   (b) 1,500A   (c) 2,000A   (d) 2,500A

70. A(n) _____ shall be considered equivalent to an overcurrent trip unit.
    (a) current transformer   (b) overcurrent relay
    (c) a and b   (d) a or b

71. A fuse or an overcurrent trip unit of a circuit breaker shall be connected in series with each ungrounded _____.
    (a) device   (b) conductor   (c) branch circuit   (d) all of these

72. Circuit breakers shall _____ all ungrounded conductors of the circuit.
    (a) open   (b) close   (c) isolate   (d) inhibit

73. Except where limited by 210.4(B), individual single-pole circuit breakers, with or without approved handle ties, shall be permitted as the protection for each ungrounded conductor of multiwire branch circuits that serve only 1Ø line-to-neutral loads.
    (a) True   (b) False

74. Single-pole breakers with approved handle ties can be used for the protection of each ungrounded conductor for line-to-line connected loads.
    (a) True   (b) False

75. No tap conductor shall supply another tap conductor.
    (a) True   (b) False

76. Overcurrent protection for tap conductors not over 25 ft is not required at the point where the conductors receive their supply providing the _____.
    (a) ampacity of the tap conductors is not less than one-third of the rating of the overcurrent device protecting the feeder conductors being tapped
    (b) tap conductors terminate in a single circuit breaker or set of fuses that limit the load to the ampacity of the tap conductors
    (c) tap conductors are suitably protected from physical damage
    (d) all of these

77. One of the requirements that permits conductors supplying a transformer to be tapped, without overcurrent protection at the tap, is that the conductors supplied by the _____ of a transformer must have an ampacity of at least one-third of the rating of the overcurrent device protecting the feeder conductors.
    (a) primary   (b) secondary   (c) tertiary   (d) none of these

78. The maximum length of an unprotected feeder tap conductor in a high bay manufacturing building over 35 ft high is _____.
    (a) 15 ft   (b) 20 ft   (c) 50 ft   (d) 100 ft

79. Industrial transformer secondary conductors can have a total length of not more than _____. The tap conductors must have an ampacity not less than the ampacity of the overcurrent protection and must terminate in a single circuit breaker or set of fuses having a rating not greater than the secondary conductor ampacity.
    (a) 8 ft   (b) 25 ft   (c) 35 ft   (d) 75 ft

80. Secondary conductors can be run up to 25 ft if installed in accordance with the following:
    (1) The secondary conductors have an ampacity that (when multiplied by the ratio of the secondary-to-primary voltage) is at least _____ percent of the rating of the overcurrent device protecting the primary of the transformer.
    (2) The secondary conductors terminate in a single circuit breaker or set of fuses that have a rating not greater then the conductor ampacity.
    (3) The secondary conductors are protected from physical damage.
    (a) 25   (b) 33   (c) 50   (d) 100

81. No overcurrent protection device shall be connected in series with any conductor that is intentionally grounded except where the overcurrent protection device opens all conductors of the circuit, including the _____ conductor, and is designed so that no pole can operate independently.
    (a) ungrounded   (b) grounding   (c) grounded   (d) none of these

82. Overcurrent protection devices shall be _____.
    (a) accessible (as applied to wiring methods)   (b) accessible (as applied to equipment)
    (c) readily accessible   (d) inaccessible to unauthorized personnel

83. Branch-circuit overcurrent protection devices are not required to be accessible to occupants of guest rooms of hotels and motels if electric maintenance is provided in a facility that is under continuous building management.
    (a) True   (b) False

84.   Overcurrent protection devices are not permitted to be located _____.
       (a) where exposed to physical damage
       (b) near easily ignitable materials, such as in clothes closets
       (c) in bathrooms of dwelling units
       (d) all of these

85.   Enclosures for overcurrent protection devices shall be mounted in a _____ position unless that is shown to be impracticable.
       (a) vertical   (b) horizontal
       (c) vertical or horizontal   (d) there are no requirements

86.   Handles or levers of circuit breakers, and similar parts that may move suddenly in such a way that persons in the vicinity are likely to be injured by being struck by them, shall be _____.
       (a) guarded   (b) isolated   (c) a and b   (d) a or b

87.   Plug fuses of 15A or less shall be identified by a(n) _____ configuration of the window, cap or other prominent part to distinguish them from fuses of higher ampere ratings.
       (a) octagonal   (b) rectangular   (c) hexagonal   (d) triangular

88.   Plug fuses of the Edison-base type have a maximum rating of _____.
       (a) 20A   (b) 30A   (c) 40A   (d) 50A

89.   Plug fuses of the Edison-base type shall be used _____.
       (a) where overfusing is necessary   (b) only as replacement in existing installations
       (c) as a replacement for Type S fuses   (d) only for 50A and above

90.   Fuseholders of the Edison-base type shall be installed only where they are made to accept _____ fuses by the use of adapters.
       (a) Edison-base   (b) medium-base   (c) heavy-duty base   (d) Type S

91.   • Which of the following statements about Type S fuses is true?
       (a) Adapters shall fit Edison-base fuseholders.
       (b) Adapters are designed to be easily removed.
       (c) Type S fuses shall be classified as not over 125V and 30A.
       (d) a and c

92.   Type _____ fuse adapters shall be designed so that once inserted in a fuseholder they cannot be easily removed.
       (a) A   (b) E   (c) S   (d) P

93.   Type S fuses, fuseholders and adapters are required to be designed so that _____ would be difficult.
       (a) installation   (b) tampering   (c) shunting   (d) b or c

94.   Dimensions of Type S fuses, fuseholders and adapters shall be standardized to permit interchangeability regardless of the _____.
       (a) model   (b) manufacturer   (c) amperage   (d) voltage

95.   Cartridge fuses and fuseholders of the 300V type are not permitted on circuits exceeding 300V _____.
       (a) between conductors   (b) to ground
       (c) or less   (d) a or c

96.   Fuseholders for cartridge fuses shall be so designed that it is difficult to put a fuse of any given class into a fuseholder that is designed for a lower _____ or a higher _____ than that of the class to which the fuse belongs.
       (a) voltage, wattage   (b) wattage, voltage   (c) voltage, current   (d) current, voltage

97.   Fuses are required to be marked with _____.
       (a) ampere and voltage rating   (b) interrupting rating where other than 10,000A
       (c) the name or trademark of the manufacturer   (d) all of these

98.   Cartridge fuses and fuseholders shall be classified according to _____ ranges.
       (a) voltage   (b) amperage   (c) voltage or amperage   (d) voltage and amperage

99. An 800A fuse rated at 600V _____ on a 250V system.
    (a) shall not be used   (b) must be used   (c) can be installed   (d) none of these

100. A(n) _____ shall be of such design that any alteration of its trip point (calibration) or the time required for its operation will require dismantling of the device or breaking of a seal for other than intended adjustments.
    (a) Type S fuse   (b) Edison-base fuse   (c) circuit breaker   (d) fuseholder

101. Circuit breakers shall be marked with their _____ rating in a manner that will be durable and visible after installation.
    (a) voltage   (b) ampere   (c) type   (d) all of these

102. Circuit breakers shall be marked with their ampere rating in a manner that will be durable and visible after installation. Such marking shall be permitted to be made visible by removal of a _____.
    (a) trim   (b) cover   (c) box   (d) a or b

103. Circuit breakers rated at _____ ampere or less and _____ volts or less shall have the ampere rating molded, stamped, etched or similarly marked into their handles or escutcheon areas.
    (a) 100, 600   (b) 600, 100   (c) 1,000, 6,000   (d) 6,000, 1,000

104. Circuit breakers having an interrupting current rating of other than _____ shall have their interrupting rating marked on the circuit breaker.
    (a) 50,000A   (b) 10,000A   (c) 15,000A   (d) 5,000A

105. Circuit breakers used as switches in 120V or 277V fluorescent-lighting circuits shall be listed and marked _____.
    (a) UL   (b) SWD or HID   (c) Amps   (d) VA

106. Circuit breakers used to switch high-intensity discharge lighting circuits shall be listed and marked as _____.
    (a) SWD   (b) HID   (c) a or b   (d) a and b

107. A circuit breaker with a _____ voltage rating, such as 240V or 480V, shall be used where the nominal voltage between any two conductors does not exceed the circuit breaker's voltage rating.
    (a) straight   (b) slash   (c) high   (d) low

108. A circuit breaker with a straight voltage rating (240V or 480V) can be used on a circuit where the nominal voltage between any two conductors does not exceed the circuit breaker's voltage rating.
    (a) True   (b) False

109. A circuit breaker with a slash rating (120/240V or 480Y/277V) can be used for a solidly grounded circuit where the nominal voltage of any one conductor to _____ does not exceed the lower of the two values, and the nominal voltage between any two conductors does not exceed the higher value.
    (a) another conductor   (b) an enclosure   (c) earth   (d) ground

Article 250 Grounding

110. A ground fault is a(an) _____ electrical connection between an ungrounded (hot) conductor and metallic enclosures, metallic raceways, metallic equipment or earth.
    (a) deliberate   (b) intentional   (c) designed   (d) unintentional

111. A ground-fault current path is an electrically conductive path from the point of a line-to-case fault extending to the _____.
    (a) ground   (b) earth   (c) electrical supply (neutral)   (d) none of these

112. An effective ground-fault current path is an intentionally constructed low-impedance path designed and intended to carry fault current from the point of a line-to-case fault on a wiring system to _____.
    (a) ground   (b) earth   (c) electrical supply (neutral)   (d) none of these

113. An effective ground-fault current path is created when all electrically conductive materials that are likely to be energized are bonded together and to the _____.
    (a) ground   (b) earth   (c) electrical supply (neutral)   (d) none of these

114.   Electrical systems that are grounded shall be connected to earth in a manner that will _____.
        (a) limit voltages due to lightning, line surges, or unintentional contact with higher voltage lines
        (b) stabilize the voltage-to-ground during normal operation
        (c) facilitate overcurrent protection device operation in case of ground faults
        (d) a and b

115.   Electrical systems such as transformers and generators shall be connected to the _____ for the purpose of limiting the
        voltage imposed by lightning, line surges or unintentional contact with higher voltage lines, by shunting the energy to
        _____.
        (a) ground   (b) earth   (c) electrical supply (neutral)   (d) none of these

116.   Electrical systems are grounded to the _____ to stabilize the system voltage.
        (a) ground   (b) earth   (c) electrical supply (neutral)   (d) none of these

117.   Non-current-carrying conductive materials enclosing electrical conductors or equipment, or forming part of such
        equipment, shall be connected to earth so as to limit the voltage-to-ground on these materials.
        (a) True   (b) False

118.   The metal parts of the electrical system in a building or structure shall be connected to the _____ for the purpose of
        limiting the voltage imposed by lightning, line surges or unintentional contact with higher voltage lines, by shunting the
        energy to the _____. Typically, this is done at the building disconnecting means in accordance with 250.24 or 250.32.
        (a) ground   (b) earth   (c) electrical supply (neutral)   (d) none of these

119.   Non-current-carrying conductive materials enclosing electrical conductors or equipment, or forming part of such
        equipment, shall be connected together to the _____ in a manner that establishes an effective ground-fault current path.
        (a) ground   (b) earth   (c) electrical supply (neutral)   (d) none of these

120.   Electrical equipment and electrically conductive material likely to become energized shall be installed in a manner that
        creates a permanent, low-impedance circuit capable of safely carrying the maximum ground-fault current likely to be
        imposed on it from where a ground fault may occur to the _____.
        (a) ground   (b) earth   (c) electrical supply (neutral)   (d) none of these

121.   Electrical equipment and wiring and other electrically conductive material likely to become energized shall be installed
        in a manner that creates a _____ likely to be imposed on it from any point on the wiring system where a ground fault
        may occur to the electrical supply source.
        (a) permanent path
        (b) low-impedance path
        (c) path capable of safely carrying the ground-fault current
        (d) all of these

122.   The earth can be used as the sole equipment grounding conductor.
        (a) True   (b) False

123.   For ungrounded systems, noncurrent-carrying conductive materials enclosing electrical conductors or equipment, or
        forming part of such equipment, shall be connected to earth in a manner that will limit the voltage imposed by lightning
        or unintentional contact with higher-voltage lines.
        (a) True   (b) False

124.   The grounding of electrical systems, circuit conductors, surge arresters and conductive non-current-carrying materials
        and equipment shall be installed and arranged in a manner that will prevent objectionable current over the grounding
        conductors or grounding paths.
        (a) True   (b) False

125.   • Currents that introduce noise or data errors in electronic equipment shall be considered objectionable currents.
        (a) True   (b) False

126.   Grounding and bonding conductors shall not be connected by _____.
        (a) pressure connections   (b) solder   (c) lugs   (d) approved clamps

127. Grounding electrode conductor fittings shall be protected from physical damage by being enclosed in _____.
(a) metal   (b) wood   (c) the equivalent of a or b (d) none of these

128. _____ on equipment to be grounded shall be removed from contact surfaces to ensure good electrical continuity.
(a) Paint   (b) Lacquer   (c) Enamel   (d) any of these

129. • When grounding service-supplied alternating-current systems, the grounding electrode conductor shall be connected (bonded) to the grounded service conductor (neutral) at _____.
(a) the load end of the service drop   (b) the meter equipment
(c) the service disconnect   (d) any of these

130. The grounding electrode conductor at the service is permitted to terminate on an equipment grounding terminal bar if a (main) bonding jumper is installed between the grounded conductor bus and the equipment grounding terminal.
(a) True   (b) False

131. A grounding connection shall not be made to any grounded circuit conductor on the _____ side of the service disconnecting means except as permitted for separately derived systems or separate buildings.
(a) supply   (b) power   (c) line   (d) load

132. Where an ac system operating at less than 1,000V is grounded at any point, the _____ conductors shall be run to each service disconnecting means and shall be bonded to each disconnect enclosure.
(a) ungrounded   (b) grounded   (c) grounding   (d) none of these

133. When service-entrance conductors exceed 1,100 kcmil for copper, the required grounded conductor for the service shall be sized not less than _____ percent of the area of the largest ungrounded service-entrance (phase) conductor.
(a) 15   (b) 19   (c) $12\frac{1}{2}$ (d) 25

134. Where the service-entrance phase conductors are installed in parallel, the size of the grounded conductor in each raceway shall be based on the size of the ungrounded service-entrance conductor in the raceway but not smaller than _____ AWG.
(a) 6   (b) 1   (c) 1/0   (d) none of these

135. A main bonding jumper shall be a _____ or similar suitable conductor.
(a) wire   (b) bus   (c) screw   (d) any of these

136. The grounding electrode conductor for a single separately derived system must connect the grounded (neutral) conductor of the derived system to the grounding electrode.
(a) True   (b) False

137. Grounding electrode taps from a separately derived system to a common grounding electrode conductor are permitted when a building or structure has multiple separately derived systems.
(a) True   (b) False

138. The grounding electrode for a separately derived system shall be as near as practicable to, and preferably in the same area as, the grounding electrode conductor connection to the system. The grounding electrode shall be the nearest one of the following:
(a) An effectively grounded metal member of the building structure.
(b) An effectively grounded metal water pipe, but only if it's within 5 ft from the point of entrance into the building.
(c) any metal structure that is effectively grounded.
(d) a or b

139. Where a grounded (neutral) conductor is installed and the neutral-to-case bond is not at the source of the separately derived system, the grounded (neutral) conductor shall be routed with the derived phase conductors and shall not be smaller than the required grounding electrode conductor specified in Table 250.66, but shall not be required to be larger than the largest ungrounded derived phase conductor.
(a) True   (b) False

140. A grounding electrode is required if a building or structure is supplied by a feeder or by more than one branch circuit.
(a) True   (b) False

141. The size of the grounding electrode conductor for a building or structure supplied by a feeder cannot be smaller than that identified in _____ based on the largest ungrounded supply conductor.
(a) 250.66   (b) 250.122   (c) Table 250.66   (d) not specified

142. Because of the increasing use of nonmetallic repairs to the interior metal water pipes, interior metal water piping located more than _____ from the point of entrance to the building shall not be used as a part of the grounding electrode system or as a conductor to interconnect electrodes that are part of the grounding electrode system.
(a) 2 ft   (b) 4 ft   (c) 5 ft   (d) 6 ft

143. The metal frame of a building that is effectively grounded is considered part of the grounding electrode system.
(a) True   (b) False

144. A bare 4 AWG copper conductor installed near the bottom of a concrete foundation or footing that is in direct contact with the earth may be used as a grounding electrode when the conductor is at least _____ in length.
(a) 25 ft   (b) 15 ft   (c) 10 ft   (d) 20 ft

145. Grounding electrodes that consist of driven rods require a minimum of _____ in contact with the soil.
(a) 10 ft   (b) 8 ft   (c) 6 ft   (d) 12 ft

146. Electrodes of pipe or conduit shall not be smaller than _____ and, where of iron or steel, shall have the outer surface galvanized or otherwise metal-coated for corrosion protection.
(a) $1/2$ in.   (b) $3/4$ in.   (c) 1 in.   (d) none of these

147. Grounding electrodes consisting of stainless-steel rods shall be listed and shall not be less than _____ in diameter.
(a) $1/2$ in.   (b) $3/4$ in.   (c) 1 in.   (d) $1 1/4$

148. A metal underground water pipe shall be supplemented by an additional electrode of a type specified in 250.52(A)(2) through (A)(7). Where the supplemental electrode is a rod, pipe or plate electrode, that portion of the bonding jumper that is the sole connection to the supplemental grounding electrode shall not be required to be larger than _____ AWG copper wire.
(a) 8   (b) 6   (c) 4   (d) 1

149. The upper end of the rod electrode shall be _____ ground level unless the aboveground end and the grounding electrode conductor attachment are protected against physical damage.
(a) above   (b) flush with   (c) below   (d) b or c

150. Supplementary electrodes for electrical equipment:
(1) Are not required to be bonded to the grounding electrode system.
(2) The bonding jumper to the supplemental electrode can be any size.
(3) The 25Ω resistance requirement of 250.56 does not apply.
(a) True   (b) False

151. • Where the resistance-to-ground of a single rod electrode exceeds 25Ω, _____.
(a) other means besides made electrodes must be used in order to provide grounding
(b) at least one additional electrode must be added
(c) no additional electrodes are required
(d) the electrode can be omitted

152. When multiple ground rods are used for a grounding electrode, they shall be separated not less than _____ apart.
(a) 6 ft   (b) 8 ft   (c) 20 ft   (d) 12 ft

153. Two or more grounding electrodes that are effectively bonded together shall be considered as a single grounding electrode system in this sense.
(a) True   (b) False

154. Where separate services supply a building and are required to be connected to a grounding electrode, the same grounding electrode shall be used. Two or more grounding electrodes that are _____ shall be considered as a single grounding electrode system in this sense.
(a) effectively bonded together   (b) spaced no more than 6 ft apart
(c) a and b   (d) none of these

155. Air terminal conductors or electrodes used for grounding air terminals _____ be used as the grounding electrodes required by 250.50 for grounding wiring systems and equipment.
(a) shall   (b) shall not   (c) can   (d) any of these

156. The grounding electrode conductor shall be made of which of the following materials?
(a) Copper   (b) Aluminum
(c) Copper-clad aluminum   (d) any of these

157. Bare aluminum or copper-clad aluminum grounding conductors shall not be used where in direct contact with masonry, the earth or where subject to corrosive conditions. Where used outside, aluminum or copper-clad aluminum grounding conductors shall not be terminated within _____ of the earth.
(a) 6 in.   (b) 12 in.   (c) 15 in.   (d) 18 in.

158. • Metal enclosures for grounding electrode conductors shall be electrically continuous from the point of attachment to cabinets or equipment to the grounding electrode.
(a) True   (b) False

159. • A service that contains 12 AWG service-entrance conductors, as permitted by 230.23(B) Ex., shall require a grounding electrode conductor sized no larger than _____.
(a) 6 AWG   (b) 4 AWG   (c) 8 AWG   (d) 10 AWG

160. • The largest size grounding electrode conductor required for any service is a _____ copper.
(a) 6 AWG   (b) 1/0 AWG   (c) 3/0 AWG   (d) 250 kcmil

161. • What size copper grounding electrode conductor is required for a service that has three sets of 500 kcmil copper conductors per phase?
(a) 1 AWG   (b) 1/0 AWG   (c) 2/0 AWG   (d) 3/0 AWG

162. In an ac system, the size of the grounding electrode conductor to a concrete-encased electrode shall not be required to be larger than _____ copper wire.
(a) 4 AWG   (b) 6 AWG   (c) 8 AWG   (d) 10 AWG

163. The connection of the grounding electrode conductor to a buried grounding electrode (driven ground rod) shall be made with a listed terminal device that is accessible .
(a) True   (b) False

164. Grounding electrode conductor connections to a concrete-encased or buried grounding electrode are required to be readily accessible.
(a) True   (b) False

165. The connection (attachment) of the grounding electrode conductor to a grounding electrode shall _____.
(a) be accessible
(b) be made in a manner that will ensure a permanent and effective grounding path
(c) a and b
(d) none of these

166. When an underground metal water-piping system is used as a grounding electrode, effective bonding shall be provided around insulated joints and sections around any equipment that is likely to be disconnected for repairs or replacement. Bonding conductors shall be made of _____ to permit removal of such equipment while retaining the integrity of the bond.
(a) stranded wire   (b) flexible conduit   (c) sufficient length   (d) none of these

167. The grounding conductor connection to the grounding electrode shall be made by _____.
(a) listed lugs
(b) exothermic welding
(c) listed pressure connectors
(d) any of these

168. A metal elbow that is installed in an underground installation of rigid nonmetallic conduit and is isolated from possible contact by a minimum cover _____ to any part of the elbow shall not be required to be grounded.
(a) of 6 in.
(b) of 12 in.
(c) of 18 in.
(d) according to Table 300.5

169. Metal enclosures and raceways for conductors added to existing installations of _____, which do not provide an equipment ground are not required to be grounded if they are less than 25 ft long and free from probable contact with grounded conductive material.
(a) nonmetallic-sheathed cable
(b) open wiring
(c) knob-and-tube wiring
(d) all of these

170. Short sections of metal enclosures or raceways used to provide support or protection of _____ from physical damage shall not be required to be grounded.
(a) conduit   (b) 600V feeders   (c) cable assemblies   (d) none of these

171. Service equipment, service raceways and service conductor enclosures shall be bonded _____.
(a) to the grounded service conductor
(b) by threaded raceways into enclosures, couplings, hubs, conduit bodies, etc
(c) by bonding-type locknuts where concentric or eccentric knockouts are not encountered
(d) any of these

172. The noncurrent-carrying metal parts of equipment, such as _____, shall be effectively bonded together.
(a) service raceways, cable trays or service cable armor
(b) service equipment enclosures containing service conductors, including meter fittings, boxes, or the like, interposed in the service raceway or armor
(c) the metallic raceway or armor enclosing a grounding electrode conductor
(d) all of these

173. Service raceways threaded into metal service equipment such as bosses (hubs) are considered to be effectively _____ to the service metal enclosure.
(a) attached   (b) bonded   (c) grounded   (d) none of these

174. • Metal raceways, cable trays, cable armor, cable sheath, enclosures, frames, fittings and other metal non-current-carrying parts that serve as the grounding conductor shall be _____ together to ensure electrical continuity and to have the capacity to conduct safely any fault current likely to be imposed.
(a) grounded   (b) effectively bonded   (c) soldered or welded   (d) any of these

175. When bonding enclosures, metal raceways, frames, fittings and other metal noncurrent-carrying parts, any nonconductive paint, enamel or similar coating shall be removed at _____.
(a) contact surfaces   (b) threads   (c) contact points   (d) all of these

176. Where required for the reduction of electric noise for electronic equipment, electrical continuity of the metal raceway is not required and the metal raceway can terminate to a(n) _____ nonmetallic fitting(s) or spacer on the electronic equipment.
(a) listed   (b) labeled   (c) identified   (d) marked

177. For circuits over 250 volts-to-ground (480Y/277V), electrical continuity can be maintained between a box or enclosure where no knockouts are encountered and a metal conduit by _____.
(a) threadless fittings for cables with metal sheath
(b) double locknuts on threaded conduit (one inside and one outside the box or enclosure)
(c) fittings that have shoulders that seat firmly against the box locknut on the inside.
(d) all of these

178. All intervening metal raceways, boxes and enclosures between Class I locations and the point of grounding for service equipment or point of grounding of a separately derived system shall be _____ according to the requirements of 250.92(B).
(a) grounded   (b) secured   (c) sealed   (d) bonded

179. A service is supplied by three metal raceways. Each raceway contains 600 kcmil ungrounded (phase) conductors. Determine the size of the service bonding jumper for each raceway.
(a) 1/0 AWG   (b) 2/0 AWG   (c) 225 kcmil   (d) 500 kcmil

180. The bonding jumper for service raceways shall be sized according to the _____.
(a) calculated load   (b) service-entrance conductor size
(c) service-drop size   (d) load to be served

181. What is the minimum size copper bonding jumper for a service raceway containing 4/0 AWG THHN aluminum conductors?
(a) 6 AWG aluminum   (b) 3 AWG copper   (c) 4 AWG aluminum   (d) 4 AWG copper

182. What is the minimum size copper equipment bonding jumper required for equipment connected to a 40A circuit?
(a) 12 AWG   (b) 14 AWG   (c) 8 AWG   (d) 10 AWG

183. The equipment bonding jumper can be installed on the outside of a raceway providing the length of the run is not more than _____ and the bonding jumper is routed with the raceway.
(a) 12 in   (b) 24 in   (c) 36 in   (d) 72 in

184. The general rule for equipment bonding jumpers is that they are not permitted to be longer than 6 ft, but an equipment bonding jumper can be longer than 6 ft at outside pole locations for the purpose of bonding or grounding isolated sections of metal raceways or elbows installed in exposed risers of metal conduit or other metal raceway.
(a) True   (b) False

185. The metal water-piping systems shall be bonded to the _____.
(a) grounded conductor at the service
(b) service equipment enclosure
(c) equipment grounding bar or bus at any panelboard within the building
(d) a and b

186. A building or structure that is supplied by a feeder must have the interior metal water-piping system bonded with a conductor sized from _____.
(a) Table 250.66   (b) Table 250.122   (c) Table 310.16   (d) none of these

187. The metal water-pipe system of a building or structure is not required to be bonded to the separately derived system neutral terminal if the water piping is bonded to the metal frame of a building or structure that serves as the grounding electrode for the separately derived system.
(a) True   (b) False

188. Metal gas piping can be considered bonded by the circuit's equipment grounding conductor of the circuit that may energize the piping.
(a) True   (b) False

189. Exposed structural steel that is interconnected to form a steel building frame, is not intentionally grounded and may become energized, must be grounded to:
(a) The service equipment enclosure.
(b) The grounded (neutral) conductor at the service.
(c) The grounding electrode conductor where of sufficient size.
(d) any of these

190. The lightning protection system grounding electrode _____ be bonded to the building grounding electrode system.
(a) shall   (b) shall not   (c) can   (d) none of these

191. Metal raceways, enclosures, frames and other noncurrent-carrying metal parts of electric equipment installed on a building equipped with a lightning protection system may require spacing from the lightning protection conductors, typically 6 ft through air or ___ through dense materials, such as concrete, brick, wood, etc.
    (a) 2 ft   (b) 3 ft   (c) 4 ft   (d) 6 ft

192. Exposed non-current-carrying metal parts likely to become energized shall be grounded where _____.
    (a) within 8 ft vertically or 5 ft horizontally of ground or grounded metal objects
    (b) located in wet or damp locations and not isolated
    (c) in electrical contact with metal
    (d) any of these

193. An electrically operated pipe organ shall have both the generator and motor frame grounded or _____.
    (a) the generator and motor shall be effectively insulated from ground
    (b) the generator and motor shall be effectively insulated from ground and from each other
    (c) the generator shall be effectively insulated from ground and from the motor driving it
    (d) both shall have double insulation

194. Permanently mounted electrical equipment and skids shall be grounded with an equipment bonding jumper sized as required by _____.
    (a) 250.50   (b) 250.66   (c) 250.122   (d) 310.15

195. Which of the following appliances installed in residential occupancies need not be grounded?
    (a) Toaster   (b) Aquarium   (c) Dishwasher   (d) Refrigerator

196. Flexible metal conduit that is not listed for grounding can be used for grounding if the length in any ground return path does not exceed 6 ft and the circuit conductors contained in the conduit are protected by overcurrent devices rated at _____ or less.
    (a) 15A   (b) 20A   (c) 30A   (d) 60A

197. The equipment grounding conductor shall be identified by _____.
    (a) a continuous outer-green finish
    (b) being bare
    (c) a continuous outer-green finish with one or more yellow stripes
    (d) any of these

198. Equipment grounding conductors shall be the same size as the circuit conductors for _____ circuits.
    (a) 15A   (b) 20A   (c) 30A   (d) all of these

199. When ungrounded conductors are increased in size to compensate for voltage drop, the equipment grounding conductor is not required to be increased because it is not a current-carrying conductor.
    (a) True   (b) False

200. • What size equipment grounding conductor is required for a nonmetallic raceway that contains the following three circuits?
    Circuit 1 - 12 AWG protected by a 20A device
    Circuit 2 - 10 AWG protected by a 30A device
    Circuit 3 - 8 AWG protected by a 40A device
    (a) 10 AWG   (b) 6 AWG   (c) 8 AWG   (d) 12 AWG

Unit 4 NEC Exam – NEC Code Order 240.8 – 250.122

1. When can breakers or fuses be used in parallel?
   (a) When they are factory assembled in parallel.
   (b) When they are listed as a unit.
   (c) a and b
   (d) a or b

2. Which of the following statements about supplementary overcurrent protection is correct?
   (a) Shall not be used in luminaires.
   (b) May be used as a substitute for a branch-circuit overcurrent protection device.
   (c) May be used to protect internal circuits of equipment.
   (d) Shall be readily accessible.

3. Circuit breakers shall _____ all ungrounded conductors of the circuit.
   (a) open   (b) close   (c) isolate   (d) inhibit

4. Except where limited by 210.4(B), individual single-pole circuit breakers, with or without approved handle ties, shall be permitted as the protection for each ungrounded conductor of multiwire branch circuits that serve only 1Ø, line-to-neutral loads.
   (a) True   (b) False

5. Single-pole breakers with approved handle ties can be used for the protection of each ungrounded conductor for line-to-line connected loads.
   (a) True   (b) False

6. Which of the following statements about Type S fuses is true?
   (a) Adapters shall fit Edison-base fuseholders.
   (b) Adapters are designed to be easily removed.
   (c) Type S fuses shall be classified as not over 125V and 30A.
   (d) a and c

7. An 800A fuse rated at 600V _____ on a 250V system.
   (a) shall not be used   (b) must be used
   (c) can be installed   (d) none of these

8. Currents that introduce noise or data errors in electronic equipment shall be considered objectionable currents.
   (a) True   (b) False

9. Grounding electrode conductor fittings shall be protected from physical damage by being enclosed in _____.
   (a) metal   (b) wood
   (c) the equivalent of a or b   (d) none of these

10. When grounding service-supplied alternating-current systems, the grounding electrode conductor shall be connected (bonded) to the grounded service conductor (neutral) at _____.
    (a) the load end of the service drop   (b) the meter equipment
    (c) the service disconnect   (d) any of these

11. Because of the increasing use of nonmetallic repairs to the interior metal water pipes, interior metal water piping located more than _____ from the point of entrance to the building shall not be used as a part of the grounding electrode system or as a conductor to interconnect electrodes that are part of the grounding electrode system.
    (a) 2 ft   (b) 4 ft   (c) 5 ft   (d) 6 ft

12. Grounding electrodes consisting of stainless-steel rods shall be listed and shall not be less than _____ in diameter.
    (a) $1/2$ in.   (b) $3/4$ in.   (c) 1 in.   (d) $1^1/4$ in.

13. Where the resistance-to-ground of a single rod electrode exceeds 25Ω, _____.
    (a) other means besides made electrodes must be used in order to provide grounding
    (b) at least one additional electrode must be added
    (c) no additional electrodes are required
    (d) the electrode can be omitted

14. Metal enclosures for grounding electrode conductors shall be electrically continuous from the point of attachment to cabinets or equipment to the grounding electrode.
    (a) True   (b) False

15. A service that contains 12 AWG service-entrance conductors, as permitted by 230.23(B) Ex., shall require a grounding electrode conductor sized no larger than _____.
    (a) 6   (b) 4   (c) 8   (d) 10

16. The largest size grounding electrode conductor required for any service is a _____ copper.
    (a) 6 AWG   (b) 1/0 AWG   (c) 3/0 AWG   (d) 250 kcmil

17. What size copper grounding electrode conductor is required for a service that has three sets of 500 kcmil copper conductors per phase?
    (a) 1 AWG   (b) 1/0 AWG   (c) 2/0 AWG   (d) 3/0 AWG

18. Grounding electrode conductor connections to a concrete-encased or buried grounding electrode are required to be readily accessible.
    (a) True   (b) False

19. Metal enclosures and raceways for conductors added to existing installations of _____, which do not provide an equipment ground are not required to be grounded if they are less than 25 ft long and free from probable contact with grounded conductive material.
    (a) nonmetallic-sheathed cable   (b) open wiring
    (c) knob-and-tube wiring   (d) all of these

20. Metal raceways, cable trays, cable armor, cable sheath, enclosures, frames, fittings and other metal non-current-carrying parts that serve as the grounding conductor must be _____ together to ensure electrical continuity and have the capacity to conduct safely any fault current likely to be imposed.
    (a) grounded   (b) effectively bonded
    (c) soldered or welded   (d) any of these

21. All intervening metal raceways, boxes and enclosures between Class I locations and the point of grounding for service equipment or point of grounding of a separately derived system must be _____ according to the requirements of 250.92(B).
    (a) grounded   (b) secured   (c) sealed   (d) bonded

22. A service is supplied by three metal raceways. Each raceway contains 600 kcmil ungrounded (phase) conductors. Determine the size of the service bonding jumper for each raceway.
    (a) 1/0 AWG   (b) 2/0 AWG   (c) 225 kcmil   (d) 500 kcmil

23. What is the minimum size copper bonding jumper for a service raceway containing 4/0 AWG THHN aluminum conductors?
    (a) 6 AWG aluminum   (b) 3 AWG copper
    (c) 4 AWG aluminum   (d) 4 AWG copper

24. When ungrounded conductors are increased in size to compensate for voltage drop, the equipment grounding conductor is not required to be increased because it is not a current-carrying conductor.
    (a) True   (b) False

25. What size equipment grounding conductor is required for a nonmetallic raceway that contains the following three circuits: Circuit 1 - 12 AWG protected by a 20A device, Circuit 2 - 10 AWG protected by a 30A device and Circuit 3 - 8 AWG protected by a 40A device?
    (a) 10 AWG   (b) 6 AWG   (c) 8 AWG   (d) 12 AWG

Unit 4 NEC Exam – Random Order 90.2 – 250.112

1. A _____ shall be considered equivalent to an overcurrent trip unit.
    (a) current transformer (b) overcurrent relay
    (c) a and b (d) a or b

2. A bare 4 AWG copper conductor installed near the bottom of a concrete foundation or footing that is in direct contact with the earth may be used as a grounding electrode when the conductor is at least _____ in length.
    (a) 25 ft (b) 15 ft (c) 10 ft (d) 20 ft

3. A building or structure that is supplied by a feeder must have the interior metal water-piping system bonded with a conductor sized from _____.
    (a) Table 250.66 (b) Table 250.122
    (c) Table 310.16 (d) none of these

4. A circuit breaker with a slash rating (120/240V or 480Y/277V) can be used for a solidly grounded circuit where the nominal voltage of any one conductor to _____ does not exceed the lower of the two values, and the nominal voltage between any two conductors does not exceed the higher value.
    (a) another conductor (b) an enclosure
    (c) earth (d) ground

5. A circuit breaker with a straight voltage rating (240V or 480V) can be used on a circuit where the nominal voltage between any two conductors does not exceed the circuit breaker's voltage rating.
    (a) True (b) False

6. A ground fault is a(n) _____ electrical connection between an ungrounded (hot) conductor and metallic enclosures, metallic raceways, metallic equipment or earth.
    (a) deliberate (b) intentional (c) designed (d) unintentional

7. A ground-fault current path is an electrically conductive path from the point of a line-to-case fault extending to the _____.
    (a) ground (b) earth
    (c) electrical supply (neutral) (d) none of these

8. A grounding connection shall not be made to any grounded circuit conductor on the _____ side of the service disconnecting means except as permitted for separately derived systems or separate buildings.
    (a) supply (b) power (c) line (d) load

9. A grounding electrode is required if a building or structure is supplied by a feeder or by more than one branch circuit.
    (a) True (b) False

10. A metal elbow that is installed in an underground installation of rigid nonmetallic conduit and is isolated from possible contact by a minimum cover of _____ to any part of the elbow shall not be required to be grounded.
    (a) 6 in. (b) 12 in.
    (c) 18 in. (d) according to Table 300.5

11. A metal underground water pipe shall be supplemented by an additional electrode of a type specified in 250.52(A)(2) through (A)(7). Where the supplemental electrode is a rod, pipe or plate electrode, that portion of the bonding jumper that is the sole connection to the supplemental grounding electrode shall not be required to be larger than _____ AWG copper wire.
    (a) 8 (b) 6 (c) 4 (d) 1

12. A(n) _____ shall be of such design that any alteration of its trip point (calibration) or the time required for its operation will require dismantling of the device or breaking of a seal for other than intended adjustments.
    (a) Type S fuse (b) Edison-base fuse (c) circuit breaker (d) fuseholder

13. Air-terminal conductors or electrodes used for grounding air terminals _____ be used as the grounding electrodes required by 250.50 for grounding wiring systems and equipment.
    (a) shall (b) shall not (c) can (d) any of these

14. An effective ground-fault current path is an intentionally constructed low-impedance path designed and intended to carry fault current from the point of a line-to-case fault on a wiring system to the _____.
    (a) ground
    (b) earth
    (c) electrical supply (neutral)
    (d) none of these

15. An effective ground-fault current path is created when all electrically conductive materials that are likely to be energized are bonded together and to the _____.
    (a) ground
    (b) earth
    (c) electrical supply (neutral)
    (d) none of these

16. Branch-circuit overcurrent protection devices are not required to be accessible to occupants of guest rooms of hotels and motels if electric maintenance is provided in a facility that is under continuous building management.
    (a) True
    (b) False

17. Cartridge fuses and fuseholders of the 300V type are not permitted on circuits exceeding 300V _____.
    (a) between conductors
    (b) to ground
    (c) or less
    (d) a or c

18. Cartridge fuses and fuseholders shall be classified according to _____ ranges.
    (a) voltage
    (b) amperage
    (c) voltage or amperage
    (d) voltage and amperage

19. Circuit breakers shall be marked with their _____ rating in a manner that will be durable and visible after installation.
    (a) voltage
    (b) ampere
    (c) type
    (d) all of these

20. Circuit breakers used to switch high-intensity discharge lighting circuits must be listed and must be marked as _____.
    (a) SWD
    (b) HID
    (c) a or b
    (d) a and b

21. Electrical equipment and electrically conductive material likely to become energized must be installed in a manner that creates a permanent, low-impedance circuit capable of safely carrying the maximum ground-fault current likely to be imposed on it from where a ground fault may occur to the _____.
    (a) ground
    (b) earth
    (c) electrical supply (neutral)
    (d) none of these

22. Electrical equipment and wiring and other electrically conductive material likely to become energized shall be installed in a manner that creates a _____ likely to be imposed on it from any point on the wiring system where a ground fault may occur to the electrical supply source.
    (a) permanent path
    (b) low-impedance path
    (c) path capable of safely carrying the ground-fault current
    (d) all of these

23. Electrical systems are grounded to the _____ to stabilize the system voltage.
    (a) ground
    (b) earth
    (c) electrical supply (neutral)
    (d) none of these

24. Electrical systems such as transformers and generators must be connected to the _____ for the purpose of limiting the voltage imposed by lightning, line surges or unintentional contact with higher voltage lines by shunting the energy to the _____.
    (a) ground
    (b) earth
    (c) electrical supply (neutral)
    (d) none of these

25. For circuits over 250 volts-to-ground (480Y/277V), electrical continuity can be maintained between a box or enclosure where no knockouts are encountered and a metal conduit by _____.
    (a) threadless fittings for cables with metal sheath
    (b) double locknuts on threaded conduit (one inside and one outside the box or enclosure)
    (c) fittings that have shoulders that seat firmly against the box locknut on the inside.
    (d) all of these

26. Fuses are required to be marked with _____.
    (a) ampere and voltage rating   (b) interrupting rating where other than 10,000A
    (c) the name or trademark of the manufacturer   (d) all of these

27. Grounding electrode taps from a separately derived system to a common grounding electrode conductor are permitted when a building or structure has multiple separately derived systems.
    (a) True   (b) False

28. Handles or levers of circuit breakers, and similar parts that may move suddenly in such a way that persons in the vicinity are likely to be injured by being struck by them, shall be _____.
    (a) guarded   (b) isolated   (c) a and b   (d) a or b

29. In an ac system, the size of the grounding electrode conductor to a concrete-encased electrode shall not be required to be larger than _____ AWG copper wire.
    (a) 4   (b) 6   (c) 8   (d) 10

30. Industrial transformer secondary conductors can have a total length of not more than _____. The tap conductors must have an ampacity not less than the ampacity of the overcurrent protection and must terminate in a single circuit breaker, or set of fuses having a rating not greater than the secondary conductor ampacity.
    (a) 8 ft   (b) 25 ft   (c) 35 ft   (d) 75 ft

31. Metal gas piping can be considered bonded by the circuit's equipment grounding conductor of the circuit that may energize the piping.
    (a) True   (b) False

32. Metal raceways, enclosures, frames and other noncurrent-carrying metal parts of electric equipment installed on a building equipped with a lightning protection system may require spacing from the lightning protection conductors, typically 6 ft through air or ___ through dense materials, such as concrete, brick, wood, etc.
    (a) 2 ft   (b) 3 ft   (c) 4 ft   (d) 6 ft

33. No tap conductor shall supply another tap conductor.
    (a) True   (b) False

34. Non-current-carrying conductive materials enclosing electrical conductors or equipment, or forming part of such equipment, shall be connected to earth so as to limit the voltage-to-ground on these materials.
    (a) True   (b) False

35. Non-current-carrying conductive materials enclosing electrical conductors or equipment, or forming part of such equipment, shall be connected together to the _____ in a manner that establishes an effective ground-fault current path.
    (a) ground   (b) earth
    (c) electrical supply (neutral) (d) none of these

36. Permanently mounted electrical equipment and skids shall be grounded with an equipment bonding jumper sized as required by _____.
    (a) 250.50   (b) 250.66   (c) 250.122   (d) 310.15

37. Secondary conductors can be run up to 25 ft if installed in accordance with the following: (1) The secondary conductors have an ampacity that (when multiplied by the ratio of the secondary-to-primary voltage) is at least _____ percent of the rating of the overcurrent device protecting the primary of the transformer. (2) The secondary conductors terminate in a single circuit breaker or set of fuses that have a rating not greater then the conductor ampacity. (3) The secondary conductors are protected from physical damage.
    (a) 25   (b) 33   (c) 50   (d) 100

38. Service equipment, service raceways and service conductor enclosures shall be bonded _____.
    (a) to the grounded service conductor
    (b) by threaded raceways into enclosures, couplings, hubs, conduit bodies, etc
    (c) by bonding-type locknuts where concentric or eccentric knockouts are not encountered
    (d) any of these

39. Service raceways threaded into metal service equipment such as bosses (hubs) are considered to be effectively _____ to the service metal enclosure.
     (a) attached   (b) bonded   (c) grounded   (d) none of these

40. Service-entrance conductors can be spliced or tapped by clamped or bolted connections at any time as long as _____.
     (a) the free ends of conductors are covered with an insulation that is equivalent to that of the conductors or with an insulating device identified for the purpose
     (b) wire connectors or other splicing means installed on conductors that are buried in the earth are listed for direct burial
     (c) no splice is made in a raceway
     (d) all of these

41. Service-entrance conductors entering, or on the exterior of, buildings or other structures shall be insulated.
     (a) True   (b) False

42. Something constructed, protected, or treated so as to prevent rain from interfering with the successful operation of the apparatus under specified test conditions is defined as _____.
     (a) raintight   (b) waterproof   (c) weathertight   (d) rainproof

43. Supplementary electrodes for electrical equipment are not required to be bonded to the grounding electrode system. The bonding jumper to the supplemental electrode can be any size. The 25Ω resistance requirement of 250.56 does not apply.
     (a) True   (b) False

44. The _____ is the point of connection between the facilities of the serving utility and the premises wiring.
     (a) service entrance   (b) service point
     (c) overcurrent protection   (d) beginning of the wiring system

45. The _____ of the circuit shall be so selected and coordinated as to permit the circuit protective devices to clear a fault without extensive damage to the electrical components of the circuit.
     (a) overcurrent protective devices
     (b) total circuit impedance
     (c) component short-circuit current ratings
     (d) all of these

46. The authority having jurisdiction for enforcement of the *Code* has the responsibility _____.
     (a) for making interpretations of the rules of the *Code*
     (b) for deciding upon the approval of equipment and materials
     (c) for waiving specific requirements in the *Code* and allowing alternate methods and material if safety is maintained
     (d) all of these

47. The authority having jurisdiction is not required to enforce any requirements of Chapter 7 (Special Conditions) or Chapter 8 (Communications Circuits) because this is not within the scope of enforcement.
     (a) True   (b) False

48. The bonding jumper for service raceways must be sized according to the _____.
     (a) calculated load   (b) service-entrance conductor size
     (c) service drop size   (d) load to be served

49. The *Code* does not cover installations in ships, watercraft, railway rolling stock, aircraft or automotive vehicles.
     (a) True   (b) False

50. The combination of all components and subsystems that convert solar energy into electrical energy is called a _____ system.
     (a) solar   (b) solar voltaic
     (c) separately derived source   (d) solar photovoltaic

CHAPTER 2
NEC Calculations and Code Questions

Scope of Chapter 2

Unit 5

Raceway, Outlet Box and Junction Box Calculations

OBJECTIVES

After reading this unit, the student should be able to briefly explain the following concepts:

Part A – Raceway Calculations	Part B – Outlet Box Calculations	Part C – Pull and Junction Box Calculations
Existing raceway calculation	Conductor equivalents	Depth of box and conduit body sizing
Raceway sizing	Sizing box – conductors all the same size	Pull and junction box size calculations
Raceway properties	Volume of box	
Understanding *NEC* Chapter 9		

After reading this unit, the student should be able to briefly explain the following terms:

Part A – Raceway Calculations		
Alternating-current conductor resistance	Lead-covered conductor	Outlet box
Bare conductors	*NEC* errors	Pigtails
Bending radius	Nipple size	Plaster rings
Compact aluminum building wire	Raceway size	Short radius conduit bodies
Conductor properties	Spare space area	Size outlet box
Conductor fill	**Part B – Outlet Box Calculations**	Strap
Conduit bodies	Cable clamps	Volume
Cross-sectional area of insulated conductors	Conductor terminating in the box	Yoke
Expansion characteristics of PVC	Conductor running through the box	**Part C – Pull and Junction Box Calculations**
Fixture wires	Conduit bodies	Angle-pull calculation
Grounding conductors	Equipment bonding jumpers	Distance between raceways
	Extension rings	Horizontal dimension
	Fixture hickey	Junction boxes
	Fixture stud	

PART A – RACEWAY FILL

5–1 UNDERSTANDING THE NEC, CHAPTER 9

Chapter 9 – Tables

Table 1 – Conductor Percent Fill

The maximum percentage of conductor fill is listed in Table 1 of Chapter 9 and is based on common conditions where the length of the conductor and number of raceway bends are within reasonable limits [Fine Print Note under Table 1]. Figure 5–1.

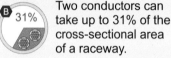

Figure 5-1
Conductor Fill – Percent of Raceway Area Permitted

Table 1 of Chapter 9, Maximum Percent Conductor Fill	
Number of Conductors	**Percent Fill Permitted**
1 conductor	53% fill
2 conductors	31% fill
3 or more conductors	40% fill
Raceway 24 inches or less	60% fill Chapter 9, Note 4

Table 1, Note 1 – Conductors all the Same Size and Insulation

When all of the conductors are the same size and insulation, the number of conductors permitted in a raceway can be determined simply by looking at the tables located in Annex C – Conduit and Tubing Fill Tables for Conductors and Fixture Wires of the Same Size.

Tables C1 through C12A are based on maximum percent fill as listed in Table 1 of Chapter 9.

Table C1 – Conductors and fixture wires in electrical metallic tubing
Table C1A – Compact conductors in electrical metallic tubing
Table C2 – Conductors and fixture wires in electrical nonmetallic tubing
Table C2A – Compact conductors in nonelectrical metallic tubing
Table C3 – Conductors and fixture wires in flexible metal conduit
Table C3A – Compact conductors in flexible metal conduit
Table C4 – Conductors and fixture wires in intermediate metal conduit
Table C4A – Compact conductors in intermediate metal conduit
Table C5 – Conductors and fixture wires in liquidtight flexible nonmetallic conduit (gray type)
Table C5A – Compact conductors in liquidtight flexible nonmetallic conduit (gray type)
Table C6 – Conductors and fixture wires in liquidtight flexible nonmetallic conduit (orange type)
Table C6A – Compact conductors in liquidtight flexible nonmetallic conduit (orange type)

Note: The appendix does not have a table for liquidtight flexible nonmetallic conduit of the black type.

Table C7 – Conductors and fixture wires in liquidtight flexible metallic conduit
Table C7A – Compact conductors in liquidtight flexible metal conduit
Table C8 – Conductors and fixture wires in rigid metal conduit

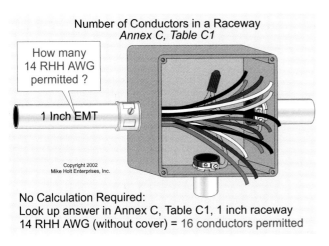

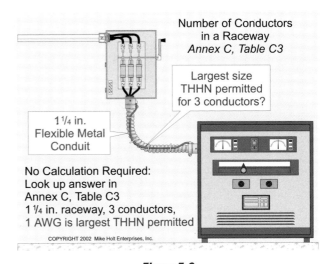

Figure 5-2
Number of Conductors in a Raceway

Figure 5-3
Number of Conductors in a Raceway

Table C8A – Compact conductors in rigid metal conduit

Table C9 – Conductors and fixture wires in rigid PVC conduit Schedule 80

Table C9A – Compact conductors in rigid PVC conduit Schedule 80

Table C10 – Conductors and fixture wires in rigid PVC conduit Schedule 40

Table C10A – Compact conductors in rigid PVC conduit Schedule 40

Table C11 – Conductors and fixture wires in Type A, Rigid PVC conduit

Table C11A – Compact conductors in Type A, PVC conduit

Table C12 – Conductors and fixture wires in Type EB, PVC conduit

Table C12A – Compact conductors in Type EB, PVC conduit

❏ **Annex C – Table C1**

How many 14 AWG RHH conductors (without cover) can be installed in 1 in. electrical metallic tubing? Figure 5–2.

(a) 25 conductors   (b) 16 conductors   (c) 13 conductors   (d) 19 conductors

     • Answer: (b) 16 conductors, Annex C, Table C1

❏ **Annex C – Table C2A – Compact Conductor**

How many compact 6 AWG XHHW conductors can be installed in $1^1/_4$ in. nonmetallic tubing?

(a) 10 conductors   (b) 6 conductors   (c) 16 conductors   (d) 13 conductors

     • Answer: (a) 10 conductors, Annex C, Table C2A

❏ **Annex C – Table C3**

If $1^1/_4$ in. flexible metal conduit has three THHN conductors (not compact), what is the largest conductor permitted to be installed? Figure 5–3.

(a) 1 AWG   (b) 1/0 AWG   (c) 2/0 AWG   (d) 3/0 AWG

     • Answer: (a) 1 AWG, Annex C, Table C3

❏ **Annex C – Table C4**

How many 4/0 AWG RHH conductors (with outer cover) can be installed in 2 in. intermediate metal conduit?

(a) 2 conductors   (b) 1 conductor   (c) 3 conductors   (d) 4 conductors

     • Answer: (c) 3 conductors, Annex C, Table C4

❏ **Annex C – Table C7 – Fixture Wire**

How many 18 AWG TFFN conductors can be installed in $^3/_4$ in. liquidtight flexible metallic conduit? Figure 5– 4.

(a) 40   (b) 26   (c) 30   (d) 39

     • Answer: (d) 39, Annex C, Table 7

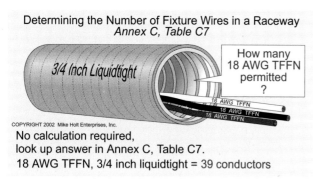

Determining the Number of Fixture Wires in a Raceway
Annex C, Table C7

How many 18 AWG TFFN permitted?

COPYRIGHT 2002 Mike Holt Enterprises, Inc.

No calculation required, look up answer in Annex C, Table C7.
18 AWG TFFN, 3/4 inch liquidtight = 39 conductors

Figure 5-4
Determining the Number of Fixture Wires in a Raceway

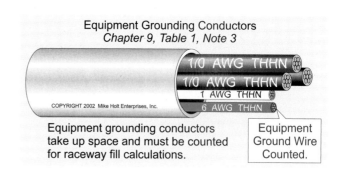

Equipment Grounding Conductors
Chapter 9, Table 1, Note 3

COPYRIGHT 2002 Mike Holt Enterprises, Inc.

Equipment grounding conductors take up space and must be counted for raceway fill calculations.

Equipment Ground Wire Counted.

Figure 5-5
Equipment Grounding Conductors

Table 1, Note 3 – Equipment Grounding Conductors

When equipment grounding conductors are installed in a raceway, the actual area of the conductor must be used when calculating raceway fill. Chapter 9, Table 5 can be used to determine the cross-sectional area of insulated conductors and Chapter 9, Table 8 can be used to determine the cross-sectional area of bare conductors [Note 8 of Table 1, Chapter 9]. Figure 5–5.

Table 1, Note 4 – Nipples, Raceways Not Exceeding 24 Inches

The cross-sectional areas of conduit and tubing can be found in Table 4 of Chapter 9. When a conduit or tubing raceway does not exceed 24 in. in length, it is called a nipple. Nipples are permitted to be filled to 60% of their total cross-sectional area. Figure 5–6.

Table 1, Note 7

When the calculated number of conductors (all of the same size and insulation) results in 0.8 or larger, the next whole number can be used. But, be careful—this only applies when the conductors are all the same size and insulation.

Table 1, Note 8

The dimensions for bare conductor are listed in Table 8 of Chapter 9.

Chapter 9, Table 4 – Conduit and Tubing Cross-Sectional Area

Table 4 of Chapter 9 lists the dimensions and cross-sectional area for conduit and tubing. The cross-sectional area of conduit or tubing is dependent on the raceway type and the maximum percentage fill as listed in Table 1 of Chapter 9.

❏ **Conduit Cross-Sectional Area**

What is the total cross-sectional area of $1^{1}/_{4}$ in. rigid metal conduit? Figure 5–7.

(a) 1.063 sq in.   (b) 1.526 sq in.
(c) 1.098 sq in.   (d) any of these
• Answer: (b) 1.526 sq in.
Chapter 9, Table 4

Chapter 9, Table 5 – Dimensions of Insulated Conductors and Fixture Wires

Table 5 of Chapter 9 lists the cross-sectional area of insulated conductors and fixture wires.

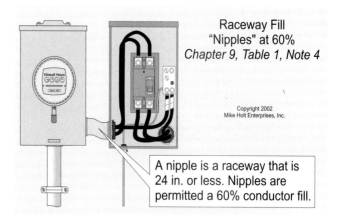

Raceway Fill "Nipples" at 60%
Chapter 9, Table 1, Note 4

Copyright 2002
Mike Holt Enterprises, Inc.

A nipple is a raceway that is 24 in. or less. Nipples are permitted a 60% conductor fill.

Figure 5-6
Raceway Fill – "Nipples" at 60%

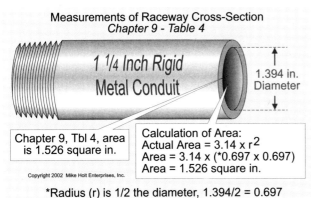

Figure 5-7

Measurements of Raceway Cross-Section

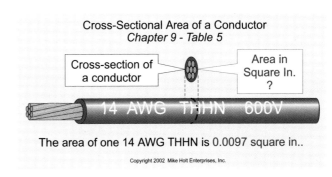

Figure 5-8

Cross-Sectional Area of a Conductor

Table 5 of Chapter 9 – Conductor Cross-Sectional Area							
	RH	RHH/RHW With *Cover*	RHH/RHW With*out* *Cover* or THW	TW	THHN THWN TFN	XHHW	BARE *Stranded* Conductors
Size AWG/kcmil	Approximate Cross-Sectional Area – Square Inches						
Column 1	Column 2	Column 3	Column 4	Column 5	Column 6	Column 7	Chapter 9, Table 8
14	.0293	.0293	.0209	.0139	.0097	.0139	**.004**
12	.0353	.0353	.0260	.0181	.0133	.0181	**.006**
10	.0437	.0437	.0333	.0243	.0211	.0243	**.011**
8	.0835	.0835	.0556	.0437	.0366	.0437	**.017**
6	.1041	.1041	.0726	.0726	.0507	.0590	**.027**
4	.1333	.1333	.0973	.0973	.0824	.0814	**.042**
3	.1521	.1521	.1134	.1134	.0973	.0962	**.053**
2	.1750	.1750	.1333	.1333	.1158	.1146	**.067**
1	.2660	.2660	.1901	.1901	.1562	.1534	**.087**
0	.3039	.3039	.2223	.2223	.1855	.1825	**.109**
00	.3505	.3505	.2624	.2624	.2233	.2190	**.137**
000	.4072	.4072	.3117	.3117	.2679	.2642	**.173**
0000	.4754	.4754	.3718	.3718	.3237	.3197	**.219**

❏ **Table 5 – THHN**

What is the cross-sectional area for one 14 AWG THHN conductor? Figure 5–8.

(a) 0.0206 sq in.   (b) 0.0172 sq in.   (c) 0.0097 sq in.   (d) 0.0278 sq in.

• Answer: (c) 0.0097 sq in.

❏ Table 5 – RHW *With Outer Cover*

What is the cross-sectional area for one 12 AWG RHW conductor *with outer cover*?

(a) 0.0206 sq in.
(b) 0.0172 sq in.
(c) 0.0353 sq in.
(d) 0.0278 sq in.
• Answer: (c) 0.0353 sq in.

❏ Table 5 – RHH *Without Outer Cover*

What is the cross-sectional area for one 10 AWG RHH (without an outer cover)?

(a) 0.0117 sq in.   (b) 0.0333 sq in.
(c) 0.0252 sq in.   (d) 0.0278 sq in.
• Answer: (b) 0.0333 sq in.

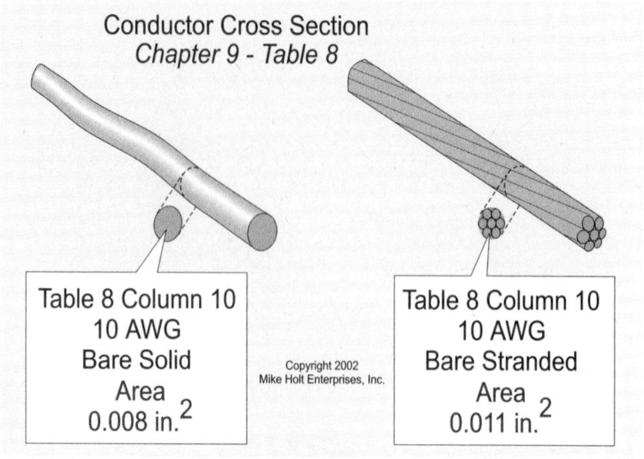

Chapter 9, Table 5A – Compact Aluminum Building Wire Nominal Dimensions and Areas

Table 5A, Chapter 9 lists the cross-sectional area for compact aluminum building wires. We will not use these tables for this unit.

Chapter 9, Table 8 – Conductor Properties

Table 8 contains conductor properties such as cross-sectional area in circular mils, number of strands per conductor, cross-sectional area in sq in. for bare conductors, and conductor's resistance at 75°C for dc for both copper and aluminum wire.

❏ Bare Conductor – Cross-Sectional Area

What is the cross-sectional area in square inches for one 10 AWG bare conductor with seven strands? Figure 5–9.

(a) 0.008   (b) 0.011   (c) 0.038   (d) a or b
• Answer: (b) 0.011.

Chapter 9, Table 9 – AC Impedance For Conductors in Conduit or Tubing

Table 9 contains the ac impedance for copper and aluminum conductors.

❏ Alternating-Current Impedance

What is the ac neutral to ohms impedance of 1/0 AWG copper installed in a metal conduit? Conductor length 1,000 ft.

(a) 0.12Ω   (b) 0.13Ω   (c) 0.14Ω   (d) 0.15Ω
• Answer: (a) 0.12Ω

Conductor Cross Section
Chapter 9 - Table 8

Table 8 Column 10
10 AWG
Bare Solid
Area
0.008 in.^{2}

Copyright 2002
Mike Holt Enterprises, Inc.

Table 8 Column 10
10 AWG
Bare Stranded
Area
0.011 in.^{2}

Figure 5-9
Bare Conductor Information

5–2 RACEWAY AND NIPPLE CALCULATIONS

Annex C – Tables 1 through 12 cannot be used to determine raceway sizing when conductors of different sizes (or types of insulation) are installed in the same raceway. The following Steps can be used to determine the raceway size and nipple size:

Step 1:   Determine the cross-sectional area (square inches) for each conductor from Table 5 of Chapter 9 for insulated conductors and Table 8 of Chapter 9 for bare conductors.

Step 2:   Determine the total cross-sectional area for all conductors.

Step 3:   Size the raceway according to the percent fill as listed in Table 1 of Chapter 9:
40% for three or more conductors.
60% for raceways 24 in. or less in length (nipples).

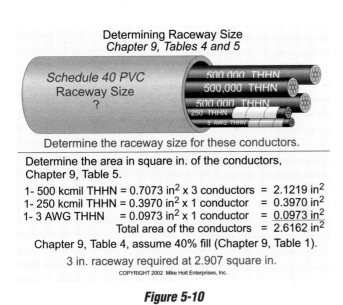

Determining Raceway Size
Chapter 9, Tables 4 and 5

Schedule 40 PVC
Raceway Size
?

500,000 THHN
500,000 THHN
500,000 THHN
250 THHN
3 AWG THHN

Determine the raceway size for these conductors.

Determine the area in square in. of the conductors,
Chapter 9, Table 5.

1- 500 kcmil THHN = 0.7073 in² x 3 conductors = 2.1219 in²
1- 250 kcmil THHN = 0.3970 in² x 1 conductor   = 0.3970 in²
1- 3 AWG THHN   = 0.0973 in² x 1 conductor   = 0.0973 in²
        Total area of the conductors = 2.6162 in²

Chapter 9, Table 4, assume 40% fill (Chapter 9, Table 1).

3 in. raceway required at 2.907 square in.

COPYRIGHT 2002 Mike Holt Enterprises, Inc.

Figure 5-10
Determining Raceway Size

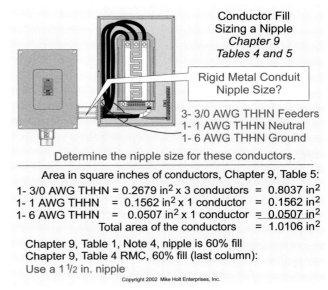

Conductor Fill
Sizing a Nipple
Chapter 9
Tables 4 and 5

Rigid Metal Conduit
Nipple Size?

3- 3/0 AWG THHN Feeders
1- 1 AWG THHN Neutral
1- 6 AWG THHN Ground

Determine the nipple size for these conductors.

Area in square inches of conductors, Chapter 9, Table 5:
1- 3/0 AWG THHN = 0.2679 in² x 3 conductors = 0.8037 in²
1- 1 AWG THHN   = 0.1562 in² x 1 conductor   = 0.1562 in²
1- 6 AWG THHN   = 0.0507 in² x 1 conductor   = 0.0507 in²
        Total area of the conductors   = 1.0106 in²

Chapter 9, Table 1, Note 4, nipple is 60% fill
Chapter 9, Table 4 RMC, 60% fill (last column):
Use a 1 1/2 in. nipple

Copyright 2002 Mike Holt Enterprises, Inc.

Figure 5-11
Conductor Fill – Sizing a Nipple

Raceway Size

A 400A feeder is installed in Schedule 40 rigid nonmetallic conduit. This raceway contains three 500 kcmil THHN conductors, one 250 kcmil THHN conductor and one 3 AWG THHN conductor. What size raceway is required for these conductors? Figure 5–10.

(a) 2 in.   (b) 2¹/₂ in.   (c) 3 in.   (d) 3¹/₂ in.

• Answer: (c) 3 in.

Step 1: Determine the cross-sectional area of the conductors, Table 5 of Chapter.
500 kcmil THHN 0.7073 sq. in. × 3 wires = 2.1219 sq in.
250 kcmil THHN 0.3970 sq. in. × 1 wire   = 0.3970 sq in.
3 AWG THHN 0.0973 sq. in. × 1 wire   = 0.0973 sq in.

Step 2: Total cross-sectional area of all conductors = 2.6162 sq in.

Step 3: Size the conduit at 40% fill [Chapter 9, Table 1] using Table 4.
3 in. Schedule 40 PVC has an cross-sectional area of 2.907 sq in. for conductors

Nipple Size

What size rigid metal nipple is required for three 3/0 AWG THHN conductors, one 1 AWG THHN conductor and one 6 AWG THHN conductor? Figure 5–11.

(a) ¹/₂ in.   (b) 1 in.   (c) 1¹/₂ in.   (d) 2 in.

• Answer: (c) 1¹/₂ in.

Step 1: Cross-sectional area of the conductors, Table 5 of Chapter 9.
3/0 AWG THHN 0.2679 sq. in. × 3 wires = 0.8037 sq in.
1 AWG THHN 0.1562 sq. in. × 1 wire   = 0.1562 sq in.
6 AWG THHN 0.0507 sq. in. × 1 wire   = 0.0507 sq in.

Step 2: Total cross-sectional area of the conductors = 1.0106 sq in..

Step 3: Size the conduit at 60% fill [Table 1, Note 4 of Chapter 9] using Table 4.
1¹/₄ in. nipple = 0.0916 sq in., too small
1¹/₂ in. nipple = 1.243 sq in., just right
2 in. nipple = 2.045 sq in., too big

5-3 EXISTING RACEWAY CALCULATIONS

There are times we need to add conductors to an existing raceway. This can be accomplished by using the following steps:

Part 1 – Determine the raceway cross-sectional spare space area.

Step 1:   Determine the raceway's cross-sectional area for conductor fill [Table 1 and Table 4 of Chapter 9].

Step 2:   Determine the area of the existing conductors [Table 5 of Chapter 9].

Step 3:   Subtract the cross-sectional area of the existing conductors (Step 2) from the area of permitted conductor fill (Step 1).

Part 2 – To determine the number of conductors permitted in the spare space area:

Step 4:   Determine the cross-sectional area of the conductors to be added [Table 5 of Chapter 9 for insulated conductors and Table 8 of Chapter 9 for bare conductors].

Step 5:   Divide the spare space area (Step 3) by the conductor's cross-sectional area (Step 4).

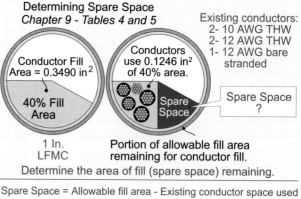

Determine the area of fill (spare space) remaining.

Spare Space = Allowable fill area - Existing conductor space used

Allowable Fill Area: Chapter 9 Table 4
1 in. LFMC, 3 or more conductors = 40% = 0.3490 sq in.

Space Used: Chapter 9, Table 5
One 10 AWG THW = 0.0333 in² x 2 conductors = 0.0666 sq in.
One 12 AWG THW = 0.0260 in² x 2 conductors = 0.0520 sq in.
One 12 AWG bare stranded, Chapter 9, Tbl 8 = 0.0060 sq in.
Allowable fill area used = 0.1246 sq in.

Spare Space = 0.3490 in² - 0.1246 in² = 0.2244 sq in. remaining fill area

Copyright 2002 Mike Holt Enterprises, Inc.

Figure 5-12
Determining Square Space

❑ **Spare Space Area**

An existing 1 in. liquidtight flexible metal conduit contains two 12 AWG THW conductors, two 10 AWG THW conductors, and one 12 AWG bare with 7 strands. What is the area remaining for additional conductors? Figure 5–12.

(a) Raceway 0.2244 sq in.
(b) Nipple 0.3994 sq in.
(c) there is no spare space
(d) a and b

  • Answer: (d) a and b

Step 1:   Conductor cross-sectional area, liquidtight flexible metal conduit [Table 1, Note 4 and Table 4 of Chapter 9].
Raceway: 0.872 = 0.349 sq in.
Nipple: 0.872 = 0.524 sq in.

Step 2:   Cross-sectional area of existing conductors.
12 AWG THW 0.0260 sq in. × 2 wires   = 0.0520 sq in.
10 AWG THW 0.0333 sq in. × 2 wires   = 0.0666 sq in.
12 AWG bare 0.006 sq in. × 1 wire   = 0.0060 sq in.
Total cross-sectional area   = 0.1246 sq in.

Note: Ground wires must be counted for raceway fill – Table 1, Note 3 of Chapter 9.

Step 3:   Subtract the area of the existing conductors from the permitted area of conductor fill.
Raceway: 0.349 sq in. – 0.1246 sq in. = 0.2244 sq in.
Nipple: 0.524 sq in. – 0.1246 sq in. = 0.3994 sq in.

❏ **Conductors in Spare Space Area**

An existing 1 in. EMT contains two 12 AWG THHN conductors, two 10 AWG THHN conductors and one 12 AWG bare 7 stranded conductor. How many additional 8 AWG THHN conductors can be added to this raceway? Figure 5–13.

(a) 7 conductors if raceway

(b) 12 conductors if nipple

(c) 15 conductors regardless of the raceway length

(d) a and b

   • Answer: (d) a and b

Step 1:   Cross-sectional area of 1 in. EMT permitted for conductor fill [Table 1, Note 4 and Table 4 of Chapter 9].
Raceway: 0.3460 sq in.
Nipple: 0.5190 sq in.

Step 2:   Cross-sectional area of existing conductors.
10 AWG THHN 0.0211 sq in. × 2 = 0.0422 sq in.
12 AWG THHN 0.0133 sq in. × 2 = 0.0266 sq in.
12 AWG bare 0.0060 sq in. × 1   = 0.0060 sq in.

Total cross-sectional area   = 0.0748 sq in.

Note: Ground wires must be counted for raceway fill. See Table 1, Note 3 of Chapter 9.

Step 3:   Subtract the area of the existing conductors from the permitted area of conductor fill.
Raceway: 0.346 sq in. – 0.0748 sq in. = 0.2712 sq in.
Nipple: 0.519 sq in. – 0.0748 sq in. = 0.4442 sq in.

Step 4:   Cross-sectional area of the conductor to be installed [Table 5 of Chapter 9].

8 AWG THHN = 0.0366 sq in.

Step 5:   Divide the spare space area (Step 3) by the conductor area.
Raceway: 0.2712 sq in./0.0366 sq in. = 7.4 or 7 conductors
Nipple: 0.4442 sq in./0.0366 sq in. = 12.1 or 12 conductors.

We must round down to 18 conductors because Note 7 of Table 1 only applies if all of the conductors are the same size and same insulation.

5–4 TIPS FOR RACEWAY CALCULATIONS

Tip 1:   Take your time.

Tip 2:   Use a ruler or straight-edge when using tables.

Tip 3:   Watch out for the different types of raceways and conductor insulation, particularly RHH/RHW with or without outer cover.

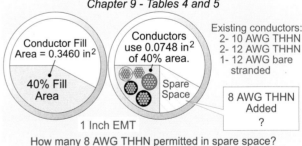

Adding Conductors to Spare Space
Chapter 9 - Tables 4 and 5

Conductor Fill Area = 0.3460 in²

40% Fill Area

Conductors use 0.0748 in² of 40% area.

Spare Space

Existing conductors:
2- 10 AWG THHN
2- 12 AWG THHN
1- 12 AWG bare stranded

8 AWG THHN Added ?

1 Inch EMT

How many 8 AWG THHN permitted in spare space?

Spare Space = Allowable fill area - Existing conductor space used

Allowable Fill Area: Chapter 9 Table 4
1 inch EMT, 3 or more conductors = 40% = 0.3460 sq in.
Space Used: Chapter 9, Table 5
One 10 AWG THHN = 0.0211 in² x 2 conductors = 0.0422 sq in.
One 12 AWG THHN = 0.0133 in² x 2 conductors = 0.0266 sq in.
One 12 AWG bare stranded, Chapter 9, Tbl 8   = 0.0060 sq in.
                         Conductor fill used = 0.0748 sq in.
Spare Space = 0.3460 in² - 0.0748 in² = 0.2712 sq in.
Chapter 9, Table 5: 8 AWG THHN = 0.0366 sq in.
0.2712 in²/0.0366 in² = 7.4 = 7- 8 AWG THHN can be added

Figure 5-13

Adding Conductors to Square Space

PART B – OUTLET BOX FILL CALCULATIONS

INTRODUCTION [314.16]

Boxes shall be of sufficient size to provide free space for all conductors. An outlet box is generally used for the attachment of devices and luminaires and has a specific amount of space (volume) for conductors, devices and fittings. The volume taken up by conductors, devices and fittings in a box must not exceed the box fill capacity. The volume of a box is the total volume of its assembled parts, including plaster rings, industrial raised covers and extension rings. The total volume includes only those fittings that are marked with its volume in cubic inches [314.16(A)].

5–5 SIZING BOX – CONDUCTORS ALL THE SAME SIZE [Table 314.16(A)]

When all of the conductors in an outlet box are the same size (insulation doesn't matter), Table 314.16(A) of the *NEC* can be used to:

(1) Determine the number of conductors permitted in the outlet box, or

(2) Determine the size outlet box required for the given number of conductors.

Note: Table 314.16(A) only applies if the outlet box contains no switches, receptacles, luminaire studs, luminaire hickeys, manufactured cable clamps, or grounding conductors (not likely).

❏ **Outlet Box Size**

What size outlet box is required for six 12 AWG THHN conductors and three 12 AWG THW conductors? Figure 5–14.

(a) $4 \times 1^1/_4$ in. square   (b) $4 \times 1^1/_2$ in. square   (c) $4 \times 1^1/_4$ in. round   (d) $4 \times 1^1/_2$ in. round

• Answer: (b) $4 \times 1^1/_2$ in. square

Table 314.16(A) permits nine 12 AWG conductors; insulation is not a factor.

❏ **Number of Conductors in Outlet Box**

Using Table 314.16(A), how many 14 AWG THHN conductors are permitted in a $4 \times 1^1/_2$ in. round box?

(a) 7 conductors   (b) 9 conductors   (c) 10 conductors   (d) 11 conductors

• Answer: (a) 7 conductors

5–6 CONDUCTOR EQUIVALENTS [314.16(B)]

Table 314.16(A) does not take into consideration the fill requirements of clamps, support fittings, devices or equipment grounding conductors within the outlet box. In no case can the volume of the box and its assembled sections be less than the fill calculation as listed below:

(1a) Conductor Terminating in the Box. Each conductor that originates outside the box and terminates or is spliced within the box is considered as one conductor. Figure 5–15.

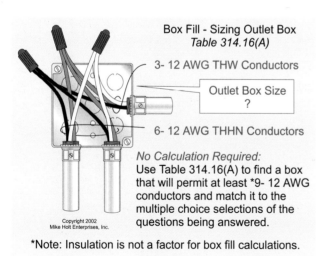

Box Fill - Sizing Outlet Box
Table 314.16(A)

3- 12 AWG THW Conductors

Outlet Box Size ?

6- 12 AWG THHN Conductors

No Calculation Required:
Use Table 314.16(A) to find a box that will permit at least *9- 12 AWG conductors and match it to the multiple choice selections of the questions being answered.

*Note: Insulation is not a factor for box fill calculations.

Copyright 2002 Mike Holt Enterprises, Inc.

Figure 5-14
Box Fill – Sizing Outlet Box

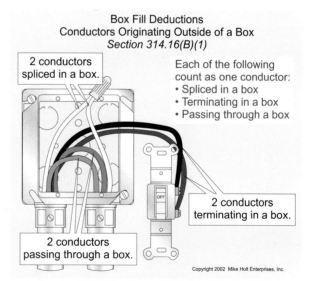

Box Fill Deductions
Conductors Originating Outside of a Box
Section 314.16(B)(1)

2 conductors spliced in a box.

Each of the following count as one conductor:
• Spliced in a box
• Terminating in a box
• Passing through a box

2 conductors terminating in a box.

2 conductors passing through a box.

Copyright 2002 Mike Holt Enterprises, Inc.

Figure 5-15
Box Fill Deductions

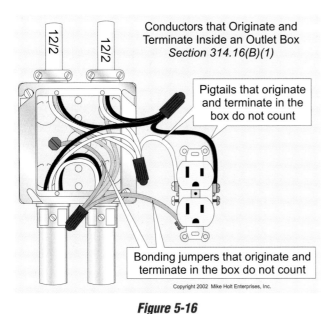

Conductors that Originate and
Terminate Inside an Outlet Box
Section 314.16(B)(1)

Pigtails that originate
and terminate in the
box do not count

Bonding jumpers that originate and
terminate in the box do not count

Copyright 2002 Mike Holt Enterprises, Inc.

Figure 5-16

Box Fill Conductors Originating Inside a Box

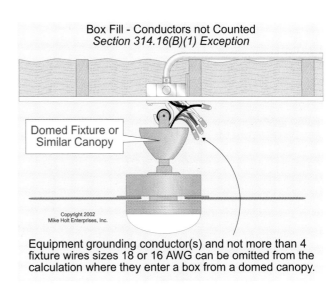

Box Fill - Conductors not Counted
Section 314.16(B)(1) Exception

Domed Fixture or
Similar Canopy

Copyright 2002
Mike Holt Enterprises, Inc.

Equipment grounding conductor(s) and not more than 4
fixture wires sizes 18 or 16 AWG can be omitted from the
calculation where they enter a box from a domed canopy.

Figure 5-17

Box Fill Conductors not Counted

(1b) Conductor Running Through the Box. Each conductor that runs through the box is considered as one conductor. Figure 5–15. Conductors, no part of which leave the box, shall not be counted. This includes equipment bonding jumpers and pigtails. Figure 5–16.

Exception. Fixture wires smaller than 14 AWG from a domed luminaire or similar canopy are not counted. Figure 5–17.

(2) Cable Fill. One or more internal cable clamps in the box are considered as one conductor volume in accordance with the volume listed in Table 314.16(B), based on the largest conductor that enters the outlet box. Figure 5–18.

Note: Small fittings such as locknuts and bushings are not counted [314.16(B)].

(3) Support Fittings Fill. One or more luminaire studs or hickeys within the box are considered as one conductor volume, based on the largest conductor that enters the outlet box. Figure 5–19.

(4) Device or Equipment Fill. Each yoke or strap containing one or more devices or equipment is considered as two conductors, based on the largest conductor that terminates on the yoke. Figure 5–20.

(5) Grounding Conductors. One or more grounding conductors are considered as one conductor volume in accordance with the volume based on the largest grounding conductor that enters the outlet box. Figure 5–21.

Note: Fixture ground wires smaller than 14 AWG from a domed luminaire or similar canopy are not counted [314.16(B)(1) Ex.].

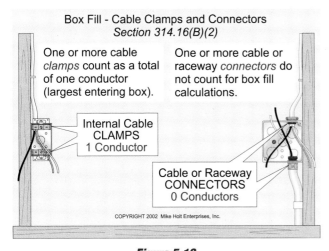

Box Fill - Cable Clamps and Connectors
Section 314.16(B)(2)

One or more cable
clamps count as a total
of one conductor
(largest entering box).

One or more cable or
raceway *connectors* do
not count for box fill
calculations.

Internal Cable
CLAMPS
1 Conductor

Cable or Raceway
CONNECTORS
0 Conductors

COPYRIGHT 2002 Mike Holt Enterprises, Inc.

Figure 5-18

Box Fill – Cable Clamps and Connectors

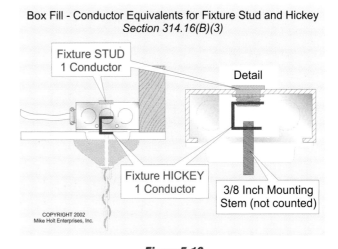

Box Fill - Conductor Equivalents for Fixture Stud and Hickey
Section 314.16(B)(3)

Fixture STUD
1 Conductor

Detail

Fixture HICKEY
1 Conductor

3/8 Inch Mounting
Stem (not counted)

COPYRIGHT 2002
Mike Holt Enterprises, Inc.

Figure 5-19

Box Fill – Conductor Equivalents for Fixture Stud and Hickey

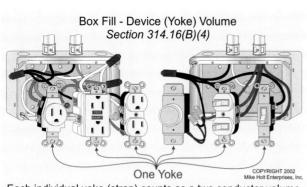

Box Fill - Device (Yoke) Volume
Section 314.16(B)(4)

One Yoke

COPYRIGHT 2002
Mike Holt Enterprises, Inc.

Each individual yoke (strap) counts as a two conductor volume, based on the largest conductor connected on that device.

Figure 5-20
Box Fill – Device Volume

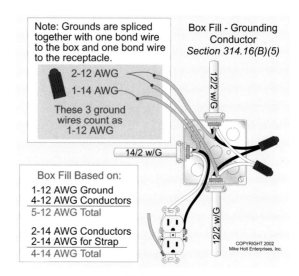

Note: Grounds are spliced together with one bond wire to the box and one bond wire to the receptacle.

Box Fill - Grounding Conductor
Section 314.16(B)(5)

2-12 AWG
1-14 AWG

These 3 ground wires count as 1-12 AWG

14/2 w/G

12/2 w/G

12/2 w/G

COPYRIGHT 2002
Mike Holt Enterprises, Inc.

Box Fill Based on:
1-12 AWG Ground
4-12 AWG Conductors
5-12 AWG Total

2-14 AWG Conductors
2-14 AWG for Strap
4-14 AWG Total

Figure 5-21
Box Fill – Grounding Conductor

What's not Counted

Wirenuts, cable connectors, raceway fittings and conductors that originate and terminate within the outlet box (such as equipment bonding jumpers and pigtails) are not counted for box fill calculations [314.16(A)].

❏ Number of Conductors

What are the total number of conductors used for box fill calculations in Figure 5–22?

(a) 5 conductors   (b) 7 conductors   (c) 9 conductors   (d) 11 conductors

• Answer: (d) 11 conductors

Switch: Five 14 AWG conductors, two conductors for device and three conductors terminating

Receptacle: Four 14 AWG conductors, two conductors for the device and two conductors terminating

Ground Wire: One conductor

Cable Clamps: One conductor

5–7 SIZING BOX – DIFFERENT SIZE CONDUCTORS [314.16(B)]

To determine the size of the outlet box when the conductors are of different sizes (insulation is not a factor), the following steps can be used:

Step 1:   Determine the number and size of conductors' equivalents in the box.

Step 2:   Determine the volume of the conductors' equivalents from Table 314.16(B).

Step 3:   Size the box by using Table 314.16(A).

Outlet Box Sizing

What size outlet box is required for 14/3 Type NM cable (with ground) that terminates on a switch, with 14/2 NM that terminates on a receptacle, if the box has internal cable clamps factory installed? Figure 5–22.

(a) 4 × 1¹/₄ square   (b) 4 × 1¹/₂ square   (c) 4 × 2¹/₈ square   (d) any of these

• Answer: (c) 4 × 2¹/₈ square

Step 1:   Determine the number and size of conductors.

14/3 NM	3 – 14 AWG
14/2 NM	2 – 14 AWG
Cable Clamps	1 – 14 AWG
Switch	2 – 14 AWG
Receptacles	2 – 14 AWG
Ground wires	1 – 14 AWG
Total	11 – 14 AWG

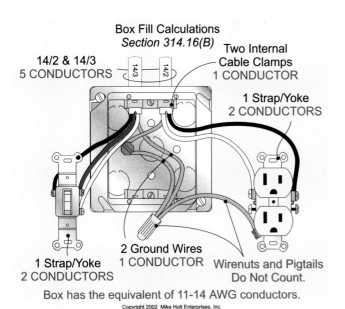

Box Fill Calculations
Section 314.16(B)

14/2 & 14/3
5 CONDUCTORS

14/3 14/2

Two Internal
Cable Clamps
1 CONDUCTOR

1 Strap/Yoke
2 CONDUCTORS

1 Strap/Yoke
2 CONDUCTORS

2 Ground Wires
1 CONDUCTOR

Wirenuts and Pigtails
Do Not Count.

Box has the equivalent of 11-14 AWG conductors.

Copyright 2002 Mike Holt Enterprises, Inc.

Figure 5-22
Box Fill Deductions

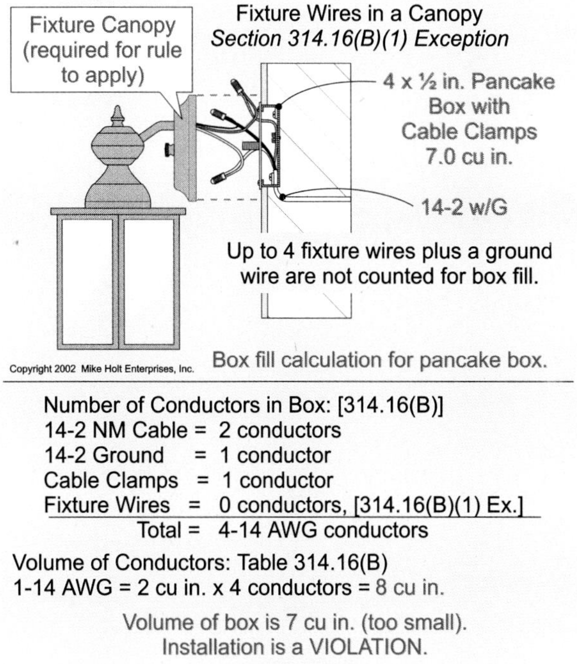

Fixture Canopy
(required for rule
to apply)

Fixture Wires in a Canopy
Section 314.16(B)(1) Exception

4 x ½ in. Pancake
Box with
Cable Clamps
7.0 cu in.

14-2 w/G

Up to 4 fixture wires plus a ground
wire are not counted for box fill.

Copyright 2002 Mike Holt Enterprises, Inc.   Box fill calculation for pancake box.

Number of Conductors in Box: [314.16(B)]
14-2 NM Cable = 2 conductors
14-2 Ground   = 1 conductor
Cable Clamps   = 1 conductor
Fixture Wires = 0 conductors, [314.16(B)(1) Ex.]
       Total = 4-14 AWG conductors

Volume of Conductors: Table 314.16(B)
1-14 AWG = 2 cu in. x 4 conductors = 8 cu in.

Volume of box is 7 cu in. (too small).
Installation is a VIOLATION.

Figure 5-23
Fixture Wires in a Canopy

Step 2:   Determine the volume of the conductors, [Table 314.16(B)].
14 AWG = 2 cu in.
14 AWG conductors volume = 11 wires × 2 cu in. = 22 cu in.

Step 3:   Select the outlet box from Table 314.16(A).
4 × 1½ square, 21 cu in. - too small
4 × 2⅛ square, 30.3 cu in. - just right

❏ **Domed Fixture Canopy [314.16(B)(1), Ex.].**

A round 4 × ½ in. box has a total volume of 7 cubic inches and has factory internal cable clamps. Can this pancake box be used with a lighting luminaire that has a domed canopy? The branch-circuit wiring is 14/2 nonmetallic-sheath cable and the luminaire has three fixture wires and one ground wire all smaller than 14 AWG. Figure 5–23.

(a) Yes   (b) No

    • Answer: (b) No

The box is limited to 7 cu in., and the conductors total 8 cu in. [314.16(B)(1) Ex.].

Step 1:   Determine the number and size of conductors within the box.
14/2 NM   2 – 14 AWG
Cable clamps 1 – 14 AWG
Ground wire   1 – 14 AWG
Total   4 – 14 AWG conductors

Step 2:   Determine the volume of the conductors [Table 314.16(B)].
14 AWG = 2 cu in.
Four 14 AWG conductors = 4 wires × 2 cu in. = 8 cu in.

❏ **Conductors Added to Existing Box**

How many 14 AWG THHN conductors can be pulled through a 4 × 2⅛ square box that has a plaster ring of 3.6 cu in.? The box already contains two receptacles, five 12 AWG THHN conductors and one 12 AWG bare grounding conductor. Figure 5–24.

(a) 4 conductors   (b) 5 conductors   (c) 6 conductors   (d) 7 conductors

    • Answer: (b) 5 conductors

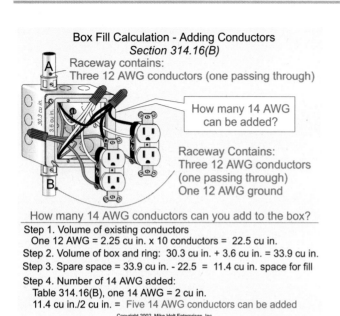

Box Fill Calculation - Adding Conductors
Section 314.16(B)
Raceway contains:
Three 12 AWG conductors (one passing through)

How many 14 AWG can be added?

Raceway Contains:
Three 12 AWG conductors (one passing through)
One 12 AWG ground

How many 14 AWG conductors can you add to the box?
Step 1. Volume of existing conductors
   One 12 AWG = 2.25 cu in. x 10 conductors = 22.5 cu in.
Step 2. Volume of box and ring: 30.3 cu in. + 3.6 cu in. = 33.9 cu in.
Step 3. Spare space = 33.9 cu in. - 22.5 = 11.4 cu in. space for fill
Step 4. Number of 14 AWG added:
   Table 314.16(B), one 14 AWG = 2 cu in.
   11.4 cu in./2 cu in. = Five 14 AWG conductors can be added
Copyright 2002 Mike Holt Enterprises, Inc.

Figure 5-24
Box Fill Calculation – Adding Conductors

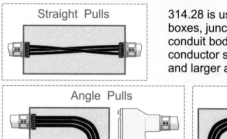
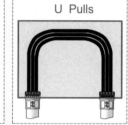

Pull and Junction Boxes - 4 AWG and Larger
Section 314.28

Straight Pulls

314.28 is used to size pull boxes, junction boxes and conduit bodies when conductor sizes 4 AWG and larger are used.

Angle Pulls

U Pulls

COPYRIGHT 2002 Mike Holt Enterprises

Figure 5-25
Pull and Junction Boxes

Step 1: Determine the number and size of the existing conductors.
   Two Receptacles   4 – 12 AWG conductors (2 yokes × 2 conductors)
   Five 12 AWG's   5 – 12 AWG conductors
   One ground   1 – 12 AWG conductor
   Total   10 – 12 AWG conductors

Step 2: Determine the volume of the existing conductors [Table 314.16(B)].
   12 AWG conductor = 2.25 cu in., 10 wires × 2.25 cu in. = 22.5 cu in.

Step 3: Determine the space remaining for the additional 14 AWG conductors.
   Remaining space = Total space less existing conductors
   Total space = 30.3 cu in. (box) [Table 314.16(A)] + 3.6 cu in. (ring) = 33.9 cu in.
   Remaining space = 33.9 cu in. – 22.5 cu in. (ten 12 AWG conductors)
   Remaining space = 11.4 cu in.

Step 4: Determine the number of 14 AWG conductors permitted in the spare space.
   Conductors added = Remaining space/added conductors volume
   Conductors added = 11.4 cu in./2 cu in. [Table 314.16(B)]
   Conductors added = 5

PART C – PULL, JUNCTION BOXES, AND CONDUIT BODIES

INTRODUCTION

Pull boxes, junction boxes, and *conduit bodies* must be sized to permit conductors to be installed so that the conductor insulation will not be damaged. For conductors 4 AWG and larger, pull boxes, junction boxes, and conduit bodies must be sized in accordance with 314.28. Figure 5–25.

5–8 PULL AND JUNCTION BOX SIZE CALCULATIONS

Straight-Pull Calculation [314.28(A)(1)]

A straight-pull calculation applies when conductors enter one side of a box and leave through the opposite wall of the box. The minimum distance from where the raceway enters to the opposite wall must not be less than eight times the trade size of the largest raceway. Figure 5–26.

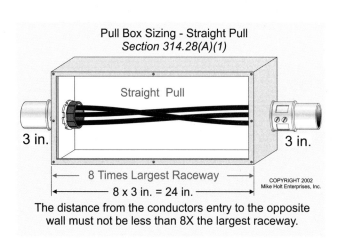

Pull Box Sizing - Straight Pull
Section 314.28(A)(1)

Straight Pull

3 in.   3 in.

8 Times Largest Raceway
8 x 3 in. = 24 in.

COPYRIGHT 2002
Mike Holt Enterprises, Inc.

The distance from the conductors entry to the opposite wall must not be less than 8X the largest raceway.

Figure 5-26
Sizing Junction/Pull Boxes for Straight Conductor Pulls

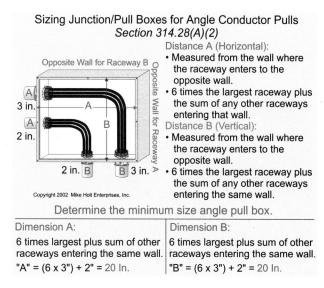

Sizing Junction/Pull Boxes for Angle Conductor Pulls
Section 314.28(A)(2)

Opposite Wall for Raceway B

Distance A (Horizontal):
• Measured from the wall where the raceway enters to the opposite wall.
• 6 times the largest raceway plus the sum of any other raceways entering that wall.

Distance B (Vertical):
• Measured from the wall where the raceway enters to the opposite wall.
• 6 times the largest raceway plus the sum of any other raceways entering the same wall.

Copyright 2002 Mike Holt Enterprises, Inc.

Determine the minimum size angle pull box.

Dimension A:	Dimension B:
6 times largest plus sum of other raceways entering the same wall.	6 times largest plus sum of other raceways entering the same wall.
"A" = (6 x 3) + 2" = 20 In.	"B" = (6 x 3") + 2" = 20 In.

Figure 5-27
Sizing Junction/Pull Boxes for Angle Conductor Pulls

Angle-Pull Calculation [314.28(A)(2)]

An angle-pull calculation applies when conductors enter one wall and leave the enclosure not opposite the wall of the conductor entry. The distance for angle-pull calculations from where the raceway enters to the opposite wall must not be less than six times the trade diameter of the largest raceway, plus the sum of the diameters of the remaining raceways on the same wall and *row*. Figure 5–27. When there is more than one row, each row shall be calculated separately and the row with the largest calculation shall be considered the minimum angle-pull dimension.

U-Pull Calculations [314.28(A)(2)]

A U-pull calculation applies when the conductors enter and leave from the same wall. The distance from where the raceways enter to the opposite wall must not be less than six times the trade diameter of the largest raceway, plus the sum of the diameters of the remaining raceways on the same wall. Figure 5–28.

Distance Between Raceways Containing The Same Conductor Calculation [314.28(A)(2)]

After sizing the pull box, the raceways must be installed so that the distance between raceways enclosing the same conductors shall not be less than six times the trade diameter of the largest raceway. This distance is measured from the nearest edge of one raceway to the nearest edge of the other raceway. Figures 5–28 and 5–29.

5–9 DEPTH OF BOX AND CONDUIT BODY SIZING [314.28(A)(2), Ex.]

When conductors enter an enclosure opposite a *removable cover*, such as the back of a pull box or conduit body, the distance from where the conductors enter to the removable cover shall not be less than the distances listed in Table 312.6(A); one wire per terminal. Figure 5–30.

❏ Depth Of Pull Or Junction Box

A 24 × 24 in. pull box has two 2 in. conduits that enter the back of the box with 4/0 AWG conductors. What is the minimum depth of the box?

(a) 4 in.   (b) 6 in.
(c) 8 in.   (d) 10 in.
  • Answer: (a) 4 in., Table 312.6(A)

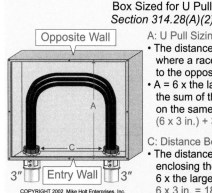

Box Sized for U Pull
Section 314.28(A)(2)

Opposite Wall

A: U Pull Sizing:
• The distance for a U Pull is from where a raceway enters a box to the opposite wall.
• A = 6 x the largest raceway plus the sum of the other raceways on the same wall.
(6 x 3 in.) + 3 in. = 21 in.

C: Distance Between Raceways
• The distance between raceways enclosing the same conductor is 6 x the largest raceway.
6 x 3 in. = 18 in.

3"   Entry Wall   3"

COPYRIGHT 2002 Mike Holt Enterprises, Inc.

Figure 5-28
Sizing Junction/Pull Boxes for U-Pulls

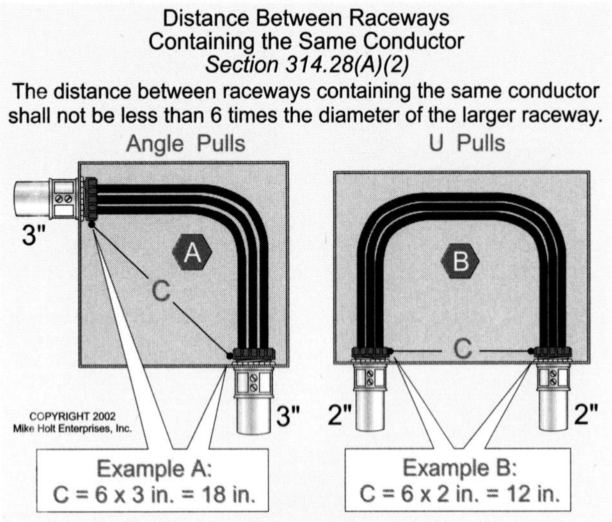

Distance Between Raceways
Containing the Same Conductor
Section 314.28(A)(2)
The distance between raceways containing the same conductor shall not be less than 6 times the diameter of the larger raceway.

Angle Pulls   U Pulls

3"

A

C

C

3"   2"   C   2"

COPYRIGHT 2002
Mike Holt Enterprises, Inc.

Example A: Example B:
C = 6 x 3 in. = 18 in.   C = 6 x 2 in. = 12 in.

Figure 5-29
Distance Between Raceways Containing the
Same Conductor

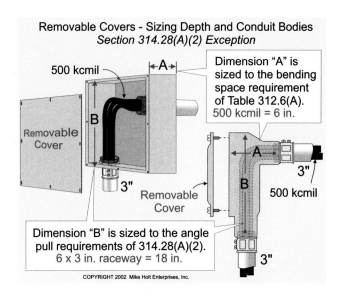

Removable Covers - Sizing Depth and Conduit Bodies
Section 314.28(A)(2) Exception

500 kcmil

A

Dimension "A" is sized to the bending space requirement of Table 312.6(A). 500 kcmil = 6 in.

B

Removable Cover

A

3"

B

3"

Removable Cover

500 kcmil

Dimension "B" is sized to the angle pull requirements of 314.28(A)(2). 6 x 3 in. raceway = 18 in.

3"

COPYRIGHT 2002 Mike Holt Enterprises, Inc.

Figure 5-30
Removable Covers – Sizing Depth and
Conduit Bodies

5–10 JUNCTION AND PULL BOX SIZING TIPS

When sizing pull and junction boxes, the following suggestions should be helpful:

Step 1:   Always draw out the problem.

Step 2:   Calculate the HORIZONTAL distance(s):
   • Left to right straight calculation
   • Left to right angle or U-pull calculation
   • Right to left straight calculation
   • Right to left angle or U-pull calculation

Step 3:   Calculate the VERTICAL distance(s):
   • Top to bottom straight calculation
   • Top to bottom angle or U-pull calculation
   • Bottom to top straight calculation
   • Bottom to top angle or U-pull calculation

5–11 PULL BOX EXAMPLES

❏ Pull Box Sizing

A junction box contains two 3 in. raceways on the left side and one 3 in. raceway on the right side. The conductors from one of the 3 in. raceways on the left wall are pulled through a 3 in. raceway on the right wall. The conductors from the other 3 in. raceways on the left wall are pulled through a 3 in. raceway at the bottom of the pull box. Figure 5–31.

▷ Horizontal Dimension

What is the horizontal dimension of this box?
   (a) 18 in.   (b) 21 in.   (c) 24 in.   (d) 30 in.
   • Answer: (c) 24 inches, 314.28. Figure 5-31, Part A

   Left to right angle pull   (6 × 3 in.) + 3 in. = 21 in.

   Left to right straight pull   8 × 3 in. = 24 in.

   Right to left angle pull   No calculation

   Right to left straight pull   8 × 3 in. = 24 in.

⇨ **Vertical Dimension**

What is the vertical dimension of this box?

(a) 18 in.   (b) 21 in.

(c) 24 in.   (d) 30 in.

- Answer: (a) 18 in.,
  314.28, Figure 5–31, Part B

Top to bottom angle	No calculation
Top to bottom straight	No calculation
Bottom to top angle	6 × 3 in. = 18 in.
Bottom to top straight	No calculation

⇨ **Distance Between Raceways**

What is the minimum distance between the two 3 in. raceways that contain the same conductors?

(a) 18 in.   (b) 21 in.

(c) 24 in.   (d) none of these

- Answer: (a) 18 in., 6 × 3 in.,
  314.28. Figure 5–31, Part C

❏ **Pull Box Sizing**

A pull box contains two 4 in. raceways on the left side and two 2 in. raceways on the top.

⇨ **Horizontal Dimension**

What is the horizontal dimension of the box?

(a) 28 in.   (b) 21 in.   (c) 24 in.   (d) none of these

- Answer: (a) 28 in., 314.28(A)(2)

Left to right angle pull	(6 × 4 in.) + 4 in. = 28 in.
Left to right straight pull	No calculation
Right to left angle pull	No calculation
Right to left straight pull	No calculation

⇨ **Vertical Dimension**

What is the vertical dimension of the box?

(a) 18 in.   (b) 21 in.   (c) 24 in.   (d) 14 in.

- Answer: (d) 14 in., 314.28(A)(2)

Top-to-bottom angle	(6 × 2 in.) + 2 in. = 14 in.
Top-to-bottom straight	No calculation
Bottom-to-top angle	No calculation
Bottom-to-top straight	No calculation

⇨ **Distance Between Raceways**

What is the minimum distance between the two 4 in. raceways that contain the same conductors?

(a) 18 in.   (b) 21 in.   (c) 24 in.   (d) none of these

- Answer: (c) 24 in., 314.28(A)(2)

6 × 4 inches = 24 in.

Pull (Junction) Box Sizing - *Section 314.28*

314.28 is used to size junction boxes for conductor sizes 4 AWG and larger and the distance between conductors containing the same conductor.

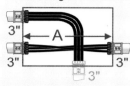

A: Horizontal Dimension
Straight Pull:
Left to Right: 8 x 3 in. = 24 in.
Right to Left: 8 x 3 in. = 24 in.
Angle Pull:
Left to Right: (6 x 3 in.) + 3 in. = 21 in.
Right to Left: No Calculation
Largest Calculation = 24 in.

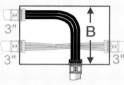

B: Vertical Dimension
Straight Pull:
Top to Bottom: No Calculation
Bottom to Top: No Calculation
Angle Pull:
Top to Bottom: (No Calculation)
Bottom to Top: 6 x 3 in. = 18 in.
Largest Calculation = 18 in.

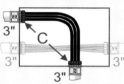

C: Distance Between Raceways
(Containing the same conductor)
Angle Pull is the only application
6 x 3 in. = 18 in.

COPYRIGHT 2002 Mike Holt Enterprises, Inc.

Figure 5-31

Pull (Junction) Box Sizing

Unit 5 – Raceway Fill, Box Fill, Junction Boxes and Conduit Bodies
Summary Questions

Calculations Questions

Part A – Raceway Fill

5–1 Understanding The National Electrical Code, Chapter 9

1. When all the conductors are the same size and insulation, the number of conductors permitted in a raceway can be determined simply by looking at the Tables listed in _____.
    (a) Chapter 9   (b) Annex B   (c) Annex C   (d) Annex D

2. When equipment grounding conductors are installed in a raceway, the actual area of the conductor shall be used when calculating raceway fill.
    (a) True   (b) False

3. When a raceway does not exceed 24 in. in length, the raceway is permitted to be filled to _____ of its cross-sectional area.
    (a) 53%   (b) 31%   (c) 40%   (d) 60%

4. How many 16 AWG TFFN conductors can be installed in $3/4$ in. electrical metallic tubing?
    (a) 40   (b) 26   (c) 30   (d) 29

5. How many 6 AWG RHH (without outer cover) can be installed in $1^1/4$ in. electrical nonmetallic tubing?
    (a) 25   (b) 16   (c) 13   (d) 7

6. How many 1/0 AWG XHHW can be installed in 2 in. flexible metal conduit?
    (a) 7   (b) 6   (c) 16   (d) 13

7. How many 12 AWG RHH (with outer cover) can be installed in a 1 in. IMC raceway?
    (a) 7   (b) 11   (c) 5   (d) 4

8. If we have a 2 in. rigid metal conduit and we want to install three THHN compact conductors, what is the largest compact conductor permitted to be installed?
    (a) 4/0 AWG   (b) 250 kcmil   (c) 300 kcmil   (d) 500 kcmil

9. The actual area of conductor fill is dependent on the raceway size and the number of conductors installed. If there are three or more conductors installed in a raceway, the total area of conductor fill is limited to _____.
    (a) 53%   (b) 31%   (c) 40%   (d) 60%

10. • What is the cross-sectional area in sq in. for 10 AWG THW?
    (a) 0.0333   (b) 0.0172   (c) 0.0252   (d) 0.0278

11. What is the area in sq in. for a 14 AWG RHW (without cover)?
    (a) 0.0209   (b) 0.0172   (c) 0.0252   (d) 0.0278

12. What is the cross-sectional area in sq in. for 10 AWG THHN?
    (a) 0.0117   (b) 0.0172   (c) 0.0252   (d) 0.0211

13. What is the cross-sectional area in sq in. for 12 AWG RHH (with outer cover)?
    (a) 0.0117   (b) 0.0353   (c) 0.0252   (d) 0.0327

14. • What is the cross-sectional area in sq in. for an 8 AWG bare solid conductor?
    (a) 0.013   (b) 0.027   (c) 0.038   (d) 0.045

5–2 Raceway and Nipple Calculations

15. The number of conductors permitted in a raceway is dependent on _____.
(a) the area of the raceway
(b) the percent area fill as listed in Chapter 9 Table 1
(c) the area of the conductors as listed in Chapter 9 Table 5 and 8
(d) all of these

16. A 200A feeder installed in Schedule 80 rigid nonmetallic conduit has three 3/0 AWG THHN, one 2 AWG THHN and one 6 AWG THHN. What size raceway is required?
(a) 2 in.   (b) $2^1/_2$ in.   (c) 3 in.   (d) $3^1/_2$ in.

17. What size rigid metal nipple is required for three 4/0 AWG THHN, one 1/0 AWG THHN, and one 4 AWG THHN?
(a) $1^1/_2$ in.   (b) 2 in.   (c) $2^1/_2$ in.   (d) none of these

5–3 Existing Raceway Calculation

18. • An existing rigid metal nipple contains four 10 AWG THHN and one 10 AWG (bare stranded) ground wire in $^3/_4$ in. rigid metal conduit. How many additional 10 AWG THHN conductors can be installed?
(a) 5   (b) 7   (c) 9   (d) 11

Part B – Outlet Box Fill Calculations

5–5 Sizing Box - Conductors All The Same Size [Table 314.16]

19. What size box is required for six 14 AWG THHN and three 14 AWG THW?
(a) $4 \times 1^1/_4$ square   (b) $4 \times 1^1/_2$ round   (c) $4 \times 1^1/_4$ round   (d) none of these

20. How many 10 AWG THHN conductors are permitted in a $4 \times 1^1/_2$ square box?
(a) 8 conductors   (b) 9 conductors   (c) 10 conductors   (d) 11 conductors

5–6 Conductor Equivalents [314.16]

21. Table 314.16(A) does not take into consideration the volume of _____.
(a) switches and receptacles   (b) luminaire studs and hickeys
(c) internal cable clamps   (d) all of these

22. When determining the number of conductors for box fill calculations, which of the following statements are true?
(a) A luminaire stud or hickey is considered as one conductor for each type, based on the largest conductor that enters the outlet box.
(b) Internal factory cable clamps are considered as one conductor for one or more cable clamps, based on the largest conductor that enters outlet box.
(c) The device yoke is considered as two conductors, based on the largest conductor that terminates on the strap (device mounting fitting).
(d) all of these

23. • When determining the number of conductors for box fill calculations, which of the following statements are true?
(a) Each conductor that runs through the box without loop (without splice) is considered as one conductor.
(b) Each conductor that originates outside the box and terminates in the box is considered as one conductor.
(c) Wirenuts, cable connectors, raceway fittings and conductors that originate and terminate within the outlet box (equipment bonding jumpers and pigtails) are not counted for box fill calculations.
(d) all of these

24. It is permitted to omit one equipment grounding conductor and not more than _____ that enter a box from a luminaire canopy.
   (a) four fixture wires   (b) four 16 AWG fixture wires
   (c) four 18 AWG fixture wires   (d) b and c

25. Can a round $4 \times 1/2$ in. box marked as 8 cu in. with manufactured cable clamps supplied with 14/2 NM be used with a luminaire that has two 18 AWG TFN and a canopy cover?
   (a) Yes   (b) No

5–7 Sizing Box - Different Size Conductors [314.16(B)]

26. What size outlet box is required for one 12/2 NM cable that terminates on a switch, one 12/3 NM cable that terminates on a receptacle, and the box has manufactured cable clamps?
   (a) $4 \times 1^1/_4$ square   (b) $4 \times 1^1/_2$ square
   (c) $4 \times 2^1/_8$ square   (d) none of these

27. • How many 14 AWG THHN conductors can be pulled through a $4 \times 1^1/_2$ square box with a plaster ring of 3.6 cu in.? The box contains two duplex receptacles, five 14 AWG THHN conductors and two grounding conductors.
   (a) 1   (b) 2   (c) 3   (d) 4

Part C – Pull, Junction Boxes, and Conduit Bodies

5–8 Pull and Junction Box Size Calculations

28. When conductors 4 AWG and larger are installed in boxes and conduit bodies we must size the enclosure according to which of the following requirements?
   (a) The minimum distance for straight pull calculations from where the conductors enter to the opposite wall must not be less than eight times the trade size of the largest raceway.
   (b) The distance for angle-pull calculations from the raceway entry to the opposite wall must not be less than six times the trade diameter of the largest raceway, plus the sum of the diameters of the remaining raceways on the same wall and row.
   (c) The distance between raceways enclosing the same conductor(s) shall not be less than six times the trade diameter of the largest raceway.
   (d) all of the above are correct

29. When conductors enter an enclosure opposite a removable cover, the distance from where the conductors enter to the removable cover shall not be less than _____.
   (a) six times the largest raceway
   (b) eight times the largest raceway
   (c) a or b
   (d) none of these

The following information applies to the next three questions.

A junction box contains two $2^1/_2$ in. raceways on the left side and one $2^1/_2$ in. raceway on the right side. The conductors from one $2^1/_2$ in. raceway (left wall) are pulled through the raceway on the right wall. The other $2^1/_2$ in. raceway conductors (on the side) are pulled through a $2^1/_2$ in. raceway at the bottom of the pull box.

30. What is the distance from the left wall to the right wall?
   (a) 18 in.   (b) 21 in.   (c) 24 in.   (d) 20 in.

31. • What is the distance from the bottom wall to the top wall?
   (a) 18 in.   (b) 21 in.   (c) 24 in.   (d) 15 in.

32. What is the distance between the raceways that contain the same conductors?
   (a) 18 in.   (b) 21 in.   (c) 24 in.   (d) 15 in.

The following information applies to the next three questions.

A junction box contains two 2 in. raceways on the left side and two 2 in. raceways on the top.

33. What is the distance from the left wall to the right wall?
    (a) 28 in.   (b) 21 in.   (c) 24 in.   (d) 14 in.

34. What is the distance from the bottom wall to the top wall?
    (a) 18 in.   (b) 21 in.   (c) 24 in.   (d) 14 in.

35. What is the distance between the 2 in. raceways that contain the same conductors?
    (a) 18 in.   (b) 21 in.   (c) 24 in.   (d) 12 in.

☆ Challenge Questions

Part A – Raceway Calculations

36. • A 3 in. Schedule 40 PVC raceway contains seven 1 AWG RHW conductors without outer cover. How many 2 AWG THW conductors may be installed in this raceway with the existing conductors?
    (a) 11   (b) 15   (c) 20   (d) 25

Part B – Box Fill Calculations

37. • Determine the minimum cubic inches required for two 10 AWG TW passing through a box, four 14 AWG THHN terminating, two 12 AWG TW terminating to a receptacle and one 12 AWG equipment bonding jumper from the receptacle to the box.
    (a) 18.5 cu in.   (b) 22 cu in.   (c) 20 cu in.   (d) 21.75 cu in.

38. • When determining the number of conductors in a box fill that has two 18 AWG fixture wires, one 14/3 nonmetallic-sheathed cable with ground, one duplex switch and two cable clamps, the count will equal _____.
    (a) 7 conductors   (b) 8 conductors   (c) 10 conductors   (d) 6 conductors

Part C – Pull and Junction Box Calculations

The Figure 5–32 applies to the next four questions.

39. The minimum horizontal dimension for the junction box shown in the diagram in Figure 5-32 is _____.
    (a) 21 in.   (b) 18 in.   (c) 24 in.   (d) 20 in.

40. • The minimum vertical dimension for the junction box shown in Figure 5-32 is _____.
    (a) 16 in.   (b) 18 in.   (c) 20 in.   (d) 24 in.

41. The minimum distance between the two 2 in. raceways that contain the same conductor C, as illustrated in Figure 5-32 is _____.
    (a) 12 in.   (b) 18 in.
    (c) 24 in.   (d) 30 in.

42. • According to Figure 5-32, if a 3 in. raceway entry (250 kcmil) is in the wall opposite of a removable cover, the distance from that wall to the cover must not be less than _____.
    (a) 4 in.   (b) 4$^1/_2$ in.
    (c) 5 in.   (d) 6 in.

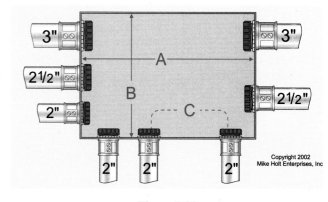

Figure 5-32

NEC Questions – Articles 250-314

Article 250 Grounding

43. When a single equipment grounding conductor is used for multiple circuits in the same raceway or cable, the single equipment grounding conductor shall be sized according to _____.
    (a) the combined rating of all the overcurrent protection devices
    (b) the largest overcurrent protection device of the multiple circuits
    (c) the combined rating of all the loads
    (d) any of these

44. Where conductors are run in parallel in multiple raceways or cables, the equipment grounding conductor, where used, shall be run in parallel in each raceway or cable.
    (a) True   (b) False

45. The terminal for the connection of the equipment grounding conductor shall be identified by a green-colored, _____.
    (a) not readily removable terminal screw with a hexagonal head.
    (b) hexagonal, not readily removable terminal nut.
    (c) pressure wire connector.
    (d) any of these

46. A grounding-type receptacle can replace a nongrounding-type receptacle at an outlet box that does not contain an equipment grounding conductor if the equipment grounding conductor is connected to the _____.
    (a) grounding electrode system as described in 250.50
    (b) grounding electrode conductor
    (c) equipment grounding terminal bar within the enclosure where the branch circuits for the receptacle or branch circuit originates
    (d) any of these

47. When considering whether equipment is effectively grounded, the structural metal frame of a building shall be permitted to be used as the required equipment grounding conductor for ac equipment.
    (a) True   (b) False

48. A grounded circuit conductor shall not be used for grounding noncurrent-carrying metal parts of equipment on the load side of _____.
    (a) the service disconnecting means
    (b) the separately derived system disconnecting means
    (c) overcurrent protection devices for separately derived systems not having a main disconnecting means
    (d) all of these

49. On the load side of the service disconnecting means, the _____ circuit conductor is permitted to ground meter enclosures if all meter enclosures are located near the service disconnecting means and no service ground-fault protection is installed.
    (a) grounding   (b) bonding   (c) grounded   (d) phase

50. An _____ shall be used to connect the grounding terminal of a grounding-type receptacle to a grounded box.
    (a) equipment bonding jumper   (b) equipment grounding jumper
    (c) a or b   (d) a and b

51. An equipment bonding jumper shall be used to connect the grounding terminal of a grounding-type receptacle to a grounded box. Where the box is surface mounted, direct metal-to-metal contact between the device yoke and the box shall be permitted to ground the receptacle to the box.
     (a) True   (b) False

52. An equipment bonding jumper for a grounding-type receptacle shall be installed between the receptacle and a flush-mounted outlet box, even when there's direct metal-to-metal contact between the metal yoke of the receptacle and the metal box.
     (a) True   (b) False

53. Contact devices or yokes designed and listed for the purpose shall be permitted in conjunction with the supporting screws to establish the grounding circuit between the device yoke and flush-type boxes.
     (a) True   (b) False

54. Equipment bonding jumpers are not required for listed wiring devices that have mounting screws that provide the grounding continuity between the metal yoke and the flush box (self-grounding receptacles).
     (a) True   (b) False

55. Where circuit conductors are spliced within a box, or terminated on equipment within or supported by a box, any separate equipment grounding conductors associated with those circuit conductors shall be spliced or joined within the box or to the box with devices suitable for the use.
     (a) True   (b) False

56. • Where one or more equipment grounding conductor enters a box, all equipment grounding conductors shall be spliced together (in the enclosure). This does not apply to insulated equipment grounding conductors for isolated ground receptacles for electronic equipment.
     (a) True   (b) False

57. When equipment grounding conductor(s) are installed in a metal box, an electrical connection is required between the equipment grounding conductor and the metal box enclosure by means of a _____.
     (a) grounding screw   (b) soldered connection
     (c) listed grounding device   (d) a or c

58. • When the dc system consists of a _____, the grounding conductor shall not be smaller than the grounded conductor.
     (a) 2-wire balancer set   (b) 3-wire balancer set
     (c) balancer winding with overcurrent protection   (d) b or c

59. The secondary circuits of current and potential instrument transformers shall be grounded where the primary windings are connected to circuits of _____ or more to ground and, where on switchboards, shall be grounded irrespective of voltage.
     (a) 300V   (b) 600V   (c) 1,000V   (d) 150V

60. • Cases or frames of instrument transformers are not required to be grounded _____.
     (a) when accessible to qualified persons only
     (b) for current transformers where the primary is not over 150 volts-to-ground and that are used exclusively to supply current to meters
     (c) for potential transformers where the primary is less than 150 volts-to-ground
     (d) a or b

61. The grounding conductor for secondary circuits of instrument transformers and for instrument cases shall not be smaller than _____ AWG copper.
     (a) 18   (b) 16   (c) 14   (d) 12

Article 280 Surge Arresters

62. A surge arrester is a protective device for limiting surge voltages by _____ or bypassing surge current.
     (a) decreasing   (b) discharging   (c) limiting   (d) derating

63. Line and ground-connecting conductors for a surge arrester shall not be smaller than _____ AWG.
    (a) 14   (b) 12   (c) 10   (d) 8

64. The conductor between a lightning surge arrester and the line and the grounding connection shall not be smaller than _____ AWG copper for installations operating at 1 kV or more.
    (a) 4   (b) 6   (c) 8   (d) 2

Article 285 Transient Voltage Surge Suppressors: TVSSs

65. Article 285 covers surge arresters.
    (a) True   (b) False

66. The scope of Article 285 applies to devices such as cord-and-plug-connected TVSS units or receptacles or appliances that have integral TVSS protection.
    (a) True   (b) False

67. A TVSS is listed to limit transient voltages by diverting or limiting surge current.
    (a) True   (b) False

68. A TVSS device must be listed.
    (a) True   (b) False

69. TVSSs shall be marked with their short-circuit current rating, and they cannot be installed where the available fault current is in excess of that rating.
    (a) True   (b) False

70. The conductors for the TVSS cannot be any longer than _____, and unnecessary bends should be avoided.
    (a) 6 in.   (b) 12 in.   (c) 18 in.   (d) none of these

71. A TVSS can be connected anywhere on the premises wiring system.
    (a) True   (b) False

Chapter 3 Wiring Methods and Materials

Article 300 Wiring Methods

72. Metric designators and trade sizes for conduit, tubing and associated fittings and accessories are designated in Table _____.
    (a) 250.66   (b) 250-122   (c) 300.1(C)   (d) 310.16

73. Unless specified elsewhere in the *Code*, Chapter 3 shall be used for voltages of _____.
    (a) 600 volts-to-ground or less
    (b) 300V between conductors or less
    (c) 600V, nominal, or less
    (d) 600V RMS

74. Single conductors are only permitted when installed as part of a wiring method as listed in Chapter _____ of the *NEC*.
    (a) 4   (b) 3   (c) 2   (d) 9

75. All conductors of a circuit, including the grounded and equipment grounding conductors, shall be contained within the same _____.
    (a) raceway   (b) cable   (c) trench   (d) all of these

76. • Circuit conductors that operate at 277V (with 600V insulation) may occupy the same enclosure or raceway with 48V dc conductors that have an insulation rating of 300V.
    (a) True   (b) False

77. Cables laid in wood notches require protection against nails or screws by using a steel plate at least _____ thick, installed before the building finish is applied.
    (a) $1/16$ in.   (b) $1/8$ in.   (c) $1/2$ in.   (d) none of these

78. Where NM cables pass through cut or drilled slots or holes in metal members, the cable needs to be protected by _____ securely covering all metal edges fastened in the opening prior to installation of the cable.
    (a) listed bushings   (b) listed grommets   (c) plates   (d) a or b

79. Where nails or screws are likely to penetrate nonmetallic-sheathed cable or electrical nonmetallic tubing installed through metal framing members, a steel sleeve, steel plate or steel clip not less than _____ in thickness shall be used to protect the cable or tubing.
    (a) $1/16$ in.   (b) $1/8$ in.   (c) $1/2$ in.   (d) none of these

80. Wiring methods installed behind panels that allow access, such as the space above a dropped ceiling, are required to be _____ according to their applicable articles.
    (a) supported   (b) painted   (c) in a metal raceway   (d) all of these

81. Where NM cable passes through factory or field openings in metal members, it must be protected by _____ bushings or _____ grommets that cover metal edges. The protection fitting shall be securely fastened in the opening prior to installation of the cable.
    (a) approved   (b) identified   (c) listed   (d) none of these

82. Where underground conductors and cables emerge from underground, they shall be protected to a point _____ above finished grade. In no case shall the protection be required to exceed 18 in. below grade.
    (a) 3 ft   (b) 6 ft   (c) 8 ft   (d) 10 ft

83. The _____ is defined as the area between the top of direct-burial cable and the finished grade.
    (a) notch   (b) cover   (c) gap   (d) none of these

84. What is the minimum cover requirement in inches for UF cable installed outdoors and underground that is supplying power to a 120V, 30A circuit?
    (a) 6   (b) 12   (c) 18   (d) 24

85. Rigid metal conduit that is directly buried outdoors must have at least _____ of cover.
    (a) 6 in.   (b) 12 in.   (c) 18 in.   (d) 24 in.

86. • When installing raceways underground in rigid nonmetallic conduit and other approved raceways, there shall be a minimum of _____ of cover.
    (a) 6 in.   (b) 12 in.   (c) 18 in.   (d) 22 in.

87. • What is the minimum cover requirement in inches for UF cable supplying power to a 120V, 15A GFCI-protected circuit outdoors under a driveway of a one-family dwelling?
    (a) 12   (b) 24   (c) 16   (d) 6

88. UF cable used with a 24V landscape lighting system is permitted to have a minimum cover of _____.
    (a) 6 in.   (b) 12 in.   (c) 18 in.   (d) 24 in.

89. Service conductors that are not encased in concrete and buried 18 in. or more below grade shall have their location identified by a warning ribbon placed in the trench at least _____ above the underground installation.
    (a) 6 in.   (b) 12 in.   (c) 18 in.   (d) none of these

90. Direct-buried conductors or cables shall be permitted to be spliced or tapped without the use of splice boxes.
    (a) True   (b) False

91. Backfill used for underground wiring must not _____.
    (a) damage the wiring method   (b) prevent compaction of the fill
    (c) contribute to the corrosion of the raceway   (d) all of these

92.   _____ are required at one or both ends of the raceway where moisture could enter a raceway and contact energized live parts.
(a) Seals   (b) Plugs   (c) Bushings   (d) a and b

93.   • When installing direct-buried cables, a _____ shall be used at the end of a conduit that terminates underground.
(a) splice kit   (b) terminal fitting   (c) bushing   (d) b or c

94.   All conductors of a circuit are required to be _____.
(a) in the same raceway   (b) in close proximity in the same trench
(c) the same size   (d) a and b

95.   Cables or raceways installed using directional boring equipment shall be _____ for this purpose. This new requirement was necessary to ensure that proper raceways are used with directional boring equipment.
(a) marked   (b) listed   (c) labeled   (d) approved

96.   Metal raceways, cable armor, boxes, cable sheathing, cabinets, elbows, couplings, fittings, supports and support hardware shall be of materials suitable for _____.
(a) corrosive locations
(b) wet locations
(c) the environment in which they are to be installed
(d) none of these

97.   Which of the following metal parts shall be protected from corrosion both inside and out?
(a) Ferrous raceways   (b) Metal elbows   (c) Boxes   (d) all of these

98.   Where corrosion protection is necessary and the conduit is threaded in the field, the threads shall be coated with a(n) _____ electrically conductive, corrosion-resistance compound.
(a) marked   (b) listed   (c) labeled   (d) approved

99.   The provisions required for mounting conduits on indoor walls or in rooms that must be hosed down frequently is _____ between the mounting surface and the electrical equipment.
(a) a permanent $1/4$ in. airspace   (b) separated by insulated bushings
(c) separated by noncombustible tubing   (d) none of these

100.   In general, areas where _____ are handled and stored may present severe corrosive conditions, particularly when wet or damp.
(a) laboratory chemicals and acids   (b) acids and alkali chemicals
(c) acids and water   (d) chemicals and water

101.   Where portions of a cable raceway or sleeve are known to be subjected to different temperatures and where condensation is known to be a problem, as in cold storage areas of buildings or where passing from the interior to the exterior of a building, the _____ shall be filled with an approved material to prevent the circulation of warm air to a colder section of the raceway or sleeve.
(a) raceways   (b) sleeve   (c) a or b   (d) none of these

102.   Raceways shall be provided with expansion fittings where necessary to compensate for thermal expansion and contraction.
(a) True   (b) False

103.   • All metal raceways, cable armor, boxes, fittings, cabinets and other metal enclosures for conductors shall be _____ joined together to form a continuous electrical conductor.
(a) electrically   (b) permanently   (c) metallically   (d) none of the above

104.   The independent support wires shall be distinguishable from fire-rated suspended-ceiling framing support wires by _____.
(a) color   (b) tagging   (c) other effective means (d) any of these

105. Ceiling-support wires used for the support of electrical raceways and cables within nonfire-rated assemblies are required to be distinguishable from the suspended-ceiling framing support wires.
(a) True   (b) False

106. Conductors in raceways shall be _____ between outlet devices and there shall be no splice or tap within a raceway itself.
(a) continuous   (b) installed   (c) copper   (d) in conduit

107. In multiwire circuits, the continuity of the _____ conductor shall not be dependent upon the device connections.
(a) ungrounded   (b) grounded   (c) grounding   (d) a and b

108. An 8 × 8 × 4 in. deep junction/splice box requires 6 in. of free conductor, measured from the point in the box where the conductors enter the enclosure. The 3 in. outside-the-box rule _____.
(a) does apply   (b) does not apply   (c) sometimes applies   (d) none of these

109. When the opening to an outlet, junction, or switch point is less than 8 inches in any dimension, each conductor shall be long enough to extend at least _____ outside the opening of the enclosure.
(a) 0 in.   (b) 3 in.   (c) 6 in.   (d) 12 in.

110. Fittings and connectors shall be used only with the specific wiring methods for which they are designed and listed.
(a) True   (b) False

111. Splices and taps are permitted in cabinets or cutout boxes if the splice or tap does not fill the wiring space at any cross-section to more than _____ percent.
(a) 20   (b) 40   (c) 60   (d) 75

112. A bushing can be used instead of a box or terminal where conductors emerge from a raceway and enter or terminate at equipment, such as open switchboards, unenclosed control equipment, or similar equipment.
(a) True   (b) False

113. A _____ shall be permitted in lieu of a box or terminal fitting at the end of a conduit where the raceway terminates behind an unenclosed switchboard or similar equipment.
(a) bushing   (b) bonding bushing   (c) coupling   (d) connector

114. The number of conductors permitted in a raceway shall be limited to _____.
(a) permit heat to dissipate
(b) prevent damage to insulation during installation
(c) prevent damage to insulation during removal of conductors
(d) all of these

115. Raceways shall be _____ between pulling points prior to the installation of conductors.
(a) mechanically completed   (b) tested for ground faults
(c) a minimum of 80 percent completed   (d) none of these

116. Prewired raceway assemblies shall be permitted only where specifically permitted elsewhere in the *Code* for the applicable wiring method.
(a) True   (b) False

117. Metal raceways shall not be _____ by welding unless the raceway is specifically designed to be, or otherwise specifically permitted to be, in the *Code*.
(a) supported   (b) terminated   (c) connected   (d) all of these

118. A 100 ft vertical run of 4/0 AWG copper requires the conductors to be supported at _____ locations.
(a) 4   (b) 5   (c) 2   (d) none of these

119. A vertical run of 4/0 AWG copper shall be supported at intervals not exceeding _____.
(a) 80 ft   (b) 100 ft   (c) 120 ft   (d) 40 ft

120. Conductors in metal raceways and enclosures shall be so arranged as to avoid heating the surrounding metal by alternating-current induction. To accomplish this, the _____ conductor(s) shall be grouped together.
(a) phase   (b) neutral   (c) ungrounded   (d) all of these

121. _____ is a nonferrous, nonmagnetic metal that has no heating due to inductive hysteresis heating.
(a) Steel   (b) Iron   (c) Aluminum   (d) all of these

122. Electrical installations in hollow spaces, vertical shafts, and ventilation or air-handling ducts shall be made so that the possible spread of fire or products of combustion will not be _____.
(a) substantially increased   (b) allowed
(c) inherent   (d) possible

123. No wiring of any type shall be installed in ducts used to transport _____.
(a) dust   (b) flammable vapors   (c) loose stock   (d) all of these

124. Equipment and devices shall be permitted within ducts or plenum chambers used to transport environmental air only if necessary for their direct action upon, or sensing of the _____.
(a) contained air   (b) air quality   (c) air temperature   (d) none of these

125. Type AC cable can be used in ducts or plenums that are used for environmental air.
(a) True   (b) False

126. One wiring method that is permitted in ducts or plenums used for environmental air is _____.
(a) flexible metal conduit of any length   (b) electrical metallic tubing
(c) armored cable (Type AC)   (d) nonmetallic-sheathed cable

127. When equipment or devices are installed in ducts or plenum chambers used to transport environmental air, and illumination is necessary to facilitate maintenance and repair, enclosed _____-type luminaires shall be permitted.
(a) screw   (b) plug   (c) gasketed   (d) neon

128. The space above a dropped ceiling used for environmental air is considered as _____ and the wiring limitations of _____ apply.
(a) a plenum, 300.22(B)   (b) other spaces, 300.22(C)
(c) a duct, 300.22(B)   (d) none of these

129. • Wiring methods permitted in the drop ceiling area used for environmental air include _____.
(a) electrical metallic tubing
(b) flexible metal conduit of any length
(c) rigid metal conduit without an overall nonmetallic covering
(d) all of these

130. The air-handling area beneath raised floors for data-processing systems is not a plenum and is not required to comply with the requirements of 300.22 but shall be permitted in accordance with Article 645.
(a) True   (b) False

131. Wiring methods and equipment installed behind panels designed to permit access (such as drop ceilings) shall be so arranged and secured to permit the removal of panels to give access to electrical equipment.
(a) True   (b) False

132. Where installed in raceways, conductors _____ AWG and larger shall be stranded.
(a) 10   (b) 6   (c) 8   (d) 4

133. In general, the minimum size phase, neutral or grounded conductor permitted for use in parallel installations is _____ AWG.
(a) 10   (b) 1   (c) 1/0   (d) 4

134. Conductors smaller than 1/0 AWG can be connected in parallel to supply control power, provided _____.
(a) they are all contained within the same raceway or cable
(b) each parallel conductor has an ampacity sufficient to carry the entire load
(c) the circuit overcurrent protection device rating does not exceed the ampacity of any individual parallel conductor
(d) all of these

135. When conductors are run in parallel, the currents should be evenly divided between the individual parallel conductors so that each conductor is evenly heated. This is accomplished by ensuring that each of the conductors within a parallel set has the same_____ and all conductors terminate in the same manner.
(a) length   (b) material   (c) cross-sectional area   (d) all of these

136. It is not the intent of 310.4 to require that conductors of one phase, neutral or grounded circuit conductor be the same as those of another phase, neutral or grounded circuit conductor to achieve _____.
(a) polarity   (b) balance   (c) grounding   (d) none of these

137. The parallel conductors in each phase or grounded conductor shall _____.
(a) be the same length and conductor material
(b) have the same circular mil area and insulation type
(c) be terminated in the same manner
(d) all of these

138. The minimum size conductor permitted in any building for branch circuits under 600V is _____ AWG.
(a) 14   (b) 12   (c) 10   (d) 8

139. Solid dielectric insulated conductors operated above 2,000V in permanent installations shall have _____ insulation and shall be shielded.
(a) ozone-resistant   (b) asbestos   (c) high-temperature   (d) perfluoro-alkoxy

140. • Insulated conductors used in wet locations shall be _____.
(a) moisture-impervious metal-sheathed
(b) RHW, TW, THW, THHW, THWN, XHHW
(c) listed for wet locations
(d) any of these

141. Insulated conductors and cables exposed to the direct rays of the sun shall be _____.
(a) listed   (b) listed and marked sunlight resistant
(c) listed for sunlight resistance   (d) b or c

142. Where conductors of different insulation are associated together, the limiting temperature of any conductor shall not be exceeded.
(a) True   (b) False

143. The _____ rating of a conductor is the maximum temperature, at any location along its length, that the conductor can withstand over a prolonged period of time without serious degradation.
(a) ambient   (b) temperature   (c) maximum withstand   (d) short-circuit

144. There are four principal determinants of conductor operating temperature, one of which is _____that is generated internally in the conductor as the result of load current flow.
(a) friction   (b) magnetism   (c) heat   (d) none of these

145. Suffixes to designate the number of conductors within a cable are _____.
(a) D - Two insulated conductors laid parallel
(b) M - Two or more insulated conductors twisted spirally
(c) T - Two or more insulated conductors twisted in parallel
(d) a and b

146. Which conductor has a temperature rating of 90°C?
(a) RH   (b) RHW   (c) THHN   (d) TW

147. TFE-insulated conductors are manufactured in sizes from 14 through _____ AWG.
(a) 2   (b) 1   (c) 2/0   (d) 4/0

148. Conductors with thermoplastic and fibrous outer braid, such as TBS insulation, are used for wiring _____.
(a) switchboards only   (b) in a dry location   (c) in a wet location   (d) luminaires

149. Lettering on conductor insulation indicates intended condition of use. THWN is rated _____.
(a) 75°C   (b) for wet locations   (c) a and b   (d) not enough information

150. THW insulation has a _____ rating when installed within electric-discharge lighting equipment, such as through fluorescent luminaires .
(a) 60°C   (b) 75°C   (c) 90°C   (d) none of these

151. The ampacities listed in the Tables of Article 310 are based on temperature alone and do not take _____ into consideration.
(a) continuous loads   (b) voltage drop   (c) insulation   (d) wet locations

152. The ampacity of a conductor can be different along the length of the conductor. The higher ampacity is permitted to be used for the lower ampacity if the lower ampacity is no more than _____ ft or no more than _____ percent of the length of the higher ampacity.
(a) 10, 20   (b) 20, 10   (c) 10, 10   (d) 15, 15

153. Where six current-carrying conductors are run in the same conduit or cable, the ampacity of each conductor shall be adjusted to a factor of _____ percent of its value.
(a) 90   (b) 60   (c) 40   (d) 80

154. Conductor derating factors shall not apply to conductors in nipples having a length not exceeding _____
(a) 12 in.   (b) 24 in.   (c) 36 in.   (d) 48 in.

155. The ampacity adjustment factors of Table 310.15(B)(2)(a) do not apply to AC or MC cable without an overall outer jacket, if which of the following conditions are met?
(a) Each cable has not more than three current-carrying conductors.
(b) The conductors are 12 AWG copper.
(c) No more than 20 current-carrying conductors are bundled or stacked.
(d) all of these

156. • When bare grounding conductors are allowed, their ampacities are limited to _____.
(a) 60°C
(b) 75°C
(c) 90°C
(d) those permitted for the insulated conductors of the same size

157. In a balanced 208Y/120V, 4-wire, 3Ø system, the grounded conductor will carry _____ amperes if the loads supplied are linear loads and no harmonic currents are present.
(a) full load   (b) zero   (c) fault-current   (d) none of these

158. On a 3Ø, 4-wire, wye circuit, where the major portion of the load consists of electrical discharge lighting, data-processing equipment, or other harmonic current inducting loads, the grounded conductor shall be counted when applying 310.15(B)(2) adjustment factors.
(a) True   (b) False

159. When determining the number of conductors that are considered as current-carrying, a grounding conductor is _____.
(a) counted as one current-carrying conductor
(b) considered to be a current-carrying conductor but not counted
(c) considered to be a noncurrent-carrying conductor and is not counted
(d) counted as one conductor for each ground wire in the raceway

160. In designing circuits, the current-carrying capacity of conductors should be corrected for heat at room temperatures above _____.
(a) 30°F   (b) 86°F   (c) 94°F   (d) 75°F

161. • Aluminum and copper-clad aluminum of the same circular mil size and insulation have _____.
(a) the same physical characteristics   (b) the same termination
(c) the same ampacity   (d) different ampacities

162. The ampacity of a single insulated 1/0 AWG THHN copper conductor in free air is _____.
(a) 260A   (b) 300A   (c) 185A   (d) 215A

163. For conductors rated 2,001 to 35,000V, thermal resistively is the reciprocal of thermal conductivity, is designated Rho and expressed in units of _____.
(a) °F-cm/volt   (b) °F-cm/watt   (c) °C-cm/volt   (d) °C-cm/watt

164. As used in the *Code*, thermal resistivity refers to the heat _____ capability through a substance by conduction.
(a) assimilation   (b) generation   (c) transfer   (d) dissipation

165. For applications where underground circuits shall be buried deeper than shown in a specific underground ampacity table or figure, the following ampacity derating factor shall be permitted to be used: _____ percent per increased foot of depth for all values of Rho.
(a) 3   (b) 6   (c) 9   (d) 12

166. Table 310.70 provides the ampacities of insulated single aluminum conductor isolated in air. If the conductor size is 8 AWG, MV-90, and the voltage range is 2,001 through 5,000, then the ampacity is _____.
(a) 64A   (b) 85A   (c) 115A   (d) 150A

167. Table 310.71 provides ampacities of an insulated three-conductor copper cable isolated in air, based on conductor temperature of 90°C (194°F) and ambient air temperature of 40°C (104°F). If the conductor size is 4/0 AWG, MV-105, and the voltage range is 2,001 through 5,000, then the ampacity is _____.
(a) 250A   (b) 285A   (c) 320A   (d) 325A

Article 312 Cabinets, Cutout Boxes and Meter Socket Enclosures

168. Cabinets or cutout boxes installed in wet locations shall be _____.
(a) waterproof   (b) raintight   (c) weatherproof   (d) watertight

169. In walls constructed of wood or other _____ material, electrical cabinets shall be flush with the finished surface or project therefrom.
(a) nonconductive   (b) porous   (c) fibrous   (d) combustible

170. Cables entering a cutout box _____.
(a) shall be secured independently to the cutout box   (b) can be sleeved through a chase
(c) shall have a maximum of two cables per connector (d) all of these

171. A switch enclosure (cabinet) shall not be used as a junction box, except where adequate space is provided so that the conductors do not fill the wiring space at any cross-section to more than 40 percent of the cross-sectional area of the space, and so that _____ do not fill the wiring space at any cross-section to more than 75 percent of the cross-sectional area of the space.
(a) splices   (b) taps   (c) conductors   (d) all of these

172. For a steel cabinet or cutout box, the metal shall not be less than _____ uncoated.
(a) 0.53 in.   (b) 0.035 in.   (c) 0.053 in.   (d) 1.35 in.

Article 314 Outlet, Device, Pull and Junction Boxes, Conduit Bodies, Fittings and Manholes

173. Round boxes shall not be used with any wiring method connector when a locknut or bushing is connected to the side of the box.
(a) True   (b) False

174. • Nonmetallic boxes are permitted for use with _____.
      (a) flexible nonmetallic conduit   (b) liquidtight nonmetallic conduit
      (c) nonmetallic cables and raceways   (d) all of these

175. Metallic boxes are required to be _____.
      (a) bonded   (b) installed   (c) grounded   (d) all of these

176. Boxes, conduit bodies and fittings installed in wet locations do not need to be listed for use in wet locations.
      (a) True   (b) False

177. According to the *NEC*, the volume of a 3 × 2 × 2 in. device box is _____
      (a) 12 cu in.   (b) 14 cu in.   (c) 10 cu in.   (d) 8 cu in.

178. When counting the number of conductors in a box, a conductor running through the box is counted as _____ conductor(s).
      (a) one   (b) two   (c) zero   (d) none of these

179. Equipment grounding conductor(s), and not more than _____ fixture wires (smaller than 14 AWG) shall be permitted to be omitted from the calculations where they enter the box from a domed luminaire or similar canopy and terminate within that box.
      (a) 2   (b) 3   (c) 4   (d) none of these

180. When determining the number of conductors in a box, and one or more factory or field-supplied internal cable clamps are present in the box, a double volume allowance for the clamps, in accordance with Table 314.16(B), shall be made based on the largest conductor present in the box.
      (a) True   (b) False

181. • For box fill calculations, a reduction of _____ conductor(s) shall be made for one hickey and two internal clamps.
      (a) 1   (b) 2   (c) 3   (d) zero

182. • What is the total volume, in cubic inches, for box fill calculations for two internal cable clamps, six 12 AWG THHN conductors, and one single-pole switch?
      (a) 12.00   (b) 13.50   (c) 14.50   (d) 20.25

183. When a box has three equipment grounding conductors in it that have originated outside the box, the three grounding conductors are counted as _____ conductor(s) when determining the number of conductors in a box for box fill calculations.
      (a) 3   (b) 6   (c) 1   (d) 0

184. Conduit bodies holding conductors larger than 6 AWG shall have a cross-sectional area at least twice that of the largest conduit to which they are connected.
      (a) True   (b) False

185. When NM cable is used with nonmetallic boxes no larger than $2^{1}/_{4}$ × 4 in., securing the cable to the box is not required if the cable is fastened within _____
      (a) 6 in.   (b) 8 in.   (c) 10 in.   (d) 12 in.

186. • In combustible walls or ceilings, the front edge of an outlet box or fitting may be set back _____ from the finished surface.
      (a) $^{3}/_{8}$ in.   (b) $^{1}/_{8}$ in.   (c) $^{1}/_{2}$ in.   (d) $^{1}/_{4}$ in.

187. Surface extensions from a flush-mounted box can be made by mounting and mechanically securing an extension ring over the flush box and attaching the extension wiring method to the extension ring.
      (a) True   (b) False

188. • Only a _____ wiring method can be used for a surface extension from a cover, and the wiring method must include an equipment grounding conductor.
      (a) solid   (b) flexible   (c) rigid   (d) cord

189.   A wood brace that is used for mounting a box shall have a cross-section not less than nominal _____
        (a) 1 × 2 in.   (b) 2 × 2 in.   (c) 2 × 3 in.   (d) 2 × 4 in.

190.   When mounting enclosures in finished surfaces, it shall be permissible to make a _____ installation when adequate support is provided by clamps, anchors or fittings identified for the application.
        (a) temporary   (b) workmanlike   (c) permanent   (d) flush

191.   Outlet boxes can be secured to suspended-ceiling framing members by mechanical means such as _____, or other means identified for the suspended-ceiling framing member(s).
        (a) bolts   (b) screws   (c) rivets   (d) all of these

192.   Outlet boxes can be secured to independent support wires which are taut and secured at both ends, if the box is supported to the independent support wires with fittings and methods identified for the purpose.
        (a) True   (b) False

193.   Enclosures that are not over _____ in size, having threaded entries and do not contain a device(s) or support a luminaire(s) or other equipment, shall be considered to be adequately supported where two or more conduits are threaded wrenchtight into the enclosure.
        (a) 50 cu in.   (b) 75 cu in.   (c) 100 cu in.   (d) 125 cu in.

194.   • Enclosures not over 100 cu in. that have threaded entries that support luminaires or contain devices shall be considered adequately supported where two or more conduits are threaded wrenchtight into the enclosure where each conduit is supported within _____ of the enclosure.
        (a) 12 in.   (b) 18 in.   (c) 24 in.   (d) 30 in.

195.   Boxes can be supported from a multiconductor cord or cable provided the conductors are protected from _____.
        (a) strain   (b) temperature   (c) sunlight   (d) abrasion

196.   • The minimum size box that is to contain a flush device shall not be less than _____ deep.
        (a) $^{15}/_{16}$ in.   (b) $^{8}/_{15}$ in.   (c) 1 in.   (d) $1^1/_2$ in.

197.   In completed installations, each outlet box shall have a _____.
        (a) cover   (b) faceplate   (c) canopy   (d) any of these

198.   A wall-mounted luminaire weighing not more than _____ can be supported to a device box with no fewer than two No. 6 or larger screws.
        (a) 4 lbs.   (b) 6 lbs.   (c) 8 lbs.   (d) 10 lbs.

199.   Luminaires shall be supported independently of the outlet box where the weight exceeds _____
        (a) 60 lbs.   (b) 50 lbs.   (c) 40 lbs.   (d) 30 lbs.

200.   A luminaire that weighs more than 50 lbs. is permitted to be supported by an outlet box or fitting that's designed and listed for the weight of the luminaire.
        (a) True   (b) False

Unit 5 NEC Exam – NEC Code Order 250.148 – 314.24

1. Where one or more equipment grounding conductors enter a box, all equipment grounding conductors shall be spliced together (in the enclosure). This does not apply to insulated equipment grounding conductors for isolated ground receptacles for electronic equipment.
   (a) True   (b) False

2. When the dc system consists of a _____, the grounding conductor shall not be smaller than the grounded conductor.
   (a) 2-wire balancer set
   (b) 3-wire balancer set
   (c) balancer winding with overcurrent protection
   (d) b or c

3. Cases or frames of instrument transformers are not required to be grounded _____.
   (a) when accessible to qualified persons only
   (b) for current transformers where the primary is not over 150 volts-to-ground and that are used exclusively to supply current to meters
   (c) for potential transformers where the primary is less than 150 volts-to-ground
   (d) a or b

4. Circuit conductors that operate at 277V (with 600V insulation) may occupy the same enclosure or raceway with 48V dc conductors that have an insulation rating of 300V.
   (a) True   (b) False

5. When installing raceways underground in rigid nonmetallic conduit and other approved raceways, there must be a minimum of _____ of cover.
   (a) 6 in.   (b) 12 in.   (c) 18 in.   (d) 22 in.

6. What is the minimum cover requirement in inches for UF cable supplying power to a 120V, 15A GFCI-protected circuit outdoors under a driveway of a one-family dwelling?
   (a) 12   (b) 24   (c) 16   (d) 6

7. When installing direct-buried cables, a _____ shall be used at the end of a conduit that terminates underground.
   (a) splice kit   (b) terminal fitting   (c) bushing   (d) b or c

8. The provisions required for mounting conduits on indoor walls or in rooms that must be hosed down frequently is _____ between the mounting surface and the electrical equipment.
   (a) a permanent $1/4$ in. airspace
   (b) separated by insulated bushings
   (c) separated by noncombustible tubing
   (d) none of these

9. All metal raceways, cable armor, boxes, fittings, cabinets and other metal enclosures for conductors must be _____ joined together to form a continuous electrical conductor.
   (a) electrically   (b) permanently   (c) metallically   (d) none of the above

10. When the opening to an outlet, junction or switch point is less than 8 inches in any dimension, each conductor shall be long enough to extend at least _____ outside the opening of the enclosure.
    (a) 0 in.   (b) 3 in.   (c) 6 in.   (d) 12 in.

11. Wiring methods permitted in the drop ceiling area used for environmental air include _____.
    (a) electrical metallic tubing
    (b) flexible metal conduit of any length
    (c) rigid metal conduit without an overall nonmetallic covering
    (d) all of these

12. The air-handling area beneath raised floors for data-processing systems is not a plenum and is not required to comply with the requirements of 300.22 but shall be permitted in accordance with Article 645.
    (a) True   (b) False

13. Insulated conductors used in wet locations shall be _____.
    (a) moisture-impervious metal-sheathed
    (b) RHW, TW, THW, THHW, THWN, XHHW
    (c) listed for wet locations
    (d) any of these

14. Where conductors of different insulation are associated together, the limiting temperature of any conductor shall not be exceeded.
    (a) True   (b) False

15. THW insulation has a _____ rating when installed within electric-discharge lighting equipment, such as through fluorescent luminaires.
    (a) 60°C   (b) 75ªC   (c) 90°C   (d) none of these

16. When bare grounding conductors are allowed, their ampacities are limited to _____.
    (a) 60°C
    (b) 75°C
    (c) 90°C
    (d) those permitted for the insulated conductors of the same size

17. Aluminum and copper-clad aluminum of the same circular mil size and insulation have _____.
    (a) the same physical characteristics   (b) the same termination
    (c) the same ampacity   (d) different ampacities

18. Nonmetallic boxes are permitted for use with _____.
    (a) flexible nonmetallic conduit   (b) liquidtight nonmetallic conduit
    (c) nonmetallic cables and raceways   (d) all of these

19. For box fill calculations, a reduction of _____ conductor(s) shall be made for one hickey and two internal clamps.
    (a) 1   (b) 2   (c) 3   (d) Zero

20. What is the total volume in cubic inches for box fill calculations for two internal cable clamps, six 12 AWG THHN conductors and one single-pole switch?
    (a) 12.00 cu in.   (b) 13.50 cu in.   (c) 14.50 cu in.   (d) 20.25 cu in.

21. When NM cable is used with nonmetallic boxes no larger than $2^1/_4 \times 4$ in., securing the cable to the box is not required if the cable is fastened within _____
    (a) 6 in.   (b) 8 in.   (c) 10 in.   (d) 12 in.

22. In combustible walls or ceilings, the front edge of an outlet box or fitting may be set back _____ from the finished surface.
    (a) $3/_8$ in.   (b) $1/_8$ in.   (c) $1/_2$ in.   (d) $1/_4$ in.

23. Only a _____ wiring method can be used for a surface extension from a cover, and the wiring method must include an equipment grounding conductor.
    (a) solid   (b) flexible   (c) rigid   (d) cord

24. Enclosures not over 100 cu in. that have threaded entries that support luminaires or contain devices shall be considered adequately supported where two or more conduits are threaded wrenchtight into the enclosure where each conduit is supported within _____ of the enclosure.
    (a) 12 in.   (b) 18 in.   (c) 24 in.   (d) 30 in.

25. The minimum size box that is to contain a flush device shall not be less than _____ deep.
    (a) $15/_{16}$ in.   (b) $8/_{15}$ in.   (c) 1 in.   (d) $1^1/_2$ in.

Unit 5 NEC Exam – Random Order 250.126 – 314.27

1. _____ are required at one or both ends of the raceway where moisture could enter a raceway and contact energized live parts.
    (a) Seals   (b) Plugs   (c) Bushings   (d) a and b

2. A 100 ft vertical run of 4/0 AWG copper requires the conductors to be supported at _____ locations.
    (a) 4   (b) 5   (c) 2   (d) none of these

3. A vertical run of 4/0 AWG copper must be supported at intervals not exceeding _____.
    (a) 80 ft   (b) 100 ft   (c) 120 ft   (d) 40 ft

4. An _____ shall be used to connect the grounding terminal of a grounding-type receptacle to a grounded box.
    (a) equipment bonding jumper   (b) equipment grounding jumper
    (c) a or b   (d) a and b

5. Cables entering a cutout box shall _____.
    (a) be secured independently to the cutout box
    (b) can be sleeved through a chase
    (c) have a maximum of two cables per connector
    (d) all of these

6. Ceiling-support wires used for the support of electrical raceways and cables within nonfire-rated assemblies are required to be distinguishable from the suspended-ceiling framing support wires.
    (a) True   (b) False

7. Conduit bodies holding conductors larger than 6 AWG shall have a cross-sectional area at least twice that of the largest conduit to which they are connected.
    (a) True   (b) False

8. Raceways shall be provided with expansion fittings where necessary to compensate for thermal expansion and contraction.
    (a) True   (b) False

9. A switch enclosure (cabinet) shall not be used as a junction box, except where adequate space is provided so that the conductors do not fill the wiring space at any cross-section to more than 40 percent of the cross-sectional area of the space, and so that _____ do not fill the wiring space at any cross-section to more than 75 percent of the cross-sectional area of the space.
    (a) splices   (b) taps   (c) conductors   (d) all of these

10. According to the *NEC*, the volume of a $3 \times 2 \times 2$ in. device box is _____
    (a) 12 cu in.   (b) 14 cu in.   (c) 10 cu in.   (d) 8 cu in.

11. All conductors of a circuit, including the grounded and equipment grounding conductors, must be contained within the same _____.
    (a) raceway   (b) cable   (c) trench   (d) all of these

12. For a steel cabinet or cutout box, the metal shall not be less than _____ uncoated.
    (a) 0.53 in.   (b) 0.035 in.   (c) 0.053 in.   (d) 1.35 in.

13. For conductors rated 2,001 to 35,000V, thermal resistively is the reciprocal of thermal conductivity, and is designated Rho and expressed in units of _____.
    (a) °F-cm/volt   (b) °F-cm/watt   (c) °C-cm/volt   (d) °C-cm/watt

14. Metallic boxes are required to be _____.
    (a) bonded   (b) installed   (c) grounded   (d) all of these

15. Rigid metal conduit that is directly buried outdoors must have at least _____ of cover.
    (a) 6 in.   (b) 12 in.   (c) 18 in.   (d) 24 in.

16. Round boxes shall not be used with any wiring method connector when a locknut or bushing is connected to the side of the box.
   (a) True   (b) False

17. Solid dielectric insulated conductors operated above 2,000V in permanent installations shall have _____ insulation and shall be shielded.
   (a) ozone-resistant   (b) asbestos
   (c) high-temperature   (d) perfluoro-alkoxy

18. The ampacity of a single insulated 1/0 AWG THHN copper conductor in free air is _____.
   (a) 260A   (b) 300A   (c) 185A   (d) 215A

19. The minimum size conductor permitted in a any building for branch circuits under 600V is _____ AWG.
   (a) 14   (b) 12   (c) 10   (d) 8

20. The terminal for the connection of the equipment grounding conductor shall be identified by a green-colored, _____.
   (a) not readily removable terminal screw with a hexagonal head.
   (b) hexagonal, not readily removable terminal nut.
   (c) pressure wire connector.
   (d) any of these

21. Type AC cable can be used in ducts or plenums that are used for environmental air.
   (a) True   (b) False

22. When determining the number of conductors that are considered as current-carrying, a grounding conductor is _____.
   (a) counted as one current-carrying conductor
   (b) considered to be a current-carrying conductor but not counted
   (c) considered to be a noncurrent-carrying conductor and is not counted
   (d) counted as one conductor for each ground wire in the raceway

23. Where six current-carrying conductors are run in the same conduit or cable, the ampacity of each conductor shall be adjusted to a factor of _____ percent of its value.
   (a) 90   (b) 60   (c) 40   (d) 80

24. A bushing can be used instead of a box or terminal where conductors emerge from a raceway and enter or terminate at equipment, such as open switchboards, unenclosed control equipment or similar equipment.
   (a) True   (b) False

25. A grounding-type receptacle can replace a nongrounding-type receptacle at an outlet box that does not contain an equipment grounding conductor if the equipment grounding conductor is connected to _____.
   (a) a grounding electrode system as described in 250.50
   (b) a grounding electrode conductor
   (c) the equipment grounding terminal bar within the enclosure where the branch circuits for the receptacle or branch circuit originates
   (d) any of these

26. A luminaire that weighs more than 50 lbs is permitted to be supported by an outlet box or fitting that's designed and listed for the weight of the luminaire.
   (a) True   (b) False

27. A TVSS can be connected anywhere on the premises wiring system.
   (a) True   (b) False

28. A TVSS device must be listed.
   (a) True   (b) False

29. A TVSS is listed to limit transient voltages by diverting or limiting surge current.
   (a) True   (b) False

30. All conductors of a circuit are required to be _____.
   (a) in the same raceway
   (b) in close proximity in the same trench
   (c) the same size
   (d) a and b

31. An 8 × 8 × 4 in.-deep junction/splice box requires 6 in. of free conductor measured from the point in the box where the conductors enter the enclosure. The 3 in. outside-the-box rule _____ apply(ies).
   (a) does   (b) does not   (c) sometimes   (d) none of these

32. An equipment bonding jumper for a grounding-type receptacle must be installed between the receptacle and a flush-mounted outlet box, even when there's direct metal-to-metal contact between the metal yoke of the receptacle and the metal box.
   (a) True   (b) False

33. An equipment bonding jumper shall be used to connect the grounding terminal of a grounding-type receptacle to a grounded box. Where the box is surface mounted, direct metal-to-metal contact between the device yoke and the box shall be permitted to ground the receptacle to the box.
   (a) True   (b) False

34. Article 285 covers surge arresters.
   (a) True   (b) False

35. Cables or raceways installed using directional boring equipment must be _____ for this purpose. This new requirement was necessary to ensure that proper raceways are used with directional boring equipment.
   (a) marked   (b) listed   (c) labeled   (d) approved

36. Conductors in metal raceways and enclosures shall be so arranged as to avoid heating the surrounding metal by alternating-current induction. To accomplish this, the _____ conductor(s) shall be grouped together.
   (a) phase   (b) neutral   (c) ungrounded   (d) all of these

37. Contact devices or yokes designed and listed for the purpose shall be permitted in conjunction with the supporting screws to establish the grounding circuit between the device yoke and flush-type boxes.
   (a) True   (b) False

38. Equipment and devices shall be permitted within ducts or plenum chambers used to transport environmental air only if necessary for their direct action upon, or sensing of, the, _____.
   (a) contained air   (b) air quality   (c) air temperature   (d) none of these

39. In a balanced 208Y/120V, 4-wire, 3Ø system, the grounded conductor will carry _____ amperes if the loads supplied are linear loads and no harmonic currents are present.
   (a) full load   (b) zero   (c) fault-current   (d) none of these

40. In designing circuits, the current-carrying capacity of conductors should be corrected for heat at room temperatures above _____.
   (a) 30°F   (b) 86°F   (c) 94°F   (d) 75°F

41. In multiwire circuits, the continuity of the _____ conductor shall not be dependent upon the device connections.
   (a) ungrounded   (b) grounded   (c) grounding   (d) a and b

42. Insulated conductors and cables exposed to the direct rays of the sun must be _____.
   (a) listed
   (c) listed for sunlight resistance
   (b) listed and marked sunlight resistant
   (d) b or c

43. Metal raceways shall not be _____ by welding unless the raceway is specifically designed to be, or otherwise specifically permitted to be, in the *Code*.
   (a) supported   (b) terminated   (c) connected   (d) all of these

44.   Metal raceways, cable armor, boxes, cable sheathing, cabinets, elbows, couplings, fittings, supports and support
       hardware shall be of materials suitable for _____.
       (a) corrosive locations
       (b) wet locations
       (c) the environment in which they are to be installed
       (d) none of these

45.   Metric designators and trade sizes for conduit, tubing, and associated fittings and accessories are designated in Table
       _____.
       (a) 250.66   (b) 250.122   (c) 300.1(C)   (d) 310.16

46.   No wiring of any type shall be installed in ducts used to transport _____.
       (a) dust   (b) flammable vapors (c) loose stock   (d) all of these

47.   Prewired raceway assemblies shall be permitted only where specifically permitted elsewhere in the *Code* for the
       applicable wiring method.
       (a) True   (b) False

48.   Service conductors that are not encased in concrete and buried 18 in. or more below grade shall have their location
       identified by a warning ribbon placed in the trench at least _____ above the underground installation.
       (a) 6 in.   (b) 12 in.   (c) 18 in.   (d) none of these

49.   Splices and taps are permitted in cabinets or cutout boxes if the splice or tap does not fill the wiring space at any cross-
       section to more than _____ percent.
       (a) 20   (b) 40   (c) 60   (d) 75

50.   Surface extensions from a flush-mounted box can be made by mounting and mechanically securing an extension ring
       over the flush box and attaching the extension wiring method to the extension ring.
       (a) True   (b) False

Unit 6

Conductor Sizing and Protection Calculations

OBJECTIVES
After reading this unit, the student should be able to briefly explain the following concepts:

Conductor allowable ampacity
Conductor ampacity
Conductor bundling derating factor, Note 8(a)
Conductor sizing summary
Conductor insulation property

Conductor size – voltage drop
Conductors in parallel
Current-carrying conductors
Equipment conductors size and protection examples
Minimum conductor size

Overcurrent protection of equipment conductors
Overcurrent protection
Overcurrent protection of conductors
Terminal ratings

After reading this unit, the student should be able to briefly explain the following terms:

Ambient temperature
American Wire Gage
Ampacity
Ampacity derating factors
Conductor bundling
Conductor properties

Continuous load
Current-carrying conductors
Fault current
Interrupting rating
Overcurrent protection device
Overcurrent protection

Parallel conductors
Temperature correction factors
Terminal ratings
Voltage drop

PART A – GENERAL CONDUCTOR REQUIREMENTS

6–1 CONDUCTOR INSULATION PROPERTY [Table 310.13]

Table 310.13 of the *NEC* provides information on conductor properties such as permitted use, maximum operating temperature and other insulation details. Figure 6–1.

The following abbreviations and explanations should be helpful in understanding Table 310.13 as well as Table 310.16.

-2   Conductor is permitted to be used at a continuous 90°C operation temperature

F   Fixture wire (solid or 7 strand) [Table 402.3]

FF   Flexible fixture wire (19 strands) [Table 402.3]

H   75°C Insulation rating

HH   90°C Insulation

N   Nylon outer cover

T   Thermoplastic insulation

W   Wet or Damp

Fixture wires, see Article 402, Table 402.3 and 402.5.

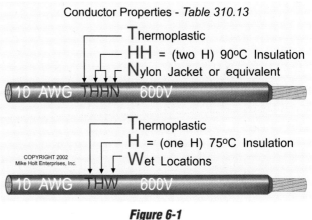

Conductor Properties - *Table 310.13*

Thermoplastic
HH = (two H) 90ºC Insulation
Nylon Jacket or equivalent

10 AWG THHN 600V

COPYRIGHT 2002
Mike Holt Enterprises, Inc.

Thermoplastic
H = (one H) 75ºC Insulation
Wet Locations

10 AWG THW 600V

Figure 6-1
Conductor Properties

TABLE 310-13 CONDUCTOR INFORMATION					
	Column 2	**Column 3**	**Column 4**	**Column 5**	**Column 6**
Type Letter	Trade Name	Maximum Operating Temperature	Applications Provisions	Sizes Available	Outer Covering
THHN	Heat resistant thermoplastic	90°C	Dry and damp locations	14 – 1,000	Nylon jacket or equivalent
THHW	Moisture- & heat-resistant thermoplastic	75°C 90°C	Wet locations Dry and damp locations	14 – 1,000	None
THW	Moisture- & heat-resistant thermoplastic	75°C 90°C	Dry, damp, and wet locations Within electrical discharge lighting equipment *See Section 410.31*	14 – 2,000	None
THWN	Moisture- & heat-resistant thermoplastic	75°C	Dry and wet locations	14 – 1,000	Nylon jacket or equivalent
TW	Moisture-resistant thermoplastic	60°C	Dry and wet locations	14 – 2,000	None
XHHW	Moisture- & heat-resistant cross-linked synthetic polymer	90°C 75°C	Dry and damp locations Wet locations	14 – 2,000	None

❏ **Table 310.13**

TW can be described as _____. Figure 6–2.

(a) thermoplastic insulation

(b) suitable for dry or wet locations

(c) maximum operating temperature of 60°C

(d) all of these

• Answer: (d) all of these

❏ **Table 402.3**

TFFN can be described as _____.

(a) stranded fixture wire

(b) thermoplastic insulation with a nylon outer cover

(c) suitable for dry and wet locations

(d) both a and b

• Answer: (d) both a and b

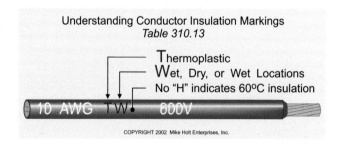

Understanding Conductor Insulation Markings
Table 310.13

Figure 6-2
Understanding Conductor Insulation Markings

6–2 CONDUCTOR ALLOWABLE AMPACITY [310.15]

The ampacity of a conductor is the current in amperes that a conductor can carry continuously without exceeding its temperature rating under specific conditions of use. Figure 6–3.

310.15 lists two ways of determining conductor ampacity:

(A) General Requirements

(1) Tables or Engineering Supervision. There are two ways to determine conductor ampacity:

- Tables 310.16 [310.15(B)]
- Engineering formula

Note: For all practical purposes, use the ampacities listed in Table 310.16.

FPN: The ampacities listed in Table 310.16 are based on temperature alone and don't take voltage drop into consideration. Voltage drop considerations are for efficiency of operation and not safety; therefore, sizing conductors for voltage drop is not a *Code* requirement. See 215.2(A)(4) FPN 2 for more details.

Conductor Ampacities
Table 310.16

Table 310.16 is based on an ambient temperature of 86°F and 3 current-carrying conductors in a raceway or cable.

COPYRIGHT 2002 Mike Holt Enterprises, Inc.

Figure 6-3
Conductor Ampacities

TABLE 310.16. ALLOWABLE AMPACITIES OF INSULATED CONDUCTORS
Based On Not More Than Three Current-Carrying Conductors and Ambient Temperature of 30°C (86°F)

Size	Temperature Rating of Conductor, See Table 310.13						Size
	60°C (40°F)	75°C (167°F)	90°C (194°F)	60°C (40°F)	75°C (167°F)	90°C (194°F)	
	TW	THHN THW THWN XHHW Wet Location	THHN THHW XHHW Dry Location	TW	THHN THW THWN XHHW Wet Location	THHN THHW WHHN Dry Location	
AWG kcmil	COPPER			ALUMINUM/COPPER-CLAD ALUMINUM			AWG kcmil
14*	20	20	25				12
12*	25	25	30	20	20	25	10
10*	30	35	40	25	30	35	8
8	40	50	55	30	40	45	8
6	55	65	75	40	50	60	6
4	70	85	95	55	65	75	4
3	85	100	110	65	75	85	3
2	95	115	130	75	90	100	2
1	110	130	150	85	100	115	1
1/0	125	150	170	100	120	135	1/0
2/0	145	175	195	115	135	150	2/0
3/0	165	200	225	130	155	175	3/0
4/0	195	230	260	150	180	205	4/0
250	215	255	290	170	205	230	250
300	240	285	320	190	230	255	300
350	260	310	350	210	250	280	350
400	280	335	380	225	270	305	400
500	320	380	430	260	310	350	500

*See 240.4(D)

(B) Table Ampacity.

The allowable ampacities of a conductor are listed in Table 310.16, based on the condition where no more than three current-carrying conductors are bundled together at an ambient temperature of 86°F.

The ampacity of a conductor is listed in Table 310.16 under the condition of no more than three current-carrying conductors bundled together in an ambient temperature of 86°F. The ampacity of a conductor changes if the ambient temperature is not 86°F or if more than three current-carrying conductors are bundled together.

CAUTION: 240.4(D) specifies that the maximum overcurrent protection device permitted for copper 14 AWG is 15A, 12 AWG is 20A, and 10 AWG is 30A. The maximum overcurrent protection device permitted for aluminum 12 AWG is 15A and 10 AWG is 20A. The overcurrent protection device limitations of 240.4(D) do not apply to the following:

Air-conditioning conductors	440.22
Capacitor conductors	Article 460
Class 1 remote-control conductors	Article 725
Cooking equipment taps	Article 725
Feeder taps	240.21
Fixture wires and taps	210.19(C) Ex.
Motor branch-circuit conductors	430.52
Motor feeder conductors	430.62
Motor control conductors	430.72
Motor taps branch circuits	430.53(D)
Motor taps feeder conductors	430.28
Power loss hazard conductors	240.4(A)
Two-wire transformers	240.4(I)
Transformer tap conductors	240.21(B)(D) and (M)
Welding conductors	Article 630

6–3 CONDUCTOR SIZING [110.6]

Conductors are sized according to the American Wire Gage (AWG) from Number 40 (No. 40) through Number 0000 (4/0 AWG). The smaller the AWG size, the larger the conductor. Conductors larger than 4/0 AWG are identified according to their circular mil area, such as 250,000, 300,000, 500,000. The circular mil size is often expressed in kcmil, such as 250 kcmil, 300 kcmil, 500 kcmil etc. Figure 6–4.

Smallest Conductor Size

The smallest size conductor permitted by the *NEC* for branch circuits, feeders or services is 14 AWG copper or 12 AWG aluminum [Table 310.5]. Some local codes require a minimum 12 AWG for commercial and industrial installations. Conductors smaller than 14 AWG are permitted for:

Class 1 remote-control circuits [402.11 Ex., and 725.27]

Fixture wire [402.5 and 410.24]

Flexible cords [400.12]

Motor control circuits [430.72]

Non-power-limited fire alarm circuits [760.23]

Power-limited fire alarm circuits [760.71(B)].

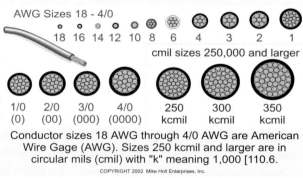

Cross-Sections and Trade Sizes of NEC Conductors
Tables 310.13 and 310.16

Conductor sizes 18 AWG through 4/0 AWG are American Wire Gage (AWG). Sizes 250 kcmil and larger are in circular mils (cmil) with "k" meaning 1,000 [110.6.

COPYRIGHT 2002 Mike Holt Enterprises, Inc.

Figure 6-4

Cross-Sections and Trade Sizes of *NEC* Conductors

Temperature Limitations of Electrical Connections
Section 110.14(C)(1)(a) - General Rule

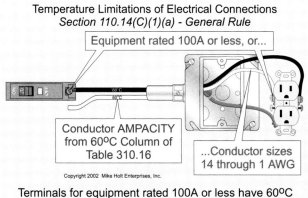

Equipment rated 100A or less, or...

Conductor AMPACITY
from 60°C Column of
Table 310.16

...Conductor sizes
14 through 1 AWG

Copyright 2002 Mike Holt Enterprises, Inc.

Terminals for equipment rated 100A or less have 60°C
terminals and require a 60°C conductor ampacity.

Figure 6-5

Temperature Limitations of Electrical Connections

Conductor Size - Terminal Ratings
Section 110.14(C)(1)(a)(2) and (3)

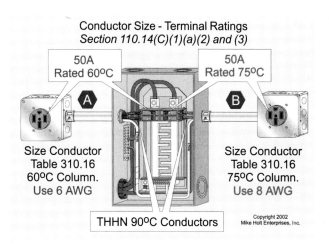

50A
Rated 60°C

50A
Rated 75°C

Size Conductor
Table 310.16
60°C Column.
Use 6 AWG

Size Conductor
Table 310.16
75°C Column.
Use 8 AWG

THHN 90°C Conductors

Copyright 2002
Mike Holt Enterprises, Inc.

Figure 6-6

Conductor Insulation & Ampacity – Terminal Ratings

6–4 TERMINAL RATINGS [110.14(C)]

Conductors are to be sized in accordance with the lowest temperature rating of any terminal, device or conductor of the circuit.

Circuits Rated 100A and Less [110.14(C)(1)(a)]

Equipment terminals rated 100A or less (and pressure connector terminals for 14 AWG through 1 AWG conductors), shall have the conductor sized no smaller than the 60°C temperature rating listed in Table 310.16, unless the terminals are marked otherwise. Figure 6–5.

❏ **Terminal Rated 60°C [110.14(C)(1)(a)(1)]**

What size THHN conductor is required for a 50A circuit listed for use at 60°C? Figure 6–6 Part A.

(a) 10 AWG   (b) 8 AWG   (c) 6 AWG   (d) any of these

• Answer: (c) 6 AWG

Conductors must be sized to the lowest temperature rating of either the equipment or the conductor. THHN insulation can be used, but the conductor size must be selected based on the 60°C terminal rating of the equipment, not the 90°C rating of the insulation. Using the 60°C column of Table 310.16, this 50A circuit requires a 6 AWG THHN conductor (rated 55A at 60°C).

❏ **Terminal Rated 75°C [110.14(C)(1)(a)(2)]**

What size THHN conductor is required for a 50A circuit listed for use at 75°C? Figure 6–6 Part B.

(a) 10 AWG   (b) 8 AWG   (c) 6 AWG   (d) any of these

• Answer: (b) 8 AWG

Conductors must be sized according to the lowest temperature rating of either the equipment or the conductor . THHN conductors can be used, but the conductor size must be selected according to the 75°C terminal rating of the equipment, not the 90°C rating of the insulation. Using the 75°C column of Table 310.16, this installation would permit 8 AWG THHN (rated 50A at 75°C) to supply the 50A load.

Circuits Over 100A [110.14(C)(1)(b)]

Terminals for equipment rated over 100A and pressure connector terminals for conductors larger than 1 AWG shall have the conductor sized according to the 75°C temperature rating listed in Table 310.16. Figure 6–7.

❏ **Over 100A [110.14(C)(1)(b)]**

What size THHN conductor is required to supply a 225A feeder?

(a) 1/0 AWG   (b) 2/0 AWG   (c) 3/0 AWG   (d) 4/0 AWG

• Answer: (d) 4/0 AWG

The conductors in this example must be sized to the lowest temperature rating of either the equipment or the conductor. THHN conductors can be used, but the conductor size must be selected according to the 75°C terminal rating of the equipment. Using the 75°C column of Table 310.16, this would require a 4/0 AWG THHN (rated 230A at 75°C) to supply the 225A load. 3/0 AWG THHN is rated 225A at 90°C, but we must size the conductor to the terminal rating at 75°C.

Minimum Conductor Size Table

When sizing conductors, the following table must always be used to determine the minimum size conductor:

Table 310.16 [110.14(C)] Terminal Size and Matching Copper Conductor		
Terminal Ampacity	60°C Terminals Wire Size	75° Terminals Wire Size
15	14	14
20	12	12
30	10	10
40	8	8
50	6	8
60	4	6
70	4	4
100	1	3
125	1/0	1
150	–	1/0
200	–	3/0
225	–	4/0
250	–	250 kcmil
300	–	350 kcmil
400	–	2 – 3/0
500	–	2 – 250 kcmil

CAUTION: When sizing conductors, we must consider conductor voltage drop, ambient temperature correction and conductor bundle adjustment factors. These subjects are covered later in this book.

What is the purpose of THHN if we can't use its higher ampacity?

In general, 90°C rated conductor ampacities cannot be used for sizing circuit conductors. However, THHN offers the opportunity of having a greater conductor ampacity for conductor ampacity adjustment. The higher ampacity of THHN can permit a conductor to be used without having to increase its size because of conductor ampacity adjustment. Remember, the advantage of THHN is not to permit a smaller conductor, but it might prevent you from having to install a larger conductor because of ampacity adjustments. Figure 6–8.

Note: This is explained in detail in Part B of this unit.

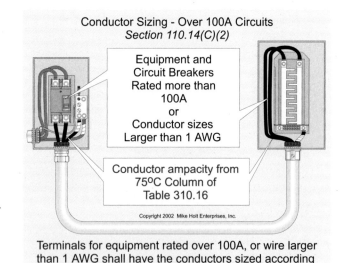

Terminals for equipment rated over 100A, or wire larger than 1 AWG shall have the conductors sized according to the 75ºC temperature rating listed in Table 310.16.

Figure 6-7
Temperature Limitations of Electrical Connections

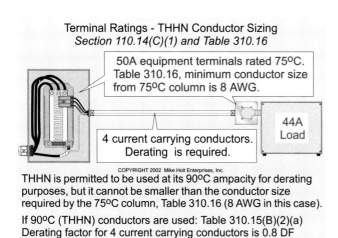

Terminal Ratings - THHN Conductor Sizing
Section 110.14(C)(1) and Table 310.16

50A equipment terminals rated 75ºC. Table 310.16, minimum conductor size from 75ºC column is 8 AWG.

44A Load

4 current carrying conductors. Derating is required.

COPYRIGHT 2002 Mike Holt Enterprises, Inc.

THHN is permitted to be used at its 90ºC ampacity for derating purposes, but it cannot be smaller than the conductor size required by the 75ºC column, Table 310.16 (8 AWG in this case).

If 90ºC (THHN) conductors are used: Table 310.15(B)(2)(a) Derating factor for 4 current carrying conductors is 0.8 DF
8 AWG THHN rated 55A x 0.8 DF = 44 ampacity (okay)

Figure 6-8

Terminal Ratings – THHN Conductor Sizing

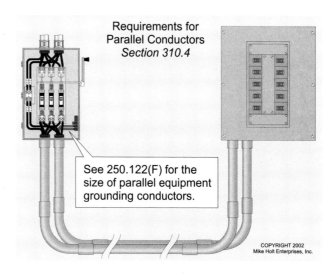

Requirements for Parallel Conductors
Section 310.4

See 250.122(F) for the size of parallel equipment grounding conductors.

COPYRIGHT 2002 Mike Holt Enterprises, Inc.

Figure 6-9

Requirements for Parallel Conductors

6–5 CONDUCTORS IN PARALLEL [310.4]

Parallel conductors (electrically joined at both ends) permit a smaller cross-sectional area per ampere. This can result in a significant cost saving for circuits over 300A. Figure 6–9. The following table demonstrates the increased circular mil area required per ampere with larger conductors.

Conductor Size	Circular Mils Chapter 9, Table 8	Ampacity 75ºC	Circular Mils Per Ampere
1/0 AWG	105,600 cm	150A	704 cm/per amp
3/0 AWG	167,600 cm	200A	838 cm/per amp
250 kcmil	250,000 cm	255A	980 cm/per amp
500 kcmil	500,000 cm	380A	1,316 cm/per amp
750 kcmil	750,000 cm	475A	1,579 cm/per amp

❏ **Sizing Parallel Conductors**

The 75ºC conductor required for a 600A service that has a demand load of 550A is _____. Figure 6–10.
  (a) 500 kcmil   (b) 750 kcmil   (c) 1,000 kcmil   (d) 1,250 kcmil
    • Answer: (d) 1,250 kcmil conductor, rated 590A [Table 310.16]

❏ **Sizing Parallel Conductors**

What size conductors would be required in one raceway (nipple) for a 600A service? The calculated demand load is 550A. Figure 6–11.
  (a) Two – 300 kcmil   (b) Two – 250 kcmil   (c) Two – 500 kcmil   (d) Two – 750 kcmil
    • Answer: (a) Two 300 kcmil conductors, each rated 285A.
      285A × 2 conductors = 570A [Table 310.16]

If we parallel the conductors, we can use two 300 kcmil conductors (total 600,000) instead of one 1,250 kcmil conductor.

Grounding Conductors in Parallel [310.4]

When equipment grounding conductors are installed with circuit conductors that are run in parallel, each raceway must have an equipment grounding conductor sized according to the overcurrent protection device rating that protects the circuit [250.122]. Figure 6–12.

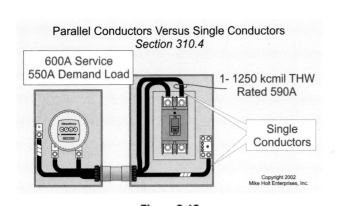

Figure 6-10
Parallel Conductors versus Single Conductors

Parallel Conductors Versus Single Conductors
Section 310.4

600A Service
550A Demand Load

1- 1250 kcmil THW
Rated 590A

Single
Conductors

Copyright 2002
Mike Holt Enterprises, Inc.

Parallel Conductors Versus Single Conductors
Section 310.4

600A Service
550A Demand Load

2- 300 kcmil THW
Rated 285A each =
570A per phase

Parallel
Conductors

Copyright 2002
Mike Holt Enterprises, Inc.

Derating for more than 3 conductors may be required if
raceway is not a nipple (over 24 in.). See 310.15(B)(2)(a).

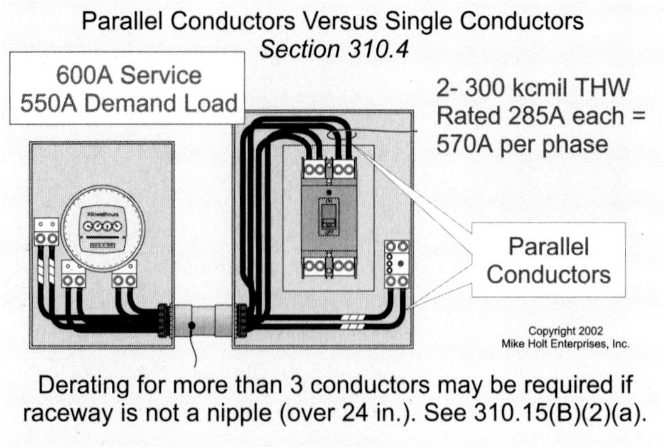

Figure 6-11
Parallel Conductors versus Single Conductors

❏ **Sizing Grounding Conductors in Parallel [250.122]**

What size equipment grounding conductor is required in each of two raceways for a 600A feeder? Figure 6–13.

(a) 3 AWG   (b) 2 AWG   (c) 1 AWG   (d) 1/0 AWG

• Answer: (c) 1 AWG

There must be a 1 AWG equipment grounding conductor in each of the two raceways.

6–6 CONDUCTOR SIZE – VOLTAGE DROP [210.19(A) FPN 4 AWG AND 215.2(A)(4) FPN 2]

The *NEC* generally does not require conductors to be sized to accommodate conductor voltage drop, but 210.19(A) FPN 4 AWG and 215.2(A)(4) FPN 2 suggest its effects should be considered.

Voltage drop is covered in detail in Unit 8 of this book.

6–7 OVERCURRENT PROTECTION [Article 240]

Overcurrent protection devices are intended to open the circuit to prevent damage to persons or property due to excessive or dangerous heat. Overcurrent protection devices have two ratings, overcurrent and ampere interrupting current (AIC).

Note. Overcurrent protection devices are also used to open the circuit to clear ground faults.

Overcurrent Rating [240.1 FPN]. Overcurrent protection for conductors and equipment is provided to open the circuit if the current reaches a value that will cause an excessive or dangerous temperature in conductors or conductor insulation. This is the actual ampere rating of the protection device, such as 15A, 20A or 30A [240.6(A)]. Figure 6–14.

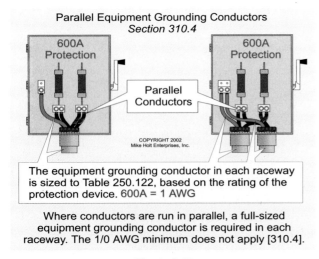

Parallel Equipment Grounding Conductors
Section 310.4

600A
Protection

600A
Protection

Parallel
Conductors

COPYRIGHT 2002
Mike Holt Enterprises, Inc.

The equipment grounding conductor in each raceway
is sized to Table 250.122, based on the rating of the
protection device. 600A = 1 AWG

Where conductors are run in parallel, a full-sized
equipment grounding conductor is required in each
raceway. The 1/0 AWG minimum does not apply [310.4].

Figure 6-12
Equipment Grounding Conductors with Parallel Conductors

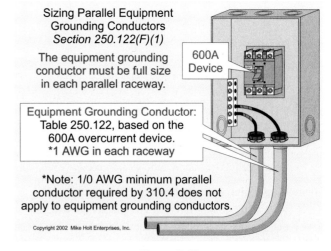

Sizing Parallel Equipment
Grounding Conductors
Section 250.122(F)(1)

The equipment grounding
conductor must be full size
in each parallel raceway.

600A
Device

Equipment Grounding Conductor:
Table 250.122, based on the
600A overcurrent device.
*1 AWG in each raceway

*Note: 1/0 AWG minimum parallel
conductor required by 310.4 does not
apply to equipment grounding conductors.

Copyright 2002 Mike Holt Enterprises, Inc.

Figure 6-13
Sizing Parallel Equipment Grounding Conductors

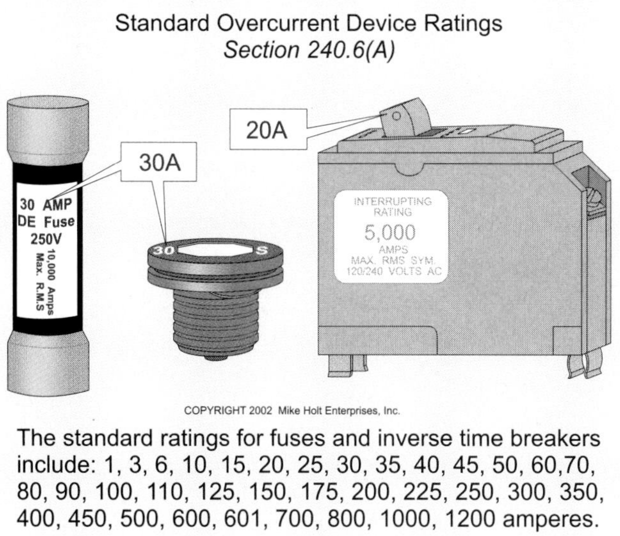

Standard Overcurrent Device Ratings
Section 240.6(A)

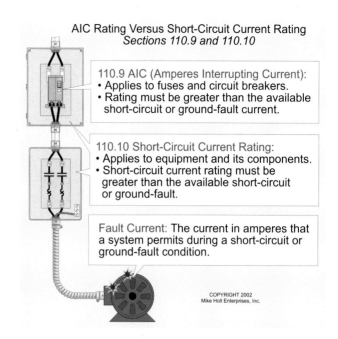

COPYRIGHT 2002 Mike Holt Enterprises, Inc.

The standard ratings for fuses and inverse time breakers include: 1, 3, 6, 10, 15, 20, 25, 30, 35, 40, 45, 50, 60,70, 80, 90, 100, 110, 125, 150, 175, 200, 225, 250, 300, 350, 400, 450, 500, 600, 601, 700, 800, 1000, 1200 amperes.

Figure 6-14
Overcurrent Device Ratings

AIC Rating Versus Short-Circuit Current Rating
Sections 110.9 and 110.10

110.9 AIC (Amperes Interrupting Current):
• Applies to fuses and circuit breakers.
• Rating must be greater than the available short-circuit or ground-fault current.

110.10 Short-Circuit Current Rating:
• Applies to equipment and its components.
• Short-circuit current rating must be greater than the available short-circuit or ground-fault.

Fault Current: The current in amperes that a system permits during a short-circuit or ground-fault condition.

COPYRIGHT 2002
Mike Holt Enterprises, Inc.

Figure 6-15
A/C Ratings Versus Short-Circuit Current Ratings

Standard Sized Protection Devices [240.6(A)]. The *NEC* lists standard sized overcurrent protection devices: 15, 20, 25, 30, 35, 40, 45, 50, 60, 70, 80, 90, 100, 110, 125, 150, 175, 200, 225, 250, 300, 350, 400, 450, 500, 600, 700, 800, 1,000, 1,200, 1,600, 2,000, 2,500, 3,000, 4,000, 5,000, and 6,000A.

Interrupting Rating [110.9]. Overcurrent protection devices, such as circuit breakers and fuses, are intended to interrupt current at fault levels and they shall have an interrupting rating sufficient for the nominal circuit voltage and the current that is available at the line terminals of the equipment.

If the overcurrent protection device is not rated for the available fault current, it could explode while attempting to clear the fault, and/or the downstream equipment could suffer serious damage causing possible hazards to people. UL, ANSI, IEEE, NEMA, manufacturers and other organizations have considerable literature on how to calculate available short-circuit current. Figure 6–15.

Note: The minimum interruption rating for a a fuse is 10,000A [240.60(C)] and a circuit breaker is 5,000A [240.83(C)] and 10,000A for fuses [240.60(C)]. Figure 6–16.

Continuous Load. Overcurrent protection devices are sized no less than 125% of the continuous load, plus 100% of the noncontinuous load [210.20(A), 215.3 and 230.42(A)]. Figure 6–17

❏ **Continuous Load**
What size protection device is required for a 100A continuous load? Figure 6–18.
(a) 150A   (b) 100A   (c) 125A   (d) 150A
• Answer: (c) 125A, 100A × 1.25 = 125A [240.6(A)]

6–8 OVERCURRENT PROTECTION OF CONDUCTORS – GENERAL REQUIREMENTS [240.4]

There are many different rules for sizing and protecting conductors and equipment. It is not simply 12 AWG wire and a 20A breaker. The general rule is that conductors must be protected according to their ampacity located at the point where the conductors receive their supply as listed in Table 310.16. Other methods of protection are permitted or required as listed in subsections (B) through (G) of this section.

(B) Next Higher Overcurrent Device Rating. The next higher protection device is permitted if all of the following conditions are met. Figure 6–19:

(1) Conductors do not supply multioutlet receptacle branch circuits for portable cord- and plug-connected loads.

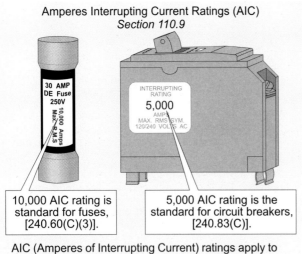

Amperes Interrupting Current Ratings (AIC)
Section 110.9

10,000 AIC rating is standard for fuses, [240.60(C)(3)].

5,000 AIC rating is the standard for circuit breakers, [240.83(C)].

AIC (Amperes of Interrupting Current) ratings apply to short-circuit and ground-fault currents, which are usually very high and must be cleared as fast as possible.

COPYRIGHT 2002 Mike Holt Enterprises, Inc.

Figure 6-16
Amperes Interrupting Current Ratings (AIC)

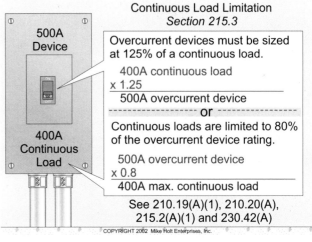

Continuous Load Limitation
Section 215.3

Overcurrent devices must be sized at 125% of a continuous load.

400A continuous load
x 1.25
500A overcurrent device

-------------------- **or** --------------------

Continuous loads are limited to 80% of the overcurrent device rating.

500A overcurrent device
x 0.8
400A max. continuous load

See 210.19(A)(1), 210.20(A), 215.2(A)(1) and 230.42(A)

COPYRIGHT 2002 Mike Holt Enterprises, Inc.

Figure 6-17
Continuous Load Limitation

(2) The ampacity of a conductor does not correspond with the standard ampere rating of a fuse or circuit breaker as listed in 240.6(A).

(3) The next size up breaker or fuse does not exceed 800A.

❏ **Overcurrent Protection of Conductors**

What size conductor is required for a 104A continuous load that is protected with a 150A breaker? Figure 6–20.

(a) 1/0 AWG   (b) 1 AWG   (c) 2 AWG   (d) any of these

• Answer: (b) 1 AWG

The conductor must be sized no less than 125% of the continuous load: 104A $\times$ 1.25 = 130A

1 AWG THHN is rated 130A at 75°C [110.14(C)(1)(a)(2)] and can be protected by a 150A protection device.

(C) Circuits with Overcurrent Protection over 800A. If the circuit overcurrent protection device exceeds 800A, the circuit conductor ampacity must not be less than the rating of the overcurrent protection device as listed in 240.6(A). Figure 6–21.

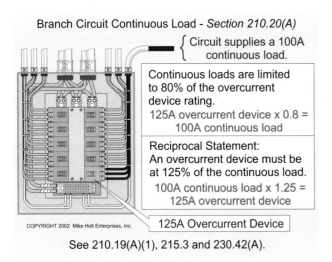

Branch Circuit Continuous Load - *Section 210.20(A)*

Circuit supplies a 100A continuous load.

Continuous loads are limited to 80% of the overcurrent device rating.
125A overcurrent device x 0.8 = 100A continuous load

Reciprocal Statement:
An overcurrent device must be at 125% of the continuous load.
100A continuous load x 1.25 = 125A overcurrent device

125A Overcurrent Device

COPYRIGHT 2002 Mike Holt Enterprises, Inc.

See 210.19(A)(1), 215.3 and 230.42(A).

Figure 6-18
Continuous Loads

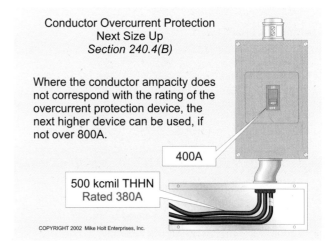

Conductor Overcurrent Protection Next Size Up
Section 240.4(B)

Where the conductor ampacity does not correspond with the rating of the overcurrent protection device, the next higher device can be used, if not over 800A.

400A

500 kcmil THHN
Rated 380A

COPYRIGHT 2002 Mike Holt Enterprises, Inc.

Figure 6-19
Overcurrent Protection – Next Size Up

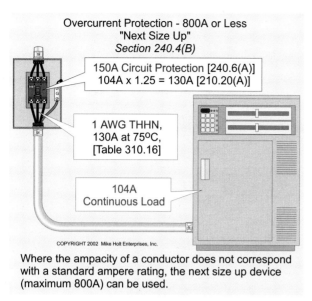

Where the ampacity of a conductor does not correspond with a standard ampere rating, the next size up device (maximum 800A) can be used.

Figure 6-20
Application of "Next Size Up" Rule

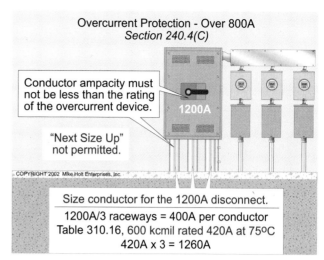

Figure 6-21
Circuits with Overcurrent Protection Over 800 Amperes

(D) Small Conductors. Unless specifically permitted in 240.4(E) through 240.4(G), overcurrent protection shall not exceed 15A for 14 AWG, 20A for 12 AWG and 30A for 10 AWG copper, or 15A for 12 AWG and 25A for 10 AWG aluminum and copper-clad aluminum after ampacity correction. Figure 6–22.

6–9 OVERCURRENT PROTECTION OF CONDUCTORS – SPECIFIC REQUIREMENTS

When sizing and protecting conductors for equipment, be sure to apply the specific *NEC* requirement.

Equipment
Air-Conditioning [440.22, and 440.32]
Appliances [422.10 and 422.11]
Cooking Appliances [210.19(A)(3), 210.21(B)(4), Note 4 of Table 220.19]
Electric Heating Equipment [424.3(B)]
Fire Protective Signaling Circuits [760.23]
Motors:
    Branch Circuits [430.22(A), and 430.52]
    Feeders [430.24, and 430.62]
    Remote Control [430.72]
Panelboard [408.16(A)]
Transformers [240.21 and 450.3]

Feeders and Services
Dwelling-Unit Feeders and Neutral [215.2, 310.15(B)(6)]
Feeder Conductor [215.2 and 215.3]
Service Conductors [230.42 and 230.90(A)]
Temporary Conductors [527.4]

Grounded (neutral) Conductor
Neutral Calculations [220.22]
Grounded Service Size [250.24(B)]

Tap Conductors
Ten feet [240.21(B)(1)]
Twenty-five feet [240.21(B)(2)]
One hundred feet [240.21(B)(4)]
Outside Feeder [240.21(B)(5)]

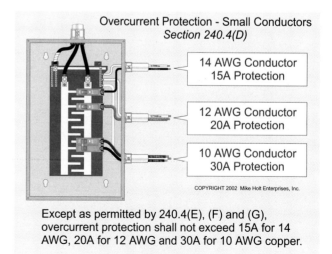

Except as permitted by 240.4(E), (F) and (G), overcurrent protection shall not exceed 15A for 14 AWG, 20A for 12 AWG and 30A for 10 AWG copper.

Figure 6-22
Protection of Small Conductors

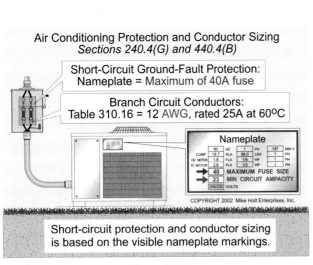

Air Conditioning Protection and Conductor Sizing
Sections 240.4(G) and 440.4(B)

Short-Circuit Ground-Fault Protection:
Nameplate = Maximum of 40A fuse

Branch Circuit Conductors:
Table 310.16 = 12 AWG, rated 25A at 60°C

Nameplate

Short-circuit protection and conductor sizing
is based on the visible nameplate markings.

Figure 6-23
Air Conditioning

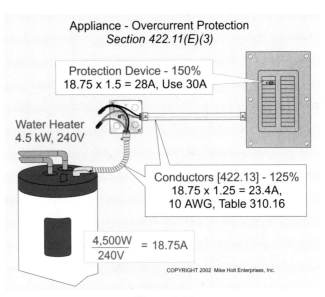

Appliance - Overcurrent Protection
Section 422.11(E)(3)

Protection Device - 150%
18.75 x 1.5 = 28A, Use 30A

Water Heater
4.5 kW, 240V

Conductors [422.13] - 125%
18.75 x 1.25 = 23.4A,
10 AWG, Table 310.16

$\dfrac{4,500W}{240V}$ = 18.75A

Figure 6-24
Water Heaters

6–10 EQUIPMENT CONDUCTORS SIZE AND PROTECTION EXAMPLES

Air-Conditioning

An air conditioner nameplate indicates the minimum circuit ampacity of 23A and maximum fuse size of 40A. What is the minimum size branch-circuit conductor and the maximum size overcurrent protection device 75°C terminals? Figure 6–23.

(a) 12 AWG, 60A fuse
(b) 12 AWG, 40A fuse
(c) 8 AWG, 50A fuse
(d) 10 AWG, 30A fuse

• Answer:(b) 12 AWG, 40A fuse

<u>Conductor:</u> The conductors must be sized based on the 60°C column of Table 310.16 and 12 AWG rated 25A.

<u>Overcurrent Protection:</u> The protection device must not be greater than a 40A fuse, either one-time or dual-element.

❑ Water Heater [422.11(E) and 422.13]

What size conductor and protection device is required for a 4,500 VA, 240V water heater? Figure 6–24.

(a) 10 AWG wire with 20A protection
(b) 10 AWG wire with 25A protection
(c) 10 AWG wire with 30A protection
(d) b or c

• Answer: (d) b or c

$I = VA/E = 4,500 VA/240V = 18.75A$

<u>Conductor Size:</u> The conductor is sized at 125% of the water heater rating [422.13].
Minimum conductor = 18.75A × 1.25 = 23.4A
Conductor is sized according to the 60°C column of Table 310.16 = 10 AWG rated 30A

<u>Overcurrent Protection:</u> Overcurrent protection device sized no more than 150% of appliance rating, [422.11(E)].
18.75A × 1.50 = 28.1A, next size up = 30A

❑ Motor [430.6(A), 430.22(A), and 430.52(C)(1)]

What size branch-circuit conductor and short-circuit protection (circuit breaker) is required for a 2-hp (12 FLC) motor rated 230V? Figure 6–25.

(a) 14 AWG with a 15A breaker
(b) 12 AWG with a 20A breaker
(c) 12 AWG with a 30A breaker
(d) 14 AWG with a 30A breaker

• Answer (d) 14 AWG with a 30A breaker

<u>Conductors:</u> Conductors are sized no less than 125% of the motor full-load current (FLC) [430.6(A) and 430.22(A)].
12A × 1.25 = 15A, Table 310.16, 14 AWG is rated 20A.

<u>Overcurrent Protection:</u> The short-circuit protection (circuit breaker) is sized at 250% of motor full-load current.
12A × 2.5 = 30A [240.6(A) and 430.52(C)(1)]

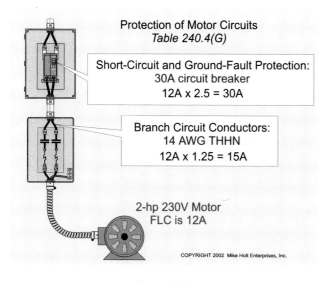

Protection of Motor Circuits
Table 240.4(G)

Short-Circuit and Ground-Fault Protection:
30A circuit breaker
12A x 2.5 = 30A

Branch Circuit Conductors:
14 AWG THHN
12A x 1.25 = 15A

2-hp 230V Motor
FLC is 12A

COPYRIGHT 2002 Mike Holt Enterprises, Inc.

Figure 6-25
Motor Protection and Conductor Sizes

Conductor Ampacities
Table 310.16

Table 310.16 is based on an ambient temperature of 86°F
and 3 current-carrying conductors in a raceway or cable.

COPYRIGHT 2002 Mike Holt Enterprises, Inc.

Figure 6-26
Conductor Ampacities

PART B – CONDUCTOR AMPACITY CALCULATIONS

6–11 CONDUCTOR AMPACITY [310.10]

The insulation temperature rating of a conductor is limited to an operating temperature that prevents serious heat damage to the conductor's insulation. If the conductor carries excessive current, the I^2R heating within the conductor can destroy the conductor insulation. To limit elevated conductor operation temperatures, the current flow (ampacity) in the conductors must be limited.

Allowable Ampacities

The ampacity of a conductor is the current the conductors can carry continuously under the specific condition of use [Article 100 definition]. The ampacity of a conductor is listed in Table 310.16 under the condition of no more than three current-carrying conductors bundled together in an ambient temperature of 86°F. The ampacity of a conductor changes if the ambient temperature is not 86°F or if more than three current-carrying conductors are bundled together in any way. Figure 6-26.

6–12 AMBIENT TEMPERATURE AMPACITY ADJUSTMENT FACTOR [Table 310.16]

The ampacity of a conductor as listed on Table 310.16 is based on the conductor operating at an ambient temperature of 86°F (30°C). When the ambient temperature is different than 86°F (30°C) for a prolonged period of time, the conductor ampacity listed in Table 310.16 must be adjusted.

In general, 90°C rated conductor ampacities cannot be used for sizing circuit conductors. However, higher insulation temperature rating offers the opportunity of having a greater conductor ampacity for conductor ampacity adjustment and a reduced temperature correction factor. The temperature adjustment factors used to determine the new conductor ampacity are listed at the bottom of Table 310.16. The following formula can be used to determine the conductor's new ampacity when the ambient temperature is not 86°F (30°C). Figure 6–27:

New Ampacity = Table 310.16 Ampacity $\times$ Ambient Temperature Adjustment Factor

Note: Conductor ampacity adjustment does not apply if the different ambient temperature exists for 10 ft or less and does not exceed 10% of the total length of the conductor [310.15(A)(2) Ex.].

❏ **Ambient Temperature Below 86°F (30°C)**

What is the ampacity of 12 AWG THHN when installed in a walk-in cooler that has an ambient temperature of 50°F? Figure 6–28.

 (a) 31A   (b) 35A   (c) 30A   (d) 20A

  • Answer: (a) 31A

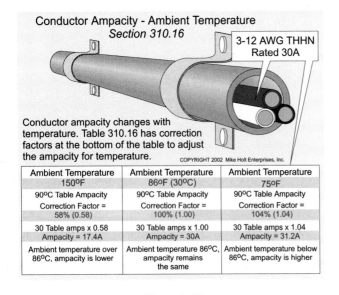

Figure 6-27

Conductor Ampacity – Ambient Temperature

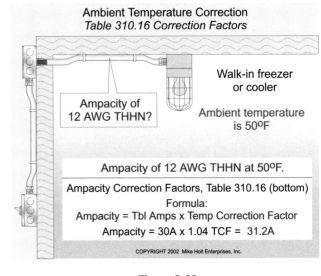

Figure 6-28

Ambient Temperature Correction

New Ampacity = Table 310.16 Ampacity × Ambient Temperature Adjustment Factor

Table 310.16 ampacity for 12 AWG THHN is 30A at 90°C.

Temperature Adjustment Factor for 90°C conductor installed in an ambient temperature of 50°F is 1.04.

New Ampacity = 30A × 1.04 = 31.2A

Note: Ampacity increases when the ambient temperature is less than 86°F (30°C).

❏ **Ambient Temperature Above 86°F (30°C)**

What is the ampacity of 6 AWG THHN when installed on a roof that has an ambient temperature of 60°C? Figure 6–29.

(a) 53A   (b) 35A   (c) 75A   (d) 60A

• Answer: (a) 53A

New Ampacity = Table 310.16 Ampacity × Ambient Temperature Adjustment Factor

Table 310.16 ampacity for 6 AWG THHN is 75A at 90°C.

Temperature Adjustment Factor, 90°C conductor rating installed at 60°C is 0.71.

New Ampacity = 75A × 0.71 = 53.25A

Note: Ampacity decreases when the ambient temperature is more than 86°F (30°C).

❏ **Conductor Size**

What size conductor is required to supply a 40A load if the conductors pass through an ambient temperature of 100°F. Figure 6–30.

(a) 10 AWG THHN   (b) 8 AWG THHN
(c) 6 AWG THHN   (d) any of these

• Answer: (b) 8 AWG THHN

The conductor to the load must have an ampacity of 40A after applying the ambient temperature adjustment factor.

New Ampacity = Table 310.16 Amperes × Ambient Temperature Adjustment Factor

10 AWG THHN = 40A × 0.91 = 36.4A

8 AWG THHN = 55A × 0.91 = 50A

Ambient Temperature Correction - *Section 310.16*

Determine ampacity of 6 AWG THHN.

Table 310.16 ampacity = 75A
Temperature correction factor, 60°C = 0.71
New Ampacity = 75A x 0.71 derating factor = 53.25A

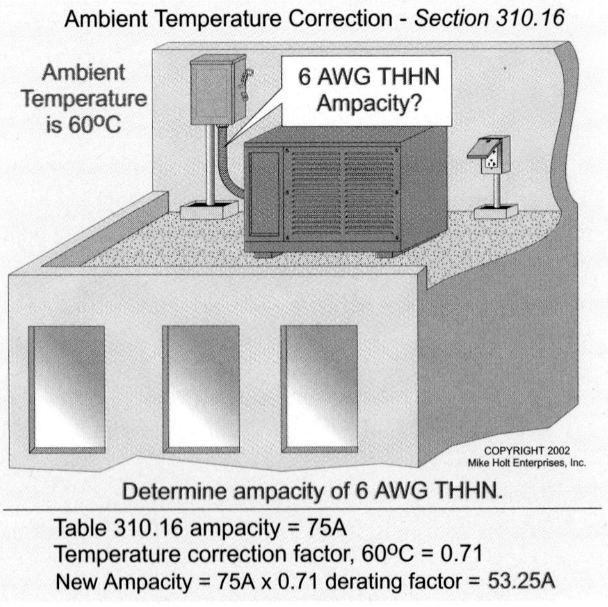

Figure 6-29

Ambient Temperature Correction

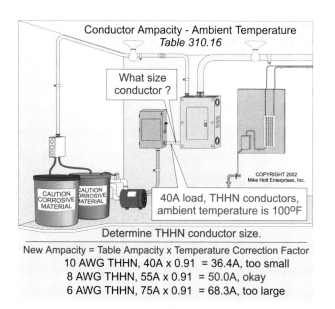

Conductor Ampacity - Ambient Temperature
Table 310.16

What size conductor ?

COPYRIGHT 2002
Mike Holt Enterprises, Inc.

40A load, THHN conductors, ambient temperature is 100°F

Determine THHN conductor size.

New Ampacity = Table Ampacity x Temperature Correction Factor
10 AWG THHN, 40A x 0.91 = 36.4A, too small
8 AWG THHN, 55A x 0.91 = 50.0A, okay
6 AWG THHN, 75A x 0.91 = 68.3A, too large

Figure 6-30
Conductor Ampacity – Ambient Temperature

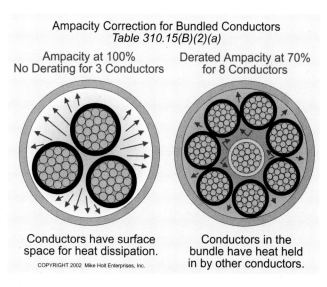

Ampacity Correction for Bundled Conductors
Table 310.15(B)(2)(a)

Ampacity at 100%
No Derating for 3 Conductors

Derated Ampacity at 70%
for 8 Conductors

Conductors have surface space for heat dissipation.

Conductors in the bundle have heat held in by other conductors.

COPYRIGHT 2002 Mike Holt Enterprises, Inc.

Figure 6-31
Ampacity Correction for Bundled Conductors

6–13 CONDUCTOR BUNDLING AMPACITY ADJUSTMENT FACTOR [Table 310.15(B)(2)]

When conductors are bundled together, the ability of the conductors to dissipate heat is reduced. The *NEC* requires that the ampacity of a conductor be reduced whenever four or more current-carrying conductors are bundled together. Figure 6–31. In general, 90°C rated conductor ampacities cannot be used for sizing a circuit conductor. However, higher insulation temperature rating offers the opportunity of having a greater conductor ampacity when used for ampacity adjustment.

The ampacity adjustment factor used to determine the new ampacity is listed in Table 310.15(B)(2)(a). The following formula can be used to determine the new conductor ampacity when more than three current-carrying conductors are bundled together:

New Ampacity = Table 310.16 Ampacity × Bundled Ampacity Adjustment Factor

Note: Conductor bundle ampacity adjustment factors do not apply to conductors in a nipple that does not exceed 24 in., see 310.15(B)(2)(a) Exception 3. Figure 6–32.

❏ **Conductor Ampacity**

What is the ampacity of four current-carrying 10 AWG THHN conductors installed in a raceway or cable? Figure 6–33.
(a) 20A   (b) 24A   (c) 32A   (d) none of these

   • Answer: (c) 32A

New Ampacity = Table 310.16 Ampacity × Bundled Ampacity Adjustment Factor

Table 310.16 ampacity for 10 AWG THHN is 40A at 90°C.

Bundle adjustment factor for four current-carrying conductors is 0.8.

New Ampacity = 40A × 0.8 = 32A

❏ **Conductor Size**

A raceway contains four current-carrying conductors. What size conductor is required to supply a 40A noncontinuous load? Figure 6–34.
(a) 10 AWG THHN   (b) 8 AWG THHN
(c) 6 AWG THHN   (d) none of these
   • Answer: (b) 8 AWG THHN

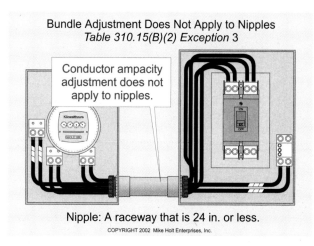

Bundle Adjustment Does Not Apply to Nipples
Table 310.15(B)(2) Exception 3

Conductor ampacity adjustment does not apply to nipples.

Nipple: A raceway that is 24 in. or less.
COPYRIGHT 2002 Mike Holt Enterprises, Inc.

Figure 6-32
Bundled Conductors in Nipples

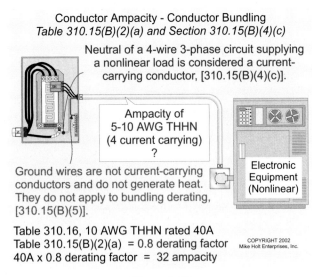

Conductor Ampacity - Conductor Bundling
Table 310.15(B)(2)(a) and Section 310.15(B)(4)(c)

Neutral of a 4-wire 3-phase circuit supplying a nonlinear load is considered a current-carrying conductor, [310.15(B)(4)(c)].

Ampacity of 5-10 AWG THHN (4 current carrying) ?

Electronic Equipment (Nonlinear)

Ground wires are not current-carrying conductors and do not generate heat. They do not apply to bundling derating, [310.15(B)(5)].

Table 310.16, 10 AWG THHN rated 40A
Table 310.15(B)(2)(a) = 0.8 derating factor
40A x 0.8 derating factor = 32 ampacity

COPYRIGHT 2002 Mike Holt Enterprises, Inc.

Figure 6-33
Conductor Ampacity – Conductor Bundling

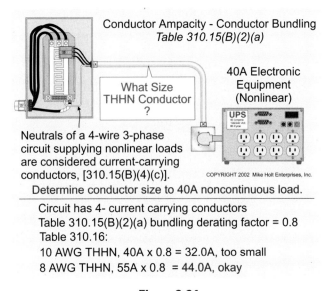

Conductor Ampacity - Conductor Bundling
Table 310.15(B)(2)(a)

What Size THHN Conductor ?

40A Electronic Equipment (Nonlinear)

UPS

Neutrals of a 4-wire 3-phase circuit supplying nonlinear loads are considered current-carrying conductors, [310.15(B)(4)(c)].

COPYRIGHT 2002 Mike Holt Enterprises, Inc.

Determine conductor size to 40A noncontinuous load.

Circuit has 4- current carrying conductors
Table 310.15(B)(2)(a) bundling derating factor = 0.8
Table 310.16:
10 AWG THHN, 40A x 0.8 = 32.0A, too small
8 AWG THHN, 55A x 0.8 = 44.0A, okay

Figure 6-34
Ampacity Correction for Bundled Conductors

The conductor must have an ampacity of 40A after applying bundle adjustment factor.

New Ampacity = Table 310.16 Ampacity × Bundled Ampacity Adjustment Factor

10 AWG THHN = 40A × 0.8 = 32A – too small

8 AWG THHN = 55A × 0.8 = 44A – just right

6 AWG THHN = 75A × 0.8 = 60A – larger than required by the *NEC*

6–14 AMBIENT TEMPERATURE AND CONDUCTOR BUNDLING ADJUSTMENT FACTORS

If the ambient temperature is different than 86°F (30°C) and there are more than three current-carrying conductors bundled together, then the ampacity listed in Table 310.16 must be adjusted for both conditions. Figure 6–35.

The following formula can be used to determine the new conductor ampacity when both ambient temperature and bundled adjustment factors apply:

Ampacity = Table 310.16 Ampacity × Temperature Factor × Bundled Factor

❏ **Conductor Ampacity**

What is the ampacity of four current-carrying 8 AWG THHN conductors installed in an ambient temperature of 100°F? Figure 6–36.

(a) 25A   (b) 40A   (c) 55A   (d) 60A

• Answer: (b) 40A

New Ampacity = Table 310.16 Ampacity × Temperature Factor × Bundled Factor

Table 310.16 ampacity of 8 AWG THHN is 55A at 90°C.

Temperature Correction factor for 90°C conductor insulation at 100°F is 0.91.

Bundled adjustment factor for four conductors is 0.8.

New Ampacity = 55A × 0.91 × 0.8 = 40A

6–15 CURRENT-CARRYING CONDUCTORS

Table 310.15(B)(2)(a) adjustment factors only apply when there are more than three current-carrying conductors bundled together. Naturally, all phase conductors are considered current-carrying, and the following should be helpful in determining which other conductors are considered current-carrying:

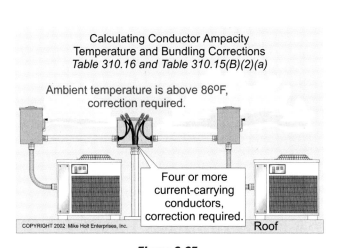

Figure 6-35

Conductor Ampacity with Temperature and Bunching

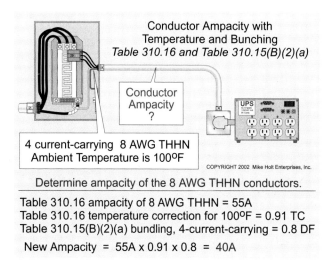

Figure 6-36

Conductor Ampacity with Temperature and Bunching

Grounded (neutral) Conductor – Balanced Circuits, 310.15(B)(4)(a)

The grounded (neutral) conductor of a balanced 3-wire circuit, or a balanced 4-wire, 3Ø wye circuit is not considered a current-carrying conductor. Figure 6–37.

Grounded (neutral) Conductor – Unbalanced 3-wire Wye Circuit, 310.15(B)(4)(b)

The grounded (neutral) conductor of a balanced 3-wire wye circuit is considered a current-carrying conductor. Figure 6–38.

This can be proven with the following formula:

$$I_{Neutral} = \sqrt{(I_{Line\ 1}^2 + I_{Line\ 2}^2) - (I_{Line\ 1} \times I_{Line\ 2})}$$

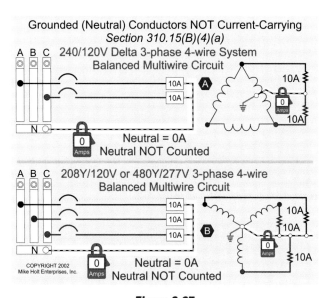

Figure 6-37

Grounded Conductors Not Current-Carrying

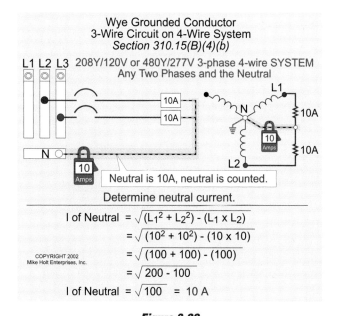

Figure 6-38

Wye Grounded Conductor

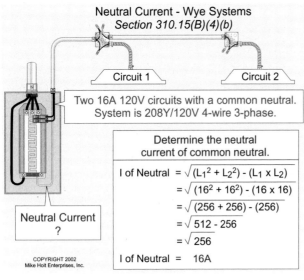

Figure 6-39

Neutral Current – Wye Systems

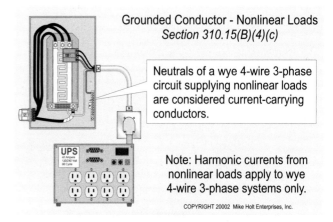

Figure 6-40

Grounded Conductor Nonlinear Loads of a
4-Wire 3-Phase System

❏ Grounded (neutral) Conductor

What is the neutral current for a balanced 16A, 3-wire, 1Ø, 208Y/120V branch circuit of a 4-wire, 3Ø wye system that supplies fluorescent lighting? Figure 6–39.

(a) 8A   (b) 16A   (c) 32A   (d) 40A

• Answer: (b) 16A

$$I_{Neutral} = \sqrt{(I_{Line\ 1}^2 + I_{Line\ 2}^2) - (I_{Line\ 1} \times I_{Line\ 2})}$$

$$I_{Neutral} = \sqrt{(16^2 + 16^2) - (16 \times 16)} = \sqrt{512 - 256} = \sqrt{256} = 16A$$

Grounded (neutral) Conductor – Nonlinear Loads, 310.15(B)(4)(c)

The grounded (neutral) conductor of a balanced 4-wire, 3Ø wye circuit that is at least 50% loaded with nonlinear loads (computers, electric-discharge lighting, etc.) is considered a current-carrying conductor. Figure 6–40.

CAUTION: Nonlinear loads produce harmonic currents that add on the neutral conductor, and the current on the neutral can be doubled. Figure 6–41.

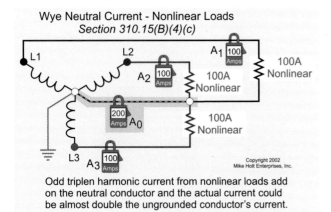

Figure 6-41

Wye-Neutral Current – Nonlinear Loads

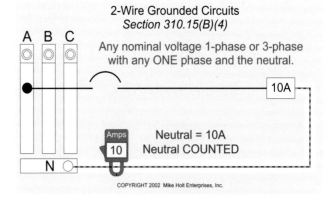

Figure 6-42

2-Wire Grounded Circuits

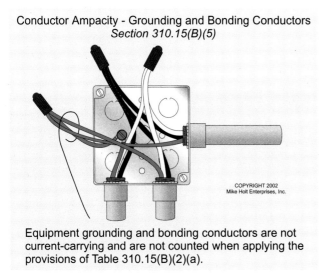

Equipment grounding and bonding conductors are not current-carrying and are not counted when applying the provisions of Table 310.15(B)(2)(a).

COPYRIGHT 2002
Mike Holt Enterprises, Inc.

Figure 6-43
Conductor Ampacity – Grounding and Bonding Conductors

Conductor Sizing Summary
Table 310.16 and Section 310.15(B)

Table 310.16 is based on an ambient temperature of 86°F and 3 current-carrying conductors in a raceway or cable.

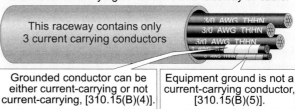

| Grounded conductor can be either current-carrying or not current-carrying, [310.15(B)(4)]. | Equipment ground is not a current-carrying conductor, [310.15(B)(5)]. |

Conductor Ampacity Correction Required

Ambient Temperature	**Conductor Bundling**
If ambient temperature is above 86°F, the conductor ampacity decreases, see the bottom of Table 310.16.	If the number of current-carrying conductors exceeds 3, the conductor ampacity decreases, Table 310.15(B)(2)(a).

COPYRIGHT 2002 Mike Holt Enterprises, Inc.

Figure 6-44
Conductor Sizing Summary

Two-wire Circuits

Both the grounded and ungrounded conductors of a 2-wire circuit carry current and both are considered current-carrying. Figure 6–42.

Grounding and Bonding Conductors, 310.15(B)(b)

Grounding and bonding conductors do not normally carry current and are not considered current-carrying. Figure 6–43.

Note: Grounding and bonding conductors are not counted when adjusting conductor ampacity for the effects of conductor bunching.

6–16 CONDUCTOR SIZING SUMMARY

The ampacity of a conductor changes with changing conditions. The factors that affect conductor ampacity are shown in Figure 6–44.

1. The allowable ampacity as listed in Table 310.16.
2. The ambient temperature correction factors, if the ambient temperature is not 86°F.
3. Conductor ampacity adjustment factors apply if four or more current-carrying conductors are bundled together.

Terminal Ratings, 110.14(C)

Equipment rated 100A or less must have the conductor sized no smaller than the 60°C column of Table 310.16. Equipment rated over 100A must have the conductors sized no smaller than the 75°C column of Table 310.16. However, higher insulation temperature rating offers the opportunity of having a greater conductor ampacity for conductor ampacity adjustment.

Unit 6 – Conductor Sizing and Protection
Summary Questions

Calculations Questions

6–1 Conductor Insulation Property [Table 310.13]

1. THHN can be described as _____.
    (a) thermoplastic insulation with a nylon outer cover
    (b) suitable for dry and wet locations
    (c) having a maximum operating temperature of 90°C
    (d) a and c

6–2 Conductor Allowable Ampacity [310.15]

2. • The maximum overcurrent protection device size for 14 AWG is 15A, 12 AWG is 20A and 10 AWG is 30A. This is a general rule, but it does not apply to motors or air conditioners according to 240.4(G).
    (a) True   (b) False

6–3 Conductor Sizing

3. • Conductor sizes are expressed in American Wire Gage (AWG) from 40 AWG through 4/0 AWG. Conductors larger than _____ are expressed in circular mils.
    (a) 1/0 AWG   (b) 1 AWG   (c) 3/0 AWG   (d) 4/0 AWG

4. The smallest size conductor permitted for branch circuits, feeders and services for residential, commercial and industrial locations is _____.
    (a) 14 AWG copper   (b) 12 AWG aluminum   (c) 12 AWG copper   (d) a and b

6–4 Terminal Ratings [110.14(C)]

5. Equipment terminals rated 100A or less (receptacles, switches, circuit breakers, fuses, etc.) and pressure connector terminals for 14 AWG through 1 AWG conductors shall have the conductor sized according to 60°C temperature rating, as listed in Table 310.16.
    (a) True   (b) False

6. • What is the minimum size THHN conductor that is permitted to terminate on a 70A circuit breaker or fuse? Be sure to comply with the requirements of 110.14(C)(1).
    (a) 8 AWG   (b) 6 AWG   (c) 4 AWG   (d) none of these

7. • What size THHN conductor is required for a 70A branch circuit if the circuit breaker and equipment are listed for 75°C terminals and the load does not exceed 65A?
    (a) 10 AWG   (b) 8 AWG   (c) 6 AWG   (d) 4 AWG

8. Terminals for equipment rated over 100A and pressure connector terminals for conductors larger than 1 AWG shall have the conductor sized according to 75°C temperature rating, as listed in Table 310.16.
    (a) True   (b) False

9. What size THHN conductor is required for an air-conditioning unit if the nameplate requires a conductor ampacity of 34A? Terminals of all the equipment and circuit breakers are rated 75°C.
    (a) 12 AWG   (b) 10 AWG   (c) 8 AWG   (d) 14 AWG

---

10. What is the minimum size THHN conductor required for a 150A circuit breaker or fuse? Be sure to comply with the requirements of 110.14(C)(2).
    (a) 1/0 AWG   (b) 2/0 AWG   (c) 3/0 AWG   (d) 4/0 AWG

11. In general, THHN (90°C) conductor ampacities cannot be used when sizing conductors; but, when more than three current-carrying conductors are bundled together, or if the ambient temperature is greater than 86°F, the allowable conductor ampacity shall be decreased. THHN offers the opportunity of having a greater ampacity for conductor adjustment purposes, thereby permitting the same conductor to be used without having to increase the conductor size.
    (a) True   (b) False

12. • What size conductor is required to supply a 190A load in a dry location? Terminals are rated 75°C.
    (a) 300 kcmil   (b) 4/0 AWG   (c) 3/0 AWG   (d) none of these

6–5 Conductors in Parallel [310.4]

13. • Phase and grounded (neutral) conductors sized 1 AWG and larger are permitted to be connected in parallel.
    (a) True   (b) False

14. To ensure that the currents are evenly distributed between the parallel conductors, each conductor within a parallel set shall be installed in the same type of raceway (metallic or nonmetallic) and shall be the same length, material, circular mils, insulation type, and must terminate in the same method.
    (a) True   (b) False

15. • When an electric relay (coil) is energized, the initial current can be very high causing significant voltage drop. The reduced voltage at the coil (because of voltage drop) can cause the coil contacts to chatter (open and close like a buzzer) or not close at all. Paralleling of control wiring conductors is often necessary and permitted by the *NEC* to reduce the effects of voltage drop for long control runs.
    (a) True   (b) False

16. When equipment grounding conductors are installed in parallel, each raceway must have a full-size equipment grounding conductor sized according to the overcurrent protection device rating of that circuit.
    (a) True   (b) False

17. All parallel equipment grounding conductors are required to be a minimum 1/0 AWG.
    (a) True   (b) False

18. What size equipment grounding conductor is required in each raceway for an 800A, 500 kcmil feeder paralleled in two raceways?
    (a) 3 AWG   (b) 2 AWG   (c) 1 AWG   (d) 1/0 AWG

19. If we have an 800A service with a calculated demand load of 750A, what size 75°C conductors would be required if paralleled in two raceways?
    (a) 4/0 AWG   (b) 250 kcmil   (c) 500 kcmil   (d) 750 kcmil

20. • What are the conductors required for a 250A feeder paralleled in two raceways?
    (a) 3 AWG   (b) 2 AWG   (c) 2/0 AWG   (d) 1/0 AWG

6–6 Conductor Size – Voltage Drop [210.19(A) FPN No. 4, and 215.2(B) FPN No. 2]

21. There is no mandatory rule in the *NEC* limiting the voltage drop on conductors, but the *Code* recommends that we consider its effect.
    (a) Truc   (b) False

6–7 Overcurrent Protection [240]

22. One of the purposes of conductor overcurrent protection is to protect the conductors against excessive or dangerous heat.
    (a) True   (b) False

23. Overcurrent devices shall be designed and rated to clear fault current and must have a short-circuit interrupting rating sufficient for the available fault levels. The minimum interruption rating for circuit breakers is _____ and _____ for fuses.
    (a) 10,000A, 10,000A   (b) 5,000A, 5,000A   (c) 5,000A, 10,000A   (d) 10,000A, 5,000A

24. Which of the following is a standard sized circuit breaker and fuse size?
    (a) 25A   (b) 90A   (c) 350A   (d) any of these

25. Where a circuit supplies continuous loads or any combination of continuous and noncontinuous loads, the rating of the overcurrent device shall not be less than the noncontinuous load plus _____ of the continuous load.
    (a) 80%   (b) 100%   (c) 125%   (d) 150%

6–8 Overcurrent Protection of Conductors - General Requirements [240.4]

26. If the ampacity of a conductor does not correspond with the standard ampere rating of a fuse or circuit breaker, the next size up protection device is permitted. This applies only if the conductors supply multioutlet receptacles for portable cord-and plug-connected loads.
    (a) True   (b) False

27. What size conductor (75°C) is required for a 70A breaker that supplies a 70A load?
    (a) 8 AWG   (b) 6 AWG   (c) 4 AWG   (d) any of these

6–11 Conductor Ampacity [310.10]

28. The temperature rating of a conductor is the maximum operating temperature the conductor insulation can withstand (without serious damage) over a prolonged period of time. The _____ provide guidance for adjusting the conductor ampacities for the different conditions.
    (a) conductor allowable ampacities
    (b) ambient temperature correction factors
    (c) correction factors
    (d) all of these

6–12 Ambient Temperature Adjustment Factor [Table 310.16]

29. The ampacities listed in Table 310.16 apply only when the ambient temperature is 40°C and there are no more than two current-carrying conductors bundled together. If the ambient temperature is not 40°C, or there are more than two current-carrying conductors in a raceway, the allowable ampacities shall be adjusted to reflect the ampacity under the condition of use.
    (a) True   (b) False

30. • What is the ampacity of an 8 AWG THHN conductor when installed in a walk-in cooler if the ambient temperature is 50°F?
    (a) 40A   (b) 50A   (c) 55A   (d) 57A

31. • What size THHN conductor is required to feed a 16A load when the conductors are in an ambient temperature of 100°F? The circuit is protected with a 20A overcurrent protection device.
    (a) 14 AWG THHN   (b) 12 AWG THHN   (c) 10 AWG THHN   (d) 8 AWG THHN

6–13 Conductor Bundle Adjustment Factor [Table 310.15(B)(2)]

32.   When four or more current-carrying conductors are bundled together for more than _____, the conductor allowable ampacity must be reduced according to the factors listed in Table 310.15(B)(2)(a).
        (a) 12 in.   (b) 24 in.   (c) 36 in.   (d) 48 in.

33.   What is the ampacity of four 1/0 AWG THHN conductors?
        (a) 111A   (b) 136A   (c) 153A   (d) 171A

34.   • A raceway contains eight current-carrying conductors. What size conductor is required to feed a 21A noncontinuous lighting load? The overcurrent protection device is rated 30A.
        (a) 14 AWG THHN   (b) 12 AWG THHN   (c) 10 AWG THHN   (d) any of these

6–14 Ambient Temperature and Conductor Bunching Adjustment Factors

35.   What is the ampacity of eight current-carrying 10 AWG THHN conductors installed in ambient temperature of 100°F?
        (a) 21A   (b) 25A   (c) 32A   (d) 40A

6–15 Current-Carrying Conductors

36.   • The neutral conductor of a balanced 3-wire delta circuit, or 4-wire, 3Ø wye circuit, is considered a current-carrying conductor for the purpose of applying the adjustment factors of Table 310.15(B)(2)(a).
        (a) True   (b) False

37.   The neutral conductor of a balanced 4-wire, 3Ø wye circuit that is at least 50 percent loaded with nonlinear loads (electric-discharge lighting, electronic ballast, dimmers, controls, computers, laboratory test equipment, medical test equipment, recording studio equipment, etc.) is not considered a current-carrying conductor for the purpose of applying bundle adjustment factors.
        (a) True   (b) False

38.   • The neutral conductor of a balanced 3-wire wye circuit is not considered a current-carrying conductor for the purpose of applying bundle adjustment factors.
        (a) True   (b) False

6–16 Conductor Sizing Summary

39.   • The ampacity of a conductor can be different along the length of the conductor. The higher calculated ampacity can be used if the length of the lower ampacity is no more than 10 ft or no more than 10 percent of the length of the circuit conductors.
        (a) True   (b) False

40.   Most terminals are rated 60°C for equipment 100A or less and 75°C for equipment terminals rated over 100A. Regardless of the conductor ampacity, conductors shall be sized no smaller than the terminal temperature rating.
        (a) True   (b) False

☆ Challenge Questions

6–7 Overcurrent Protection [240]

41.   • A continuous load of 27A requires the circuit overcurrent protection device to be sized at _____.
        (a) 20A   (b) 30A   (c) 40A   (d) 35A

42.   • What size overcurrent protection device is required for a 45A continuous load? The circuit is in a raceway with 14 current-carrying conductors.
        (a) 45A   (b) 50A   (c) 60A   (d) 70A

43.   • A 65A continuous load requires a _____ overcurrent protection device.
      (a) 60A   (b) 70A   (c) 75A   (d) 90A

44.   • A department store (continuous load) feeder supplies a lighting load of 103A. The minimum size overcurrent
      protection device permitted for this feeder is _____.
      (a) 110A   (b) 125A   (c) 150A   (d) 175A

6–12 Ambient Temperature Adjustment Factor [Table 310.16]

45.   • A 2 AWG TW conductor is installed in a location where the ambient temperature is expected to be 102°F. The
      temperature correction factor for conductor ampacity in this location is _____.
      (a) 0.96   (b) 0.88   (c) 0.82   (d) 0.71

46.   • If the ambient temperature is 71°C, the minimum insulation that a conductor must have and still have the capacity to
      carry current is _____.
      (a) 60°C   (b) 105°C   (c) 90°C   (d) any of these

6–13 Conductor Bundle Adjustment Factors [Table 310.15(B)(2)]

47.   • The ampacity of six current-carrying 4/0 AWG XHHW aluminum conductors installed in a ground floor slab (wet
      location) is _____.
      (a) 135A   (b) 185A   (c) 144A   (d) 210A

6–14 Ambient Temperature and Conductor Bundle Adjustment Factors

48.   • The ampacity of 15 current-carrying 10 AWG RHW aluminum conductors in an ambient temperature of 75°F is
      _____.
      (a) 30A   (b) 22A   (c) 16A   (d) 12A

49.   • A(n) _____ AWG THHN conductor is required for a 19.7A load if the ambient temperature is 75°F and there are nine
      current-carrying conductors in the raceway.
      (a) 14   (b) 12   (c) 10   (d) 8

50.   • The ampacity of nine current-carrying 10 AWG THW conductors installed in a 20 in.-long raceway is _____.
      (a) 25A   (b) 30A   (c) 35A   (d) none of these

51.   • The ampacity of 10 current-carrying 6 AWG THHW conductors installed in an 18 in. long conduit in a dry location having an ambient temperature of 39°C is _____.
       (a) 47A   (b) 68A   (c) 66A   (d) 75A

6–15 Current-Carrying Conductors

52.   A raceway contains the following: one 4-wire multiwire branch circuit that supplies a balanced incandescent 120V lighting load; one 4-wire multiwire branch circuit that supplies a balanced 120V fluorescent lighting load; two conductors that supply a receptacle; and one equipment grounding conductor. The system is 3Ø, 208Y/120V. Taking these factors into consideration, how many of these conductors are considered current-carrying?
       (a) 7 conductors   (b) 8 conductors   (c) 9 conductors   (d) 11 conductors

53.   • There is a total of nine 10 AWG THW conductors in a raceway. The system voltage is 3Ø, 208Y/120V. One conductor is an equipment grounding conductor, four conductors supply a 4-wire multiwire branch circuit for balanced electric-discharge luminaires and the remaining conductors supply a 4-wire multiwire branch circuit for balanced incandescent luminaires. Taking all of these factors into consideration, how many of these conductors are considered current-carrying?
       (a) 6 conductors   (b) 7 conductors   (c) 9 conductors   (d) 10 conductors

NEC Questions – Articles 314-354

Article 314 Outlet, Device, Pull and Junction Boxes, Conduit Bodies, Fittings, and Manholes

54.   Where a box is used as the sole support of a ceiling-suspended (paddle) fan, the box shall be listed for the application and for the weight of the fan to be supported.
       (a) True   (b) False

55.   When sizing a pull box in a straight run which contains conductors of 4 AWG or larger, the length of the box shall not be less than _____ for systems not over 600V.
       (a) 8 times the diameter of the largest raceway
       (b) 6 times the diameter of the largest raceway
       (c) 48 times the outside diameter of the largest shielded conductor
       (d) 36 times the largest conductor

56.   Pull boxes or junction boxes that have any dimension over _____ shall have all conductors cabled or racked up in an approved manner.
       (a) 3 ft   (b) 6 ft   (c) 9 ft   (d) 12 ft

57.   Which of the following are required to be accessible?
       (a) outlet boxes   (b) junction boxes   (c) pull boxes   (d) all of these

58.   Listed boxes designed for underground installation can be directly buried when covered by _____.
       (a) concrete   (b) gravel
       (c) noncohesive granulated soil   (d) b and c

59.   Sheet steel boxes not over 100 cu in. in size shall be made from steel not less than _____ thick.
       (a) 0.0625 in.   (b) 0.0757 in.   (c) 0.075 in.   (d) 0.025 in.

60.   Metal boxes over _____ in size shall be constructed so as to be of ample strength and rigidity.
       (a) 50 cu in.   (b) 75 cu in.   (c) 100 cu in.   (d) 125 cu in.

61.   All boxes and conduit bodies, covers, extension rings, plaster rings and the like shall be durably and legibly marked with the manufacturer's name or trademark.
       (a) True   (b) False

62. For systems over 600V, the length of a pull box in a straight run shall not be less than _____ entering the box.
    (a) 18 times the diameter of the largest raceway
    (b) 48 times the diameter of the largest raceway
    (c) 48 times the outside diameter of the largest shielded conductor or cable
    (d) 36 times the largest conductor

63. For straight pulls, the length of a pull box shall not be less than _____ times the outside diameter, over sheath, of the largest conductor or cable entering the box on systems over 600V.
    (a) 18   (b) 16   (c) 36   (d) 48

64. • For angle or U-pulls, the distance between the conductor entry (for systems over 600V, nominal) and the opposite wall of the box shall not be less than _____ times the outside diameter, over sheath, of the largest cable or conductor.
    (a) 6   (b) 12   (c) 24   (d) 36

65. The distance between a cable or conductor entry and its exit from the box shall be not less than _____ times the outside diameter, over sheath, of that cable or conductor on a 1,000V system.
    (a) 16   (b) 18   (c) 36   (d) 40

Article 320 Armored Cable: Type AC

66. • The use of AC cable is permitted in _____ installations.
    (a) wet   (b) cable tray   (c) exposed   (d) b and c

67. Armored cable is limited or not permitted _____.
    (a) in commercial garages   (b) where subject to physical damage
    (c) motion picture studios   (d) all of these

68. Exposed runs of AC cable shall closely follow the surface of the building finish or of running boards. Exposed runs shall also be permitted to be installed on the underside of joists where supported at each joist and located so as not to be subject to physical damage.
    (a) True   (b) False

69. AC cable installed through, or parallel to, framing members must be protected against physical damage from penetration by screws or nails.
    (a) True   (b) False

70. Where run across the top of floor joists, or within 7 ft of floor or floor joists, across the face of rafters or studding in attics and roof spaces that are accessible, AC cable shall be protected by substantial guard strips that are _____.
    (a) at least as high as the cable   (b) constructed of metal
    (c) made for the cable   (d) none of these

71. When AC cable is run across the top of a floor joist in an attic without permanent ladders or stairs, substantial guard strips within _____ of the scuttle hole shall protect the cable.
    (a) 7 ft   (b) 6 ft   (c) 5 ft   (d) 3 ft

72. When armored cable is run on the side of rafters, studs or floor joists in an accessible attic, protection is required for the cable with running boards.
    (a) True   (b) False

73. The radius of the curve of the inner edge of any bend shall not be less than _____ for AC cable.
    (a) five times the largest conductor within the cable
    (b) three times the diameter of the cable
    (c) five times the diameter of the cable
    (d) six times the outside diameter of the conductors

74. AC cable shall be supported at intervals not exceeding $4^1/_2$ ft and the cable shall be secured within _____ of cable termination.
    (a) 4 in.   (b) 8 in.   (c) 9 in.   (d) 12 in.

75. Armored cable used for the connection of recessed luminaires or equipment within an accessible ceiling does not need to be secured for lengths up to _____.
   (a) 2 ft   (b) 3 ft   (c) 4 ft   (d) 6 ft

Article 322 Flat Cable Assemblies: Type FC

76. FC cable is an assembly of parallel conductors formed integrally with an insulating material web specifically designed for field installation in surface metal raceway.
   (a) True   (b) False

78. Flat cable assemblies are suitable to supply tap devices for _____ loads. The rating of the branch circuit shall not exceed 30A.
   (a) lighting   (b) small appliance   (c) small power   (d) all of these

78. Flat cable assemblies shall not be installed outdoors or in wet or damp locations unless _____ for use in wet locations.
   (a) special permission is granted   (b) approved   (c) identified   (d) none of these

79. Flat cable assemblies shall not be installed _____.
   (a) where subject to corrosive vapors unless suitable for the application
   (b) in hoistways
   (c) in any hazardous (classified) location
   (d) all of these

80. All extensions from flat cable assemblies shall be made by approved wiring methods within the _____ that is/are installed at either end of the flat cable assembly runs.
   (a) end-caps   (b) junction boxes
   (c) surface metal raceway   (d) underfloor metal raceway

81. Tap devices used in FC assemblies shall be rated at not less than _____ or more than 300 volts-to-ground, and they shall be color-coded in accordance with the requirements of 322.120(C).
   (a) 20A   (b) 15A   (c) 30A   (d) 40A

82. Flat cable assemblies shall consist of _____ conductors.
   (a) 2   (b) 3   (c) 4   (d) any of these

83. Flat cable assemblies shall have conductors of _____ AWG special stranded copper wires.
   (a) 14   (b) 12   (c) 10   (d) all of these

Article 324 Flat Conductor Cable: Type FCC

84. A field-installed wiring system for branch circuits designed for installation under carpet squares is defined as _____.
   (a) underfloor wiring   (b) undercarpet wiring
   (c) flat conductor cable   (d) underfloor conductor cable

85. FCC cable consists of _____ copper conductors placed edge-to-edge and separated and enclosed within an insulating assembly.
   (a) 3 or more square   (b) 2 or more round   (c) 3 or more flat   (d) 2 or more flat

86. FCC systems shall be permitted both for general-purpose and appliance branch circuits; they shall not be permitted for individual branch circuits.
   (a) True   (b) False

87. The maximum voltage permitted between ungrounded conductors of flat conductor cable systems is _____.
   (a) 600V   (b) 300V   (c) 250V   (d) 150V

88. General-use branch circuits using flat conductor cable shall not exceed _____.
   (a) 15A   (b) 20A   (c) 30A   (d) 40A

89. Use of FCC cable systems shall be permitted on wall surfaces in _____.
    (a) surface metal raceways   (b) cable trays
    (c) busways   (d) any of these

90. Use of FCC systems in damp locations shall be _____.
    (a) restricted
    (b) permitted
    (c) permitted provided the system is encased in concrete
    (d) approved by special permission

91. Floor-mounted-type flat conductor cable and fittings shall be covered with carpet squares no larger than _____.
    (a) 36 square inches   (b) 36 inches square   (c) 30 square inches   (d) 24 inches square

92. Flat conductor cable cannot be installed in _____.
    (a) residential units   (b) schools   (c) hospitals   (d) any of these

93. No more than _____ layers of flat conductor cable can cross at any one point.
    (a) 2   (b) 3   (c) 4   (d) none of these

94. All bare FCC cable end fittings shall be _____.
    (a) sealed   (b) insulated   (c) listed   (d) all of these

95. The top shield installed over all floor-mounted FCC cable shall completely _____ all cable runs, corners, connectors and ends.
    (a) cover   (b) encase   (c) protect   (d) none of these

96. Receptacles, receptacle housings and self-contained devices used with flat conductor cable systems shall be _____.
    (a) rated a minimum of 20A   (b) rated a minimum of 15A
    (c) identified for this use   (d) none of these

97. Each FCC transition assembly shall incorporate means for _____.
    (a) facilitating the entry of the FCC cable into the assembly
    (b) connecting the FCC cable to grounded conductors
    (c) electrically connecting the assembly to the metal cable shields and grounding conductors
    (d) all of these

98. Metal shields for flat conductor cable must be electrically continuous to the _____.
    (a) floor   (b) cable
    (c) equipment grounding conductor   (d) none of these

99. FCC cable shall be clearly and durably marked _____.
    (a) on the top side at intervals not exceeding 30 in.
    (b) on both sides at intervals not exceeding 24 in.
    (c) with conductor material, maximum temperature and ampacity
    (d) b and c

Article 326 Integrated Gas Spacer Cable: Type IGS

100. IGS cable is a factory assembly of one or more conductors, each individually insulated and enclosed in a loose-fit, nonmetallic flexible conduit as an integrated gas spacer cable rated _____ volts.
    (a) 0 through 6,000   (b) 600 through 6,000   (c) 0 through 600   (d) 3,000 through 6,000

Article 328 Medium Voltage Cable: Type MV

101. MV is defined as a single or multiconductor solid dielectric insulated cable rated _____ or higher.
    (a) 601V   (b) 1,001V   (c) 2,001V   (d) 6,001V

Article 330 Metal-Clad Cable: Type MC

102. MC cable shall not be used where exposed to the following destructive corrosive conditions, unless the metallic sheath is suitable for the conditions or is protected by material suitable for the conditions:
(a) Direct burial in the earth
(b) In concrete
(c) In cinder fill
(d) all of these

103. MC cable installed through or parallel to framing members must be protected against physical damage from penetration by screws or nails by $1^1/_4$ in. separation or protected by a suitable metal plate.
(a) True
(b) False

104. Smooth-sheath MC cable that has an external diameter of not greater than 1 in. shall have a bending radius of not more than _____ times the cable external diameter.
(a) 5
(b) 10
(c) 12
(d) 13

105. MC cable shall be supported and secured at intervals not exceeding _____.
(a) 3 ft
(b) 6 ft
(c) 4 ft
(d) 2 ft

106. MC cable can be unsupported where it is:
(a) Fished between concealed access points in finished buildings or structures and support is impracticable.
(b) Not more than 2 ft in length at terminals where flexibility is necessary.
(c) Not more than 6 ft from the last point of support within an accessible ceiling for the connection of luminaires.
(d) any of these

107. The minimum size conductor permitted for MC cable is _____ AWG.
(a) 18 copper
(b) 14 copper
(c) 12 copper
(d) none of these

108. The metallic sheath of metal-clad cable shall be continuous and _____.
(a) flame-retardant
(b) weatherproof
(c) close fitting
(d) all of these

Article 332 Mineral-Insulated, Metal-Sheathed Cable: Type MI

109. The outer sheath of MI cable is made of _____.
(a) aluminum
(b) steel alloy
(c) copper
(d) b or c

110. Which of the following statements about MI cable is correct?
(a) It may be used in any hazardous location.
(b) A single run of cable shall not contain more than the equivalent of four quarter bends.
(c) It shall be securely supported at intervals not exceeding 10 ft.
(d) none of these

111. Bends in MI cable shall be made so that the cable will not be _____.
(a) damaged
(b) shortened
(c) a and b
(d) none of these

112. The radius of the inner edge of any bend in MI cable shall not be less than five times the external diameter of the metallic sheath for cable and not more than _____ in external diameter.
(a) $^1/_2$ in.
(b) $^1/_4$ in.
(c) $^5/_8$ in.
(d) $1^1/_2$ in.

113. The radius of the inner edge of any bend in MI cable shall not be less than _____ times the cable external diameter for any cable having a diameter greater than $^3/_4$ in., but not more than 1 in.
(a) 6
(b) 3
(c) 8
(d) 10

114. MI cable shall be securely supported at intervals not exceeding _____.
(a) 3 ft
(b) $3^1/_2$ ft
(c) 5 ft
(d) 6 ft

115. Where single-conductor MI cables are used, all ungrounded (phase) conductors and, when used, the _____ conductor, shall be grouped together to minimize induced voltage on the metal sheath.
(a) larger
(b) neutral
(c) grounding
(d) largest

116. Where MI cable terminates, a _____ shall be provided immediately after stripping to prevent the entrance of moisture into the insulation.

(a) bushing   (b) connector   (c) fitting   (d) seal

117. MI cable conductors shall be made of _____ with a resistance corresponding to standard AWG and kcmil sizes.

(a) solid copper   (b) solid or stranded copper

(c) stranded copper   (d) solid copper or aluminum

118. Where the outer sheath of MI cable is made of copper, it shall provide an adequate path for equipment grounding purposes.

(a) True   (b) False

119. The conductor insulation of MI cable shall be a highly-compressed refractory mineral that will provide proper _____ for all conductors.

(a) covering   (b) spacing   (c) resistance   (d) none of these

Article 334 Nonmetallic-Sheathed Cable: Types NM, NMC and NMS

120. NM cable shall be _____.

(a) marked   (b) approved   (c) identified   (d) listed

121. NM, NMC and NMS nonmetallic-sheathed cables shall be permitted to be used in _____.

(a) one-family dwellings   (b) multifamily dwellings

(c) other structures   (d) all of these

122. NM and NMC cables shall not be used in one-and two-family dwellings exceeding three floors above grade.

(a) True   (b) False

123. NM cable can be installed in multifamily dwellings of Types III, IV and V construction except as prohibited in 334.

(a) True   (b) False

124. NM cable can be installed as open runs in dropped or suspended ceilings in other than one- and two-family and multifamily dwellings.

(a) True   (b) False

125. NM cable shall not be used _____.

(a) in commercial buildings

(b) in the air void of masonry block not subject to excessive moisture

(c) for exposed work

(d) embedded in poured cement, concrete or aggregate

126. NM cable must closely follow the surface of the building finish or running boards when run exposed.

(a) True   (b) False

127. Where NMC cable is run at angles with joists in unfinished basements, it shall be permissible to secure cables not smaller than _____ conductors directly to the lower edges of the joist.

(a) two, 6 AWG   (b) three, 8 AWG   (c) three, 10 AWG   (d) a or b

128. NM cable installed through, or parallel to, framing members shall be protected against physical damage from penetration by screws or nails. Grommets or bushings for the protection of NM cables shall be _____ for the purpose, and they must remain in place.

(a) marked   (b) approved   (c) identified   (d) listed

129. NM cables run horizontally through framing are considered supported and secured where such support does not exceed $4\frac{1}{2}$ ft intervals and the NM cable is securely fastened in place by an approved means within 12 in. of each box, cabinet, conduit body or other NM cable termination.

(a) True   (b) False

130. NM cable shall be secured in place within _____ of every cabinet, box or fitting.

(a) 6 in.   (b) 10 in.   (c) 12 in.   (d) 18 in.

131. Two conductor NM cables cannot be stapled on edge.
    (a) True   (b) False

132. NM cable, installed within accessible ceilings for the connections to luminaires and equipment, does not need to be secured within 12 in. from the luminaire or equipment when the free length does not exceed ___.
    (a) $4^1/_2$ ft   (b) $2^1/_2$ ft   (c) $3^1/_2$ ft   (d) any of these

133. Switch, outlet and tap devices of insulating material shall be permitted to be used without boxes in exposed NMC cable.
    (a) True   (b) False

134. The ampacity of NM cable shall be that of 60°C conductors, as listed in 310.15. However, the 90°C rating can be used for ampacity adjustment purposes provided the final derated ampacity does not exceed that of a _____ rated conductor.
    (a) 120°C   (b) 60°C   (c) 90°C   (d) none of these

135. Insulation rating of ungrounded conductors in NM cable shall be _____.
    (a) 60°C   (b) 75°C   (c) 90°C   (d) any of these

136. • The difference in the construction specifications between NM cable and NMC cable is that NMC cable is _____.
    (a) corrosion-resistant   (b) flame-retardant
    (c) fungus-resistant   (d) a and c

Article 336 Power and Control Tray Cable: Type TC

137. TC cable shall be permitted to be used _____.
    (a) for power and lighting circuits   (b) in cable trays in hazardous locations
    (c) in Class 1 control circuits   (d) all of these

138. TC tray cable shall not be installed _____.
    (a) where it will be exposed to physical damage   (b) as open cable on brackets or cleats
    (c) direct buried unless identified for such use   (d) all of these

Article 338 Service-Entrance Cable: Types SE and USE

139. _____ is a type of multiconductor cable permitted for use as an underground service-entrance cable.
    (a) SE   (b) NMC   (c) UF   (d) USE

140. • USE or SE cable must have a minimum of _____ conductors (including the uninsulated one) in order for one of the conductors to be uninsulated.
    (a) one   (b) two   (c) three   (d) four

Article 340 Underground Feeder and Branch-Circuit Cable: Type UF

141. UF cable shall not be used where subjected to physical damage. When this cable is subject to physical damage, it shall be protected by a suitable method such as a raceway.
    (a) True   (b) False

142. UF cable shall not be used in _____.
    (a) motion picture studios   (b) storage battery rooms
    (c) hoistways   (d) all of these

143. UF cable shall not be used _____.
    (a) in any hazardous (classified) location
    (b) embedded in poured cement, concrete, or aggregate
    (c) where exposed to direct rays of the sun, unless identified as sunlight-resistant
    (d) all of these

144. The maximum size underground UF cable is _____.
    (a) 14   (b) 10   (c) 1/0   (d) 4/0

145. The overall covering of UF cable shall be _____.
    (a) flame retardant   (b) moisture, fungus and corrosion resistant
    (c) suitable for direct burial in the earth   (d) all of these

Article 342 Intermediate Metal Conduit: Type IMC

146. IMC can be installed in or under cinder fill that is subjected to permanent moisture _____.
    (a) where the conduit is not less than 18 in. under the fill
    (b) when protected on all sides by 2 in. of concrete
    (c) where protected by corrosion protection judged suitable
    (d) any of these

147. Materials such as straps, bolts, screws, etc. that are associated with the installation of IMC in concrete in wet locations are required to be _____.
    (a) weatherproof   (b) weathertight   (c) corrosion-resistant   (d) none of these

148. When practical, contact of dissimilar metals shall be avoided anywhere in a raceway system to prevent _____.
    (a) corrosion   (b) galvanic action   (c) shorts   (d) none of these

149. One-inch IMC raceway containing three or more conductors can be conductor filled to _____ percent.
    (a) 53   (b) 31   (c) 40   (d) 60

150. The cross-sectional area of 1 in. IMC is approximately _____
    (a) 1.22 sq in.   (b) 0.62 sq in.   (c) 0.96 sq in.   (d) 2.13 sq in.

151. A run of IMC shall not contain more than the equivalent of _____ quarter bends including the offsets located immediately at the outlet or fitting.
    (a) 1   (b) 2   (c) 3   (d) 4

152. IMC shall be firmly fastened within _____ of each outlet box, junction box, device box, fitting, cabinet or other conduit termination.
    (a) 12 in.   (b) 18 in.   (c) 2 ft   (d) 3 ft

153. One-inch IMC shall be supported every _____.
    (a) 8 ft   (b) 10 ft   (c) 12 ft   (d) 14 ft

154. For industrial machinery, straight exposed vertical risers of IMC with threaded couplings are permitted if supported at the top and bottom no more than _____ apart.
    (a) 10 ft   (b) 12 ft   (c) 15 ft   (d) 20 ft

155. Horizontal runs of IMC supported by openings through framing members at intervals not exceeding 10 ft and securely fastened within 3 ft of termination points shall be permitted.
    (a) True   (b) False

156. Threadless couplings approved for use with IMC in wet locations shall be the _____ type.
    (a) rainproof   (b) raintight   (c) moistureproof   (d) concrete-tight

157. Threadless couplings and connectors must not be used on threaded IMC ends unless the fittings are listed for the purpose.
    (a) True   (b) False

158. Running threads of IMC shall not be used on conduit for connection at couplings.
    (a) True   (b) False

Article 344 Rigid Metal Conduit: Type RMC

159. RMC shall be permitted to be installed in concrete, in direct contact with the earth or in areas subject to severe corrosive influences when protected by _____ and judged suitable for the condition.
    (a) ceramic   (b) corrosion protection   (c) backfill   (d) a natural barrier

160. RMC can be installed in or under cinder fill that is subjected to permanent moisture when protected on all sides by a layer of noncinder concrete not less than _____ thick.
(a) 2 in.   (b) 4 in.   (c) 6 in.   (d) 18 in.

161. Materials such as straps, bolts, etc., associated with the installation of RMC in a wet location are required to be _____.
(a) weatherproof   (b) weathertight   (c) corrosion-resistant   (d) none of these

162. Aluminum fittings and enclosures shall be permitted to be used with _____ conduit.
(a) steel rigid metal   (b) aluminum rigid metal (c) rigid nonmetallic   (d) a and b

163. The minimum radius for a bend of 1 in. rigid conduit is _____, when using a one-shot bender.
(a) $10^1/_2$ in.   (b) $11^1/_2$ in.   (c) $5^3/_4$ in.   (d) $9^1/_2$ in.

164. The minimum radius of a field bend on $1^1/_4$ in. RMC is _____.
(a) 7 in.   (b) 8 in.   (c) 14 in.   (d) 10 in.

165. Two-inch RMC shall be supported every _____.
(a) 10 ft   (b) 12 ft   (c) 14 ft   (d) 15 ft

166. Straight runs of 1 in. RMC using threaded couplings may be secured at intervals not exceeding _____.
(a) 5 ft   (b) 10 ft   (c) 12 ft   (d) 14 ft

167. Exposed vertical risers for industrial machinery or fixed equipment can be supported at intervals not exceeding _____ if the conduit is made up with threaded couplings, firmly supported at the top and bottom of the riser and no other means of support is available.
(a) 6 ft   (b) 10 ft   (c) 20 ft   (d) none of these

168. Horizontal runs of RMC supported by openings through _____ at intervals not exceeding 10 ft and securely fastened within 3 ft of termination points shall be permitted.
(a) walls   (b) trusses   (c) rafters   (d) framing members

169. When threadless couplings and connectors used in the installation of RMC are buried in masonry or concrete, they shall be of the _____ type.
(a) raintight   (b) wet and damp location
(c) nonabsorbent   (d) concrete-tight

170. Threadless couplings and connectors used with RMC and installed in wet locations shall be of the _____ type.
(a) raintight   (b) wet and damp location
(c) nonabsorbent   (d) weatherproof

171. Each length of RMC shall be clearly and durably identified every _____.
(a) 3 ft   (b) 5 ft   (c) 10 ft   (d) none of these

172. RMC as shipped shall _____.
(a) be in standard lengths of 10 ft   (b) include a coupling on each length
(c) be threaded on each end   (d) all of these

Article 348 Flexible Metal Conduit: Type FMC

173. Which of the following conductor types are required to be used when FMC is installed in a wet location?
(a) THWN   (b) XHHW   (c) THW   (d) any of these

174. FMC cannot be installed _____.
(a) underground   (b) embedded in poured concrete
(c) where subject to physical damage   (d) all of these

175. FMC can be installed exposed or concealed where not subject to physical damage.
(a) True   (b) False

176. The largest size THHN conductor permitted in $^3/_8$ in. FMC is _____ AWG.
(a) 12   (b) 16   (c) 14   (d) 10

177. How many 12 AWG XHHW conductors, not counting a bare ground wire, are allowed in $^3/_8$ in. FMC (maximum of 6 ft) with outside fittings?
      (a) 4   (b) 3   (c) 2   (d) 5

178. FMC shall be secured _____.
      (a) at intervals not exceeding $4^1/_2$ ft
      (b) within 12 in. on each side of a box where fished
      (c) where fished
      (d) at intervals not exceeding 3 ft at motor terminals

179. In a concealed FMC installation, _____ connectors shall not be used.
      (a) straight   (b) angle   (c) grounding-type   (d) none of these

Article 350 Liquidtight Flexible Metal Conduit: Type LFMC

180. The use of listed and marked LFMC shall be permitted for _____.
      (a) direct burial where listed and marked for the purpose
      (b) exposed work
      (c) concealed work
      (d) all of these

181. • LFMC smaller than _____ shall not be used, except as permitted in 348.20(A).
      (a) $^3/_8$ in.   (b) $^1/_2$ in.   (c) $1^1/_2$ in.   (d) $1^1/_4$ in.

182. The maximum number of 14 THHN permitted in $^3/_8$ in. LFMC with outside fittings is _____.
      (a) 4   (b) 7   (c) 5   (d) 6

183. Where flexibility is necessary, securing LFMC is not required for lengths not exceeding _____ at terminals.
      (a) 2 ft   (b) 3 ft   (c) 4 ft   (d) 6 ft

184. When LFMC is used as a fixed raceway, it shall be secured within _____ in. on each side of the box and shall be at intervals not exceeding _____ ft.
      (a) 12, $4^1/_2$ (b) 18, 3   (c) 12, 3   (d) 18, 4

185. When LFMC is used to connect equipment requiring flexibility, a separate _____ conductor shall be installed.
      (a) bond jumper   (b) bonding   (c) equipment grounding   (d) none of these

Article 352 Rigid Nonmetallic Conduit: Type RNC

186. Extreme _____ may cause RNCs to become brittle and therefore more susceptible to damage from physical contact.
      (a) sunlight   (b) corrosive conditions   (c) heat   (d) cold

187. RNC and fittings can be used in areas of dairies, laundries, canneries, or other wet locations and in locations where walls are frequently washed, however, the entire conduit system including boxes and _____ shall be installed & equipped to prevent water from entering the conduit.
      (a) luminaries   (b) fittings   (c) supports   (d) all of these

188. RNC can be installed exposed in buildings _____, where not subject to physical damage.
      (a) three floors   (b) twelve floors   (c) six floors   (d) of any height

189. RNC can be used to support nonmetallic conduit bodies, but the conduit bodies shall not contain devices, luminaires, or other equipment.
      (a) True   (b) False

190. RNC shall not be used _____.
      (a) in hazardous (classified) locations
      (b) for the support of luminaires or other equipment
      (c) where subject to physical damage unless identified for such use
      (d) all of these

191. When installing RNC _____.
      (a) all cut ends shall be trimmed inside and outside to remove rough edges
      (b) there shall be a support within 2 ft of each box and cabinet
      (c) all joints shall be made by an approved method
      (d) a and c

192. RNC shall be securely fastened within _____ of each box.
      (a) 6 in.   (b) 24 in.   (c) 12 in.   (d) 36 in.

193. One-inch RNC shall be supported every _____.
      (a) 2 ft   (b) 3 ft   (c) 4 ft   (d) 6 ft

194. Expansion fittings for RNC shall be provided to compensate for thermal expansion and contraction when the length change in a straight run between securely mounted boxes, cabinets, elbows or other conduit terminations is expected to be _____ or greater.
      (a) $1/4$ in.   (b) $1/2$ in.   (c) 1 in.   (d) none of these

195. An equipment grounding conductor is not required in the conduit if the grounded conductor is used to ground equipment as permitted in 250.142.
      (a) True   (b) False

196. RNC and fittings used above ground shall be resistant to _____.
      (a) moisture and chemical atmospheres   (b) low temperatures and sunlight
      (c) distortion from heat   (d) all of these

Article 354 Nonmetallic Underground Conduit with Conductors: Type NUCC

197. NUCC and its associated fittings shall be _____ for the purpose.
      (a) listed   (b) approved   (c) identified   (d) none of these

198. The use of NUCC shall be permitted _____.
      (a) for direct-burial underground installation   (b) to be encased or embedded in concrete
      (c) in cinder fill   (d) all of these

199. NUCC shall not be used _____.
      (a) in exposed locations   (b) inside buildings
      (c) in hazardous (classified) locations   (d) all of these

200. NUCC larger than _____ shall not be used.
      (a) 1 in.   (b) 2 in.   (c) 3 in.   (d) 4 in.

Unit 6 NEC Exam – NEC Code Order 314.71 – 352.28

1. For angle or U-pulls, the distance between the conductor entry (for systems over 600V, nominal) and the opposite wall of the box shall not be less than _____ times the outside diameter, over sheath, of the largest cable or conductor.
    (a) 6   (b) 12   (c) 24   (d) 36

2. The use of AC cable is permitted in _____ installations.
    (a) wet   (b) cable tray   (c) exposed   (d) b and c

3. Tap devices used in FC assemblies shall be rated at not less than _____ or more than 300 volts-to-ground, and they shall be color-coded in accordance with the requirements of 322.120(C).
    (a) 20A   (b) 15A   (c) 30A   (d) 40A

4. MV is defined as a single or multiconductor solid dielectric insulated cable rated _____ or higher.
    (a) 601V   (b) 1,001V   (c) 2,001V   (d) 6,001V

5. MC cable shall not be used where exposed to the following destructive corrosive conditions unless the metallic sheath is suitable for the conditions or is protected by material suitable for the conditions:
    (a) Direct burial in the earth   (b) In concrete
    (c) In cinder fill   (d) all of these

6. Smooth-sheath MC cable that has an external diameter of not greater than 1 in. shall have a bending radius of not more than _____ times the cable external diameter.
    (a) 5   (b) 10   (c) 12   (d) 13

7. Which of the following statements about MI cable is correct?
    (a) It may be used in any hazardous location.
    (b) A single run of cable shall not contain more than the equivalent of four quarter bends.
    (c) It shall be securely supported at intervals not exceeding 10 ft.
    (d) none of these

8. NM, NMC and NMS nonmetallic-sheathed cables shall be permitted to be used in _____.
    (a) one-family dwellings   (b) multifamily dwellings
    (c) other structures   (d) all of these

9. Insulation rating of ungrounded conductors in NM cable must be _____.
    (a) 60 °C   (b) 75 °C   (c) 90 °C   (d) any of these

10. The difference in the construction specifications between NM cable and NMC cable is that NMC cable is _____.
    (a) corrosion-resistant   (b) flame-retardant
    (c) fungus-resistant   (d) a and c

11. USE or SE cable must have a minimum of _____ conductors (including the uninsulated one) in order for one of the conductors to be uninsulated.
    (a) one   (b) two   (c) three   (d) four

12. The maximum size underground UF cable is _____.
    (a) 14   (b) 10   (c) 1/0   (d) 4/0

13. One-inch IMC raceway containing three or more conductors can be conductor filled to _____ percent.
    (a) 53   (b) 31   (c) 40   (d) 60

14. The cross-sectional area of 1 in. IMC is approximately _____
    (a) 1.22 sq in.   (b) 0.62 sq in.   (c) 0.96 sq in.   (d) 2.13 sq in.

15. For industrial machinery, straight exposed vertical risers of IMC with threaded couplings are permitted if supported at the top and bottom no more than _____ apart.
    (a) 10 ft   (b) 12 ft   (c) 15 ft   (d) 20 ft

16. Horizontal runs of IMC supported by openings through framing members at intervals not exceeding 10 ft and securely fastened within 3 ft of termination points shall be permitted.
    (a) True   (b) False

17. Aluminum fittings and enclosures shall be permitted to be used with _____ conduit.
    (a) steel rigid metal   (b) aluminum rigid metal
    (c) rigid nonmetallic   (d) a and b

18. The minimum radius of a field bend on $1^1/_4$ in. RMC is _____
    (a) 7 in.   (b) 8 in.   (c) 14 in.   (d) 10 in.

19. Two-inch RMC shall be supported every _____.
    (a) 10 ft   (b) 12 ft   (c) 14 ft   (d) 15 ft

20. The largest size THHN conductor permitted in $3/_8$ in. FMC is _____ AWG.
    (a) 12   (b) 16   (c) 14   (d) 10

21. The use of listed and marked LFMC shall be permitted for _____.
    (a) direct burial where listed and marked for the purpose
    (b) exposed work
    (c) concealed work
    (d) all of these

22. LFMC smaller than _____ shall not be used except as permitted in 348.20(A).
    (a) $3/_8$ in.   (b) $1/_2$ in.   (c) $1^1/_2$ in.   (d) $1^1/_4$ in.

23. When LFMC is used to connect equipment requiring flexibility, a separate _____ conductor must be installed.
    (a) bond jumper   (b) bonding
    (c) equipment grounding   (d) none of these

24. RNC can be installed exposed in buildings of _____ where not subject to physical damage.
    (a) three floors   (b) twelve floors
    (c) six floors   (d) any height

25. When installing RNC, _____.
    (a) all cut ends shall be trimmed inside and outside to remove rough edges
    (b) there shall be a support within 2 ft of each box and cabinet
    (c) all joints shall be made by an approved method
    (d) a and c

Unit 6 NEC Exam – Random Order 90.2 – 354.6

1. A field-installed wiring system for branch circuits designed for installation under carpet squares is defined as _____.
   (a) underfloor wiring   (b) undercarpet wiring
   (c) flat conductor cable   (d) underfloor conductor cable

2. Each FCC transition assembly shall incorporate means for _____.
   (a) facilitating the entry of the FCC cable into the assembly
   (b) connecting the FCC cable to grounded conductors
   (c) electrically connecting the assembly to the metal cable shields and grounding conductors
   (d) all of these

3. Expansion fittings for RNC shall be provided to compensate for thermal expansion and contraction when the length change in a straight run between securely mounted boxes, cabinets, elbows, or other conduit terminations is expected to be _____ or greater.
   (a) $1/4$ in.   (b) $1/2$ in.   (c) 1 in.   (d) none of these

4. Extreme _____ may cause RNC to become brittle and therefore more susceptible to damage from physical contact.
   (a) sunlight   (b) corrosive conditions
   (c) heat   (d) cold

5. FCC cable shall be clearly and durably marked _____.
   (a) on the top side at intervals not exceeding 30 in.
   (b) on both sides at intervals not exceeding 24 in.
   (c) with conductor material, maximum temperature and ampacity
   (d) b and c

6. Flat cable assemblies shall consist of _____ conductors.
   (a) 2   (b) 3   (c) 4   (d) any of these

7. Flat cable assemblies shall not be installed outdoors or in wet or damp locations unless _____ for use in wet locations.
   (a) special permission is granted   (b) approved
   (c) identified   (d) none of these

8. Floor-mounted-type flat conductor cable and fittings shall be covered with carpet squares no larger than _____.
   (a) 36 square inches   (b) 36 inches square
   (c) 30 square inches   (d) 24 inches square

9. AC cable installed through, or parallel to, framing members must be protected against physical damage from penetration by screws or nails.
   (a) True   (b) False

10. AC cable shall be supported at intervals not exceeding $4^1/2$ ft and the cable shall be secured within _____ of cable termination.
    (a) 4 in.   (b) 8 in.   (c) 9 in.   (d) 12 in.

11. Armored cable used for the connection of recessed luminaires or equipment within an accessible ceiling does not need to be secured for lengths up to _____.
    (a) 2 ft   (b) 3 ft   (c) 4 ft   (d) 6 ft

12. Exposed vertical risers for industrial machinery or fixed equipment can be supported at intervals not exceeding _____ if the conduit is made up with threaded couplings, firmly supported at the top and bottom of the riser and no other means of support is available.
    (a) 6 ft   (b) 10 ft   (c) 20 ft   (d) none of these

13. FCC systems shall be permitted both for general-purpose and appliance branch circuits. They shall not be permitted for individual branch circuits.
    (a) True   (b) False

14. MC cable can be unsupported where it is:
    (a) Fished between concealed access points in finished buildings or structures and support is impracticable.
    (b) Not more than 2 ft in length at terminals where flexibility is necessary.
    (c) Not more than 6 ft from the last point of support within an accessible ceiling for the connection of luminaires.
    (d) any of these

15. MC cable installed through, or parallel to, framing members must be protected against physical damage from penetration by screws or nails by $1^1/_4$ in. separation or protected by a suitable metal plate.
    (a) True　　　　　　　　(b) False

16. MI cable shall be securely supported at intervals not exceeding _____.
    (a) 3 ft　　　　　　(b) $3^1/_2$ ft　　　　　　(c) 5 ft　　　　　　(d) 6 ft

17. NM and NMC cables shall not be used in one-and two-family dwellings exceeding three floors above grade.
    (a) True　　　　　　　　(b) False

18. NM cable can be installed as open runs in dropped or suspended ceilings in other than one- and two-family and multifamily dwellings.
    (a) True　　　　　　　　(b) False

19. NM cable can be installed in multifamily dwellings of Types III, IV and V construction except as prohibited in 334.
    (a) True　　　　　　　　(b) False

20. NM cable installed through, or parallel to, framing members must be protected against physical damage from penetration by screws or nails. Grommets or bushings for the protection of NM cables must be _____ for the purpose and they must remain in place.
    (a) marked　　　　　　(b) approved　　　　　　(c) identified　　　　　　(d) listed

21. NM cable must be _____.
    (a) marked　　　　　　(b) approved　　　　　　(c) identified　　　　　　(d) listed

22. NM cable, installed within accessible ceilings for the connections to luminaires and equipment, does not need to be secured within 12 in. from the luminaire or equipment when the free length does not exceed _____.
    (a) $4^1/_2$ ft　　　　　　(b) $2^1/_2$ ft　　　　　　(c) $3^1/_2$ ft　　　　　　(d) any of these

23. No more than _____ layers of flat conductor cable can cross at any one point.
    (a) 2　　　　　　(b) 3　　　　　　(c) 4　　　　　　(d) none of these

24. NUCC and its associated fittings must be _____ for the purpose.
    (a) listed　　　　　　(b) approved　　　　　　(c) identified　　　　　　(d) none of these

25. RNC and fittings can be used in areas of dairies, laundries, canneries or other wet locations and in locations where walls are frequently washed; however, the entire conduit system including boxes and _____ shall be installed and equipped to prevent water from entering the conduit.
    (a) luminaries　　　　　　(b) fittings　　　　　　(c) supports　　　　　　(d) all of these

26. RNC shall not be used _____.
    (a) in hazardous (classified) locations
    (b) for the support of luminaires or other equipment
    (c) where subject to physical damage unless identified for such use
    (d) all of these

27. Straight runs of 1 in. RMC using threaded couplings may be secured at intervals not exceeding _____.
    (a) 5 ft　　　　　　(b) 10 ft　　　　　　(c) 12 ft　　　　　　(d) 14 ft

28. The _____ rating of a conductor is the maximum temperature at any location along its length that the conductor can withstand over a prolonged period of time without serious degradation.
    (a) ambient　　　　　　　　　　　(b) temperature
    (c) maximum withstand　　　　　　(d) short-circuit

29. The ampacities listed in the Tables of Article 310 are based on temperature alone and do not take _____ into consideration.
(a) continuous loads
(b) voltage drop
(c) insulation
(d) wet locations

30. The ampacity adjustment factors of Table 310.15(B)(2)(a) do not apply to AC or MC cable without an overall outer jacket, if which of the following conditions are met?
(a) Each cable has not more than three current-carrying conductors.
(b) The conductors are 12 AWG copper.
(c) No more than 20 current-carrying conductors are bundled or stacked.
(d) all of these

31. The ampacity of NM cable shall be that of 60°C conductors, as listed in 310.15. However, the 90°C rating can be used for ampacity adjustment purposes, provided the final derated ampacity does not exceed that of a _____ rated conductor.
(a) 120°C
(b) 60°C
(c) 90°C
(d) none of these

32. The conductor between a lightning surge arrester and the line and the grounding connection shall not be smaller than _____ AWG copper (for installations operating at 1 kV or more).
(a) 4
(b) 6
(c) 8
(d) 2

33. The conductors for the TVSS cannot be any longer than _____, and unnecessary bends should be avoided.
(a) 6 in.
(b) 12 in.
(c) 18 in.
(d) none of these

34. The connected load to which the demand factors of Table 220.32 apply shall include 3 VA per _____ for general lighting and general-use receptacles.
(a) inch
(b) foot
(c) square inch
(d) square foot

35. The dedicated equipment space for electrical equipment that is required for panelboards is measured from the floor to a height of _____ above the equipment, or to the structural ceiling, whichever is lower.
(a) 3 ft
(b) 6 ft
(c) 12 ft
(d) 30 ft

36. The dimension of working clearance for access to live parts operating at 300V, nominal-to-ground, where there are exposed live parts on both sides of the workspace (not guarded) with the operator between, is _____ according to Table 110.26(A).
(a) 3 ft
(b) 3½ ft
(c) 4 ft
(d) 4½ ft

37. The disconnecting means is not required to be located at the building or structure, where documented safe switching procedures are established and maintained and where the installation is monitored by _____ persons.
(a) maintenance
(b) management
(c) service
(d) qualified

38. The earth can be used as the sole equipment grounding conductor.
(a) True
(b) False

39. The following systems must be installed in accordance with the *NEC*:
(a) Signaling
(b) Communications
(c) Power and Lighting
(d) all of these

40. The general rule for equipment bonding jumpers is that they are not permitted to be longer than 6 ft, but an equipment bonding jumper can be longer than 6 ft at outside pole locations for the purpose of bonding or grounding isolated sections of metal raceways or elbows installed in exposed risers of metal conduit or other metal raceway.
(a) True
(b) False

41. The grounded conductor of a 3-wire branch circuit supplying a household electric range shall be permitted to be smaller than the ungrounded conductors when the maximum demand of an 8.75 kW range has been computed according to Column C of Table 220.19. However, the ampacity of the neutral conductor shall not be less than _____ percent of the branch-circuit rating and not be smaller than _____ AWG.
(a) 50, 6
(b) 70, 6
(c) 50, 10
(d) 70, 10

42. The grounding electrode conductor for a single separately derived system must connect the grounded (neutral) conductor of the derived system to the grounding electrode.
(a) True　　　　　　　　(b) False

43. The grounding electrode for a separately derived system shall be as near as practicable to, and preferably in the same area as, the grounding electrode conductor connection to the system. The grounding electrode shall be the nearest one of the following:
(a) An effectively grounded metal member of the building structure.
(b) An effectively grounded metal water pipe, but only if it's within 5 ft from the point of entrance into the building.
(c) Any metal structure that is effectively grounded.
(d) a or b

44. The independent support wires shall be distinguishable from fire-rated suspended-ceiling framing support wires by _____.
(a) color　　　　　　　　　　　　(b) tagging
(c) other effective means　　　　　(d) any of these

45. The lightning protection system grounding electrode _____ be bonded to the building grounding electrode system.
(a) shall　　　　(b) shall not　　　　(c) can　　　　(d) none of these

46. The maximum distance the required receptacle for a dwelling unit countertop surface can be above a dwelling unit kitchen counter surface is _____
(a) 10 in.　　　　(b) 12 in.　　　　(c) 18 in.　　　　(d) 20 in.

47. The metal parts of the electrical system in a building or structure must be connected to the _____ for the purpose of limiting the voltage imposed by lightning, line surges or unintentional contact with higher voltage lines by shunting the energy to the _____. Typically, this is done at the building disconnecting means in accordance with 250.24 or 250.32.
(a) ground　　　　　　　　　　　　(b) earth
(c) electrical supply (neutral)　　　　(d) none of these

48. The metal water-pipe system of a building or structure is not required to be bonded to the separately derived system neutral terminal if the water piping is bonded to the metal frame of a building or structure that serves as the grounding electrode for the separately derived system.
(a) True　　　　　　　　(b) False

49. The metal water-piping systems must be bonded to the _____.
(a) grounded conductor at the service
(b) service equipment enclosure
(c) equipment grounding bar or bus at any panelboard within the building
(d) a and b

50. The metallic sheath of metal-clad cable shall be continuous and _____.
(a) flame-retardant　　　　　　　(b) weatherproof
(c) close fitting　　　　　　　　　(d) all of these

Unit 7

Motor Calculations

OBJECTIVES

After reading this unit, the student should be able to briefly explain the following concepts:

Branch circuit	Feeder conductor size	Motor VA calculations
Short circuit	Highest-rated motor	Motor calculation steps
Ground-fault protection	Motor overcurrent protection	Overload protection
Feeder protection	Motor branch-circuit conductors	

After reading this unit, the student should be able to briefly explain the following terms:

Code letter	Motor nameplate current rating	Service factor (SF)
Dual-element fuse	Next smaller device size	Short circuit
Ground fault	One-time fuse	Temperature rise
Inverse-time breaker	Overcurrent protection	
Heaters	Overcurrent	
Motor FLC	Overload	

INTRODUCTION

When sizing conductors and overcurrent protection for motors, we must comply with the requirements of the *NEC*; specifically Article 430 [240.4(G)].

7–1 MOTOR BRANCH-CIRCUIT CONDUCTORS [430.22(A)]

Branch-circuit conductors to a single motor must have an ampacity of not less than 125 percent of the motor's full-load current (FLC) as listed in Tables 430.147 through 430.150 [430.6(A)]. The actual conductor size must be selected from Table 310.16 according to the terminal temperature rating (60° or 75°C) of the equipment [110.14(C)]. Figure 7–1.

❏ **Motor Branch-Circuit Conductors**

What size THHN conductor is required for a 2-hp, 230V, 1Ø motor? Figure 7–2.

(a) 14 AWG   (b) 12 AWG

(c) 10 AWG   (d) 8 AWG

    • Answer: (a) 14 AWG

Motor FLC – Table 430.148: 2-hp, 230V 1Ø = 12A

Conductor sized no less than 125% of motor FLC:

12A × 1.25 = 15A, Table 310.16,

14 AWG THHN rated 20A at 60°C

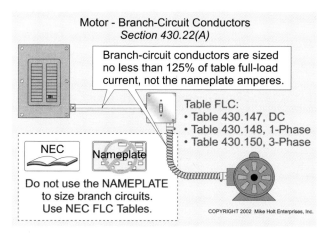

Figure 7-1

Sizing Motor Branch-Circuit Conductors

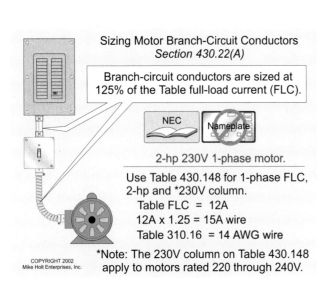

Sizing Motor Branch-Circuit Conductors
Section 430.22(A)

Branch-circuit conductors are sized at 125% of the Table full-load current (FLC).

NEC   Nameplate

2-hp 230V 1-phase motor.

Use Table 430.148 for 1-phase FLC, 2-hp and *230V column.
Table FLC = 12A
12A x 1.25 = 15A wire
Table 310.16 = 14 AWG wire

*Note: The 230V column on Table 430.148 apply to motors rated 220 through 240V.

COPYRIGHT 2002
Mike Holt Enterprises, Inc.

Figure 7-2
Sizing Motor Branch Circuit Conductors

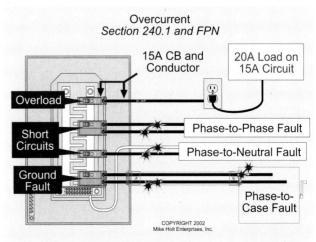

Overcurrent
Section 240.1 and FPN

15A CB and Conductor   20A Load on 15A Circuit

Overload
Short Circuits
Ground Fault

Phase-to-Phase Fault
Phase-to-Neutral Fault
Phase-to-Case Fault

COPYRIGHT 2002
Mike Holt Enterprises, Inc.

Overcurrent: (Article 100 Definition) Any current in excess of the rated current of equipment or materials. Causes of overcurrent are overloads, short circuits and ground faults.

Figure 7-3
Overcurrent

Note: The minimum size conductor permitted for building wiring is 14 AWG [310.5]; however, local codes and many industrial facilities have requirements that 12 AWG be used as the smallest branch-circuit wire.

7–2 MOTOR OVERCURRENT PROTECTION

Motors and their associated equipment must be protected against overcurrent (overload, short circuit or ground fault) [Article 100]. Figure 7–3. Due to the special characteristics of induction motors, overcurrent protection is generally accomplished by having the overload protection separated from the short-circuit and ground-fault protection device. Figure 7–4. Article 430, Part C contains the requirements for motor overload protection and Part D of Article 430 contains the requirements for motor short-circuit and ground-fault protection.

Overload Protection

Overload is the condition where current exceeds the equipment ampere rating which can result in equipment damage due to dangerous overheating [Article 100]. Overload protection devices, sometimes called heaters, are intended to protect the motor, the motor control equipment and the branch-circuit conductors from excessive heating due to motor overload [430.31].

Overload protection is not intended to protect against short-circuits or ground-fault currents. Figure 7–5.

If overload protection is to be accomplished by the use of fuses, a fuse must be installed to protect each ungrounded conductor [430.36 and 430.55].

Short-Circuit and Ground-Fault Protection

Branch-circuit short-circuit and ground-fault protection devices are intended to protect the motor, the motor control apparatus and the conductors against short circuit or ground faults, but they are not intended to protect against an overload [430.51]. Figure 7–6.

Note: The short-circuit and ground-fault protection device required for motor circuits is not the type required for personnel [210.8], feeders [215.9 and 240.13], services [230.95] or temporary wiring for receptacles [527.6].

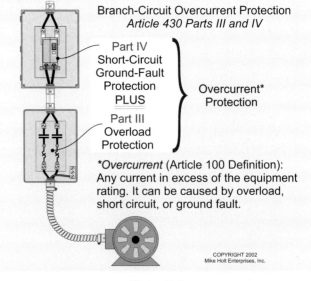

Branch-Circuit Overcurrent Protection
Article 430 Parts III and IV

Part IV
Short-Circuit
Ground-Fault
Protection
PLUS
Part III
Overload
Protection

Overcurrent*
Protection

Overcurrent (Article 100 Definition): Any current in excess of the equipment rating. It can be caused by overload, short circuit, or ground fault.

COPYRIGHT 2002
Mike Holt Enterprises, Inc.

Figure 7-4
Branch Circuit Overcurrent Protection

Overload Types
Section 430.31

Overloads protect the motor, conductor and associated equipment from excessive heat due to motor overloads. They are not intended to protect against short circuits and ground faults.

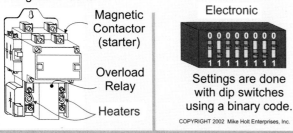

Magnetic Contactor (starter)

Overload Relay

Heaters

Electronic

Settings are done with dip switches using a binary code.

COPYRIGHT 2002 Mike Holt Enterprises, Inc.

Fuses can provide overload protection as well as short-circuit and ground-fault protection, see 430.55.

Figure 7-5
Types of Overloads

Short-Circuit and Ground-Fault Protection
Section 430.51

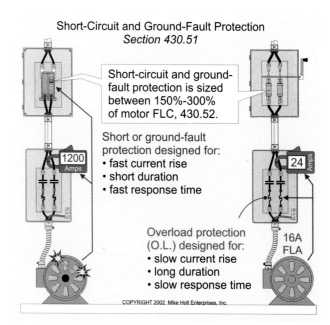

Short-circuit and ground-fault protection is sized between 150%-300% of motor FLC, 430.52.

Short or ground-fault protection designed for:
• fast current rise
• short duration
• fast response time

Overload protection (O.L.) designed for:
• slow current rise
• long duration
• slow response time

COPYRIGHT 2002 Mike Holt Enterprises, Inc.

1200 Amps

24 Amps

16A FLA

Figure 7-6
Short-Circuit and Ground-Fault Protection

7–3 OVERLOAD PROTECTION [430.32(A)]

In addition to short-circuit and ground-fault protection, motors must be protected against overload. Generally, the motor overload device is part of the motor starter; however, a separate overload device like a dual-element fuse can be used. Motors rated more than 1-hp without integral thermal protection, and motors 1-hp or less (automatically started) [430.32(C)], must have an overload device sized in response to the motor nameplate current rating [430.6(A)]. The overload device must be sized no larger than the requirements of 430.32.

Service Factor

Motors with a nameplate service factor (S.F.) rating of 1.15 or more shall have the overload protection device sized no more than 125 percent of the motor nameplate current rating.

❏ Service Factor

If a dual-element fuse is used for overload protection, what size fuse is required for a 5-hp, 230V, 1Ø motor, service factor 1.16, if the motor nameplate current rating is 28A? Figure 7–7.

    (a) 25A   (b) 30A   (c) 35A   (d) 40A

    • Answer: (c) 35A

Overload protection is sized to motor nameplate current rating [430.6(A), 430.32(A)(1), and 430.55].

28A × 1.25 = 35A [240.6(A)]

Temperature Rise

Motors with a nameplate temperature rise rating not over 40°C shall have the overload protection device sized no more than 125 percent of motor nameplate current rating.

❏ Temperature Rise

If a dual-element fuse is used for the overload protection, what size fuse is required for a 50-hp, 460V, 3Ø motor, temperature rise 39°C, motor nameplate current rating of 60A (FLA)? Figure 7–8.

    (a) 40A   (b) 50A   (c) 60A   (d) 70A

    • Answer: (d) 70A

Overloads are sized according to the motor nameplate current rating, not the motor FLC rating.

60A × 1.25 = 75A, 70A [240.6(A) and 430.32(A)(1)]

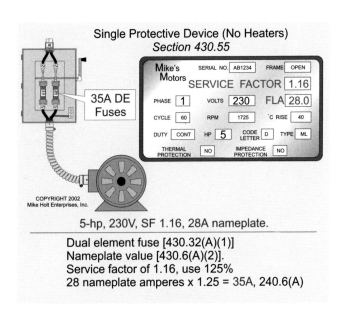

Single Protective Device (No Heaters)
Section 430.55

5-hp, 230V, SF 1.16, 28A nameplate.

Dual element fuse [430.32(A)(1)]
Nameplate value [430.6(A)(2)].
Service factor of 1.16, use 125%
28 nameplate amperes x 1.25 = 35A, 240.6(A)

Figure 7-7
Sizing Fuses for Overcurrent Protection

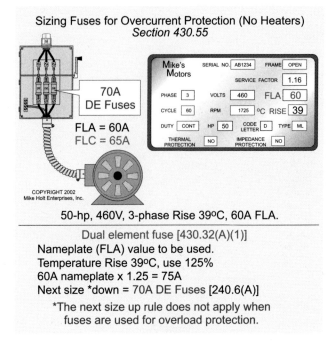

Sizing Fuses for Overcurrent Protection (No Heaters)
Section 430.55

50-hp, 460V, 3-phase Rise 39°C, 60A FLA.

Dual element fuse [430.32(A)(1)]
Nameplate (FLA) value to be used.
Temperature Rise 39°C, use 125%
60A nameplate x 1.25 = 75A
Next size *down = 70A DE Fuses [240.6(A)]

*The next size up rule does not apply when
fuses are used for overload protection.

Figure 7-8
Sizing Fuses for Overcurrent Protection

All Other Motors

Motors that do not have a service factor rating of 1.15 and up, or a temperature rise rating of 40°C and less, must have the overload protection device sized at not more than 115 percent of the motor nameplate ampere rating.

Number of Overloads [430.37]

An overload protection device must be installed in each ungrounded conductor according to the requirements of Table 430.37. If a fuse is used for overload protection, a fuse must be installed in each ungrounded conductor [430.36].

7–4 BRANCH-CIRCUIT SHORT-CIRCUIT GROUND-FAULT PROTECTION [430.52(C)(1)]

In addition to overload protection, each motor and its accessories require short-circuit and ground-fault protection. *NEC* 430.52(C) requires the motor branch-circuit short-circuit and ground-fault protection (except torque motors) to be sized no greater than the percentages listed in Table 430.52. When the short-circuit ground-fault protection device value determined from Table 430.52 does not correspond with the standard rating or setting of overcurrent protection devices as listed in 240.6(A), the next higher protection device size may be used [430.52(C)(1) Ex. 1]. Figure 7–9.

To determine the percentage from Table 430.52 to be used to size the motor branch-circuit short-circuit ground-fault protection device, the following steps should be helpful:

Step 1:   Locate the motor type on Table 430.52

Step 2:   Select the percentage from Table 430.52 according to the type of protection device such as nontime delay (one-time) fuse, dual-element fuse or inverse-time circuit breaker.

NEC Table 430.52			
	Percent of FLC (FLC) Tables 430.147, 148 and 150		
Type of Motor	**One-Time Fuse**	**Dual-Element Fuse**	**Circuit Breaker**
Direct-Current and Wound-Rotor Motors	150	150	150
All Other Motors	300	175	250

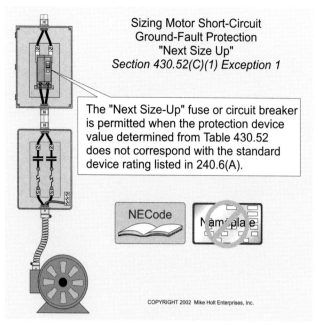

Figure 7-9
Sizing Motor Short-Circuit Ground-Fault Protection

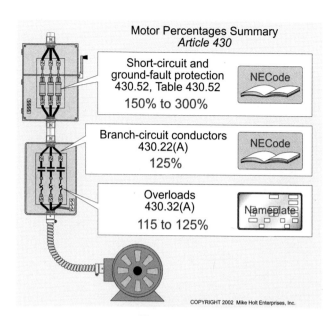

Figure 7-10
Motor Percentages Summary

Note: Where the protection device rating determined by Table 430.52 does not correspond with the standard size or rating of fuses or circuit breakers listed in 240.6(A), use of the next higher standard size or rating shall be permitted [430.52(C)(1) Ex. 1].

❑ **Branch Circuit**

Which of the following statements are true?

(a) The branch-circuit short-circuit protection (nontime delay fuse) for a 3-hp, 115V, 1Ø motor shall not exceed 110A.

(b) The branch-circuit short-circuit protection (dual-element fuse) for a 5-hp, 230V, 1Ø, motor shall not exceed 50A.

(c) The branch-circuit short-circuit protection (inverse-time breaker) for a 25-hp, 460V, 3Ø synchronous motor shall not exceed 70A.

(d) all of these are truc

• Answer: (d) all of these are true

Short-circuit and ground-fault protection, 430.53(C)(1) Ex. 1 and Table 430.52:

Table 430.148 – 34A × 3.00 = 102A, next size up permitted, 110A

Table 430.148 – 28A × 1.75 = 49A, next size up permitted, 50A

Table 430.150 – 26A × 2.50 = 65A, next size up permitted, 70A

WARNING: Conductors are sized at 125% of the motor FLC [430.22(A)], overloads are sized from 115% to 125% of the motor nameplate current rating [430.32(A)(1)], and the short-circuit ground-fault protection device is sized from 150% to 300% of the motor FLC [Table 430.52]. There is no relationship between the branch-circuit conductor ampacity (125%) and the short-circuit ground-fault protection device (150% up to 300%). Figure 7–10.

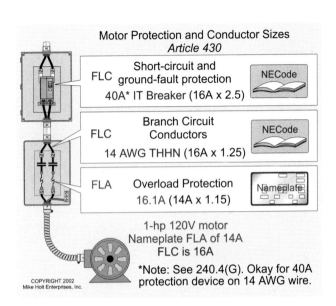

Figure 7-11
Motor Protection and Conductor Sizes

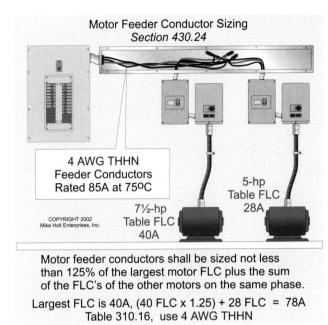

Figure 7-12
Motor Feeder Conductor Sizing

❑ **Branch Circuit**

Which of the following statements are true for a 1-hp, 120V motor, nameplate current rating of 14A? Figure 7–11.

(a) The branch-circuit conductors can be 14 AWG THHN.

(b) Overload protection is from 16.1A.

(c) Short-circuit and ground-fault protection is permitted to be a 40A circuit breaker.

(d) all of these are true
  • Answer: (d) all of these are true
  Conductor Size [430.22(A)]
  16A × 1.25 = 20A, 14 AWG at 60°C, Table 310.16
  Overload Protection Size [430.32(A)(1)]
  14A (nameplate) × 1.15 = 16.1A
  Short-Circuit and Ground-Fault Protection [430.52(C)(1), Tables 430.52 and 240.6(A)]
  16A × 2.50 = 40A circuit breaker

This bothers many electrical people, but the 14 AWG THHN conductors and motor are protected against overcurrent by the 16A overload protection device and the 40A short-circuit protection device.

7–5 FEEDER CONDUCTOR SIZE [430.24]

Conductors that supply several motors must have an ampacity of not less than:

(1) 125% of the highest-rated motor FLC [430.17], plus

(2) The sum of the FLCs of the other motors (on the same phase) [430.6(A)].

❑ **Feeder Conductor Size**

What size feeder conductor is required for two motors; a 7$\frac{1}{2}$-hp, 230V (40A), 1Ø and a 5-hp, 230V (28A), 1Ø? Terminals rated for 75°C. Figure 7–12.

(a) 50A   (b) 60A   (c) 70A   (d) 80A
  • Answer: (d) 80A, (40A × 1.25) + 28A = 78A

Note: 4 AWG conductor at 75°C, Table 310.16 is rated for 85A.

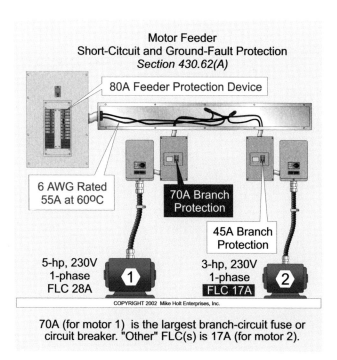

Motor Feeder
Short-Citcuit and Ground-Fault Protection
Section 430.62(A)

80A Feeder Protection Device

6 AWG Rated
55A at 60ºC

70A Branch
Protection

45A Branch
Protection

5-hp, 230V
1-phase
FLC 28A ①

3-hp, 230V
1-phase
FLC 17A ②

COPYRIGHT 2002 Mike Holt Enterprises, Inc.

70A (for motor 1) is the largest branch-circuit fuse or
circuit breaker. "Other" FLC(s) is 17A (for motor 2).

Figure 7-13
Sizing Feeder Protection

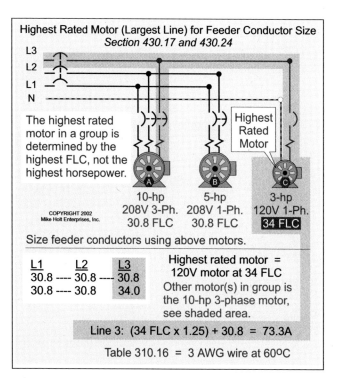

Highest Rated Motor (Largest Line) for Feeder Conductor Size
Section 430.17 and 430.24

L3
L2
L1
N

The highest rated
motor in a group is
determined by the
highest FLC, not the
highest horsepower.

Highest
Rated
Motor

COPYRIGHT 2002
Mike Holt Enterprises, Inc.

10-hp
208V 3-Ph.
30.8 FLC

5-hp
208V 1-Ph.
30.8 FLC

3-hp
120V 1-Ph.
34 FLC

Size feeder conductors using above motors.

L1	L2	L3
30.8	30.8	30.8
30.8	30.8	34.0

Highest rated motor =
120V motor at 34 FLC
Other motor(s) in group is
the 10-hp 3-phase motor,
see shaded area.

Line 3: (34 FLC x 1.25) + 30.8 = 73.3A

Table 310.16 = 3 AWG wire at 60ºC

Figure 7-14
Highest-Rated Motor

7–6 FEEDER PROTECTION [430.62(A)]

The feeder shall be provided with a protective device having a rating or setting not greater than the largest rating or setting of the branch-circuit short-circuit and ground-fault protective device, plus the sum of the full-load currents of the other motors of the group.

Motor feeder conductors must have protection against short circuits and ground faults but not overload.

❏ **Feeder Protection**

What size feeder protection (inverse-time breaker) is required for a 5-hp, 230V, 1Ø motor and a 3-hp, 230V, 1Ø motor? Figure 7–13.

(a) 30A breaker　　　　(b) 40A breaker　　　　(c) 50A breaker　　　　(d) 80A breaker

　• Answer: (d) 80A breaker

Motor FLC [Table 430.148]
5-hp motor FLC = 28A
3-hp motor FLC = 17A
Branch-Circuit Protection [430.52(C)(1), Table 430.52 and 240.6(A)]
5-hp: 28A × 2.5 = 70A
3-hp: 17A × 2.5 = 42.5A, next size up, 45A
Feeder Conductor [430.24(A)]
(28A × 1.25) + 17A = 52A, 6 AWG rated 55A at 60ºC, Table 310.16
Feeder Protection [430.62]
Not greater than 70A protection, plus 17A = 87A, next size down, 80A

7-7 HIGHEST-RATED MOTOR [430.17]

When selecting the feeder conductors or the feeder short-circuit ground-fault protection device, the highest-rated motor shall be the highest-rated motor FLC, (not the highest-rated horsepower) [430.17].

❏ **Highest-Rated Motor**

Which is the highest-rated motor of the following? Figure 7–14.

(a) 10-hp, 208V, 3Ø   (b) 5-hp, 208V, 1Ø

(c) 3-hp, 120V, 1Ø   (d) any of these

• Answer: (c) 3-hp, 120V, 1Ø, FLC of 34A

10-hp = 30.8A [Table 430.150]

5-hp = 30.8A [Table 430.148]

3-hp = 34.0A [Table 430.148]

7-8 MOTOR CALCULATIONS STEPS

Steps & *NEC* Rules	M1	M2	M3
Step 1: Motor FLC Tables 430.147, 148 & 150	_____ FLC	_____ FLC	_____ FLC
Step 2: Overload Protection Standard 430.32(A)(1)	____ × 1. ____ = ____	____ × 1. ____ = ____	____ × 1. ____ = ____
Step 3: Branch-Circuit Conductors 430.22(A), Table 310.16	_____ × 1.25 = _____	_____ × 1.25 = _____	_____ × 1.25 = _____
Step 4: Branch-Circuit Protection Table 430.52, 430.52(C) 240.6(A)	_____ × _____ = _____ Next Size Up	_____ × _____ = _____ Next Size Up	_____ × _____ = _____ Next Size Up
Step 5: Feeder Conductor 430.24 and Table 310.16	_____ × 1.25 + _____ + _____ + _____ = _____ Table 310.16, Use _____ (60 or 75°C?)		
Step 6: Feeder Protection 430.62, Table 430.52, and 240.6(A)	_____ + _____ + _____ + _____ = _____ Next Size Down		

❏ Motor Calculation Steps Example

Given: One 10-hp, 208V, 3Ø motor and three 1-hp, 120V, 1Ø motors. Determine the branch-circuit conductor (THHN) and short-circuit ground-fault protection device (inverse-time breaker) for all motors, then determine the feeder conductor and protection size. Fig. 7–15.

	M1	**M2**
Step 1: Motor FLC Tables 430.147, 148 or 150	Table 430.150 30.8 FLC	Table 430.148 16 FLC
Step 2: Overload Protection Nameplate 430.32(A)(1)	No Nameplate, Use FLC 30.8A × 1.15 = 35.4A	No Nameplate, Use FLC 16A × 1.15 = 18.4A
Step 3: Branch-Circuit Conductors 125 Percent of FLC 110.14(C), 430.22(A), Table 310-16	30.8A × 1.25 = 38.5A Table 310.16, 8 AWG THHN 60°C	16A × 1.25 = 20A Table 310.16, 14 AWG THHN 60°C
Step 4: Branch-Circuit Protection FLC × Table 430.52 percent 430.52(C), 240.6(A)	30.8A × 2.5 = 77A Next size up 240.6(A) = 80A	16A × 2.5 = 40A 240.6(A) = 40A
Step 5: Feeder Conductor FLC × 125 percent + FLC others 430.24, Table 310.16	(30.8A × 1.25) + 16A = 54.5A Table 310.16, Use 6 AWG THHN, rated 55A 60°C	
Step 6: Feeder Protection Largest Branch protection + FLCs of other motors 430.62, Table 430.52, 240.6(A)	Inverse-Time Breaker 80A + 16A = 96A, next size down, 90A	

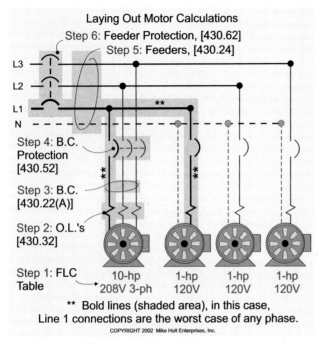

Laying Out Motor Calculations

Step 6: Feeder Protection, [430.62]
Step 5: Feeders, [430.24]

L3
L2
L1
N

**

Step 4: B.C.
Protection
[430.52]

Step 3: B.C.
[430.22(A)]

Step 2: O.L.'s
[430.32]

Step 1: FLC
Table

| 10-hp | 1-hp | 1-hp | 1-hp |
| 208V 3-ph | 120V | 120V | 120V |

** Bold lines (shaded area), in this case,
Line 1 connections are the worst case of any phase.

COPYRIGHT 2002 Mike Holt Enterprises, Inc.

Figure 7-15
Laying Out Motor Calculations

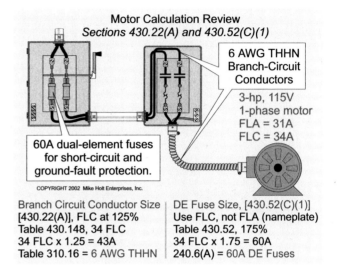

Motor Calculation Review
Sections 430.22(A) and 430.52(C)(1)

6 AWG THHN
Branch-Circuit
Conductors

3-hp, 115V
1-phase motor
FLA = 31A
FLC = 34A

60A dual-element fuses
for short-circuit and
ground-fault protection.

COPYRIGHT 2002 Mike Holt Enterprises, Inc.

| Branch Circuit Conductor Size [430.22(A)], FLC at 125% Table 430.148, 34 FLC 34 FLC x 1.25 = 43A Table 310.16 = 6 AWG THHN | DE Fuse Size, [430.52(C)(1)] Use FLC, not FLA (nameplate) Table 430.52, 175% 34 FLC x 1.75 = 60A 240.6(A) = 60A DE Fuses |

Figure 7-16
Motor Calculation Review

7-9 MOTOR CALCULATION REVIEW

❏ Branch Circuit

Size the branch-circuit conductors (THHN) and short-circuit ground-fault protection device for a 3-hp, 115V, 1Ø motor. The motor nameplate full-load amperes are 31A and dual-element fuses are to be used for short-circuit and ground-fault protection. Figure 7–16.

Branch-Circuit Conductors [430.22(A)]

Branch-circuit conductors to a single motor must have an ampacity of not less than 125% of the motor FLC as listed in Tables 430.147 through 430.150 [430.6(A)].

34A × 125% = 43A

Table 310.16 60°C terminals: The conductor must be a 6 AWG THHN rated 55A [110.14(C)(1)].

Branch-Circuit Short-Circuit Protection [430.52(C)(1)]

The branch-circuit short-circuit and ground-fault protection device protects the motor, the motor control apparatus and the conductors against overcurrent due to short-circuits or ground-faults, but not overload [430.51]. The branch-circuit short-circuit and ground-fault protection devices are sized by considering the type of motor and the type of protection device according to the motor FLC listed in Table 430.52. When the protection device values determined from Table 430.52 do not correspond with the standard rating of overcurrent protection devices as listed in 240.6(A), the next higher overcurrent protection device must be installed.

34A × 175% = 60A dual-element fuse [240.6(A)]

See Example No. D8 in Annex D of the *NEC*.

❏ Feeder

Size the feeder conductor (THHN) and protection device (inverse-time breakers 75°C terminal rating) for the following motors: Three 1-hp, 120V, 1Ø motors; three 5-hp, 208V, 1Ø motors; one wound-rotor 15-hp, 208V, 3Ø motor. Figure 7–17.

Branch-circuit short-circuit protection [240.6(A), 430.52(C)(1), and Table 430.52]

15-hp: 46.2A × 150% (wound-rotor) = 69A: Next size up = 70A

5-hp: 30.8A × 250% = 77A: Next size up = 80A

1-hp: 16A × 250% = 40A

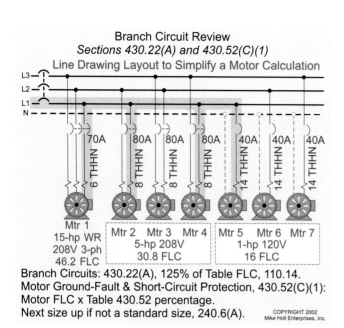

Branch Circuit Review
Sections 430.22(A) and 430.52(C)(1)
Line Drawing Layout to Simplify a Motor Calculation

Branch Circuits: 430.22(A), 125% of Table FLC, 110.14.
Motor Ground-Fault & Short-Circuit Protection, 430.52(C)(1):
Motor FLC x Table 430.52 percentage.
Next size up if not a standard size, 240.6(A).

COPYRIGHT 2002
Mike Holt Enterprises, Inc.

Figure 7-17
Branch Circuit Review

Feeder Conductor Review - Section 430.24
1/0 AWG THHN rated 150A at 75ºC.

COPYRIGHT 2002 Mike Holt Enterprises, Inc.

	Line 1	Line 2	Line 3
Mtr 1, 15-hp	46.2	46.2	46.2
Mtr 2, 5-hp	30.8	30.8	
Mtr 3, 5-hp		30.8	30.8
Mtr 4, 5-hp	30.8		30.8
Mtr 5, 6, & 7, 1-hp	16.0	16.0	16.0
Largest Line	123.8	123.8	123.8

(46.2 x 1.25) + 30.8 + 30.8 + 16.0 = 136A for feeder
Table 310.16, 1/0 AWG THHN rated 150A at 75ºC

Figure 7-18
Feeder Conductor Review

Feeder Conductor [430.24]

Conductors that supply several motors must have an ampacity of not less than 125% of the highest-rated motor FLC [430.17], plus the sum of the other motor FLCs [430.6(A)]. Figure 7–18

(46.2A × 1.25) + 30.8A + 30.8A + 16A = 136A

Table 310.16, 1/0 AWG THHN, rated 150A.

Note: When sizing the feeder conductor, be sure to only include the motors that are on the same phase. For that reason, only four motors are used in this calculation.

Feeder Protection [430.62]

Feeder conductors must be protected against short-circuits and ground-faults and sized not to be greater than the maximum branch-circuit short-circuit ground-fault protection device [430.52(C)(1)], plus the sum of the FLCs of the other motors (on the same phase).

80A + 30.8A + 46.2A + 16A = 174A

Next size down, 150A circuit breaker [240.6(A)]. See Example D8 in Annex D of the *NEC*. Figure 7–19.

Note: When sizing the feeder protection, be sure to only include the motors that are on the same phase. For that reason, only four motors are used in this calculation.

7-10 MOTOR VA CALCULATIONS

The input VA of a motor is determined by multiplying the motor volts by the motor amperes. To determine the *motor VA rating*, the following formulas can be used:

Motor VA (1Ø) = Motor Volt Rating × Motor Ampere Rating
Motor VA (3Ø) = Motor Volt Rating × Motor Ampere Rating × $\sqrt{3}$

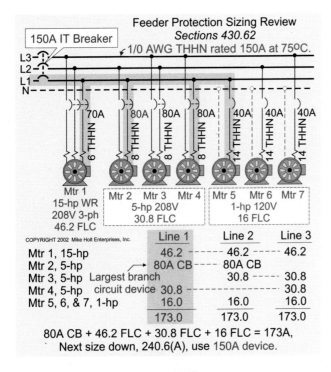

Feeder Protection Sizing Review
Sections 430.62

150A IT Breaker

1/0 AWG THHN rated 150A at 75°C.

Mtr 1
15-hp WR
208V 3-ph
46.2 FLC

Mtr 2 Mtr 3 Mtr 4
5-hp 208V
30.8 FLC

Mtr 5 Mtr 6 Mtr 7
1-hp 120V
16 FLC

COPYRIGHT 2002 Mike Holt Enterprises, Inc.

	Line 1	Line 2	Line 3
Mtr 1, 15-hp	46.2	46.2	46.2
Mtr 2, 5-hp	80A CB	80A CB	
Mtr 3, 5-hp Largest branch		30.8	30.8
Mtr 4, 5-hp circuit device	30.8		30.8
Mtr 5, 6, & 7, 1-hp	16.0	16.0	16.0
	173.0	173.0	173.0

80A CB + 46.2 FLC + 30.8 FLC + 16 FLC = 173A,
Next size down, 240.6(A), use 150A device.

Figure 7-19
Feeder Protection Sizing Review

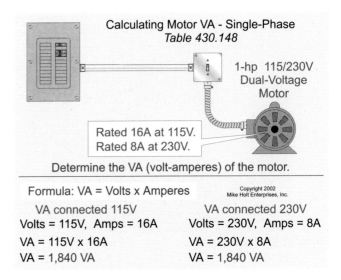

Calculating Motor VA - Single-Phase
Table 430.148

1-hp 115/230V
Dual-Voltage
Motor

Rated 16A at 115V.
Rated 8A at 230V.

Determine the VA (volt-amperes) of the motor.

Formula: VA = Volts x Amperes   Copyright 2002 Mike Holt Enterprises, Inc.

VA connected 115V	VA connected 230V
Volts = 115V, Amps = 16A	Volts = 230V, Amps = 8A
VA = 115V x 16A	VA = 230V x 8A
VA = 1,840 VA	VA = 1,840 VA

Figure 7-20
Calculating Motor VA – Single Phase

❏ Motor VA – Single-Phase

What is the input VA for a 1-hp motor rated 115/230V? Figure 7–20.

(a) 1,840 VA at 115V   (b) 1,840 VA at 230V   (c) a and b   (d) none of these

• Answer: (c) a and b, Table 430.148

Motor VA = Volts $\times$ FLC

VA at 230V = 230V $\times$ 8A = 1,840 VA

VA at 115V = 115V $\times$ 16A = 1,840 VA

Note: Many people believe that a 230V motor consumes less power than a 115V motor (I thought that), but both motors consume the same amount of power.

❏ Motor VA – Three-Phase

What is the input VA for a 5-hp, 230V, 3Ø motor? Figure 7–21.

(a) 6,055 VA   (b) 3,730 VA

(c) 6,440 VA   (d) 8,050 VA

• Answer: (a) 6,055 VA, Table 430.150

Motor VA = Volts $\times$ FLC $\times \sqrt{3}$

Table 430.150 FLC = 15.2A

Motor VA = 230V $\times$ 15.2A $\times$ 1.732

Motor VA = 6,055 VA

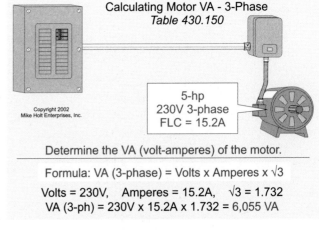

Calculating Motor VA - 3-Phase
Table 430.150

Copyright 2002
Mike Holt Enterprises, Inc.

5-hp
230V 3-phase
FLC = 15.2A

Determine the VA (volt-amperes) of the motor.

Formula: VA (3-phase) = Volts x Amperes x √3

Volts = 230V,   Amperes = 15.2A,   √3 = 1.732
VA (3-ph) = 230V x 15.2A x 1.732 = 6,055 VA

Figure 7-21
Calculating Motor VA – 3-Phase

Unit 7 – Motor Calculations
Summary Questions

Calculations Questions

Introduction

1. When sizing conductors, Table 240.4(G) allows us to use Article 430 for sizing overcurrent protection of motors.
    (a) True   (b) False

7–1 Motor Branch-Circuit Conductors [430.22(A)]

2. What size THHN conductor is required for a 5-hp, 230V, 1Ø motor? Terminals are rated 75°C.
    (a) 14 AWG   (b) 12 AWG   (c) 10 AWG   (d) 8 AWG

7–2 Motor Overcurrent Protection

3. Motors and their associated equipment shall be protected against overcurrent (overload, short-circuit or ground-fault), but because of the special characteristics of induction motors, overcurrent protection is generally accomplished by having the overload protection separate from the short-circuit and ground-fault protection.
    (a) True   (b) False

4. • Which parts of Article 430 contain the requirements for motor overcurrent protection?
    (a) Motor and Branch-Circuit Overload Protection - Part III
    (b) Motor Branch-Circuit, Short-Circuit and Ground-Fault Protection - Part IV
    (c) a and b
    (d) none of these

5. • Overload is the condition where current is greater than the equipment ampacity rating, resulting in equipment damage due to dangerous overheating [Article 100]. Overload protection devices, sometimes called heaters, are intended to protect the _____ from dangerous overheating.
    (a) motor   (b) motor control equipment
    (c) branch-circuit conductors   (d) all of these

6. The branch-circuit short-circuit and ground-fault protection device is intended to protect the motor, the motor control apparatus and the conductors against overcurrent due to _____.
    (a) short circuits   (b) ground faults   (c) overloads   (d) a and b

7–3 Overload Protection [430.32(A)]

7. • The *NEC* requires motor overload protection devices to be sized according to the motor full-load current rating as listed in Tables 430.147, 148, or 150.
    (a) True   (b) False

8. If the motor overload protection relay sized according to 430.32 is not capable of carrying the motor starting and running current, the next size up overload can be used if sized according to the requirements of 430.32(C).
    (a) True   (b) False

9. Motors with a nameplate service factor (S.F.) rating of 1.15 or more must have the overload protection device sized at no more than _____ percent of the motor nameplate current rating.
    (a) 100   (b) 115   (c) 125   (d) 135

10. Motors with a nameplate temperature rise rating not over 40°C must have the overload protection device sized at no more than _____ percent of motor nameplate current rating.
    (a) 100   (b) 115   (c) 125   (d) 135

11. Motors that have a service factor of 1.10 must have the overload protection device sized at not more than _____ percent of the motor nameplate ampere rating.
    (a) 100   (b) 115   (c) 125   (d) 135

12. If a dual-element fuse is used for overload protection, what size fuse is required for a 5-hp, 208V, 3Ø motor with a service factor of 1.16 and a motor nameplate current rating of 16A (FLA)?
    (a) 20A   (b) 25A   (c) 30A   (d) 35A

13. If a dual-element fuse is used for the overload protection, what size fuse is required for a 30-hp, 460V, 3Ø synchronous motor, with a temperature rise of 39°C?
    (a) 20A   (c) 30A   (b) 25A   (d) 40A

7–4 Branch Circuit Short-Circuit Ground-Fault Protection [430.52(C)(1)]

14. In addition to overload protection, each motor and its accessories require short-circuit and ground-fault protection according to 430.52. When sizing the branch-circuit protection device, we must consider which of the following factors?
    (a) The motor type, such as induction, synchronous, wound-rotor, etc.
    (b) The motor Code letter starting characteristics.
    (c) The type of protection device to be used; fuse or breaker.
    (d) all of these

15. The *NEC* requires motor branch-circuit short-circuit and ground-fault protection to be sized not greater than the percentages listed in Table 430.52. When the short-circuit, ground-fault protection device value determined from Table 430.52 does not correspond with the standard rating of overcurrent protection devices as listed in 240.6(A), the next _____ device size shall be used.
    (a) smaller   (b) larger   (c) a or b   (d) none of these

16. To determine the percentage of the motor FLC from Table 430.52 that is to be used to size the motor branch-circuit short-circuit and ground-fault protection device, which of the following steps should be used?
    (a) Locate the motor type on Table 430.52, such as dc, wound rotor, high-reactance, autotransformer start or all other motors.
    (b) Locate the motor starting conditions, such as Code letter or no Code letter.
    (c) Select the percentage from Table 430.52 according to the type of protection device, such as one-time fuse, dual-element fuse or circuit breaker.
    (d) all of these

17. If the branch-circuit short-circuit and ground-fault protection dual-element (time delay) fuse selected (sized not greater than the percentages listed in Table 430.52) is not capable of carrying the load, the next larger size dual-element fuse can be used. The next size dual-element fuse cannot exceed _____ of the motor full-load current rating.
    (a) 125%   (b) 150%   (c) 175%   (d) 225%

18. Conductors are sized at _____ of the motor full-load currents [430.22(A)], overloads from _____ and the motor short-circuit, ground-fault protection device (inverse-time circuit breaker) is sized up to _____.
    (a) 125%, 115%, 250%   (b) 100%, 125%, 150%
    (c) 125%, 125%, 125%   (d) 100%, 100%, 100%

19. Which of the following statements are true for a 10-hp, 208V, 3Ø motor with a nameplate current of 29A?
    (a) The branch-circuit conductors can be 8 AWG THHN.
    (b) Overload protection is 33A.
    (c) Short-circuit and ground-fault protection can be an 80A circuit breaker.
    (d) all of these

7–5 Feeder Conductor Size [430.24]

20. Feeder conductors that supply several motors must have an ampacity of not less than _____.
    (a) 125% of the highest-rated motor's FLC
    (b) the sum of the full-load currents of the other motors on the same phase
    (c) a or b
    (d) a and b

21. Motor feeder conductors (sized according to 430.24) must have a feeder protection device to protect against short circuits and ground faults (not overloads), sized not greater than the _____.
    (a) largest branch-circuit short-circuit ground-fault protection device [430.52] of any motor of the group
    (b) sum of full-load currents of the other motors on the same time phase
    (c) a or b
    (d) a and b

7–6 Feeder Protection [430.62(A)]

22. • Which of the following statements about a 30-hp, 460V, 3Ø synchronous motor and a 10-hp 460V, 3Ø motor are true?
    (a) The 30-hp motor has 8 AWG THHN with an 80A breaker.
    (b) The 10-hp motor has 14 AWG THHN with a 35A breaker.
    (c) The feeder conductors must be 6 AWG THHN with a 90A breaker.
    (d) all of these

7–7 Highest Rated Motor [430.17]

23. When selecting the feeder conductors and short-circuit ground-fault protection device, the highest-rated motor shall be the highest-rated _____.
    (a)-hp   (b) full-load current
    (c) nameplate current   (d) any of these

24. Which is the highest-rated motor of the following?
    (a) 25-hp, synchronous, 3Ø, 460V   (b) 20-hp, 3Ø, 460V
    (c) 15-hp, 3Ø, 460V   (d) 3-hp, 120V

7–10 Motor VA Calculations

25. What is the VA input of a dual voltage 5-hp, 3Ø motor rated 460/230V?
    (a) 3,027 VA at 460V   (b) 6,055 VA at 230V
    (c) 6,055 VA at 460V   (d) b and c

26. What is the input VA of a 3-hp, 208V, 1Ø motor?
    (a) 3,890 VA   (b) 6,440 VA   (c) 6,720 VA   (d) none of these

☆ Challenge Questions

7–1 Motor Branch Circuit Conductors [430.22(A)]

27. • The branch-circuit conductors of a 5-hp, 230V motor with a nameplate rating of 25A shall have an ampacity of not less than _____. Note: The motor is used for intermittent duty and, due to the nature of the apparatus it drives, it cannot run for more than five minutes at any one time.
    (a) 33A   (b) 37A   (c) 21A   (d) 23A

7–3 Overload Protection [430.32(A)]

28. The standard overload protection device for a 2-hp, 115V motor that has a full-load current rating of 24A and a nameplate rating of 21.5A shall not exceed _____.
     (a) 20.6A   (b) 24.7A   (c) 29.9A   (d) 33.8A

29. • The maximum overload protective device relay for a 2-hp, 115V motor with a nameplate rating of 22A is _____. The Service Factor is 1.2.
     (a) 30.8A   (b) 33.8A   (c) 33.6A   (d) 22.6A

Ultimate Trip Setting [430.32(A)(2)]

30. The ultimate trip overload device of a thermally protected $1^1/_2$-hp, 120V motor would be rated no more than _____.
     (a) 31.2A   (b) 26A   (c) 28A   (d) 23A

7–4 Branch-Circuit Short-Circuit Ground-Fault Protection [430.52(C)(1)]

31. A 2-hp, 120V motor requires a _____ branch-circuit short-circuit protection device. Note: Use an inverse-time breaker for protection.
     (a) 20A   (b) 30A   (c) 40A   (d) 60A

32. • The branch-circuit short-circuit protection device for a 10-hp, 230V, 1Ø motor shall not exceed _____. Note: Use an inverse-time breaker for protection.
     (a) 125A   (b) 50A   (c) 75A   (d) 80A

33. The branch-circuit protection (circuit breaker) for a 125-hp, 240V, dc motor is _____.
     (a) 400A   (b) 600A   (c) 700A   (d) 800A

7–5 Feeder Conductor Size [430.24]

34. • The motor feeder conductor size for three 15-hp, 208V, 3Ø motors; three 3-hp, 208V, 1Ø motors; and three 1-hp, 120V, 1Ø motors would be _____.
     (a) 2/0 AWG   (b) 3/0 AWG   (c) 4/0 AWG   (d) 250 kcmil

7–6 Feeder Protection [430.62(A)]

35. • There are three motors: one 5-hp, 230V, 1Ø motor with a service factor of 1.2 and two $1^1/_2$-hp, 120V, 1Ø motors. Using an inverse-time breaker, the 3-wire feeder conductor protection device after balancing all three motors would be _____.
     (a) 60A   (b) 70A   (c) 80A   (d) 90A

36. • If an inverse-time breaker is used for the feeder short-circuit protection, what size protection is required for the following 3Ø motors?
     (1) Motor 1 = 40-hp 52 FLC
     (2) Motor 2 = 20-hp 27 FLC
     (3) Motor 3 = 10-hp 14 FLC
     (4) Motor 4 = 5-hp 7.6 FLC
     (a) 225A   (b) 200A   (c) 125A   (d) 175A

37. • If dual-element fuses are used to protect a 3-wire, 115/230V feeder conductor for twenty-two $^1/_2$-hp, 115V, 1Ø motors, the fuse size selected should not be greater than _____.
     (a) 125A   (b) 90A   (c) 100A   (d) 110A

38. • The feeder protection for one 25-hp, 208V, 3Ø motor and three 3-hp, 120V, 1Ø motors would be _____ after balancing. Note: Use inverse-time breakers (ITB).
     (a) 225A   (b) 200A   (c) 300A   (d) 250A

NEC Questions – Articles 354-406

Article 354 Nonmetallic Underground Conduit with Conductors: Type NUCC

39.   In order to _____ NUCC, the conduit shall be trimmed away from the conductors or cables using an approved method that will not damage the conductor or cable insulation or jacket.
    (a) facilitate installing
    (b) enhance the appearance of the installation of
    (c) terminate
    (d) provide safety to the persons installing

40.   NUCC shall be capable of being supplied on reels without damage or _____ and shall be of sufficient strength to withstand abuse, such as impact or crushing in handling and during installation, without damage to conduit or conductors.
    (a) distortion   (b) breakage   (c) shattering   (d) all of these

ARTICLE 356 Liquidtight Flexible Nonmetallic Conduit: Type LFNC

41.   Type LFNC-B shall be permitted to be installed in lengths longer than _____ where secured in accordance with 356.30.
    (a) 2 ft   (b) 3 ft   (c) 6 ft   (d) 10 ft

ARTICLE 358 Electrical Metallic Tubing: Type EMT

42.   The minimum and maximum size of EMT is _____, except for special installations.
    (a) $5/16$ and 3 in.   (b) $3/8$ and 4 in.   (c) $1/2$ and 3 in.   (d) $1/2$ and 4 in.

43.   A run of EMT between outlet boxes shall not exceed _____ offsets close to the box.
    (a) 360 degrees plus
    (b) 360 degrees total including
    (c) four quarter bends plus
    (d) 180 degrees total including

44.   EMT shall not be threaded.
    (a) True   (b) False

45.   • EMT shall be supported within 3 ft of each coupling.
    (a) True   (b) False

46.   Couplings and connectors used with EMT shall be made up _____.
    (a) of metal
    (b) in accordance with industry standards
    (c) tight
    (d) none of these

ARTICLE 360 Flexible Metallic Tubing: Type FMT

47.   Flexible metallic tubing shall be permitted to be used _____.
    (a) in dry and damp locations
    (b) for direct burial
    (c) in lengths over 6 ft
    (d) for a maximum of 1,000V

48.   The maximum size flexible metallic tubing permitted is _____
    (a) $3/8$ in.   (b) $1/2$ in.   (c) $3/4$ in.   (d) 1 in.

49.   Where $3/8$ in. flexible metallic tubing has a fixed bend for installation purposes and not flexed for service, the minimum radius of the bend shall not be less than _____
    (a) 8 in.   (b) $12^1/_2$ in.   (c) $3^1/_2$ in.   (d) 4 in.

ARTICLE 362 Electrical Nonmetallic Tubing: Type ENT

50.   ENT is composed of a material that is resistant to moisture, chemical atmospheres and is _____.
    (a) flexible   (b) flame-retardant   (c) fireproof   (d) flammable

51. When ENT is installed concealed in walls, floors and ceilings of buildings exceeding three floors above grade, a thermal barrier shall be provided having a minimum _____ minute finish rating as listed for fire-rated assemblies.
    (a) 5   (b) 10   (c) 15   (d) 30

52. When a building is supplied with a(n) _____ fire sprinkler system, ENT can be installed exposed or concealed in buildings of any height.
    (a) listed   (b) identified   (c) approved   (d) none of these

53. ENT can be installed above a suspended ceiling if the suspended ceiling provides a thermal barrier having a _____-minute finish rating as identified in listings of fire-rated assemblies.
    (a) 5   (b) 10   (c) 15   (d) none of these

54. When a building is supplied with an approved fire sprinkler system(s), ENT can be installed above any suspended ceiling.
    (a) True   (b) False

55. ENT and fittings shall be permitted to be _____ in a concrete slab on grade where the tubing is placed upon sand or approved screenings.
    (a) encased   (b) embedded   (c) either a or b   (d) none of these

56. ENT is not permitted in hazardous (classified) locations except for intrinsically safe applications.
    (a) True   (b) False

57. ENT is permitted for direct earth burial when used with fittings listed for this purpose.
    (a) True   (b) False

58. ENT is not permitted in places of assembly unless it is encased in at least _____ of concrete.
    (a) 1 in.   (b) 2 in.   (c) 3 in.   (d) 4 in.

59. In electrical nonmetallic tubing, the maximum number of bends between pull points cannot exceed _____ degrees, including any offsets.
    (a) 320   (b) 270   (c) 360   (d) unlimited

60. All cut ends of ENT shall be trimmed inside and _____ to remove rough edges.
    (a) outside   (b) tapered   (c) filed   (d) beveled

61. ENT shall be secured in place every _____.
    (a) 12 in.   (b) 18 in.   (c) 24 in.   (d) 36 in.

62. Bushings or adapters are required at ENT terminations to protect the conductors from abrasion.
    (a) True   (b) False

63. The _____ of conductors used in prewired ENT manufactured assemblies shall be identified by means of a printed tag or label attached to each end of the manufactured assembly.
    (a) type   (b) size   (c) quantity   (d) all of these

ARTICLE 366 Auxiliary Gutters

64. An auxiliary gutter is permitted to contain _____.
    (a) conductors   (b) overcurrent devices (c) busways   (d) none of these

65. Because an auxiliary gutter is used to supplement wiring space, it is not a raceway and shall not extend a distance greater than _____ beyond the equipment that it supplements.
    (a) 50 ft   (b) 30 ft   (c) 10 ft   (d) 25 ft

66. When conductor ampacity adjustment factors of 310.15(B)(2)(a) are used, an auxiliary gutter shall not contain more than _____ at any cross section.
    (a) 25 conductors   (b) 40 current-carrying conductors
    (c) 20 conductors   (d) no limit on the number of conductors

67. The continuous current-carrying capacity of $1^1/_2$ sq in. copper busbar mounted in an unventilated enclosure is _____.
    (a) 500A   (b) 750A   (c) 650A   (d) 1,500A

68. The maximum ampere rating of a 4 in. $\times$ $1/_2$ in. busbar that is 4 ft long and installed in an auxiliary gutter is _____.
    (a) 500A   (b) 750A   (c) 650A   (d) 2,000A

69. Bare conductors shall be securely and rigidly supported so that the minimum clearance between bare current-carrying metal parts of different potential mounted on the same surface will not be less than 2 in., nor less than _____. for parts that are held free in the air.
    (a) $1/_4$ in   (b) $1/_2$ in   (c) 1 in   (d) 2 in

70. Conductors, including splices and taps, shall not fill the auxiliary gutter to more than _____ percent of its cross-sectional area.
    (a) 30   (b) 40   (c) 60   (d) 75

71. Auxiliary gutters shall be constructed and installed so as to maintain _____ continuity.
    (a) mechanical   (b) electrical   (c) a or b   (d) a and b

Article 368 Busways

72. Busways shall not be installed _____.
    (a) where subject to severe physical damage
    (b) outdoors or in wet or damp locations unless identified for such use
    (c) in hoistways
    (d) all of these

73. Busways shall be securely supported, unless otherwise designed and marked as such, at intervals not to exceed _____.
    (a) 10 ft   (b) 5 ft   (c) 3 ft   (d) 8 ft

74. It shall be permissible to extend busways vertically through dry floors if totally enclosed (unventilated) where passing through, and for a minimum distance of _____ above the floor to provide adequate protection from physical damage.
    (a) 6 ft   (b) $6^1/_2$ ft   (c) 8 ft   (d) 10 ft

75 When a vertical busway penetrates two or more floors (in other than industrial establishments), a minimum 4 in.-high curb shall be installed around the busway floor opening to prevent liquids from entering the vertical busway. The curb shall be installed within _____ of the floor opening for the busway and electrical equipment shall be located so that liquids retained by the 4 in. curb will not damage equipment.
    (a) 12 in.   (b) 6 in.   (c) 12 ft   (d) 6 ft

76. • When devices or plug-in connections for tapping off feeder or branch circuits from busways include an externally operable fusible switch that is out of reach, _____ shall be provided for operation of the disconnecting means from the floor.
    (a) ropes   (b) chains   (c) hook sticks   (d) any of these

77. Bus runs with nominal voltage levels exceeding 600V, having sections located both inside and outside of buildings, shall have a _____ at the building wall to prevent interchange of air between indoor and outdoor sections.
    (a) waterproof rating   (b) vapor seal   (c) fire seal   (d) b and c

78. When bus enclosures terminate at machines cooled by flammable gas, _____ shall be provided to prevent accumulation of flammable gas within the bus enclosures.
    (a) seal-off bushings   (b) baffles   (c) a or b   (d) none of these

Article 370 Cablebus

79. Cablebus is ordinarily assembled at the _____.
     (a) point of installation   (b) manufacturer's location
     (c) distributor's location   (d) none of these

80. The cablebus assembly is designed to carry _____ current and to withstand the magnetic forces of such current.
     (a) service   (b) load   (c) fault   (d) grounded

81. Cablebus shall not be permitted for _____.
     (a) a service   (b) branch circuits   (c) exposed work   (d) concealed work

82. Cablebus framework that is _____ shall be permitted as the equipment grounding conductor for branch circuits and feeders.
     (a) bonded   (b) welded   (c) protected   (d) galvanized

83. In a cablebus, the size and number of conductors shall be that for which the cablebus is designed, and in no case smaller than _____ AWG.
     (a) 1/0   (b) 2/0   (c) 3/0   (d) 4/0

84. The individual conductors in a cablebus shall be supported at intervals not greater than _____ for vertical runs.
     (a) $\frac{1}{2}$ ft   (b) 1 ft   (c) $1\frac{1}{2}$ ft   (d) 2 ft

85. • A cablebus system shall include approved fittings for dead ends.
     (a) True   (b) False

86. Each section of cablebus shall be marked with the manufacturer's name or trade designation and the minimum diameter, number, voltage rating, and ampacity of the conductors to be installed. Markings shall be so located as to be visible after installation.
     (a) True   (b) False

Article 372 Cellular Concrete Floor Raceways

87. In cellular concrete floor raceways, a cell is defined as a single, enclosed _____ space in a floor made of precast cellular concrete slabs, the direction of the cell being parallel to the direction of the floor member.
     (a) circular   (b) oval   (c) tubular   (d) hexagonal

88. A transverse metal raceway for electrical conductors, furnishing access to predetermined cells of precast cellular concrete floors, which permits installation of electrical conductors from a distribution center to the floor cells is usually known as a(n) _____.
     (a) cell   (b) header   (c) open-bottom raceway   (d) none of these

89. The header on a cellular concrete floor raceway shall be installed _____ to the cells.
     (a) in a straight line   (b) at right angles   (c) a and b   (d) none of these

90. Connections from cellular concrete floor raceway headers to cabinets shall be made by means of _____.
     (a) listed metal raceways   (b) PVC raceways
     (c) listed fittings   (d) a and c

91. For cellular concrete floor raceways, junction boxes shall be _____ the floor grade and sealed against the free entrance of water or concrete.
     (a) leveled to   (b) above   (c) below   (d) perpendicular to

92. In cellular concrete floor raceways a grounding conductor shall connect the insert receptacle to a _____.
     (a) negative ground connection provided in the raceway
     (b) negative ground connection provided on the header
     (c) positive ground connection provided on the header
     (d) grounded terminal located within the insert

93. When an outlet is _____ from a cellular concrete floor raceway, the sections of circuit conductors supplying the outlet shall be removed from the raceway.
     (a) discontinued   (b) abandoned   (c) removed   (d) all of these

Article 374 Cellular Metal Floor Raceways

94. A _____ is defined as a single, enclosed tubular space in a cellular metal floor member.
     (a) cellular metal floor raceway   (b) cell
     (c) header   (d) none of these

95. Without special permission, no conductor larger than _____ AWG shall be installed in a cellular metal floor raceway.
     (a) 1/0   (b) 4/0   (c) 1   (d) no restriction

96. Loop wiring shall _____ in a cellular metal raceway.
     (a) not be permitted   (b) not be considered a splice or tap
     (c) be considered a splice or tap when used   (d) be permitted on conductor sizes 10 or less

97. A junction box used with a cellular metal floor raceway shall be _____.
     (a) level with the floor grade
     (b) sealed against the entrance of water or concrete
     (c) metal and electrically continuous with the raceway
     (d) all of these

98. Inserts for cellular metal floor raceways shall be leveled to the floor grade and sealed against the entrance of _____.
     (a) concrete   (b) water   (c) moisture   (d) all of these

Article 376 Metal Wireways

99. Metal wireways are sheet metal troughs with _____ that are used to form a raceway system.
     (a) removable covers   (b) hinged covers   (c) a or b   (d) none of these

100. Metal wireways can be installed either exposed or concealed under all conditions.
     (a) True   (b) False

100. Wireways shall be permitted for _____.
     (a) exposed work   (b) concealed work
     (c) wet locations if listed for the purpose   (d) a and c

101. Wireways can pass transversely through a wall _____.
     (a) if the length passing through the wall is unbroken
     (b) if the wall is not fire rated
     (c) in hazardous locations
     (d) if the wall is fire rated

102. Conductors larger than that for which the wireway is designed may be installed in any wireway.
     (a) True   (b) False

103. Where a wireway is used as a pull box for insulated conductors 4 AWG or larger, the distance between raceway and cable entries enclosing the same conductor shall not be less than that required in 314.28(A)(1) for straight pulls and 314.28(A)(2) for angle pulls.
     (a) True   (b) False

104. Wireways shall be supported where run horizontally at each end and at intervals not to exceed _____, or for individual lengths longer than _____ at each end or joint, unless listed for other support intervals.
     (a) 5 ft   (b) 10 ft   (c) 3 ft   (d) 6 ft

105. Splices and taps shall be permitted within a wireway provided they are accessible. The conductors, including splices and taps, shall not fill the wireway to more than _____ percent of its area at that point.
     (a) 25   (b) 80   (c) 125   (d) 75

106. Extensions from wireways are not permitted.
     (a) True   (b) False

Article 382 Nonmetallic Extensions

107. Nonmetallic surface extensions are permitted in _____.
     (a) buildings not over three stories   (b) buildings over four stories
     (c) all buildings   (d) none of these

108. Nonmetallic extensions shall be secured in place by approved means at intervals not exceeding _____
     (a) 6 in.   (b) 8 in.   (c) 10 in.   (d) 16 in.

109. Each run of nonmetallic extensions shall terminate in a fitting that covers the _____.
     (a) device   (b) box   (c) end of the extension   (d) end of the assembly

Article 386 Surface Metal Raceways

110. It is permissible to run unbroken lengths of surface metal raceways through dry _____.
     (a) walls   (b) partitions   (c) floors   (d) all of these

111. In general, the voltage limitation between conductors in a surface metal raceway shall not exceed _____ unless the metal has a thickness of not less than 0.040 in., nominal.
     (a) 300V   (b) 150V   (c) 600V   (d) 1,000V

112. The adjustment factors of 310.15(B)(2)(a), (Notes to Ampacity Tables of 0 through 2,000V), shall not apply to conductors installed in surface metal raceways where _____.
     (a) the cross-sectional area exceeds 4.2 in. (square inches)
     (b) the current-carrying conductors do not exceed 30 in number
     (c) the total cross-sectional area of all conductors does not exceed 20 percent of the interior cross-sectional area of the raceway
     (d) all of these

113. • The maximum number of conductors permitted in any surface raceway shall be _____.
     (a) no more than 30 percent of the inside diameter
     (b) no greater than the number for which it was designed
     (c) no more than 75 percent of the cross-sectional area
     (d) that which is permitted in the Table 312.6(A)

114. The conductors, including splices and taps, in a metal surface raceway shall not fill the raceway to more than _____ percent of its cross-sectional area at that point.
     (a) 75   (b) 40   (c) 38   (d) 53

115. Surface metal raceway enclosures providing a transition from other wiring methods shall have a means for connecting a(n) _____.
     (a) grounded conductor   (b) ungrounded conductor
     (c) equipment grounding conductor   (d) all of these

116. Where combination surface metal raceways are used for both signaling and for lighting and power circuits, the different systems shall be run in separate compartments identified by _____ of the interior finish, and the same relative position of compartments shall be maintained throughout the premises.
     (a) the brilliant colors   (b) etching   (c) identification   (d) the contrasting colors

117. Surface metal raceways and their fittings shall be so designed that the sections can be _____.
     (a) electrically coupled together
     (b) mechanically coupled together
     (c) installed without subjecting the wires to abrasion
     (d) all of these

Article 388 Surface Nonmetallic Raceways

118. • The use of surface nonmetallic raceways shall be permitted _____.
     (a) in dry locations   (b) where concealed   (c) in hoistways   (d) all of these

Article 390 Underfloor Raceways

119. Flat-top underfloor raceways over 4 in. but not over 8 in. wide with a minimum of 1 in. spacing between raceways shall be covered with concrete to a depth of not less than 1 in. Raceways spaced less than 1 in. apart shall be covered with concrete to a depth of _____
     (a) 1 in.   (b) 4 in.   (c) $1^1/_2$ in.   (d) 2 in.

120. The combined area of all conductors installed at any point of an underfloor raceway shall not exceed _____ percent of the internal cross-sectional area.
     (a) 75   (b) 60   (c) 40   (d) 30

121. Loop wiring in an underfloor raceway _____ to be a splice or tap.
     (a) is considered   (b) shall not be considered
     (c) is permitted   (d) none of these

122. When an outlet from an underfloor raceway is discontinued, the circuit conductors supplying the outlet _____.
     (a) may be spliced
     (b) may be reinsulated
     (c) may be handled like abandoned outlets on loop wiring
     (d) shall be removed from the raceway

123. Underfloor raceways shall be laid so that a straight line from the center of one _____ to the center of the next _____ will coincide with the centerline of the raceway system.
     (a) termination point   (b) junction box   (c) receptacle   (d) panelboard

124. Inserts set in fiber underfloor raceways after the floor is laid shall be _____ into the raceway.
     (a) taped   (b) glued   (c) screwed   (d) mechanically secured

Article 392 Cable Trays

125. A cable tray is a unit or assembly of units or sections and associated fittings forming a _____ system used to securely fasten or support cables and raceways.
     (a) structural   (b) flexible   (c) movable   (d) secure

126. Cable trays can be used as a support system for _____.
     (a) services, feeders and branch circuits   (b) communications circuits
     (c) control and signaling circuits   (d) all of these

127. • The intent of Article 392 is to limit the use of cable trays to industrial establishments only.
     (a) True   (b) False

128. Where exposed to direct rays of the sun, insulated conductors and jacketed cables shall be _____ as being sunlight resistant.
     (a) listed   (b) approved   (c) identified   (d) none of these

129. Cable trays and their associated fittings shall be _____ for the intended use.
     (a) listed   (b) approved   (c) identified   (d) none of these

130. Cable trays can be used in corrosive areas and in areas requiring voltage isolation.
     (a) True   (b) False

131. Cable tray systems shall not be used _____.
     (a) in hoistways   (b) where subject to severe physical damage
     (c) in hazardous locations   (d) a and b

132. Cable trays shall _____.
    (a) include fittings for changes in direction and elevation
    (b) have side rails or equivalent structural members
    (c) be made of corrosion-resistant material
    (d) all of these

133. Nonmetallic cable trays shall be made of _____ material.
    (a) fire-resistant   (b) waterproof   (c) corrosive   (d) flame-retardant

134. Each run of cable tray shall be _____ before the installation of cables.
    (a) tested for 25 ohms resistance   (b) insulated
    (c) completed   (d) all of these

135. Supports for cable trays shall be provided in accordance with _____.
    (a) installation instructions   (b) the *NEC*
    (c) a or b   (d) none of these

136. In industrial facilities where conditions of maintenance and supervision ensure that only qualified persons will service the installation, cable tray systems can be used to support _____.
    (a) raceways   (b) cables
    (c) boxes and conduit bodies   (d) all of these

137 For raceways terminating at the tray, a(n) _____ cable tray clamp or adapter shall be used to securely fasten the raceway to the cable tray system.
    (a) listed   (b) approved   (c) identified   (d) none of these

138. Steel or aluminum cable tray systems can be used as an equipment grounding conductor provided the cable tray sections and fittings are identified for _____ purposes.
    (a) grounding   (b) special   (c) industrial   (d) all

139. Aluminum cable trays shall not be used as an equipment grounding conductor for circuits with ground-fault protection and above _____.
    (a) 2,000A   (b) 300A   (c) 500A   (d) 1,200A

140. Steel cable trays shall not be used as equipment grounding conductors for circuits with ground-fault protection above _____.
    (a) 200A   (b) 300A   (c) 600A   (d) 800A

141. Steel or aluminum cable tray systems can be used as equipment grounding conductors, provided the cable tray sections and fittings have been _____ marked to show the cross-sectional area of metal in channel cable trays, or cable trays of one-piece construction.
    (a) legibly   (b) durably   (c) a or b   (d) a and b

142. Cable _____ made and insulated by approved methods shall be permitted to be located within a cable tray provided they are accessible and do not project above the side rails.
    (a) connections   (b) jumpers   (c) splices   (d) conductors

143. A box shall not be required where cables or conductors from cable trays are installed in bushed conduit and tubing used as support or protection against _____.
    (a) abuse   (b) unauthorized access   (c) physical damage   (d) tampering

144. Where single conductor cables comprising each phase or grounded conductor of a circuit are connected in parallel in a cable tray, the conductors shall be installed _____ to prevent current unbalance in the paralleled conductors due to inductive reactance.
    (a) in groups consisting of not more than three conductors per phase or neutral
    (b) in groups consisting of not more than one conductor per phase or neutral
    (c) as individual conductors securely bound to the cable tray
    (d) in separate groups

145. Where a solid-bottom cable tray having a usable inside depth of 6 in. or less contains multiconductor control and/or signal cables only, the sum of the cross-sectional areas of all cables at any cross-section shall not exceed _____ percent of the interior cross-sectional area of the cable tray.
(a) 25   (b) 30   (c) 35   (d) 40

Article 394 Concealed Knob-and-Tube Wiring

146. Concealed knob-and-tube wiring shall be permitted to be used in commercial garages, theaters and similar locations, motion picture studios, hazardous (classified) locations or in the hollow spaces of walls, ceilings and attics where such spaces are insulated by loose, rolled or foamed-in-place insulating material that envelops the conductors.
(a) True   (b) False

147. Where knob-and-tube conductors pass through wood cross members in plastered partitions, conductors shall be protected by noncombustible, nonabsorbent, insulating tubes extending not less than _____ beyond the wood member.
(a) 2 in.   (b) 3 in.   (c) 4 in.   (d) 6 in.

148. Supports for concealed knob-and-tube wiring shall be installed within _____ in. of each side of each tap or splice and at intervals not exceeding _____ ft.
(a) 3, $2^1/_2$ (b) 2, $3^1/_2$ (c) 6, $4^1/_2$ (d) 4, $6^1/_2$

149. Where solid knobs are used, conductors shall be securely tied to them by _____ equivalent to that of the conductor.
(a) wires having insulation   (b) wires having an AWG
(c) nonconductive material   (d) none of these

Article 396 Messenger-Supported Wiring

150. An exposed wiring support system using a messenger wire to support insulated conductors is known as _____.
(a) open wiring   (b) messenger-supported wiring
(c) field wiring   (d) none of these

151. _____ shall be permitted to be installed in messenger-supported wiring.
(a) Multiconductor service-entrance cable
(b) Mineral Insulated (Type MI) cable
(c) Multiconductor underground feeder cable
(d) all of these

152. The conductors supported by messenger shall be permitted to come into contact with the messenger supports or any structural members, walls or pipes.
(a) True   (b) False

153. The messenger shall be supported at dead ends and at intermediate locations so as to eliminate _____ on the conductors.
(a) static   (b) magnetism   (c) tension   (d) induction

Article 398 Open Wiring on Insulators

154. Open wiring on insulators is a(n) _____ wiring method using cleats, knobs, tubes and flexible tubing for the protection and support of single insulated conductors run in or on buildings and not concealed by the building structure.
(a) temporary   (b) acceptable   (c) enclosed   (d) exposed

155. Open conductors entering or leaving locations subject to dampness, wetness or corrosive vapors shall have _____ formed on them and shall then pass upward and inward from the outside of the buildings, or from the damp, wet or corrosive location, through noncombustible, nonabsorbent insulating tubes.
(a) weather heads   (b) drip loops   (c) identification   (d) blisters

156. • Open wiring on insulators within _____ from the floor shall be considered exposed to physical damage.
(a) 4 ft   (b) 2 ft   (c) 7 ft   (d) none of these

157. Where open conductors cross ceiling joists and wall studs, and are exposed to physical damage, they shall be protected by a substantial running board. Running boards shall extend at least _____ in. outside the conductors, but not more than _____ in., and the protecting sides shall be at least 2 in. high and at least 1 in., nominal, in thickness.
(a) $1/_2$, 1   (b) $1/_2$, 2   (c) 1, $2^1/_2$ (d) 1, 2

158. A conductor used for open wiring that must penetrate a wall, floor, or other framing member, shall be carried through a _____.
(a) separate sleeve or tube   (b) weatherproof tube
(c) tube of absorbent material   (d) grounded metallic tube

159. Open conductors shall be separated by at least _____ from metal raceways, piping, or other conducting material, and from any exposed lighting, power, or signaling conductor, or shall be separated by a continuous and firmly fixed nonconductor in addition to the insulation of the conductor.
(a) 2 in.   (b) $2^1/_2$ in.   (c) 3 in.   (d) $3^1/_2$ in.

160. Conductors smaller than 8 AWG for open wiring on insulators shall be supported within _____ of a tap or splice.
(a) 6 in.   (b) 8 in.   (c) 10 in.   (d) 12 in.

161. When nails are used to mount knobs for the support of open wiring on insulators, they shall not be smaller than _____-penny.
(a) six   (b) eight   (c) ten   (d) none of these

162. When screws are used to mount knobs for the support of open wiring on insulators, they shall be of a length sufficient to penetrate the wood to a depth equal to at least _____ the height of the knob and the full thickness of the cleat.
(a) one-eighth   (b) one-quarter   (c) one-third   (d) one-half

163. Conductors that are _____ AWG or larger, supported on solid knobs, shall be securely tied to the knobs by tie wires having an insulation equivalent to that of the open wire.
(a) 14   (b) 12   (c) 10   (d) 8

Chapter 4 Equipment for General Use

Article 400 Flexible Cords and Cables

164. HPD cord is permitted for _____ usage.
(a) not hard   (b) hard   (c) extra hard   (d) all of these

165. TPT and TST cords shall be permitted in lengths not exceeding _____ when attached directly, or by means of a special type of plug, to a portable appliance rated 50W or less.
(a) 8 ft   (b) 10 ft   (c) 15 ft   (d) can't be used at all

166. In no case shall conductors within flexible cords and cables be associated together in such a way with respect to the kind of circuit, the wiring method used, or the number of conductors such that the _____ temperature of the conductors is exceeded.
(a) operating   (b) governing   (c) ambient   (d) limiting

167. A 3-conductor 16 AWG, SJE cable (one conductor is used for grounding) shall have a maximum ampacity of _____ for each conductor.
(a) 13A   (b) 12A   (c) 15A   (d) 8A

168. Flexible cords and cables can be used for _____.
(a) wiring of luminaires
(b) connection of portable lamps or appliances
(c) connection of utilization equipment to facilitate frequent interchange
(d) all of these

169. Flexible cords shall not be used as a substitute for _____ wiring.
(a) temporary   (b) fixed   (c) overhead   (d) none of these

170. Unless specifically permitted in 400.7, flexible cords and cables shall not be used where _____.
(a) run through holes in walls, ceilings or floors
(b) run through doorways, windows or similar openings
(c) attached to building surfaces
(d) all of these

171. Flexible cords and cables shall not be concealed behind building _____, or run through doorways, windows or similar openings.
(a) structural ceilings   (b) suspended or dropped ceilings
(c) floors or walls   (d) all of these

172. Repair of hard-service cord that is _____ AWG and larger shall be permitted if conductors are spliced in accordance with 110.14(B) and the completed splice retains the insulation, outer sheath properties and usage characteristics of the cord being spliced.
(a) 16   (b) 15   (c) 14   (d) 12

173. Flexible cords and cables shall be connected to devices and to fittings so that tension will not be transmitted to joints or terminal screws. This shall be accomplished by _____.
(a) knotting the cord   (b) winding with tape
(c) fittings designed for the purpose   (d) all of these

174. Flexible cords and cables shall be protected by _____ where passing through holes in covers, outlet boxes or similar enclosures.
(a) bushings   (b) fittings   (c) a or b   (d) none of these

175. One conductor of flexible cords intended to be used as a(n) _____ circuit conductor shall have a continuous marker readily distinguishing it from the other conductor or conductors.
(a) grounded   (b) equipment   (c) ungrounded   (d) all of these

176. The flexible cord conductor identification required in the *Code* shall consist of one of five methods. One method is a tracer in a braid of any color contrasting with that of the braid and _____ in the braid of the other conductor or conductors.
(a) a solid color   (b) no tracer   (c) two tracers   (d) none of these

177. A flexible cord conductor intended to be used as a(n) _____ conductor shall have a continuous identifying marker readily distinguishing it from the other conductor or conductors.
(a) ungrounded   (b) equipment grounding (c) service   (d) high-leg

Article 402 Fixture Wires

178. Fixture wires 18 AWG TFFN are rated for _____.
(a) 14A   (b) 10A   (c) 8A   (d) 6A

179. The smallest size fixture wire permitted in the *NEC* is _____ AWG.
(a) 22   (b) 20   (c) 18   (d) 16

180. Fixture wires shall be permitted for connecting luminaires to the _____ conductors supplying the luminaires.
(a) service   (b) branch-circuit   (c) supply   (d) none of these

181. Fixture wires shall be permitted for installation in luminaires and in similar equipment where enclosed or protected and not subject to _____ in use or for connecting luminaires to the branch-circuit conductors supplying the luminaires.
(a) bending or twisting   (b) knotting
(c) stretching or straining   (d) none of these

Article 404 Switches

182. Switches or circuit breakers shall not disconnect the grounded conductor of a circuit unless the switch or circuit breaker _____.
    (a) can be opened and closed by hand levers only
    (b) simultaneously disconnects all conductors of the circuit
    (c) opens the grounded conductor before it disconnects the ungrounded conductors
    (d) none of these

183. Switch or circuit-breaker enclosures can be used as a junction box or raceway for conductors feeding through splices, or taps when installed in accordance with 312.8.
    (a) True   (b) False

184. Which of the following switches must indicate whether they are in the (open) OFF or (closed) ON position?
    (a) General-use switches.   (b) Motor-circuit switches.
    (c) Circuit breakers.   (d) all of these

185. All switches and circuit breakers used as switches shall be installed so that they may be operated from a readily accessible place. They shall be installed so that the center of the grip of the operating handle of the switch or circuit breaker, when in its highest position, is not more than 6 ft 7 in. above the floor or working platform.
    (a) True   (b) False

186. Snap switches shall not be grouped or ganged in enclosures with other _____ if the voltage between adjacent devices exceeds 300V, unless they are installed in enclosures equipped with permanently installed barriers between adjacent devices.
    (a) snap switches   (b) receptacles   (c) similar devices   (d) all of these

187. Snap switches shall not be grouped or ganged in enclosures unless they can be arranged so that the voltage between adjacent switches does not exceed _____, or unless they are installed in enclosures equipped with permanently installed barriers between adjacent switches.
    (a) 100V   (b) 200V   (c) 300V   (d) 400V

188. All snap switches, including dimmer and similar control switches, shall be effectively grounded so that they can provide a means to ground metal faceplates, whether or not a metal faceplate is installed.
    (a) True   (b) False

189. A faceplate for a flush-mounted snap switch shall not be less than _____ thick when made of a nonferrous metal.
    (a) 0.03 in.   (b) 0.04 in.   (c) 0.003 in.   (d) 0.004 in.

190. Snap switches installed in recessed boxes must have the _____ seated against the finished wall surface.
    (a) mounting yoke   (b) body   (c) toggle   (d) all of these

191. Knife switches rated for more than 1,200A at 250V or less _____.
    (a) are used only as isolating switches
    (b) may be opened under load
    (c) should be placed so that gravity tends to close them
    (d) should be connected in parallel

192. A form of general-use snap switch, suitable only for use on ac circuits, can control _____.
    (a) resistive and inductive loads that do not exceed the ampere and voltage rating of the switch
    (b) tungsten-filament lamp loads that do not exceed the ampere rating of the switch at 120V
    (c) motor loads that do not exceed 80 percent of the ampere and voltage rating of the switch
    (d) all of these

193. A form of general-use snap switch, suitable for use on either ac or dc circuits, may be used for control of inductive loads not exceeding _____ percent of the ampere rating of the switch at the applied voltage.
    (a) 75   (b) 90   (c) 100   (d) 50

194. Snap switches rated _____ or less directly connected to aluminum conductors shall be listed and marked CO/ALR.
(a) 15A   (b) 20A   (c) 25A   (d) 30A

195. General-use _____ switches shall be used only to control permanently installed incandescent luminaires unless otherwise listed for control of other loads.
(a) dimmer   (b) fan speed control   (c) timer   (d) all of these

196. Where in the off position, a switching device with a marked "OFF" position must completely disconnect all _____ conductors of the load it controls.
(a) grounded   (b) ungrounded   (c) grounding   (d) all of these

197. A fused switch shall not have fuses _____ except as permitted in 240.8.
(a) in series   (b) in parallel   (c) less than 100A   (d) over 15A

Article 406 Receptacles, Cord Connectors, and Attachment Plugs (Caps)

198. Isolated ground receptacles installed in nonmetallic boxes shall be covered with a nonmetallic faceplate because a metal faceplate cannot be grounded to the circuit equipment grounding conductor.
(a) True   (b) False

199. • When replacing an ungrounded type receptacle in a bedroom of a dwelling unit and no grounding means exists in the receptacle enclosure, you must use a _____.
(a) nongrounding receptacle   (b) grounding receptacle
(c) GFCI-type receptacle   (d) a or c

200. Receptacles connected to circuits having different voltages, frequencies or types of current (ac or dc) on the _____ shall be of such design that the attachment plugs used on these circuits are not interchangeable.
(a) building   (b) interior   (c) same premises   (d) exterior

Unit 7 NEC Exam – NEC Code Order 358.28 – 406.3

1. EMT shall not be threaded.
    (a) True   (b) False

2. EMT must be supported within 3 ft of each coupling.
    (a) True   (b) False

3. Flexible metallic tubing shall be permitted to be used _____.
    (a) in dry and damp locations   (b) for direct burial
    (c) in lengths over 6 ft   (d) for a maximum of 1,000V

4. ENT is permitted for direct earth burial when used with fittings listed for this purpose.
    (a) True   (b) False

5. ENT is not permitted in places of assembly unless it is encased in at least _____ of concrete.
    (a) 1 in.   (b) 2 in.   (c) 3 in.   (d) 4 in.

6. The continuous current-carrying capacity of a $1^1/_2$-sq in. copper busbar mounted in an unventilated enclosure is _____.
    (a) 500A   (b) 750A   (c) 650A   (d) 1,500A

7. The maximum ampere rating of a 4 in. x $^1/_2$ in. busbar that is 4 ft long and installed in an auxiliary gutter is _____.
    (a) 500A   (b) 750A   (c) 650A   (d) 2,000A

8. Conductors, including splices and taps, shall not fill the auxiliary gutter to more than _____ percent of its cross-sectional area.
    (a) 30   (b) 40   (c) 60   (d) 75

9. Auxiliary gutters shall be constructed and installed so as to maintain _____ continuity.
    (a) mechanical   (b) electrical   (c) a or b   (d) a and b

10. It shall be permissible to extend busways vertically through dry floors if totally enclosed (unventilated) where passing through, and for a minimum distance of, _____ above the floor to provide adequate protection from physical damage.
    (a) 6 ft   (b) $6^1/_2$ ft   (c) 8 ft   (d) 10 ft

11. When devices or plug-in connections for tapping off feeder or branch circuits from busways include an externally operable fusible switch that is out of reach, _____ shall be provided for operation of the disconnecting means from the floor.
    (a) ropes   (b) chains   (c) hook sticks   (d) any of these

12. A cablebus system shall include approved fittings for dead ends.
    (a) True   (b) False

13. Each section of cablebus shall be marked with the manufacturer's name or trade designation and the minimum diameter, number, voltage rating and ampacity of the conductors to be installed. Markings shall be so located as to be visible after installation.
    (a) True   (b) False

14. Loop wiring shall _____ in a cellular metal raceway.
    (a) not be permitted
    (b) not be considered a splice or tap
    (c) be considered a splice or tap when used
    (d) be permitted on conductor sizes 10 or less

15. Metal wireways can be installed either exposed or concealed under all conditions.
    (a) True   (b) False

16. Wireways shall be permitted for _____.
    (a) exposed work
    (b) concealed work
    (c) wet locations if listed for the purpose
    (d) a and c

17. The maximum number of conductors permitted in any surface raceway shall be _____.
    (a) no more than 30 percent of the inside diameter
    (b) no greater than the number for which it was designed
    (c) no more than 75 percent of the cross-sectional area
    (d) that which is permitted in the Table 312.6(A)

18. The use of surface nonmetallic raceways shall be permitted _____.
    (a) in dry locations   (b) where concealed
    (c) in hoistways   (d) all of these

19. The intent of Article 392 is to limit the use of cable trays to industrial establishments only.
    (a) True   (b) False

20. Aluminum cable trays shall not be used as an equipment grounding conductor for circuits with ground-fault protection and above _____.
    (a) 2,000A   (b) 300A   (c) 500A   (d) 1,200A

21. Open wiring on insulators within _____ from the floor shall be considered exposed to physical damage.
    (a) 4 ft   (b) 2 ft   (c) 7 ft   (d) none of these

22. In no case shall conductors within flexible cords and cables be associated together in such a way with respect to the kind of circuit, the wiring method used or the number of conductors such that the _____ temperature of the conductors is exceeded.
    (a) operating   (b) governing   (c) ambient   (d) limiting

23. Flexible cords and cables shall be protected by _____ where passing through holes in covers, outlet boxes or similar enclosures.
    (a) bushings   (b) fittings   (c) a or b   (d) none of these

24. Fixture wires shall be permitted for installation in luminaires and in similar equipment where enclosed or protected and not subject to _____ in use or for connecting luminaires to the branch-circuit conductors supplying the luminaires.
    (a) bending or twisting   (b) knotting
    (c) stretching or straining   (d) none of these

25. When replacing an ungrounded type receptacle in a bedroom of a dwelling unit and no grounding means exists in the receptacle enclosure, you must use a _____.
    (a) nongrounding receptacle   (b) grounding receptacle
    (c) GFCI-type receptacle   (d) a or c

Unit 7 NEC Exam – Random Order 90.3 – 406.2

1. A form of general-use snap switch, suitable for use on either ac or dc circuits, may be used for control of inductive loads not exceeding _____ percent of the ampere rating of the switch at the applied voltage.
    (a) 75   (b) 90   (c) 100   (d) 50

2. A 3-conductor 16 AWG, SJE cable (one conductor is used for grounding) shall have a maximum ampacity of _____ for each conductor.
    (a) 13A   (b) 12A   (c) 15A   (d) 8A

3. A flexible cord conductor intended to be used as a(n) _____ conductor shall have a continuous identifying marker readily distinguishing it from the other conductor or conductors.
    (a) ungrounded   (b) equipment grounding
    (c) service   (d) high-leg

4. A junction box used with a cellular metal floor raceway shall be _____.
    (a) level with the floor grade
    (b) sealed against the entrance of water or concrete
    (c) metal and electrically continuous with the raceway
    (d) all of these

5. Connections from cellular concrete floor raceway headers to cabinets shall be made by means of _____.
    (a) listed metal raceways   (b) PVC raceways
    (c) listed fittings   (d) a and c

6. Couplings and connectors used with EMT shall be made up _____.
    (a) of metal
    (b) in accordance with industry standards
    (c) tight
    (d) none of these

7. It is permissible to run unbroken lengths of surface metal raceways through dry _____.
    (a) walls   (b) partitions   (c) floors   (d) all of these

8. Knife switches rated for more than 1,200A at 250V or less _____.
    (a) are used only as isolating switches
    (b) may be opened under load
    (c) should be placed so that gravity tends to close them
    (d) should be connected in parallel

9. Nonmetallic cable trays shall be made of _____ material.
    (a) fire-resistant   (b) waterproof   (c) corrosive   (d) flame-retardant

10. Nonmetallic surface extensions are permitted in _____.
    (a) buildings not over three stories   (b) buildings over four stories
    (c) all buildings   (d) none of these

11. Repair of hard-service cord that is _____ AWG and larger shall be permitted if conductors are spliced in accordance with 110.14(B) and the completed splice retains the insulation, outer sheath properties, and usage characteristics of the cord being spliced.
    (a) 16   (b) 15   (c) 14   (d) 12

12. Steel cable trays shall not be used as equipment grounding conductors for circuits with ground-fault protection above _____.
    (a) 200A   (b) 300A   (c) 600A   (d) 800A

13. Surface metal raceway enclosures providing a transition from other wiring methods shall have a means for connecting a(n) _____.
(a) grounded conductor
(b) ungrounded conductor
(c) equipment grounding conductor
(d) all of these

14. The combined area of all conductors installed at any point of an underfloor raceway shall not exceed _____ percent of the internal cross-sectional area.
(a) 75
(b) 60
(c) 40
(d) 30

15. The flexible cord conductor identification required in the *Code* shall consist of one of five methods. One method is a tracer in a braid of any color contrasting with that of the braid and _____ in the braid of the other conductor or conductors.
(a) a solid color
(b) no tracer
(c) two tracers
(d) none of these

16. When conductor ampacity adjustment factors of 310.15(B)(2)(a) are used, an auxiliary gutter shall not contain more than _____ at any cross-section.
(a) 25 conductors
(b) 40 current-carrying conductors
(c) 20 conductors
(d) no limit on the number of conductors

17. Wireways can pass transversely through a wall _____.
(a) if the length passing through the wall is unbroken
(b) if the wall is not fire rated
(c) in hazardous locations
(d) if the wall is fire rated

18. Wireways shall be supported where run horizontally at each end and at intervals not to exceed _____, or for individual lengths longer than _____ at each end or joint, unless listed for other support intervals.
(a) 5 ft
(b) 10 ft
(c) 3 ft
(d) 6 ft

19. A box shall not be required where cables or conductors from cable trays are installed in bushed conduit and tubing used as support or protection against _____.
(a) abuse
(b) unauthorized access
(c) physical damage
(d) tampering

20. A cable tray is a unit or assembly of units or sections and associated fittings forming a _____ system used to securely fasten or support cables and raceways.
(a) structural
(b) flexible
(c) movable
(d) secure

21. An auxiliary gutter is permitted to contain _____.
(a) conductors
(b) overcurrent devices
(c) busways
(d) none of these

22. Bus runs with nominal voltage levels exceeding 600V, having sections located both inside and outside of buildings, shall have a _____ at the building wall to prevent interchange of air between indoor and outdoor sections.
(a) waterproof rating
(b) vapor seal
(c) fire seal
(d) b and c

23. Cable trays and their associated fittings must be _____ for the intended use.
(a) listed
(b) approved
(c) identified
(d) none of these

24. Cable trays can be used as a support system for _____.
(a) services, feeders, branch circuits
(b) communications circuits
(c) control and signaling circuits
(d) all of these

25. Cable trays can be used in corrosive areas and in areas requiring voltage isolation.
(a) True
(b) False

26. Each run of nonmetallic extensions shall terminate in a fitting that covers the _____.
    (a) device   (b) box
    (c) end of the extension   (d) end of the assembly

27. ENT and fittings shall be permitted to be _____ in a concrete slab on grade where the tubing is placed upon sand or approved screenings.
    (a) encased   (b) embedded   (c) either a or b   (d) none of these

28. ENT can be installed above a suspended ceiling if the suspended ceiling provides a thermal barrier having a _____-minute finish rating as identified in listings of fire-rated assemblies.
    (a) 5   (b) 10   (c) 15   (d) none of these

29. ENT is not permitted in hazardous (classified) locations except for intrinsically safe applications.
    (a) True   (b) False

30. Flexible cords and cables shall not be concealed behind building _____ or run through doorways, windows or similar openings.
    (a) structural ceilings
    (b) suspended or dropped ceilings
    (c) floors or walls
    (d) all of these

31. For raceways terminating at the tray, a(n) _____ cable tray clamp or adapter must be used to securely fasten the raceway to the cable tray system.
    (a) listed   (b) approved   (c) identified   (d) none of these

32. General-use _____ switches must be used only to control permanently installed incandescent luminaires unless otherwise listed for control of other loads.
    (a) dimmer   (b) fan speed control
    (c) timer   (d) all of these

33. In industrial facilities where conditions of maintenance and supervision ensure that only qualified persons will service the installation, cable tray systems can be used to support _____.
    (a) raceways   (b) cables
    (c) boxes and conduit bodies   (d) all of these

34. Isolated ground receptacles installed in nonmetallic boxes must be covered with a nonmetallic faceplate because a metal faceplate cannot be grounded to the circuit equipment grounding conductor.
    (a) True   (b) False

35. Loop wiring in an underfloor raceway _____ to be a splice or tap.
    (a) is considered   (b) shall not be considered
    (c) is permitted   (d) none of these

36. Snap switches shall not be grouped or ganged in enclosures with other _____ if the voltage between adjacent devices exceeds 300V, unless they are installed in enclosures equipped with permanently installed barriers between adjacent devices.
    (a) snap switches   (b) receptacles   (c) similar devices   (d) all of these

37. Supports for cable trays must be provided in accordance with _____.
    (a) installation instructions   (b) the *NEC*
    (c) a or b   (d) none of these

38. The _____ of conductors used in prewired ENT manufactured assemblies shall be identified by means of a printed tag or label attached to each end of the manufactured assembly.
    (a) type   (b) size   (c) quantity   (d) all of these

39. The maximum size flexible metallic tubing permitted is _____
    (a) $3/8$ in.   (b) $1/2$ in.   (c) $3/4$ in.   (d) 1 in.

40. The minimum headroom for working spaces about service equipment, switchboards, panelboards or motor control centers shall be 6 ft 6 in. except for service equipment or panelboards in existing dwelling units that do not exceed 200A.
    (a) True   (b) False

41. The minimum headroom of working spaces about motor control centers shall be _____.
    (a) 3 ft   (b) 5 ft   (c) 6 ft   (d) 6 ft 6 in.

42. The *NEC* requires series-rated installations to be field-marked to indicate the maximum level of fault-current for which the system has been installed.
    (a) True   (b) False

43. The outer sheath of MI cable is made of _____.
    (a) aluminum   (b) steel alloy   (c) copper   (d) b or c

44. The parallel conductors in each phase or grounded conductor shall _____.
    (a) be the same length and conductor material
    (b) have the same circular mil area and insulation type
    (c) be terminated in the same manner
    (d) all of these

45. The requirements in "Annexes" must be complied with.
    (a) True   (b) False

46. The scope of Article 285 applies to devices such as cord-and-plug-connected TVSS units or receptacles or appliances that have integral TVSS protection.
    (a) True   (b) False

47. The size of the grounding electrode conductor for a building or structure supplied by a feeder cannot be smaller than that identified in _____ based on the largest ungrounded supply conductor.
    (a) 250.66   (b) 250.122
    (c) Table 250.66   (d) not specified

48. The standard ampere ratings for fuses and inverse-time circuit breakers are listed in 240.6. Additional standard ratings for fuses shall be _____.
    (a) 1A   (b) 6A   (c) 601A   (d) all of these

49. The term "luminaire" replaces the terms "fixture" and "lighting fixture" throughout the 2002 *NEC*.
    (a) True   (b) False

50. The total rating of utilization equipment fastened in place shall not exceed _____ percent of the branch- circuit ampere rating, where the circuit supplies receptacles for cord-and-plug-connected equipment.
    (a) 50   (b) 75   (c) 100   (d) 125

Unit 8

Voltage Drop Calculations

OBJECTIVES

After reading this unit, the student should be able to briefly explain the following concepts:

Alternating-current resistance as compared to dc	Limiting current to limit voltage drop	Sizing conductors to prevent excessive voltage drop
Conductor resistance – ac circuits	Limiting conductor length to limit voltage drop	Voltage-drop considerations
Conductor resistance – dc circuits		Voltage-drop recommendations
Determining circuit voltage drop	Resistance	
Extending circuits		

After reading this unit, the student should be able to briefly explain the following terms:

American Wire Gage	E_{VD} = Conductor voltage drop expressed in volts	Q = ac adjustment factor
CM = circular mils		R = resistance
Conductor	I = load in ampere at 100 percent	Skin effect
Cross-sectional area	K = dc constant	Temperature coefficient
D = distance	Ohm's Law method	VD = volts dropped
Eddy currents		

PART A – CONDUCTOR RESISTANCE CALCULATIONS

8–1 CONDUCTOR RESISTANCE

Metals intended to carry electric current are called conductors or wires, and by their nature, they oppose the flow of electrons. Conductors can be solid or stranded, and they can be made from copper, aluminum, silver or even gold. The conductor's opposition to the flow of current is impacted by the conductor material (copper/aluminum), its cross-sectional area (wire size), its length and its operating temperature.

Material

Silver is the best conductor because it has the lowest resistance, but its high cost limits its use to special applications. *Aluminum* is often used when weight or cost are important considerations, but *copper* is the most common type of metal used for electrical conductors. Figure 8–1.

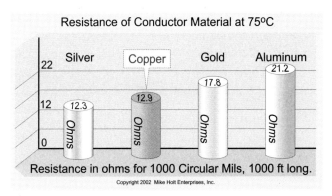

Resistance of Conductor Material at 75°C

Resistance in ohms for 1000 Circular Mils, 1000 ft long.

Copyright 2002 Mike Holt Enterprises, Inc.

Figure 8-1
Resistance of Conductor Material at 75°C

Cross-Sectional Area

The *cross-sectional area* of a conductor is the conductor's surface area expressed in circular mils. The greater the conductor cross-sectional area (the larger the conductor), the greater the number of available electron paths and the lower the conductor resistance. Conductors are sized according to the American Wire Gage (AWG), which ranges from a 40 AWG to 4/0 AWG. Conductors larger than 4/0 are identified in circular mils such as 250,000, 500,000, etc.

Conductor resistance varies inversely with the conductor's diameter; that is, the smaller the wire size, the greater the resistance, and the larger the wire size, the lower the resistance. Figure 8–2.

"Resistance" is the total opposition of current flow in a dc circuit, measured in ohms. "Impedance" is the total opposition to current flow in an ac circuit, measured in ohms.

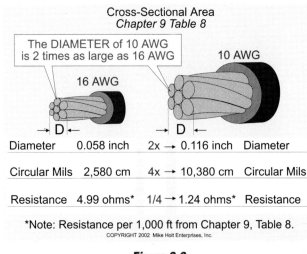

Cross-Sectional Area
Chapter 9 Table 8

The DIAMETER of 10 AWG is 2 times as large as 16 AWG

16 AWG　　10 AWG

Diameter	0.058 inch	2x → 0.116 inch	Diameter
Circular Mils	2,580 cm	4x → 10,380 cm	Circular Mils
Resistance	4.99 ohms*	1/4 → 1.24 ohms*	Resistance

*Note: Resistance per 1,000 ft from Chapter 9, Table 8.
COPYRIGHT 2002 Mike Holt Enterprises, Inc.

Figure 8-2
Cross-Sectional Area

Conductor Length

The resistance of a conductor is directly proportional to its length. The following table provides examples of conductor resistance and *circular mils area* for conductor lengths of 1,000 ft. Naturally, longer or shorter lengths will result in different conductor resistances.

Conductor Properties – *NEC* Chapter 9, Table 8			
Conductor Size American Wire Gage	**Conductor Resistance Per 1,000 Feet at 75°C**	**Conductor Diameter Inches**	**Conductor Area Circular Mils**
14 AWG	3.140Ω (stranded)	0.073	4,110
12 AWG	1.980Ω (stranded)	0.092	6,530
10 AWG	1.240Ω (stranded)	0.116	10,380
8 AWG	0.778Ω (stranded)	0.146	16,510
6 AWG	0.491Ω (stranded)	0.184	26,240

Temperature

The resistance of a conductor changes with changing temperature; this is called *temperature coefficient*. Temperature coefficient describes the effect that temperature has on the resistance of a conductor. Positive temperature coefficient indicates that as the temperature rises, the conductor resistance will also rise. Examples of conductors that have a positive temperature coefficient are silver, copper, gold and aluminum conductors. Negative temperate coefficient means that as the temperature increases, the conductor resistance decreases.

The conductor resistances listed in the *NEC* Tables 8 and 9 of Chapter 9 are based on an operating temperature of 75°C. A three-degree change in temperature results in a one percent change in conductor resistance for both copper and aluminum conductors. The formula to determine the change in conductor resistance with changing temperature is listed at the bottom of Table 8. For example, the resistance of copper at 90°C is about five percent more than at 75°C.

8–2 CONDUCTOR RESISTANCE – DIRECT-CURRENT CIRCUITS, [Chapter 9, Table 8]

The *NEC* lists the resistance and area circular mils for both dc and ac circuit conductors. Direct current circuit conductor resistances are listed in Chapter 9, Table 8 and ac circuit conductor resistances are listed in Chapter 9, Table 9.

The *dc conductor resistances* listed in Chapter 9, Table 8 apply to conductor lengths of 1,000 ft. The following formula can be used to determine the conductor resistance for conductor lengths other than 1,000 ft:

$$\text{DC Conductor Resistance} = \frac{\text{Conductor Resistance Ohms}}{1,000 \text{ ft}} \times \text{Conductor Length}$$

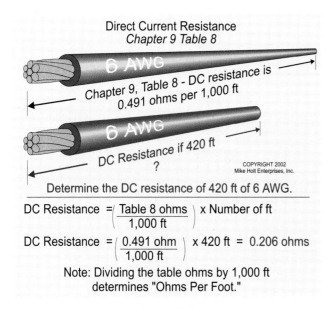

Figure 8-3
Direct-Current Resistance

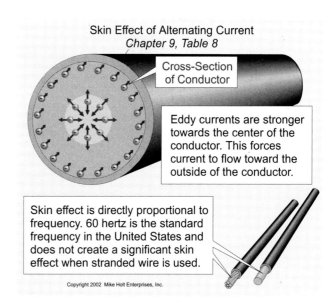

Figure 8-4
Skin Effect of Alternating Current

❏ **Conductor Resistance Copper**

What is the dc resistance of 420 ft. of 6 AWG copper? Figure 8–3.

(a) 0.49Ω   (b) 0.29Ω   (c) 0.72Ω   (d) 0.21Ω

• Answer: (d) 0.21Ω

The dc resistance of 6 AWG copper 1,000 ft long is 0.491Ω, Chapter 9, Table 8.

The dc resistance of 420 ft is: (0.491Ω/1,000 ft) × 420 ft = 0.206Ω, rounded to 0.21

❏ **Conductor Resistance Aluminum**

What is the resistance of 1,490 ft of 3 AWG aluminum?

(a) 0.60Ω   (b) 0.29Ω   (c) 0.72Ω   (d) 0.21Ω

• Answer: (a) 0.60Ω

The resistance of 3 AWG aluminum 1,000 ft long is 0.403Ω, Chapter 9, Table 8.

The resistance of 1,490 ft is: (0.403Ω/1,000 ft) × 1,490 ft = 0.60Ω

8–3 CONDUCTOR IMPEDANCE – ALTERNATING-CURRENT CIRCUITS

In dc circuits, the only property that opposes the flow of electrons is resistance. In ac circuits, the expanding and collapsing magnetic field within the conductor induces an electromotive force that opposes the flow of ac. This opposition to the flow of ac is called *inductive reactance*.

In addition, ac flowing through a conductor generates small, erratic, independent currents called *eddy currents*. Eddy currents are greatest in the center of the conductors and repel the flowing electrons toward the conductor surface; this is known as *skin effect*. Figure 8–4.

Because of skin effect, the effective cross-sectional area of an ac conductor is reduced, which results in an increased opposition to current flow. The total opposition to the movement of electrons in an ac circuit (resistance and inductive reactance) is called *impedance*.

8–4 ALTERNATING-CURRENT IMPEDANCE AS COMPARED TO DIRECT-CURRENT RESISTANCE

The opposition to current flow is greater for ac (impedance) as compared to dc circuits (resistance) because of inductive reactance, eddy currents and skin effect. The following two tables give examples of the difference between ac impedance as compared to dc resistance. Figure 8–5.

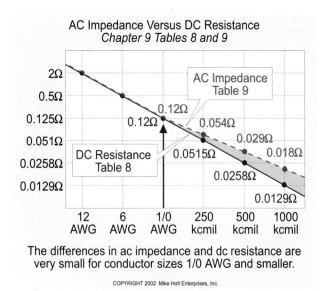

The differences in ac impedance and dc resistance are very small for conductor sizes 1/0 AWG and smaller.

COPYRIGHT 2002 Mike Holt Enterprises, Inc.

Figure 8-5
Conductor Resistance

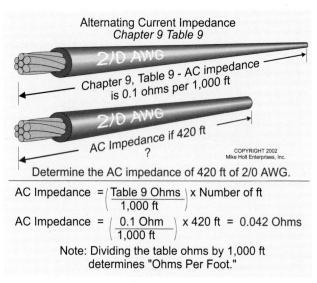

Determine the AC impedance of 420 ft of 2/0 AWG.

$$\text{AC Impedance} = \left(\frac{\text{Table 9 Ohms}}{1,000 \text{ ft}} \right) \times \text{Number of ft}$$

$$\text{AC Impedance} = \left(\frac{0.1 \text{ Ohm}}{1,000 \text{ ft}} \right) \times 420 \text{ ft} = 0.042 \text{ Ohms}$$

Note: Dividing the table ohms by 1,000 ft determines "Ohms Per Foot."

Figure 8-6
Alternating-Current Impedance

COPPER – Alternating-Current Impedance versus Direct-Current Resistance at 75°C			
Conductor Size	**AC Impedance Chapter 9, Table 9**	**DC Resistance Chapter 9, Table 8**	**AC impedance greater than dc resistance by %**
250,000	0.054Ω per 1,000 ft	0.0515Ω per 1,000 ft	4.85%
500,000	0.029Ω per 1,000 ft	0.0258Ω per 1,000 ft	12.40%
1,000,000	0.018Ω per 1,000 ft	0.0129Ω per 1,000 ft	39.50%
ALUMINUM – Alternating-Current Impedance versus Direct-Current Resistance at 75°C			
Conductor Size	**AC Impedance Chapter 9, Table 9**	**DC Resistance Chapter 9, Table 8**	**AC impedance greater than dc resistance by %**
250,000	0.086Ω per 1,000 ft	0.0847Ω per 1,000 ft	1.5%
500,000	0.045Ω per 1,000 ft	0.0424Ω per 1,000 ft	6.13%
1,000,000	0.025Ω per 1,000 ft	0.0212Ω per 1,000 ft	17.12%

8–5 IMPEDANCE [Chapter 9, Table 9 of The NEC]

An alternating-current conductor's opposition to current flow (resistances and reactance) is listed in Chapter 9, Table 9 of the *NEC*. The total opposition to current flow in an ac circuit is called impedance and this is dependent on the conductor's material (copper or aluminum) and on the magnetic property of the raceway or cable they are installed within.

❏ **Alternating-Current Ohms-to-Neutral Impedance Per 1,000 Ft**

What is the ac ohms-to-neutral impedance of 250,000 cm that is 1,000 ft long?

Copper conductor in nonmetallic raceway	= 0.052Ω
Copper conductor in aluminum raceway	= 0.057Ω
Copper conductor in steel raceway	= 0.054Ω
Aluminum conductor in nonmetallic raceway	= 0.085Ω
Aluminum conductors in aluminum raceway	= 0.090Ω
Aluminum conductors in steel raceway	= 0.086Ω

Alternating-Current Conductor Impedance Formula

The following formula can be used to determine conductor impedance:

$$\text{Alternating-Current Impedance} = \frac{\text{Conductor Ohms-to-Neutral Impedance}}{1,000 \text{ ft}} \times \text{Conductor Length}$$

❏ Ohms-to-Neutral Impedance

What is the ac ohms-to-neutral impedance of 420 ft of 2/0 AWG copper installed in a metal raceway? Figure 8–6.

(a) 0.069Ω　　　　　(b) 0.042Ω　　　　　(c) 0.072Ω　　　　　(d) 0.021Ω

• Answer: (b) 0.042Ω

The ohms-to-neutral impedance of 2/0 AWG copper is 0.1Ω per 1,000 ft, Chapter 9, Table 9.

　The ohms-to-neutral impedance of 420 ft of 2/0 AWG is: (0.1Ω/1,000 ft) × 420 ft = 0.042Ω

What is the ac ohms-to-neutral impedance of 169 ft of 500 kcmil aluminum conductors installed in aluminum conduit?

(a) 0.0049Ω　　　　　(b) 0.0029Ω　　　　　(c) 0.0081Ω　　　　　(d) 0.0021Ω

• Answer: (c) 0.0081Ω

The ohms-to-neutral impedance of 500 kcmil installed in aluminum conduit is 0.048Ω per 1,000 ft.

　Ohms-to-neutral impedance of 169 ft of 500 kcmil in aluminum conduit: (0.048Ω/1,000 ft) × 169 ft = 0.0081Ω

Converting Copper to Aluminum or Aluminum to Copper

When requested to determine the replacement conductor for copper or aluminum, the following steps should be helpful:

Step 1:　　Determine the ohms-to-neutral impedance of the existing conductor using Table 9, Chapter 9 for 1,000 ft.

Step 2:　　Using Table 9, Chapter 9, locate a replacement conductor that has an ohms-to-neutral impedance of not more than the existing conductors.

Step 3:　　Verify that the replacement conductor has an ampacity [Table 310.16] sufficient for the load.

❏ Aluminum to Copper

A 240V, 100A, 1Ø load is wired with 2/0 AWG aluminum conductors in a steel raceway. What size copper wire can we use to replace the aluminum wires and not have a greater voltage drop?

Note: The wire selected must have an ampacity of at least 100A. Figure 8–7.

(a) 1/0 AWG　　　　　(b) 1 AWG　　　　　(c) 2 AWG　　　　　(d) 3 AWG

　• Answer: (b) 1 AWG copper

The ohms-to-neutral impedance of 2/0 AWG aluminum (steel raceway) is 0.16Ω per 1,000 ft.

The ohms-to-neutral impedance of 1 AWG copper (steel raceway) is 0.16Ω per 1,000 ft (ampacity of 130A at 75°C).

Determining the Resistance of Parallel Conductors

The total resistance of a parallel circuit is always less than the smallest resistor. The equal resistors' formula can be used to determine the resistance total of parallel conductors:

$$\text{Resistance Total} = \frac{\text{Resistance of One Conductor*}}{\text{Number of Parallel Conductors}}$$

*Resistance according to Chapter 9, Table 8 or Impedance Chapter 9, Table 9 of the *NEC*, assuming 1,000 ft unless specified otherwise.

❏ DC Resistance of Parallel Conductors

What is the dc resistance for two 500 kcmil conductors in parallel? Figure 8–8.

(a) 0.0129Ω　　　　　(b) 0.0258Ω　　　　　(c) 0.0518Ω　　　　　(d) 0.0347V

　• Answer: (a) 0.0129Ω

$$\text{Resistance Total} = \frac{\text{Resistance of One Conductor}}{\text{Number of Parallel Conductors}} = \frac{0.0258\Omega}{2 \text{ conductors}} = 0.0129\Omega$$

　Note: The resistance of 1,000 kcmil in Chapter 9, Table 8, is 0.0129Ω.

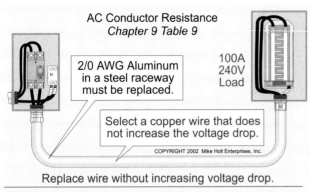

AC Conductor Resistance
Chapter 9 Table 9

2/0 AWG Aluminum in a steel raceway must be replaced.

100A
240V
Load

Select a copper wire that does not increase the voltage drop.

COPYRIGHT 2002 Mike Holt Enterprises, Inc.

Replace wire without increasing voltage drop.

Calculation not required. Use Chapter 9 Table 9 to find a copper resistance equal to or slightly more than aluminum.

2/0 AWG aluminum, steel raceway = 0.16 ohm per 1,000 ft
Closest copper in a steel raceway = 1 AWG = 0.16 ohms per 1,000 ft

Table 310.16, 1 AWG rated 130A at 75°C is okay for 100A load and will not increase voltage drop.

Figure 8-7
AC Conductor Resistance

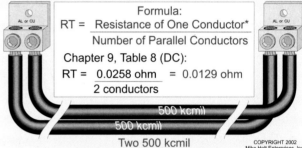

Direct Current Resistance of Parallel Conductors
Chapter 9 Table 8

Formula:
$$RT = \frac{\text{Resistance of One Conductor*}}{\text{Number of Parallel Conductors}}$$

Chapter 9, Table 8 (DC):
$$RT = \frac{0.0258 \text{ ohm}}{2 \text{ conductors}} = 0.0129 \text{ ohm}$$

500 kcmil
500 kcmil
Two 500 kcmil

COPYRIGHT 2002
Mike Holt Enterprises, Inc.

*NOTE: Since the length of the conductors is not specified, assume 1,000 ft.

Figure 8-8
Direct Current Resistance of Parallel Conductors

PART B – VOLTAGE-DROP CALCULATIONS

8-6 VOLTAGE-DROP CONSIDERATIONS

The voltage drop of a circuit is in direct proportion to the conductor's resistance and the magnitude (size) of the current. The longer the conductor, the greater the conductor resistance, the greater the conductor voltage drop; or the greater the current, the greater the conductor voltage drop.

Note: Undervoltage for inductive loads can cause overheating, inefficiency and a shorter life span for electrical equipment. This is especially true in such solid-state equipment as TVs, data-processing equipment (e.g., computers) and other similar equipment. When a conductor resistance causes the voltage to drop below an acceptable point, the conductor size should be increased. Figure 8–9.

8–7 NEC VOLTAGE-DROP RECOMMENDATIONS

Contrary to many beliefs, the *NEC* generally does not require conductors to be increased in size to accommodate voltage drop. However, it does recommend that we consider the effects of conductor voltage drop when sizing conductors.

See some of these recommendations in the Fine Print Notes to Sections 210.19(A), 215.2(A)(4), 230.31(C) and 310.15(A)(1). Please be aware that Fine Print Notes in the *NEC* are recommendations, *not* requirements [90.5(C)]. The *Code* recommends that the maximum combined voltage drop for both the feeder and branch circuit should not exceed five percent, and the maximum on the feeder or branch circuit should not exceed three percent. Figure 8–10.

❏ **NEC Voltage-Drop Recommendation**

What are the minimum *NEC* recommended operating volts for a 115V rated load that is connected to a 120V source? Figure 8–11.

(a) 120V   (b) 115V   (c) 114V   (d) 116V

• Answer: (c) 114V

The maximum conductor voltage-drop recommended for both the feeder and branch circuit is five percent of the voltage source (120V). The total conductor voltage drop (feeder and branch circuit) should not exceed 120V × 0.05 = 6V. The operating voltage at the load is calculated by subtracting the conductor's voltage drop from the voltage source:
120V – 6V = 114V.

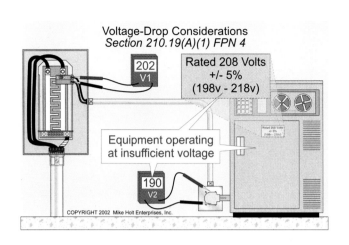

Figure 8-9

Voltage-Drop Considerations

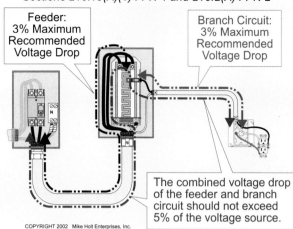

Figure 8-10

Maximum Overall Voltage Drop

8–8 DETERMINING CIRCUIT CONDUCTORS VOLTAGE DROP

When the circuit conductors have already been installed, the voltage drop of the conductors can be determined by the Ohm's Law method or by the formula method:

Ohm's Law Method – 1Ø Only

$$VD = I \times R$$

VD = Conductor voltage drop expressed in volts.

I = The load in amperes at 100%, not at 125%, for motors or continuous loads.

R* = Conductor Resistance, Chapter 9, Table 8 for dc or Chapter 9, Table 9 for ac.

*For conductors 1/0 AWG and smaller, the difference in resistance between dc and ac circuits is so little that it can be ignored. In addition, you can ignore the small difference in resistance between stranded and solid wires.

❏ **Voltage Drop 120V**

What is the voltage drop of two 12 AWG THHN conductors that supply a 16A, 120V load located 100 ft from the power supply? Figure 8–12.

(a) 3.2V   (b) 6.4V   (c) 9.6V   (d) 12.8V

• Answer: (b) 6.4V

  $VD = I \times R$

  I = 16A

  R = 2Ω per 1,000 ft, Chapter 9, Table 9: (2Ω/1,000 ft) × 200 ft = 0.4Ω

  $VD = 16A \times 0.4\Omega = 6.4V$

❏ **Voltage Drop 240V**

A 240V, 24A, 1Ø load is located 160 ft from the panelboard and is wired with 10 AWG THHN. What is the voltage drop of the circuit conductors? Figure 8–13.

(a) 4.53V   (b) 9.22V   (c) 3.64V   (d) 5.54V

• Answer: (b) 9.22V

  $VD = I \times R$

  I = 24A

  R = 1.2Ω per 1,000 ft, Chapter 9, Table 9: (1.2Ω/1,000 ft) × 320 ft = 0.384Ω

  $VD = 24A \times 0.384\Omega = 9.216 VD$

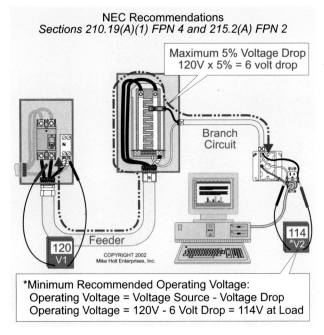

Figure 8-11
Overall Voltage Drop

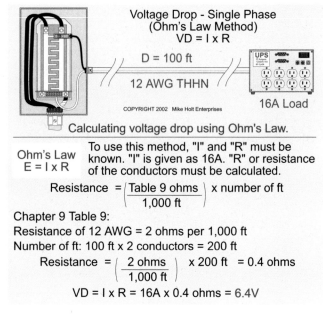

Figure 8-12
Voltage Drop Using Ohm's Law

Voltage Drop Using the Formula Method

In addition to the Ohm's Law method, the following formula can be used to determine the conductor voltage drop:

$$VD (1\emptyset) = \frac{2 \times (K \times Q) \times I \times D}{CM} \qquad VD (3\emptyset) = \frac{\sqrt{3} \times (K \times Q) \times I \times D}{CM}$$

Note: $\sqrt{3} = 1.732$

VD = Volts Dropped: The voltage drop of the circuit expressed in volts. The *NEC* recommends a maximum 3% voltage drop for either the branch circuit or feeder.

K = Direct-Current Constant: This constant K represents the dc resistance for a 1,000-circular mils conductor that is 1,000 ft long, at an operating temperature of 75°C. The constant K value is 12.9Ω for copper and 21.2Ω for aluminum.

Q = Alternating-Current Adjustment Factor: For ac circuits with conductors 2/0 AWG and larger, the dc resistance constant K must be adjusted for the effects of self-induction (eddy currents). The "Q" Adjustment Factor is calculated by dividing the ac ohms-to-neutral impedance listed in Chapter 9, Table 9 by the dc resistance listed in Chapter 9, Table 8 in the *NEC*. For all practical exam purposes, this resistance adjustment factor can be ignored because exams rarely give ac voltage drop questions with conductors larger than 1/0 AWG.

I = Amperes: The load in amperes at 100% (not at 125% for motors or continuous loads).

D = Distance: The distance the load is from the power supply.

CM = Circular-Mils: The circular mils of the circuit conductor as listed in *NEC* Chapter 9, Table 8.

❏ Voltage Drop – 1Ø

A 24A, 240V load is located 160 ft from a panelboard and is wired with 10 AWG THHN. What is the approximate voltage drop of the branch-circuit conductors? Figure 8–14.

(a) 4.25V   (b) 9.5V   (c) 3%   (d) 5%

    • Answer: (b) 9.5V

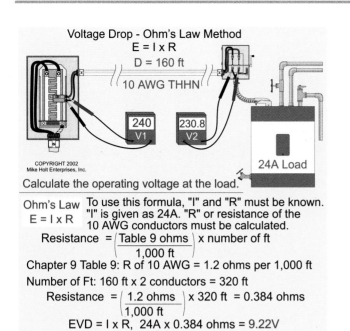

Voltage Drop - Ohm's Law Method
E = I x R
D = 160 ft
10 AWG THHN

240 V1   230.8 V2

24A Load

COPYRIGHT 2002
Mike Holt Enterprises, Inc.

Calculate the operating voltage at the load.

Ohm's Law
E = I x R

To use this formula, "I" and "R" must be known. "I" is given as 24A. "R" or resistance of the 10 AWG conductors must be calculated.

Resistance $= \left(\dfrac{\text{Table 9 ohms}}{1,000 \text{ ft}}\right)$ x number of ft

Chapter 9 Table 9: R of 10 AWG = 1.2 ohms per 1,000 ft

Number of Ft: 160 ft x 2 conductors = 320 ft

Resistance $= \left(\dfrac{1.2 \text{ ohms}}{1,000 \text{ ft}}\right)$ x 320 ft = 0.384 ohms

EVD = I x R, 24A x 0.384 ohms = 9.22V

V2 = 240V source - 9.22 volts dropped = 230.8V

Figure 8-13
Voltage Drop Ohm's Law Method

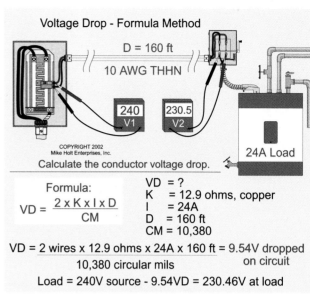

Voltage Drop - Formula Method
D = 160 ft
10 AWG THHN

240 V1   230.5 V2

24A Load

COPYRIGHT 2002
Mike Holt Enterprises, Inc.

Calculate the conductor voltage drop.

Formula:
$VD = \dfrac{2 \times K \times I \times D}{CM}$

VD = ?
K = 12.9 ohms, copper
I = 24A
D = 160 ft
CM = 10,380

VD = 2 wires x 12.9 ohms x 24A x 160 ft = 9.54V dropped on circuit
10,380 circular mils

Load = 240V source - 9.54VD = 230.46V at load

Figure 8-14
Voltage Drop Formula Method

$$VD = \frac{2 \times K \times I \times D}{CM}$$

K = 12.9Ω, copper

I = 24A

D = 160 ft

CM = 10,380 Chapter 9, Table 8 (10 AWG)

$$VD = \frac{2 \text{ wires} \times 12.9\ \Omega \times 24A \times 160 \text{ ft}}{10,380 \text{ circular mils}} = 9.54 \text{ VD}$$

❏ **Voltage Drop – 3Ø**

A 3Ø, 36 kVA load rated 208V is located 80 ft from the panelboard and is wired with 1 AWG THHN aluminum. What is the approximate voltage drop of the feeder circuit conductors? Figure 8–15.

(a) 3.5V   (b) 7V   (c) 3%   (d) 5%

• Answer: (a) 3.5V

$$VD = \frac{\sqrt{3} \times K \times I \times D}{CM}$$

$\sqrt{3}$ = 1.732

K = 21.2Ω, aluminum

$$I = \frac{VA}{(E \times \sqrt{3})} = \frac{36,000 \text{ VA}}{(208\ V \times 1.732)} = 100A$$

D = 80 ft

CM = 83,690 Chapter 9, Table 8

$$VD = \frac{1.732 \times 21.2\ \Omega \times 100A \times 80 \text{ ft}}{83,690 \text{ circular mils}} = 3.51 \text{ VD}$$

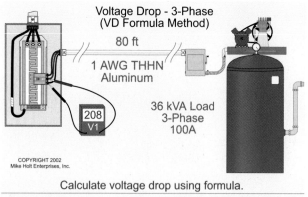

Calculate voltage drop using formula.

Formula	
VD = $\dfrac{\sqrt{3} \times K \times I \times D}{CM}$	$\sqrt{3}$ = 1.732 K = 21.2 ohms, aluminum I = 100A D = 80 ft CM = 83,690

VD = $\dfrac{1.732 \times 21.2 \text{ ohms} \times 100A \times 80 \text{ ft}}{83,690 \text{ circular mils}}$ = 3.5V

Figure 8-15

3-Phase Voltage Drop Formula Method

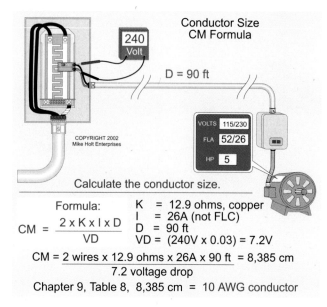

Calculate the conductor size.

Formula:	K = 12.9 ohms, copper
CM = $\dfrac{2 \times K \times I \times D}{VD}$	I = 26A (not FLC) D = 90 ft VD = (240V x 0.03) = 7.2V

CM = $\dfrac{2 \text{ wires} \times 12.9 \text{ ohms} \times 26A \times 90 \text{ ft}}{7.2 \text{ voltage drop}}$ = 8,385 cm

Chapter 9, Table 8, 8,385 cm = 10 AWG conductor

Figure 8-16

Conductor Size CM Method

8–9 SIZING CONDUCTORS TO PREVENT EXCESSIVE VOLTAGE DROP

The size of a conductor (actually its resistance) affects voltage drop. If we want to decrease the voltage drop of a circuit, we can increase the size of the conductor (reduce its resistance). When sizing conductors to prevent excessive voltage drop, use the following formulas:

$$\textbf{CM (1Ø)} = \frac{2 \times K \times I \times D}{\textbf{Volts Dropped}} \qquad \textbf{CM (3Ø)} = \frac{\sqrt{3} \times K \times I \times D}{\textbf{Volts Dropped}}$$

❏ **Size Conductor – 1Ø**

A 5-hp motor is located 90 ft from a 120/240V, 1Ø panelboard. What size conductor should be used if the motor nameplate indicates 26A at 230V? Terminals rated for 75°C. Figure 8–16.

(a) 10 AWG   (b) 8 AWG   (c) 6 AWG   (d) 4 AWG

• Answer: (a) 10 AWG

$$CM = \frac{2 \times K \times I \times D}{VD}$$

K = 12.9Ω, copper

I = 26A at 230V

D = 90 ft

VD = 240V × 0.03 = 7.2V

$$CM = \frac{2 \text{ wires} \times 12.9\ \Omega \times 26A \times 90 \text{ ft}}{7.2V} = 8,385, 10 \text{ AWG, Chapter 9, Table 8}$$

Note: 430.22(A) requires that the motor conductors be sized not less than 125% of the motor FLCs as listed in Table 430.148. The motor FLC is 26A and the conductor must be sized at: 26A × 1.25 = 32.5A. The 10 AWG THHN required for voltage drop is rated for 35A at 75°C according to Table 310.16 and 110.14(C).

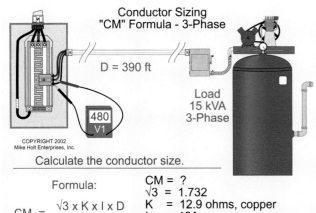

Conductor Sizing
"CM" Formula - 3-Phase

D = 390 ft

Load
15 kVA
3-Phase

480
V1

COPYRIGHT 2002
Mike Holt Enterprises, Inc.

Calculate the conductor size.

Formula:

$$CM = \frac{\sqrt{3} \times K \times I \times D}{\text{allowable VD}}$$

CM = ?
√3 = 1.732
K = 12.9 ohms, copper
I   = 18A
D = 390 ft
VD = (480v x 0.03) = 14.4V

$$CM = \frac{1.732 \times 12.9 \text{ ohms} \times 18A \times 390 \text{ ft}}{14.4 \text{ voltage drop}} = 10,892 \text{ cm}$$

Chapter 9, Table 8, 10,892 cm = 8 AWG conductors

Figure 8-17
Conductor Sizing CM Method

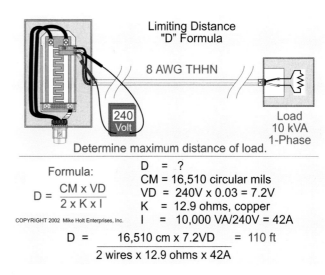

Limiting Distance
"D" Formula

8 AWG THHN

240
Volt

Load
10 kVA
1-Phase

Determine maximum distance of load.

Formula:

$$D = \frac{CM \times VD}{2 \times K \times I}$$

D = ?
CM = 16,510 circular mils
VD = 240V x 0.03 = 7.2V
K = 12.9 ohms, copper
I = 10,000 VA/240V = 42A

COPYRIGHT 2002 Mike Holt Enterprises, Inc.

$$D = \frac{16,510 \text{ cm} \times 7.2VD}{2 \text{ wires} \times 12.9 \text{ ohms} \times 42A} = 110 \text{ ft}$$

Figure 8-18
Limiting Distance "D" Formula

❏ **Size Conductor – 3Ø**

A 3Ø, 15 kVA load rated 480V is located 390 ft from the panelboard. What size conductor is required to prevent the voltage drop from exceeding three percent? Figure 8–17.

(a) 10 AWG   (b) 8 AWG   (c) 6 AWG   (d) 4 AWG

• Answer: (b) 8 AWG

$$CM = \frac{\sqrt{3} \times K \times I \times D}{VD}$$

K = 12.9Ω, copper

$$I = 18A, \frac{VA}{(E \times 1.732)} = \frac{15,000 \text{ VA}}{(480V \times 1.732)} = 18A$$

D = 390 ft

VD = 480V × 0.03 = 14.4V

$$CM = \frac{1.732 \times 12.9 \text{ Ω} \times 18 \text{ A} \times 390 \text{ ft}}{14.4V} = 10,892, \text{ 8 AWG, Chapter 9, Table 8}$$

8–10 LIMITING CONDUCTOR LENGTH TO LIMIT VOLTAGE DROP

Voltage drop can also be reduced by limiting the length of the conductors. The following formulas can be used to help determine the maximum conductor length to limit the voltage drop to *NEC* suggestions:

$$D (1Ø) = \frac{CM \times VD}{2 \times K \times I} \qquad D (3Ø) = \frac{CM \times VD}{\sqrt{3} \times K \times I}$$

❏ **Distance – 1Ø**

What is the maximum distance a 240V, 1Ø, 10 kVA load can be located from the panelboard so the voltage drop does not exceed three percent? The load is wired with 8 AWG THHN, Fig 8–18.

(a) 55 ft   (b) 110 ft   (c) 165 ft   (d) 220 ft

• Answer: (b) 110 ft

$$D = \frac{CM \times VD}{2 \times K \times I}$$

CM = 16,510 (8 AWG), Chapter 9, Table 8

VD = 240V $\times$ 0.03 = 7.2V

K = 12.9Ω, copper

I = VA/E = 10,000 VA/240V = 42A

$$D = \frac{16,510 \text{ circular mils} \times 7.2 \text{ V}}{2 \text{ wires} \times 12.9 \ \Omega \times 42 \text{ A}} = 110 \text{ ft}$$

❏ **Distance – 3Ø**

What is the maximum distance a 480V, 3Ø, 37.5 kVA transformer, wired with 6 AWG THHN, can be located from the panelboard so that the voltage drop does not exceed three percent? Figure 8–19.

(a) 275 ft   (b) 325 ft

(c) 375 ft   (d) 425 ft

    • Answer: (c) 375 ft

$$D = \frac{CM \times VD}{\sqrt{3} \times K \times I}$$

CM = 26,240, 6 AWG Chapter 9, Table 8

VD = 480V $\times$ 0.03 = 14.4V

K = 12.9Ω, copper

$$I = \frac{VA}{(\text{Volts} \times 1.732)} = \frac{37,500 \text{ VA}}{(480V \times 1.732)} = 45A$$

$$D = \frac{26,240 \text{ circular mils} \times 14.4V}{1.732 \times 12.9\Omega \times 45A} = 376 \text{ ft}$$

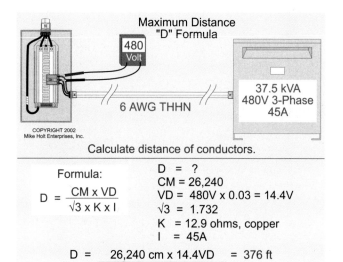

Maximum Distance "D" Formula

6 AWG THHN

37.5 kVA 480V 3-Phase 45A

Calculate distance of conductors.

Formula:

$$D = \frac{CM \times VD}{\sqrt{3} \times K \times I}$$

D = ?
CM = 26,240
VD = 480V x 0.03 = 14.4V
√3 = 1.732
K = 12.9 ohms, copper
I = 45A

$$D = \frac{26,240 \text{ cm} \times 14.4VD}{1.732 \times 12.9 \text{ ohms} \times 45A} = 376 \text{ ft}$$

Figure 8-19
Maximum Distance "D" Formula

8–11 LIMITING CURRENT TO LIMIT VOLTAGE DROP

Sometimes the only method of limiting the circuit voltage drop is to limit the load on the conductors. The following formulas can be used to determine the maximum load.

$$I \ (1\text{Ø}) = \frac{CM \times VD}{2 \times K \times D} \qquad I \ (3\text{Ø}) = \frac{CM \times VD}{\sqrt{3} \times K \times D}$$

❏ **Maximum Load – 1Ø**

An existing installation contains 1/0 AWG THHN aluminum conductors in a nonmetallic raceway to a panelboard located 220 ft from a 230V power source. What is the maximum load that can be placed on the panelboard so that the *NEC* recommendation for voltage drop is not exceeded? Figure 8–20.

(a) 51A   (b) 78A   (c) 94A   (d) 115A

    • Answer: (b) 78A

$$I = \frac{CM \times VD}{2 \times K \times D}$$

CM = 105,600 (1/0 AWG), Chapter 9, Table 8

VD = 230V $\times$ 0.03 = 6.9V

K = 21.2Ω, aluminum

D = 220 ft

$$I = \frac{105,600 \text{ circular mils} \times 6.9 \text{ VD}}{2 \text{ wires} \times 21.2\Omega \times 220 \text{ ft}} = 78A$$

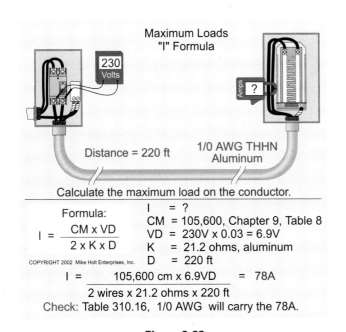

Maximum Loads
"I" Formula

Distance = 220 ft   1/0 AWG THHN
                            Aluminum

Calculate the maximum load on the conductor.

Formula: I   = ?
                  CM   = 105,600, Chapter 9, Table 8
I = CM x VD   VD   = 230V x 0.03 = 6.9V
    ‾‾‾‾‾‾‾   K   = 21.2 ohms, aluminum
    2 x K x D   D   = 220 ft
COPYRIGHT 2002 Mike Holt Enterprises, Inc.

I =   105,600 cm x 6.9VD   = 78A
       2 wires x 21.2 ohms x 220 ft
Check: Table 310.16, 1/0 AWG will carry the 78A.

Figure 8-20
Maximum Loads "I" Formula

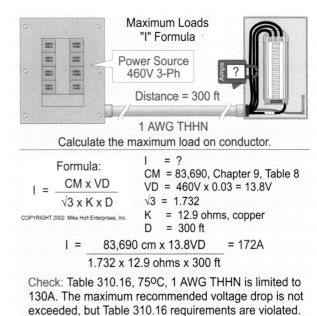

Maximum Loads
"I" Formula

Power Source
460V 3-Ph

Distance = 300 ft

1 AWG THHN
Calculate the maximum load on conductor.

Formula: I   = ?
                  CM   = 83,690, Chapter 9, Table 8
I = CM x VD   VD   = 460V x 0.03 = 13.8V
    ‾‾‾‾‾‾‾   √3   = 1.732
    √3 x K x D   K   = 12.9 ohms, copper
COPYRIGHT 2002 Mike Holt Enterprises, Inc.   D   = 300 ft

I =   83,690 cm x 13.8VD   = 172A
       1.732 x 12.9 ohms x 300 ft

Check: Table 310.16, 75ºC, 1 AWG THHN is limited to
130A. The maximum recommended voltage drop is not
exceeded, but Table 310.16 requirements are violated.

Figure 8-21
Maximum Loads "I" Formula

Note: The maximum load permitted on 1/0 AWG THHN Aluminum at 75°C terminals is 120A [Table 310.16].

❑ **Maximum Load – 3Ø**

An existing installation contains 1 AWG THHN conductors in an aluminum raceway to a panelboard located 300 ft from a 3Ø, 460/230V power source. What is the maximum load the conductors can carry so that the *NEC* recommendation for voltage drop is not exceeded? Figure 8–21.

(a) 170A   (b) 190A   (c) 210A   (c) 240A
  • Answer: (a) 170A

$$I = \frac{CM \times VD}{\sqrt{3} \times K \times D}$$

CM = 83,690 (1 AWG), Chapter 9, Table 8

VD = 460V $\times$ 0.03 = 13.8V

K = 12.9Ω, copper

D = 300 ft

$$I = \frac{(83,690 \text{ circular mils} \times 13.8V)}{(1.732 \times 12.9\Omega \times 300 \text{ ft})} = 172A$$

Note: The maximum load permitted on 1 AWG THHN at 75°C is 130A [110.14(C) and Table 310.16].

8–12 EXTENDING CIRCUITS

If you want to extend an existing circuit and you want to limit the voltage drop, follow these steps:

Step 1:   Determine the voltage drop of the existing conductors.

$$VD\ (1\emptyset) = \frac{2 \times K \times I \times D}{CM}$$

$$VD\ (3\emptyset) = \frac{\sqrt{3} \times K \times I \times D}{CM}$$

Step 2:   Determine the voltage drop permitted for the extension by subtracting the voltage drop of the existing conductors from the permitted voltage drop.

Step 3:   Determine the extended conductor size.

$$CM\ (1\emptyset) = \frac{2 \times K \times I \times D}{VD}$$

$$CM\ (3\emptyset) = \frac{\sqrt{3} \times K \times I \times D}{VD}$$

❏ **Extending Circuits**

An existing junction box is located 55 ft from the panelboard and contains 4 AWG THW aluminum conductors. This circuit is to be extended 65 ft and supply a 50A, 240V load. What size copper conductors must be used for the extension? Figure 8–22.

(a) 8 AWG   (b) 6 AWG
(c) 4 AWG   (d) 5 AWG

  • Answer: (b) 6 AWG

Step 1:   Determine the voltage drop of the existing conductors:

$$\text{Single-Phase VD} = \frac{2 \times K \times I \times D}{CM}$$

$K = 21.2\Omega$, aluminum
$I = 50A$
$D = 55$ ft
$CM = 4$ AWG (41,740 circular mils, Chapter 9, Table 8)

$$VD = \frac{2 \text{ wires} \times 21.2\ \Omega \times 50A \times 55 \text{ ft}}{41,740 \text{ circular mils}} = 2.79V$$

Step 2:   Determine the voltage drop permitted for the extension by subtracting the voltage drop of the existing conductors from the total permitted voltage drop.

Total permitted voltage drop = 240V $\times$ 0.03 = 7.2V $-$ 2.79V = 4.41V

Step 3:   Determine the extended conductor size.

$$CM = \frac{2 \times K \times I \times D}{VD}$$

$K = 12.9\Omega$, copper
$I = 50A$
$D = 65$ ft
$VD = 4.41V$

$$CM = \frac{2 \text{ wires} \times 12.9\Omega \times 50A \times 65 \text{ ft}}{4.41 \text{ VD}} = 19,014, 6 \text{ AWG, Chapter 9, Table 8}$$

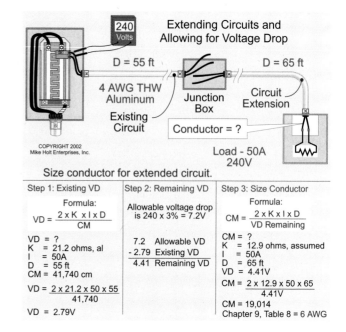

Extending Circuits and Allowing for Voltage Drop

240 Volts

D = 55 ft   D = 65 ft

4 AWG THW Aluminum   Junction Box   Circuit Extension

Existing Circuit   Conductor = ?

COPYRIGHT 2002
Mike Holt Enterprises, Inc.

Load - 50A 240V

Size conductor for extended circuit.

Step 1: Existing VD	Step 2: Remaining VD	Step 3: Size Conductor
Formula: $VD = \frac{2 \times K \times I \times D}{CM}$	Allowable voltage drop is 240 x 3% = 7.2V	Formula: $CM = \frac{2 \times K \times I \times D}{VD \text{ Remaining}}$
VD = ? K = 21.2 ohms, al I = 50A D = 55 ft CM = 41,740 cm	7.2 Allowable VD - 2.79 Existing VD 4.41 Remaining VD	CM = ? K = 12.9 ohms, assumed I = 50A D = 65 ft VD = 4.41V
$VD = \frac{2 \times 21.2 \times 50 \times 55}{41,740}$		$CM = \frac{2 \times 12.9 \times 50 \times 65}{4.41V}$
VD = 2.79V		CM = 19,014 Chapter 9, Table 8 = 6 AWG

Figure 8-22
Extending Circuits

Unit 8 – Voltage-Drop Summary Questions

Calculations Questions

8–1 Conductor Resistance

1. The _____ the number of free electrons, the better the conductivity of the conductor.
    (a) greater   (b) fewer

2. Conductor resistance is determined by the _____.
    (a) material type   (b) cross-sectional area
    (c) conductor length   (d) all of these

3. ___ is the best conductor, better than gold, but the high cost limits its use to special applications, such as fuse elements and some switch contacts.
    (a) Silver   (b) Copper   (c) Aluminum   (d) None of these

4. _____ conductors are often used when weight or cost are important considerations.
    (a) Silver   (b) Copper   (c) Aluminum   (d) None of these

5. • Conductor cross-sectional area is expressed in _____.
    (a) sq in.   (b) mils   (c) circular mils   (d) none of these

6. The resistance of a conductor is directly proportional to its length.
    (a) True   (b) False

7. The resistance of a conductor changes with temperature. Temperature coefficient describes the effect that temperature has on the resistance of a conductor. Conductors with a _____ temperature coefficient have an increase in resistance with an increase in temperature.
    (a) positive   (b) negative   (c) neutral   (d) none of these

8–2 Conductor Resistance – Direct Current [Chapter 9, Table 8]

8. The *National Electrical Code* lists the resistance and area in circular mils for both dc and ac conductors. DC conductor resistances are listed in Chapter 9 Table _____ and ac conductor resistances are listed in Chapter 9 Table _____.
    (a) 1, 5   (b) 3, 4   (c) 9, 8   (d) 8, 9

9. What is the dc resistance of 400 ft of 6 AWG copper conductor?
    (a) 0.2Ω   (b) 0.3Ω   (c) 0.4Ω   (d) 0.5Ω

10. What is the dc resistance of 200 ft of 1 AWG copper conductor?
    (a) 0.0308Ω   (b) 0.0311Ω   (c) 0.0423Ω   (d) 0.0564Ω

11. What is the dc resistance of 1,000 ft of 3 AWG aluminum conductor?
    (a) 0.231Ω   (b) 0.313Ω   (c) 0.422Ω   (d) 0.403Ω

12. What is the dc resistance of 800 ft of 1/0 AWG aluminum conductor?
    (a) 0.23Ω   (b) 0.16Ω   (c) 0.08Ω   (d) 0.56Ω

13. What is the dc resistance of 100 ft of 1 AWG copper conductor?
    (a) 0.233Ω   (b) 0.162Ω   (c) 0.0154Ω   (d) 0.561Ω

14. What is the dc resistance of 500 ft of 3 AWG aluminum conductor?
    (a) 0.2331Ω   (b) 0.1614Ω   (c) 0.0231Ω   (d) 0.2015Ω

8–3 Conductor Impedance – Alternating Current Circuits

15. The intensity of the magnetic field is dependent on the intensity of ac. The greater the current flow, the greater the overall magnetic field.
    (a) True   (b) False

16. • The expanding and collapsing magnetic field within the conductor exerts a force on the moving electrons. This force is called counter-electromotive force (CEMF).
    (a) True   (b) False

17. _____ currents are small independent currents that are produced as a result of the expanding and collapsing magnetic field. They flow erratically within the conductor opposing current flow and consuming power.
    (a) Lenz   (b) Ohm's   (c) Eddy   (d) Kerchoff's

18. The expanding and collapsing magnetic field induces a counter voltage within the conductors, which repels the flowing electrons towards the conductor surface. This is known as _____ effect.
    (a) inductive   (b) skin   (c) surface   (d) watt

8–4 Alternating-Current Impedance as Compared to Direct-Current Resistance

19. The opposition to current flow is greater for ac circuits than for dc circuits because of _____.
    (a) eddy currents   (b) skin effect   (c) CEMF   (d) all of these

8–5 Conductor Ohms to Neutral Impedance – Alternating Current [Chapter 9, Table 9]

20. The ac conductor resistances listed in Chapter 9 Table 9 of the *NEC* are different for copper and aluminum and for nonmagnetic and magnetic raceways.
    (a) True   (b) False

21. What is the ac ohms-to-neutral resistance for 300 ft of 2/0 AWG copper conductor?
    (a) 0.03Ω   (b) 0.04Ω   (c) 0.05Ω   (d) 0.06Ω

22. What is the ac ohms-to-neutral resistance for 1,000 ft of 500 kcmil copper conductor installed in an aluminum raceway?
    (a) 0.032Ω   (b) 0.027Ω   (c) 0.029Ω   (d) 0.030Ω

23. What is the ac ohms-to-neutral resistance for 100 ft of 3 AWG copper conductor?
    (a) 0.012Ω   (b) 0.025Ω   (c) 0.33Ω   (d) 0.43Ω

24. What is the ac ohms-to-neutral resistance of 400 ft of 500 kcmil copper conductor installed in PVC conduit?
    (a) 0.0108Ω   (b) 0.0204Ω   (c) 0.0333Ω   (d) 0.0431Ω

25. A 2-wire circuit supplies a 36A load that is located 100 ft from the panelboard. The load is wired with 1 AWG THHN aluminum in PVC conduit. What is the total ac ohms-to-neutral resistance of the circuit conductors?
    (a) 0.05Ω   (b) 0.25Ω   (c) 0.50Ω   (d) 0.62Ω

26. What size copper conductors in a steel raceway can be used to replace 1/0 AWG aluminum that supplies a 110A load? Note: We do not want to increase the circuit voltage drop.
    (a) 3 AWG   (b) 2 AWG   (c) 1 AWG   (d) 1/0 AWG

27. What is the dc resistance in ohms for three 300 kcmil conductors in a parallel run of 1,000 ft in length?
    (a) 0.014Ω   (b) 0.026Ω   (c) 0.052Ω   (d) 0.047Ω

28. What is the ac ohms-to-neutral resistance for three 1/0 AWG THHN aluminum conductors in parallel for 1,000 ft?
    (a) 0.114Ω   (b) 0.231Ω   (c) 0.413Ω   (d) 0.067Ω

8–6 Voltage-Drop Considerations

29. Because of the great demand for electricity, utilities sometimes are required to reduce their output voltage. In addition, the utility and customer transformers, services, feeders and branch-circuit conductors oppose the flow of current. The opposition to current flow results in voltage drop. All circuits have voltage drop, simply because all conductors have resistance.
    (a) True   (b) False

30. When sizing conductors for feeders and branch circuits, the *NEC* _____ that we take voltage drop into consideration.
    (a) permits   (b) suggests   (c) requires   (d) demands

31. _____ equipment such as motors and electromagnetic ballasts can overheat at reduced voltage. This results in reduced equipment operating life and inconvenience to the customer.
    (a) Inductive   (b) Electronic   (c) Resistive   (d) all of these

32. • _____ equipment such as computers, laser printers, copy machines, etc., can suddenly power down because of reduced voltage, resulting in data losses.
    (a) Inductive   (b) Electronic   (c) Resistive   (d) all of these

33. Resistive loads such as incandescent luminaires and electric space heating can have their power output decreased by the square of the voltage.
    (a) True   (b) False

34. What is the power consumed by a 4.5 kW, 230V water heater operating at 200V?
    (a) 2,700W   (b) 3,400W   (c) 4,500W   (d) 5,500W

35. How can conductor voltage drop be reduced?
    (a) Reduce conductor resistance   (b) Increase conductor size
    (c) Decrease conductor length   (d) all of these

8–7 NEC Voltage-Drop Recommendations

36. If the branch circuit supply voltage is 208V, the maximum recommended voltage drop of the circuit should not be more than _____.
    (a) 3.6V   (b) 6.24V   (c) 6.9V   (d) 7.2V

37. If the feeder supply voltage is 240V, the maximum recommended voltage drop of the feeder should not be more than _____.
    (a) 3.6V   (b) 6.24V   (c) 6.9V   (d) 7.2V

8–8 Determining Circuit Conductors Voltage Drop

38. What is the voltage drop of two 12 AWG THHN conductors supplying a 12A continuous load? Note: The continuous load is located 100 ft from the power supply.
    (a) 3.2V   (b) 4.75V   (c) 4V   (d) 12.8V

39. A 240V, 24A, 1Ø load is located 160 ft from the panelboard. The load is wired with 10 AWG THHN. What is the approximate voltage drop of the branch-circuit conductors?
    (a) 4.25V   (b) 9.5V   (c) 3.2V   (d) 5.9V

40. A 208V, 36 kVA, 3Ø load is located 100 ft from the panelboard and is wired with 1 AWG THHN aluminum. What is the approximate voltage drop of the circuit conductors?
    (a) 3.5V   (b) 5V   (c) 3V   (d) 4.4V

8–9 Sizing Conductors to Prevent Excessive Voltage Drop

41. A 1Ø, 5-hp motor is located 110 ft from a panelboard. The nameplate indicates that the voltage is 115/230 and the FLA is 52/26A. What size conductor is required if the motor windings are connected in parallel and operate at 115V? Apply *NEC* recommended voltage-drop limits.
    (a) 10 AWG THHN　　　　(b) 8 AWG THHN　　　　(c) 6 AWG THHN　　　　(d) 3 AWG THHN

42. A 1Ø, 5-hp motor is located 110 ft from a panelboard. The nameplate indicates that the voltage is 115/230 and the FLA is 52/26A. What size conductor is required if the motor windings are connected in series and operate at 230V? Apply *NEC* recommended voltage-drop limits.
    (a) 10 AWG THHN　　　　(b) 8 AWG THHN　　　　(c) 6 AWG THHN　　　　(d) 4 AWG THHN

43. A 480V, 15 kW, 3Ø load is located 300 ft from the panelboard. What size copper conductor is required to prevent the voltage drop from exceeding 3 percent?
    (a) 10 AWG THHN　　　　(b) 8 AWG THHN　　　　(c) 6 AWG THHN　　　　(d) 4 AWG THHN

8–10 Limiting Conductor Length to Limit Voltage Drop

44. What is the approximate distance that a 240V, 7.5 kVA, 1Ø load can be located from the panelboard so the voltage drop does not exceed 3 percent? The load is wired with 8 AWG THHN copper.
    (a) 55 ft　　　　(b) 110 ft　　　　(c) 145 ft　　　　(d) 220 ft

45. What is the approximate distance a 460V, 37.5 kVA, 3Ø transformer, wired with 6 AWG THHN copper can be located from the panelboard so the voltage drop does not exceed 3 percent?
    (a) 250 ft　　　　(b) 300 ft　　　　(c) 345 ft　　　　(d) 400 ft

8–11 Limiting Current to Limit Voltage Drop

46. An existing installation consists of 1/0 AWG THHN copper conductors in a nonmetallic raceway to a panelboard located 200 ft from a 240V, 1Ø power source. What is the maximum load that can be placed on the panelboard so that the *NEC* recommendations for voltage drop are not exceeded?
    (a) 94A　　　　(b) 109A　　　　(c) 71A　　　　(d) 147A

47. An existing installation contains 2 AWG THHN aluminum conductors in an aluminum raceway to a panelboard located 300 ft from a 460/230V, 3Ø power source. What is the maximum load the conductors can carry without exceeding the *NEC* recommendation for voltage drop?
    (a) 83A　　　　(b) 85A　　　　(c) 64A　　　　(d) 49A

8–12 Extending Circuits

48. An existing junction box is located 65 ft from the panelboard and contains 4 AWG THHN aluminum conductors. What size copper conductor can be used to extend this circuit 85 ft and supply a 50A, 208V load? Apply *NEC* recommended voltage drop limits.
    (a) 8 AWG THHN　　　　(b) 6 AWG THHN　　　　(c) 4 AWG THHN　　　　(d) 10 AWG THHN

49. What is the circuit voltage if the conductor voltage drop is 3.3V? Assume 3 percent voltage drop.
    (a) 110V　　　　(b) 115V　　　　(c) 120V　　　　(d) none of these

☆ Challenge Questions

8–1 Conductor Resistance

50. The resistance of a conductor is affected by temperature change. This is called the _____.
    (a) temperature correction factor　　　　(b) temperature coefficient
    (c) ambient temperature factor　　　　(d) none of these

8–3 Conductor Ohms-to-Neutral Impedance – Alternating Current Circuits

51. The total opposition to current flow in an ac circuit is expressed in ohms and is called _____.
     (a) impedance   (b) conductance   (c) reluctance   (d) resistance

8–5 Alternating-Current Conductor Ohms-to-Neutral Impedance [Chapter 9, Table 9]

52. A 240V, 40A, 1Ø load is located 150 ft from an existing junction box. The junction box is located 50 ft from the panelboard and is wired with 4 AWG THHN aluminum wire. The total resistance of the two 4 AWG conductors from the panelboard to the junction box is approximately _____.
     (a) 0.03Ω   (b) 0.09Ω   (c) 0.05Ω   (d) 0.04Ω

53. A load is located 100 ft from a 230V power supply and is wired with 4 AWG THHN aluminum conductors. What size copper conductor can be used to replace the aluminum conductors and not increase the conductor voltage drop?
     (a) 6 AWG   (b) 8 AWG   (c) 1/0 AWG   (d) 2 AWG

8–7 NEC Voltage-Drop Recommendations

54. A 40A, 240V rated, 1Ø load is wired 150 ft from a junction box and the junction box is located 50 ft from a panelboard (for a total of 200 ft). If the voltage at the panelboard is 240V, what is the minimum voltage recommended by the *NEC* at the 40A load?
     (a) 228.2V   (b) 232.8V   (c) 236.2V   (d) 117.7V

8–8 Determining Circuit Conductors Voltage Drop

55. What is the voltage drop of two 4 AWG aluminum conductors that supply a 5-hp, 208V, 1Ø motor that has a nameplate rating of 55A? The motor is located 95 ft from the power supply.
     (a) 3.25V   (b) 5.31V   (c) 6.24V   (d) 7.26V

8–9 Sizing Conductors to Prevent Excessive Voltage Drop

56. A 240V, 40A, 1Ø load is located 150 ft from an existing junction box. The junction box is located 50 ft from the panelboard. When the 40A load is on, the voltage at the junction box would be calculated to be 236V. The *NEC* recommends voltage drop for this branch circuit not exceed 3 percent of the 240 voltage source (7.2V). What size copper conductor could be installed from the junction box to the load and still meet *NEC* recommendations?
     (a) 3 AWG   (b) 1 AWG   (c) 1/0 AWG   (d) 6 AWG

8–10 Limiting Conductor Length to Limit Voltage Drop

57. How far can a 230V, 50A, 3Ø load be located from the panel if fed with 3 AWG THHN and still meet *NEC* recommendations for voltage drop?
     (a) 275 ft   (b) 300 ft   (c) 325 ft   (d) 350 ft

8–11 Limiting Current to Limit Voltage Drop

58. Two 8 AWG THHN copper conductors supply a 120V load that is located 225 ft from the panelboard. What is the maximum load in amperes that can be applied to these conductors without exceeding the *NEC* recommendation on conductor voltage drop?
     (a) 0A   (b) 5A   (c) 10A   (d) 15A

8–13 Miscellaneous Voltage-Drop Questions

59. An 8Ω resistor is connected to a 120V power supply. Using a voltmeter, we measure 112V across the resistor. What is the current of the 8Ω resistor in amperes?
     (a) 14A   (b) 13A   (c) 15A   (d) 19A

60. A 480V, 1Ø feeder carries 400A and has a 7.2 voltage drop. What is the total resistance of the conductors in this circuit?
    (a) 0.1880Ω   (b) 0.1108Ω   (c) .0190Ω   (d) .0180Ω

61. An 8Ω resistor operates at 112V and is connected to a 115V power supply. The voltage drop of this circuit is _____ percent of the voltage source?
    (a) 2.6   (b) 2.3   (c) 3.5   (d) 3.3

NEC Questions – Articles 406-430

ARTICLE 406 Receptacles, Cord Connectors, and Attachment Plugs (Caps)

62. Receptacles mounted in boxes that are set back of the wall surface shall be installed so that the mounting _____ of the receptacle is held rigidly at the surface of the wall.
    (a) screws or nails   (b) yoke or strap   (c) face plate   (d) none of these

63. Receptacles mounted to and supported by a cover shall be secured by more than one screw.
    (a) True   (b) False

64. Receptacle faceplate covers made of insulating material shall be noncombustible and not less than _____ in thickness.
    (a) 0.10 in.   (b) 0.04 in.   (c) 0.01 in.   (d) 0.22 in.

65. Receptacles, cord connectors, and attachment plugs shall be constructed so that the receptacle or cord connectors will not accept an attachment plug with a different _____ or current rating than that for which the device is intended.
    (a) voltage   (b) amperage   (c) heat   (d) all of these

66. A receptacle shall be considered to be in a location protected from the weather when located under roofed open porches, canopies, marquees, and the like, and will not be subjected to _____.
    (a) spray from a hose   (b) a direct lightning hit
    (c) beating rain or water runoff   (d) falling or wind-blown debris

67. Receptacles installed outdoors in a location protected from the weather or other damp locations, shall be in an enclosure that is _____ when the receptacle is covered.
    (a) raintight   (b) weatherproof   (c) rainproof   (d) weathertight

68. A receptacle installed outdoors in a location protected from the weather or in other damp locations must have an enclosure for the receptacle that is weatherproof when the receptacle is _____.
    (a) covered   (b) enclosed   (c) protected   (d) none of these

69. A receptacle shall be considered to be in a location protected from the weather (damp location) where _____.
    (a) located under a roofed open porch
    (b) not subjected to beating rain or water runoff
    (c) a or b
    (d) a and b

70. _____, 125 and 250V receptacles installed outdoors in a wet location must have an enclosure that is weatherproof.
    (a) 15A   (b) 20A   (c) a and b   (d) none of these

71. Which of the following statements are true for receptacle covers in a wet location?
    (1.) The receptacle cover shall be listed as weatherproof while the attachment plug is inserted for stationary or fixed loads that are intended to have an attachment plug inserted into the receptacle.
    (2.) The receptacle cover shall be listed as weatherproof while the attachment plug is not inserted for portable loads such as appliances and power tools.
    (a) 1 only   (b) 2 only   (c) 1 and 2   (d) none of these

72. An enclosure that is weatherproof only when the receptacle cover is closed can be used for receptacles in a wet location when the receptacle is used for _____ while attended.
    (a) portable equipment   (b) portable tools
    (c) fixed equipment   (d) a and b

73. The enclosure for a receptacle installed in an outlet box flush-mounted on a wall surface in a damp or wet location shall be made weatherproof by means of a weatherproof faceplate assembly that provides a _____ connection between the plate and the wall surface.
    (a) sealed   (b) weathertight
    (c) sealed and protected   (d) watertight

74. Receptacles and cord connectors having grounding terminals must have those terminals effectively _____.
    (a) grounded   (b) bonded   (c) labeled   (d) listed

75. Grounding-type attachment plugs shall be used only with a cord having a(n) _____ conductor.
    (a) equipment grounding   (b) isolated
    (c) computer circuit   (d) insulated

Article 408 Switchboards and Panelboards

76. Conductors and busbars on a switchboard, panelboard or control board shall be located so as to be free from _____ and shall be held firmly in place.
    (a) obstructions   (b) physical damage   (c) a and b   (d) none of these

77. Barriers shall be placed in all service switchboards to isolate the service _____ and terminals from the remainder of the switchboard.
    (a) busbars   (b) conductors   (c) cables   (d) none of these

78. Each switchboard or panelboard that is used as service equipment shall be provided with a main bonding jumper within the panelboard, or one of the sections of the switchboard, for connecting the grounded service conductor on its _____ side to the switchboard or panelboard frame.
    (a) load   (b) supply   (c) phase   (d) high-leg

79. Panelboards supplied by a 3Ø, 4-wire, delta-connected system shall have that phase having the higher voltage-to-ground (high-leg) connected to the _____ phase.
    (a) A   (b) B   (c) C   (d) any of these

80. All panelboard circuits and circuit _____ shall be legibly identified as to purpose or use on a circuit directory located on the face or inside of the panel doors.
    (a) manufacturers   (b) conductors   (c) feeders   (d) modifications

81. Switchboards that have any exposed live parts shall be located in permanently _____ locations and then only where under competent supervision and accessible only to qualified persons.
    (a) dry   (b) mounted   (c) supported   (d) all of these

82. Switchboards shall be placed so as to reduce the probability of communicating _____ to adjacent combustible materials.
    (a) sparks   (b) backfeed   (c) fire   (d) all of these

83. A space of _____ or more shall be provided between the top of any switchboard and any combustible ceiling.
    (a) 12 in.   (b) 18 in.   (c) 2 ft   (d) 3 ft

84. An insulated conductor used within a switchboard shall be _____.
    (a) listed
    (b) flame-retardant
    (c) rated for the highest voltage it may contact
    (d) all of these

85. Conduit or raceways, including their end fittings, shall not rise more than _____ above the bottom of a switchboard enclosure.
    (a) 3 in.   (b) 4 in.   (c) 5 in.   (d) 6 in.

86. Noninsulated busbars shall have a minimum space of _____ between the bottom of enclosure and busbar.
    (a) 6 in.   (b) 8 in.   (c) 10 in.   (d) 12 in.

87. To qualify as a lighting and appliance branch-circuit panelboard, the number of circuits rated at 30A or less and having a neutral conductor shall be _____of the total.
    (a) more than 10 percent   (b) 10 percent
    (c) 20 percent   (d) 40 percent

88. A lighting and appliance branch-circuit panelboard shall be provided with physical means to prevent the installation of more _____ devices than that number for which the panelboard was designed, rated and approved.
    (a) overcurrent   (b) equipment   (c) breaker   (d) all of these

89. A lighting and appliance branch-circuit panelboard contains six 3-pole breakers and eight 2-pole breakers. The maximum allowable number of single-pole breakers that can be added to this panelboard is _____.
    (a) 8   (b) 16   (c) 28   (d) 42

90. A lighting and appliance branch-circuit panelboard is considered protected by the _____ conductor protection device if the protection device rating is not greater than the panelboard rating.
    (a) grounded   (b) feeder   (c) branch circuit   (d) none of these

91. Panelboards equipped with snap switches rated at 30A or less shall have overcurrent protection not in excess of _____.
    (a) 30A   (b) 50A   (c) 100A   (d) 200A

92. When equipment grounding conductors are installed in panelboards, a _____ is required for the proper termination of the equipment grounding conductors.
    (a) neutral   (b) grounded terminal bar
    (c) grounding terminal bar   (d) none of these

93. Each _____ conductor must terminate within the panelboard in an individual terminal that is not also used for another conductor.
    (a) grounded   (b) ungrounded   (c) grounding   (d) all of these

94. Pilot lights, instruments, potential transformers, current transformers and other switchboard devices with potential coils shall be supplied by a circuit that is protected by overcurrent devices rated _____.
    (a) 15A or more   (b) 15A or less   (c) 20A or less   (d) 10A or less

95. The minimum spacing of busbars of opposite polarity that are held in free air in a panelboard is _____ when not over 125V, nominal.
    (a) $^1/_2$ in.   (b) 1 in.   (c) 2 in.   (d) 4 in.

Article 410 Luminaires (Lighting Fixtures), Lampholders and Lamps

96. Luminaires, lampholders and receptacles shall have no live parts normally exposed to contact, but cleat-type lampholders and receptacles located at least _____ above the floor shall be permitted to have exposed contacts.
    (a) 3 ft   (b) 6 ft   (c) 8 ft   (d) none of these

97. A luminaire marked "Suitable for Damp Locations" _____ be used in a wet location.
    (a) can   (b) cannot

98. Luminaires can be installed in a cooking hood if the luminaire is identified for use within a _____ cooking hood.
    (a) nonresidential   (b) commercial   (c) multifamily   (d) all of these

99.   No part of cord-connected luminaires, hanging luminaires, lighting track, pendants or suspended-ceiling fans shall be located within a zone measured 3 ft horizontally and _____ vertically from the top of the bathtub rim or shower stall threshold.
(a) 4 ft   (b) 6 ft   (c) 8 ft   (d) none of these

100.   Lampholders installed over highly combustible material shall be of the _____ type.
(a) industrial   (b) switched   (c) unswitched   (d) residential

101.   Unless an individual switch is provided for each luminaire located over combustible material, lampholders shall be located at least _____ above the floor, or shall be located or guarded so that the lamps cannot be readily removed or damaged.
(a) 3 ft   (b) 6 ft   (c) 8 ft   (d) 10 ft

102.   The *NEC* requires a lighting outlet in clothes closets.
(a) True   (b) False

103.   Incandescent luminaires that have open lamps, and pendant-type luminaires, can be installed in clothes closets where proper clearance is maintained from combustible products.
(a) True   (b) False

104.   Surface-mounted fluorescent luminaires in clothes closets can be installed on the wall above the door or on the ceiling provided there is a minimum clearance of _____ between the luminaire and the nearest point of a storage space.
(a) 3 in.   (b) 6 in.   (c) 9 in.   (d) 12 in.

105.   In clothes closets, recessed incandescent luminaires with a completely enclosed lamp shall be permitted to be installed in the wall or on the ceiling, provided there is a minimum clearance of _____ between the luminaire and the nearest point of a storage space.
(a) 3 in.   (b) 6 in.   (c) 9 in.   (d) 12 in.

106.   Coves for luminaires shall have adequate space and shall be located so that the lamps and equipment can be properly installed and _____.
(a) maintained   (b) protected from physical damage
(c) tested   (d) inspected

107.   When an electric-discharge luminaire is mounted directly over an outlet box, the luminaire must provide access to the conductor wiring within the outlet box.
(a) True   (b) False

108.   The maximum weight of a luminaire that may be mounted by the screw-shell of a brass socket is _____
(a) 2 lbs.   (b) 6 lbs.   (c) 3 lbs.   (d) 50 lbs.

109.   The handhole of metal luminaire poles can be omitted for metal poles _____ or less in height above finished grade. This is only permitted if the pole is provided with a hinged base and the grounding terminal is accessible within the hinged base.
(a) 8 ft   (b) 18 ft   (c) 20 ft   (d) none of these

110.   Metal poles that support luminaires must meet the following requirements: _____.
(a) They must have an accessible handhole (sized 2 × 4 in.) with a raintight cover
(b) An accessible grounding terminal must be installed accessible from the handhole
(c) a and b
(d) none of these

111.   Metal poles used for the support of luminaires shall be bonded to a(n) _____.
(a) grounding electrode   (b) grounded conductor
(c) equipment grounding conductor   (d) any of these

112.   Luminaires attached to the suspended ceiling framing shall be secured to the framing member with screws, bolts, rivets or clips _____ and identified for use with the type of ceiling framing member(s) and luminaires involved.
(a) marked   (b) labeled   (c) identified   (d) listed

113. Trees can be used to support luminaires.
     (a) True   (b) False

114. Exposed conductive parts of luminaires shall be _____.
     (a) grounded   (b) painted   (c) bonded   (d) a and b

115. Luminaires shall be wired so that the _____ of lampholders will be connected to the same fixture, circuit conductor or terminal.
     (a) conductor   (b) neutral   (c) base   (d) screw-shells

116. Luminaires shall be wired with conductors having insulation suitable for the environmental conditions and _____ to which the conductors will be subjected.
     (a) temperature   (b) voltage   (c) current   (d) all of these

117. Fixture wires used for luminaires shall not be smaller than _____ AWG.
     (a) 22   (b) 18   (c) 16   (d) 14

118. Splices and taps shall not be located within a luminaire (fixture) _____.
     (a) arms or stems   (b) bases or screw-shells   (c) a and b   (d) a or b

119. No _____ splices or taps shall be made within or on a luminaire (fixture).
     (a) unapproved   (b) untested   (c) uninspected   (d) unnecessary

120. _____ conductors shall be used for wiring on luminaire (fixture) chains and on other movable or flexible parts.
     (a) Solid   (b) Covered   (c) Insulated   (d) Stranded

121. Wiring on fixture chains and other movable parts shall be _____.
     (a) rated for 110°C (b) stranded   (c) hard usage rated   (d) none of these

122. Luminaires that require adjustment or aiming after installation can be cord-connected without an attachment plug.
     (a) True   (b) False

123. A listed luminaire or a listed fixture assembly shall be permitted to be cord-connected if located _____ the outlet box and the cord is continuously visible for its entire length outside the fixture and is not subject to strain or physical damage.
     (a) within   (b) directly below (c) directly above   (d) adjacent to

124. Luminaires designed for end-to-end connection to form a continuous assembly, or luminaires connected together by recognized wiring methods, shall be permitted to contain the conductors of a 2-wire branch circuit, or one _____ branch circuit, supplying the connected luminaires and need not be listed as a raceway.
     (a) small appliance (b) appliance   (c) multiwire   (d) industrial

125. Branch-circuit conductors within _____ of a ballast shall have an insulation temperature rating not lower than 90°C (194°F) unless supplying a luminaire that is listed and marked as suitable for a different insulation temperature.
     (a) 1 in.   (b) 3 in.   (c) 6 in.   (d) None of these

126. All luminaires requiring ballasts or transformers shall be plainly marked with their electrical _____ and the manufacturer's name, trademark or other suitable means of identification.
     (a) characteristics (b) rating   (c) frequency   (d) none of these

127. Tubing having cut threads and used as arms or stems on luminaires shall have a wall thickness not less than _____
     (a) 0.020 in.   (b) 0.025 in.   (c) 0.040 in.   (d) 0.015 in.

128. Portable lamps shall be wired with _____ recognized by 400.4 and have an attachment plug of the polarized or grounding type.
     (a) flexible cable   (b) flexible cord
     (c) nonmetallic flexible cable   (d) nonmetallic flexible cord

129. Lampholders with Edison-base screw-shells are designed for lampholders and screw-in receptacle adapters.
     (a) True   (b) False

130. Lampholders installed in wet or damp locations shall be of the _____ type.
      (a) waterproof   (b) weatherproof   (c) moistureproof   (d) moisture-resistant

131. Switched lampholders shall be of such construction that the switching mechanism interrupts the electrical connection to the _____.
      (a) lampholder   (b) center contact   (c) branch circuit   (d) all of these

132. A 1,000W incandescent lamp requires a _____ base.
      (a) mogul   (b) standard   (c) medium   (d) copper

133. A recessed incandescent luminaire (fixture) shall be installed so that adjacent combustible material will not be subjected to temperatures in excess of _____°C.
      (a) 75   (b) 90   (c) 125   (d) 150

134. Recessed incandescent luminaires shall have _____ protection and shall be identified as thermally protected.
      (a) physical   (b) corrosion   (c) thermal   (d) all of these

135. The minimum distance that an outlet box containing tap supply conductors can be placed from a recessed luminaire (fixture) is _____.
      (a) 1 ft   (b) 2 ft   (c) 3 ft   (d) 4 ft

136. The raceway or cable for tap conductors to recessed luminaires shall have a minimum length of _____
      (a) 6 in.   (b) 12 in.   (c) 18 in.   (d) 24 in.

137. Incandescent-type luminaires shall be marked to indicate the maximum allowable _____ of lamps.
      (a) voltage   (b) amperage   (c) rating   (d) wattage

138. Ballasts for fluorescent or electric-discharge lighting installed indoors must have _____ protection.
      (a) short-circuit   (b) overcurrent   (c) integral thermal   (d) none of these

139. Surface-mounted luminaires with a ballast must have a minimum clearance of _____ from combustible low-density cellulose fiberboard, unless the fixture is marked "Suitable for Surface Mounting on Combustible Low-Density Cellulose Fiberboard."
      (a) $1/_2$ in.   (b) 1 in.   (c) $1^1/_2$ in.   (d) 2 in.

140. Auxiliary equipment not installed as part of a luminaire (lighting fixture) assembly shall be enclosed in accessible, permanently installed _____.
      (a) nonmetallic cabinets   (b) enclosures
      (c) metal cabinets   (d) all of these

141. An autotransformer, which is used as part of a ballast for supplying lighting units and raises the voltage to more than 300V, shall be supplied by a(n) _____ system.
      (a) system   (b) grounded   (c) listed   (d) identified

142. Electric-discharge luminaires having an open-circuit voltage exceeding _____ shall not be installed in or on dwelling occupancies.
      (a) 120V   (b) 250V   (c) 600V   (d) 1,000V

143. Lighting track is a manufactured assembly and its length may not be altered by the addition or subtraction of sections of track.
      (a) True   (b) False

144. Lighting track is a manufactured assembly designed to support and _____ luminaires that are capable of being readily repositioned on the track.
      (a) connect   (b) protect   (c) energize   (d) all of these

145. Lighting track fittings shall be permitted to be equipped with general-purpose receptacles.
      (a) True   (b) False

146. Lighting track shall not be installed _____.
    (a) where subject to physical damage   (b) in wet or damp locations
    (c) a and b   (d) none of these

147. Lighting track shall not be installed less than _____ above the finished floor except where protected from physical damage or track operating at less than 30V rms, open-circuit voltage.
    (a) 4 ft   (b) 5 ft   (c) 5½ ft   (d) 6 ft

148. Track lighting shall not be installed within the zone measured 3 ft horizontally and _____ vertically from the top of the bathtub rim.
    (a) 2 ft   (b) 3 ft   (c) 4 ft   (d) 8 ft

149. • _____ identified for use on lighting track shall be designed specifically for the track on which they are to be installed.
    (a) Fittings   (b) Receptacles   (c) Devices   (d) all of these

Article 422 Appliances

150. Branch-circuit conductors to individual appliances shall not be sized _____ than required by the appliance markings or instructions.
    (a) larger   (b) smaller

151. Individual appliances that are continuously loaded shall have the branch-circuit rating sized no less than _____ percent of the appliance marked ampere rating.
    (a) 150   (b) 100   (c) 125   (d) 80

152. If a protective device rating is marked on an appliance, the branch-circuit overcurrent protection device rating shall not be greater than _____ percent of the protective device rating marked on the appliance.
    (a) 100   (b) 50   (c) 80   (d) 115

153. Infrared lamps for industrial heating appliances shall have overcurrent protection not exceeding _____.
    (a) 30A   (b) 40A   (c) 50A   (d) 60A

154. The rating or setting of an overcurrent protection device for a 16.3A single nonmotor-operated appliance should not exceed _____.
    (a) 15A   (b) 35A   (c) 25A   (d) 45A

155. Central heating equipment, other than fixed electric space-heating equipment, shall be supplied by a(n) _____ branch circuit.
    (a) multiwire   (b) individual
    (c) multipurpose   (d) small-appliance branch circuit

156. Water heaters having a capacity of _____ gallons or less shall have a branch-circuit rating not less than 125 percent of the nameplate rating of the water heater.
    (a) 60   (b) 75   (c) 90   (d) 120

157. A waste disposal can be cord-and-plug-connected, but the cord must not be less than 18 in. or more than _____ and it shall be protected from physical damage.
    (a) 30 in.   (b) 36 in.   (c) 42 in.   (d) 48 in.

158. The cord for a dishwasher and trash compactor shall not be longer than _____ measured from the back of the appliance.
    (a) 2 ft   (b) 4 ft   (c) 6 ft   (d) 8 ft

159. Ceiling-suspended (paddle) fans that do not exceed _____ in weight, with or without accessories, shall be supported by outlet boxes identified for such use and supported in accordance with 314.23 and 314.27.
    (a) 20 lbs.   (b) 25 lbs.   (c) 30 lbs.   (d) 35 lbs.

160. Listed outlet boxes, or outlet box systems that are identified for the purpose shall be permitted to support ceiling-suspended fans that weigh no more than _____
    (a) 50 lbs.   (b) 60 lbs.   (c) 70 lbs.   (d) all of these

161. The maximum allowable-hp rating of a permanently-connected appliance, when the branch-circuit overcurrent protection device is used as the appliance disconnecting means, is _____ or 300 VA.
(a) $^1/_8$-hp   (b) $^1/_4$-hp   (c) $^1/_2$-hp   (d) $^3/_4$-hp

162. For permanently connected appliances rated over _____, or $^1/_8$-hp, the branch-circuit switch or circuit breaker shall be permitted to serve as the disconnecting means where the switch or circuit breaker is within sight from the appliance or is capable of being locked in the open position.
(a) 200 VA   (b) 300 VA   (c) 400 VA   (d) 500 VA

163. Appliances that have a unit switch with an _____ setting that disconnects all the ungrounded conductors can serve as the disconnecting means for the appliance.
(a) on   (b) off   (c) on/off   (d) all of these

164. Electric heaters of the cord-and-plug-connected immersion type shall be constructed and installed so that current-carrying parts are effectively _____ from electrical contact with the substance in which they are immersed.
(a) isolated   (b) protected   (c) insulated   (d) all of these

165. Electrically heated smoothing irons shall be equipped with an identified _____ means.
(a) disconnecting   (b) temperature-limiting
(c) current-limiting   (d) none of these

166. An infrared heating lamp used in a medium-base lampholder shall be rated _____ or less.
(a) 150W   (b) 300W   (c) 600W   (d) 750W

167. Each electric appliance shall be provided with a(n) _____ giving the identifying name and the rating in volts and amperes, or in volts and watts.
(a) pamphlet   (b) nameplate   (c) auxiliary statement   (d) owner's manual

Article 424 Fixed Electric Space-Heating Equipment

168. The branch-circuit conductor and overcurrent protection device for fixed electric space-heating equipment loads shall not be smaller than _____ percent of the total load.
(a) 80   (b) 100   (c) 125   (d) 150

169. Permanently installed electric baseboard heaters equipped with factory-installed receptacle outlets can be used as the outlets required by 210.50(B).
(a) True   (b) False

170. Fixed electric space-heating equipment requiring supply conductors with insulation rated over _____ shall be clearly and permanently marked.
(a) 75°C   (b) 60°C   (c) 90°C   (d) all of these

171. Fixed electric space-heating equipment shall be installed to provide the _____ spacing between the equipment and adjacent combustible material, unless it has been found to be acceptable where installed in direct contact with combustible material.
(a) required   (b) minimum   (c) maximum   (d) safest

172. If the disconnect is not within sight of the fixed electric space heater (without supplementary overcurrent protection devices), it shall be capable of being _____.
(a) locked   (b) locked in the closed position
(c) locked in the open position   (d) within sight

173. Resistance-type heating elements in electric space-heating equipment shall be protected at not more than _____.
(a) 95 percent of the nameplate value   (b) 60A
(c) 48A   (d) 150 percent of the rated current

174. • Electric heating appliances employing resistance-type heating elements rated more than _____ shall have the heating elements subdivided.
(a) 60A   (b) 50A   (c) 48A   (d) 35A

175. Electric space-heating cables shall be furnished with factory-assembled nonheating leads that are a minimum of _____ in length.
 (a) 6 in.   (b) 18 in.   (c) 3 ft   (d) 7 ft

176. On space-heating cables, blue leads indicate a cable rated at _____, nominal.
 (a) 120V   (b) 240V   (c) 208V   (d) 277V

177. Conductors located above a heated ceiling shall be considered as operating in an ambient temperature of _____.
 (a) 86°C   (b) 30°C   (c) 50°C   (d) 20°C

178. Electric space-heating cables shall not extend beyond the room or area in which they _____.
 (a) provide heat   (b) originate   (c) terminate   (d) are connected

179. Electric space-heating cables shall not be installed over cabinets whose clearance from the ceiling is less than the minimum _____ dimension of the cabinet to the nearest cabinet edge that is open to the room or area.
 (a) horizontal   (b) vertical   (c) overall   (d) depth

180. The minimum clearance between an electric space-heating cable and an outlet box used for surface luminaires shall not be less than _____
 (a) 8 in.   (b) 14 in.   (c) 18 in.   (d) 6 in.

181. GFCI protection for personnel shall be provided for electrically heated floors in _____ locations.
 (a) bathroom   (b) spa   (c) hot tub   (d) all of these

182. When installing duct heaters, sufficient clearance shall be maintained to permit replacement and adjustment of controls and heating elements.
 (a) True   (b) False

183. A boiler employing resistance-type immersion heating elements contained in an ASME-rated and stamped vessel, and rated at more than 120A, shall have the heating elements subdivided into loads not exceeding _____.
 (a) 70A   (b) 100A   (c) 120A   (d) 150A

184. The size of the branch-circuit overcurrent protective devices and conductors for an electrode-type boiler, rated less than 50 kW and not greater than 600V, shall be calculated on the basis of _____.
 (a) 125 percent of the total load excluding motors
 (b) 125 percent of the total load including motors
 (c) 150 percent of the nameplate value
 (d) 100 percent of the nameplate value

185. For electrode-type boilers, each boiler shall be designed so that in normal operation there is no change in state of the heat transfer medium and it shall be equipped with a temperature-sensitive _____.
 (a) protective device   (b) limiting means   (c) shut-off device   (d) all of these

Article 426 Fixed Outdoor Electric Deicing and Snow-Melting Equipment

186. A heating panel is a complete assembly provided with a junction box or a length of flexible conduit for connection to a(n) _____.
 (a) wiring system   (b) service   (c) branch circuit   (d) approved conductor

187. Examples of snow-melting resistance heaters include _____.
 (a) tubular heaters and strip heaters
 (b) immersion heaters and heating blankets
 (c) heating cables or heating tape
 (d) all of these

188. Resistance heating elements of deicing _____ shall not be installed where they bridge expansion joints unless adequately protected from expansion and contraction.
 (a) heating cables   (b) units   (c) panels   (d) all of these

189. Embedded deicing and snow-melting equipment cables, units, and panels shall not be installed where they bridge _____, unless provisions are made for expansion and contraction.
   (a) roads   (b) over water spans   (c) runways   (d) expansion joints

190. Exposed elements of impedance heating systems shall be physically guarded, isolated, or thermally insulated with a _____ jacket to protect against contact by personnel in the area.
   (a) corrosion-resistant   (b) waterproof
   (c) weatherproof   (d) flame-retardant

191. An impedance heating system that is operating at a _____ greater than 30, but not more than 80, shall be grounded at a designated point(s).
   (a) voltage   (b) amperage   (c) wattage   (d) temperature

192. All fixed outdoor deicing and snow-melting equipment shall be provided with a means for disconnecting all _____ conductors.
   (a) grounded   (b) grounding   (c) ungrounded   (d) all of these

Article 427 Fixed Electric Heating Equipment for Pipelines and Vessels

193. Types of pipeline resistive heaters are heating _____.
   (a) blankets   (b) tape   (c) coils   (d) a and b

194. The ampacity of branch-circuit conductors and the rating or setting of overcurrent protective devices supplying fixed electric heating equipment for pipelines and vessels shall be not less than _____ percent of the total load of the heaters.
   (a) 75   (b) 100   (c) 125   (d) 150

195. External surfaces of pipeline and vessel heating equipment that operate at temperatures exceeding _____ shall be physically guarded, isolated or thermally insulated to protect against contact by personnel in the area.
   (a) 110°F   (b) 120°F   (c) 130°F   (d) 140°F

196. Ground-fault protection of equipment shall be provided for electric heat tracing and heating panels furnished for pipeline and vessels.
   (a) True   (b) False

197. For a skin-effect heating installation complying with Article 426 or 427, the provisions of 300.20 shall apply to the installation of a single conductor in a ferromagnetic envelope (metal enclosure).
   (a) True   (b) False

Article 430 Motors, Motor Circuits and Controllers

198. For general motor applications, the motor branch-circuit short-circuit and ground-fault protection device shall be sized based on the _____ amperes.
   (a) motor nameplate   (b) NEMA standards   (c) *NEC* Table   (d) Factory Mutual

199. The motor _____ current as listed in Tables 430.147 through 430.150 shall be used for sizing motor branch-circuit conductors and short-circuit, ground-fault protection devices.
   (a) nameplate   (b) full-load   (c) load   (d) none of these

200. A motor for general use shall be marked with a time rating of _____.
   (a) continuous   (b) 30 or 60 minutes   (c) 5 or 15 minutes   (d) any of these

Unit 8 NEC Exam – NEC Code Order 406.4 – 427.47

1. Receptacles mounted in boxes that are set back of the wall surface shall be installed so that the mounting _____ of the receptacle is held rigidly at the surface of the wall.
    (a) screws or nails
    (b) yoke or strap
    (c) face plate
    (d) none of these

2. Receptacle faceplate covers made of insulating material shall be noncombustible and not less than _____ inches in thickness.
    (a) 0.10
    (b) 0.04
    (c) 0.01
    (d) 0.22

3. Receptacles installed outdoors in a location protected from the weather or other damp locations shall be in an enclosure that is _____ when the receptacle is covered.
    (a) raintight
    (b) weatherproof
    (c) rainproof
    (d) weathertight

4. The enclosure for a receptacle installed in an outlet box flush-mounted on a wall surface in a damp or wet location shall be made weatherproof by means of a weatherproof faceplate assembly that provides a _____ connection between the plate and the wall surface.
    (a) sealed
    (b) weathertight
    (c) sealed and protected
    (d) watertight

5. When equipment grounding conductors are installed in panelboards, a _____ is required for the proper termination of the equipment grounding conductors.
    (a) neutral
    (b) grounded terminal bar
    (c) grounding terminal bar
    (d) none of these

6. Pilot lights, instruments, potential transformers, current transformers and other switchboard devices with potential coils shall be supplied by a circuit that is protected by overcurrent devices rated _____.
    (a) 15A or more
    (b) 15A or less
    (c) 20A or less
    (d) 10A or less

7. Incandescent luminaires that have open lamps, and pendant-type luminaires, can be installed in clothes closets where proper clearance is maintained from combustible products.
    (a) True
    (b) False

8. Surface-mounted fluorescent luminaires in clothes closets can be installed on the wall above the door or on the ceiling provided there is a minimum clearance of _____ between the luminaire and the nearest point of a storage space.
    (a) 3 in.
    (b) 6 in.
    (c) 9 in.
    (d) 12 in.

9. Trees can be used to support luminaires.
    (a) True
    (b) False

10. Splices and taps shall not be located within a luminaire (fixture) _____.
    (a) arms or stems
    (b) bases or screw-shells
    (c) a and b
    (d) a or b

11. Luminaires that require adjustment or aiming after installation can be cord-connected without an attachment plug.
    (a) True
    (b) False

12. Tubing having cut threads and used as arms or stems on luminaires shall have a wall thickness not less than _____
    (a) 0.020 in.
    (b) 0.025 in.
    (c) 0.040 in.
    (d) 0.015 in.

13. Surface-mounted luminaires with a ballast must have a minimum clearance of _____ from combustible low-density cellulose fiberboard, unless the fixture is marked "Suitable for Surface Mounting on Combustible Low-Density Cellulose Fiberboard."
    (a) $1/_2$ in.
    (b) 1 in.
    (c) $1^1/_2$ in.
    (d) 2 in.

14. Electric-discharge luminaires having an open-circuit voltage exceeding _____ shall not be installed in or on dwelling occupancies.
    (a) 120V
    (b) 250V
    (c) 600V
    (d) 1,000V

15.   _____ identified for use on lighting track shall be designed specifically for the track on which they are to be installed.
       (a) Fittings   (b) Receptacles   (c) Devices   (d) all of these

16.   Individual appliances that are continuously loaded shall have the branch-circuit rating sized no less than _____ percent
       of the appliance marked ampere rating.
       (a) 150   (b) 100   (c) 125   (d) 80

17.   The rating or setting of an overcurrent protection device for a 16.3A single nonmotor-operated appliance should not
       exceed _____.
       (a) 15A   (b) 35A   (c) 25A   (d) 45A

18.   Electric heaters of the cord-and-plug-connected immersion type shall be constructed and installed so that current-
       carrying parts are effectively _____ from electrical contact with the substance in which they are immersed.
       (a) isolated   (b) protected   (c) insulated   (d) all of these

19.   Resistance-type heating elements in electric space-heating equipment shall be protected at not more than _____.
       (a) 95 percent of the nameplate value   (b) 60A
       (c) 48A   (d) 150 percent of the rated current

20.   Electric heating appliances employing resistance-type heating elements rated more than _____ shall have the heating
       elements subdivided.
       (a) 60A   (b) 50A   (c) 48A   (d) 35A

21.   The minimum clearance between an electric space-heating cable and an outlet box used for surface luminaires shall not
       be less than _____
       (a) 8 in.   (b) 14 in.   (c) 18 in.   (d) 6 in.

22.   When installing duct heaters, sufficient clearance shall be maintained to permit replacement and adjustment of controls
       and heating elements.
       (a) True   (b) False

23.   The size of the branch-circuit overcurrent protective devices and conductors for an electrode-type boiler, rated less than
       50 kW and not greater than 600V, shall be calculated on the basis of _____.
       (a) 125 percent of the total load excluding motors
       (b) 125 percent of the total load including motors
       (c) 150 percent of the nameplate value
       (d) 100 percent of the nameplate value

24.   Exposed elements of impedance heating systems shall be physically guarded, isolated or thermally insulated with a
       _____ jacket to protect against contact by personnel in the area.
       (a) corrosion-resistant   (b) waterproof
       (c) weatherproof   (d) flame-retardant

25.   For a skin-effect heating installation complying with Article 426 or 427, the provisions of 300.20 shall apply to the
       installation of a single conductor in a ferromagnetic envelope (metal enclosure).
       (a) True   (b) False

Unit 8 NEC Exam – Random Order 100 – 427.22

1. A 1,000W incandescent lamp requires a _____ base.
    (a) mogul　　　(b) standard　　　(c) medium　　　(d) copper

2. A boiler employing resistance-type immersion heating elements contained in an ASME-rated and stamped vessel, and rated at more than 120A, shall have the heating elements subdivided into loads not exceeding _____.
    (a) 70A　　　(b) 100A　　　(c) 120A　　　(d) 150A

3. A heating panel is a complete assembly provided with a junction box or a length of flexible conduit for connection to a(n) _____.
    (a) wiring system
    (c) branch circuit
    (b) service
    (d) approved conductor

4. A listed luminaire or a listed fixture assembly shall be permitted to be cord-connected if located _____ the outlet box and the cord is continuously visible for its entire length outside the fixture and is not subject to strain or physical damage.
    (a) within　　　(b) directly below　　　(c) directly above　　　(d) adjacent to

5. A luminaire marked "Suitable for Damp Locations" _____ be used in a wet location.
    (a) can　　　(b) cannot

6. All luminaires requiring ballasts or transformers shall be plainly marked with their electrical _____ and the manufacturer's name, trademark, or other suitable means of identification.
    (a) characteristics
    (c) frequency
    (b) rating
    (d) none of these

7. Auxiliary equipment not installed as part of a luminaire (lighting fixture) assembly, shall be enclosed in accessible, permanently installed _____.
    (a) nonmetallic cabinets
    (c) metal cabinets
    (b) enclosures
    (d) all of these

8. Branch-circuit conductors within _____ of a ballast shall have an insulation temperature rating not lower than 90°C (194°F) unless supplying a luminaire that is listed and marked as suitable for a different insulation temperature.
    (a) 1 in.　　　(b) 3 in.　　　(c) 6 in.　　　(d) none of these

9. Conductors located above a heated ceiling shall be considered as operating in an ambient temperature of _____.
    (a) 86 °C　　　(b) 30 °C　　　(c) 50 °C　　　(d) 20 °C

10. For electrode-type boilers, each boiler shall be designed so that in normal operation there is no change in state of the heat transfer medium and it shall be equipped with a temperature-sensitive _____.
    (a) protective device
    (c) shut-off device
    (b) limiting means
    (d) all of these

11. Infrared lamps for industrial heating appliances shall have overcurrent protection not exceeding _____.
    (a) 30A　　　(b) 40A　　　(c) 50A　　　(d) 60A

12. Lampholders installed over highly combustible material shall be of the _____ type.
    (a) industrial　　　(b) switched　　　(c) unswitched　　　(d) residential

13. Lighting track shall not be installed _____.
    (a) where subject to physical damage
    (c) a and b
    (b) in wet or damp locations
    (d) none of these

14. Luminaires can be installed in a cooking hood if the luminaire is identified for use within a _____ cooking hood.
    (a) nonresidential　　　(b) commercial　　　(c) multifamily　　　(d) all of these

15. Panelboards equipped with snap switches rated at 30A or less shall have overcurrent protection not in excess of _____.
    (a) 30A　　　(b) 50A　　　(c) 100A　　　(d) 200A

16. _____, 125 and 250V receptacles installed outdoors in a wet location must have an enclosure that is weatherproof.
     (a) 15A   (b) 20A   (c) a and b   (d) none of these

17. A lighting and appliance branch-circuit panelboard is considered protected by the _____ conductor protection device if the protection device rating is not greater than the panelboard rating.
     (a) grounded   (b) feeder   (c) branch circuit   (d) none of these

18. A receptacle installed outdoors in a location protected from the weather or in other damp locations must have an enclosure for the receptacle that is weatherproof when the receptacle is _____.
     (a) covered   (b) enclosed   (c) protected   (d) none of these

19. A receptacle shall be considered to be in a location protected from the weather (damp location) where _____.
     (a) located under a roofed open porch
     (b) not subjected to beating rain or water runoff
     (c) a or b
     (d) a and b

20. Ballasts for fluorescent or electric-discharge lighting installed indoors must have _____ protection.
     (a) short-circuit   (b) overcurrent
     (c) integral thermal   (d) none of these

21. Each _____ conductor must terminate within the panelboard in an individual terminal that is not also used for another conductor.
     (a) grounded   (b) ungrounded   (c) grounding   (d) all of these

22. GFCI protection for personnel shall be provided for electrically heated floors in _____ locations.
     (a) bathroom   (b) spa   (c) hot tub   (d) all of these

23. Ground-fault protection of equipment shall be provided for electric heat tracing and heating panels furnished for pipeline and vessels.
     (a) True   (b) False

24. If the disconnect is not within sight of the fixed electric space heater (without supplementary overcurrent protection devices), it must be capable of being _____.
     (a) locked   (b) locked in the closed position
     (c) locked in the open position   (d) within sight

25. Incandescent-type luminaires shall be marked to indicate the maximum allowable _____ of lamps.
     (a) voltage   (b) amperage   (c) rating   (d) wattage

26. Listed outlet boxes, or outlet box systems that are identified for the purpose, shall be permitted to support ceiling-suspended fans that weigh no more than _____
     (a) 50 lbs.   (b) 60 lbs.   (c) 70 lbs.   (d) all of these

27. Luminaires attached to the suspended ceiling framing shall be secured to the framing member with screws, bolts, rivets or clips _____ and identified for use with the type of ceiling framing member(s) and luminaires involved.
     (a) marked   (b) labeled   (c) identified   (d) listed

28. On space-heating cables, blue leads indicate a cable rated at _____, nominal.
     (a) 120V   (b) 240V   (c) 208V   (d) 277V

29. Receptacles mounted to and supported by a cover shall be secured by more than one screw.
     (a) True   (b) False

30. The cord for a dishwasher and trash compactor shall not be longer than _____ measured from the back of the appliance.
     (a) 2 ft   (b) 4 ft   (c) 6 ft   (d) 8 ft

31. The upper end of the rod electrode shall be _____ ground level unless the aboveground end and the grounding electrode conductor attachment are protected against physical damage.
    (a) above   (b) flush with   (c) below   (d) b or c

32. The working space in front of the electric equipment shall not be less than _____ wide or less than the width of the equipment, whichever is greater.
    (a) 15 in.   (b) 30 in.   (c) 40 in.   (d) 60 in.

33. There shall be no more than _____ disconnects installed for each service, or for each set of service-entrance conductors as permitted in 230.2 and 230.40.
    (a) two   (b) four   (c) six   (d) none of these

34. There shall be no more than _____ disconnects installed for each supply.
    (a) two   (b) four   (c) six   (d) none of these

35. Threadless couplings and connectors must not be used on threaded IMC ends unless the fittings are listed for the purpose.
    (a) True   (b) False

36. 208Y/120V or 480Y/227V, 3Ø, 4-wire, wye systems used to supply nonlinear loads such as personal computers, energy-efficient electronic ballasts, electronic dimming, etc., cause distortion of the phase and neutral currents; this can produce high, unwanted, and potentially hazardous harmonic neutral currents. The *Code* cautions us that the system design should allow for the possibility of high harmonic neutral currents.
    (a) True   (b) False

37. Track lighting shall not be installed within the zone measured 3 ft horizontally and _____ vertically from the top of the bathtub rim.
    (a) 2 ft   (b) 3 ft   (c) 4 ft   (d) 8 ft

38. TVSSs must be marked with their short-circuit current rating and they cannot be installed where the available fault current is in excess of that rating.
    (a) True   (b) False

39. Two 20A small-appliance branch circuits can supply more than one kitchen of a dwelling.
    (a) True   (b) False

40. UF cable shall not be used where subjected to physical damage. When this cable is subject to physical damage, it shall be protected by a suitable method such as a raceway.
    (a) True   (b) False

41. Unless identified for use in the operating environment, no conductors or equipment shall be _____ having a deteriorating effect on the conductors or equipment.
    (a) located in damp or wet locations
    (b) exposed to fumes, vapors or gases
    (c) exposed to liquids or excessive temperatures
    (d) all of these

42. Unless specifically permitted in 240.4(E) through 240.4(G), the overcurrent protection shall not exceed _____ after any correction factors for ambient temperature and the number of conductors have been applied.
    (a) 15A for 14 AWG copper   (b) 20A for 12 AWG copper
    (c) 30A for 10 AWG copper   (d) all of these

43. Unless specified elsewhere in the *Code*, Chapter 3 shall be used for voltages of _____.
    (a) 600 volts-to-ground or less   (b) 300V between conductors or less
    (c) 600V, nominal, or less   (d) 600V RMS

44. Utilization equipment is equipment that utilizes electricity for _____.
    (a) chemicals   (b) heating   (c) lighting   (d) any of these

45. What is the minimum cover requirement in inches for UF cable installed outdoors and underground that is supplying power to a 120V, 30A circuit?
    (a) 6   (b) 12   (c) 18   (d) 24

46. What is the minimum size copper equipment bonding jumper required for equipment connected to a 40A circuit?
    (a) 12 AWG   (b) 14 AWG   (c) 8 AWG   (d) 10 AWG

47. When a building is supplied with a(n) _____ fire sprinkler system, ENT can be installed exposed or concealed in buildings of any height.
    (a) listed   (b) identified   (c) approved   (d) none of these

48. When a building is supplied with an approved fire sprinkler system(s), ENT can be installed above any suspended ceiling.
    (a) True   (b) False

49. When computations result in a fraction of an ampere that is less than_____, such fractions shall be permitted to be dropped.
    (a) 0.49   (b) 0.50   (c) 0.51   (d) none of these

50. When conductors are run in parallel, the currents should be evenly divided between the individual parallel conductors so that each conductor is evenly heated. This is accomplished by ensuring that each of the conductors within a parallel set has the same_____ and all conductors terminate in the same manner.
    (a) length   (b) material
    (c) cross-sectional area   (d) all of these

2002 Code Changes

Mike's all new Illustrated Changes to the NEC, 2002 Edition is a detailed review of the most important changes to the 2002 NEC. Printed in full-color, Illustrated Changes to the NEC, 2002 Edition contains over 200 pages with more than 175 detailed graphics. This library is a must have for everyone in the electrical industry. In addition to the Illustrated Code Changes book, you'll receive 9 hours of video in both high-quality DVD format and VHS video tape, the interactive Code Change CD-Rom and Mike's 2002 Code Tabs for your Code book.

Call Today 1.888.NEC.CODE or visit us online at www.NECcode.com for the latest information and pricing.

Unit 9

Single-Family Dwelling Unit Load Calculations

OBJECTIVES

After reading this unit, the student should be able to briefly explain the following concepts:

Air-conditioning versus heat
Appliance (small) circuits
Appliance demand load
Clothes dryer demand load
Cooking equipment calculations

Dwelling unit calculations – Optional method
Dwelling unit calculations – Standard method

Laundry circuit
Lighting and receptacles

After reading this unit, the student should be able to briefly explain the following terms:

General lighting
General–use receptacles
Optional method

Rounding
Standard method
Unbalanced demand load

Voltages

PART A - GENERAL REQUIREMENTS

9–1 GENERAL REQUIREMENTS

Article 220 provides the requirement for residential branch circuit, feeders, and service calculations. Other important articles include Branch Circuits – 210, Feeders – 215, Services – 230, Overcurrent Protection – 240, Wiring Methods – 300, Conductors – 310, Appliances – 422, Electric Space-Heating Equipment – 424, Motors – 430, and Air-Conditioning – 440.

9–2 VOLTAGES [220.2(A)]

Unless other voltages are specified, branch-circuit, feeder, and service loads shall be computed at nominal system voltages of 120, 120/240, 208Y/120, or 240. Figure 9–1.

9–3 FRACTION OF AN AMPERE [220.2(B)]

The rules for rounding an ampere require that when a calculation results in a fraction of an ampere of 0.50 and more, we "round up" to the next ampere. If the calculation is 0.49 of an ampere or less, we "round down" to the next lower ampere.

❏ **Rounding**

What size THHN conductor is required to supply a 230V, 9.6 kW, 1Ø fixed space heater that has a 3A blower motor [424.3(B)]? Terminal rating of 75°C. Figure 9–2.

   (a) 8 AWG   (b) 6 AWG   (c) 4 AWG   (d) 2AWG

   • Answer: (b) 6 AWG

   According to 424.3(B), the conductors and overcurrent protection device to electric space-heating equipment shall be sized no less than 125% of the total load (heat plus motors).

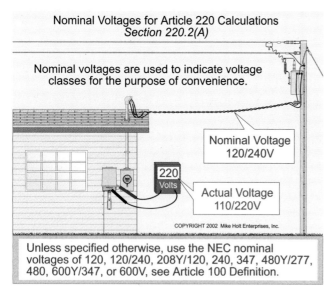

Figure 9-1
Nominal Voltages for Article 220 Calculations

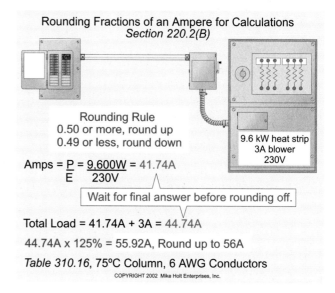

Figure 9-2
Rounding Fractions of an Ampere for Calculations

Space Heater Amperes = Watts/Volts = 9,600W/230V = 41.74A

Total Load (heat + motor) = 41.74A + 3A = 44.74A

Conductor Sized 44.74A × 1.25 = 55.92A, round up to 56A

6 AWG THHN is rated 65A at 75°C terminals [Table 310.16].

9–4 APPLIANCE (SMALL) CIRCUITS [210.11(C)(1)]

A minimum of two 20A *small-appliance branch circuits* are required for receptacle outlets in the kitchen, dining room, breakfast room, pantry or similar dining areas. In general, receptacles or lighting outlets can not be connected to these 20A small-appliance branch circuits [210.52(B)(2) Exceptions]. Figure 9–3.

Feeder and Service

When sizing the feeder or service, each dwelling unit shall have two 20A small-appliance branch circuits with a feeder load of 1,500 VA for each circuit [220.16(A)].

Other Related *Code* Sections

Areas that the small-appliance circuits supply [210.52(B)(1)].
Receptacle outlets required for kitchen countertops [210.52(C)].
15 or 20A receptacles can be used on 20A circuit [210.21(B)(3)].

9–5 COOKING EQUIPMENT – BRANCH CIRCUIT [Table 220.19, Note 4]

To determine the branch-circuit demand load for cooking equipment, we must comply with the requirements of Table 220.19, Note 4. The branch-circuit demand load for one range shall be according to the demand loads listed in Table 220.19, but a minimum 40A circuit is required for 8.75 kW and larger ranges [210.19(A)(3)].

❏ **Less Than 12 kW Column C**

What is the branch-circuit demand load (in amperes) for one 9 kW range? Figure 9–4.

(a) 21A   (b) 27A   (c) 38A   (d) 33A

• Answer: (d) 33A

The demand load for one range in Table 220.19 Column C is 8 kW.

This can be converted to amperes by dividing the power by the voltage, $I = \dfrac{P}{E} = \dfrac{8 \text{ kW} \times 1,000}{240\text{V}^*} = 33A$

*Assume 120/240V, 1Ø for all calculations unless the question gives a specific voltage and system.

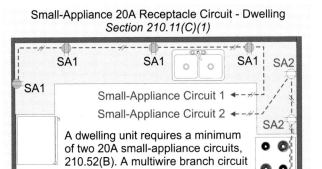

Small-Appliance 20A Receptacle Circuit - Dwelling
Section 210.11(C)(1)

SA1   SA1   SA1 SA2
SA1
    Small-Appliance Circuit 1
    Small-Appliance Circuit 2
                                    SA2
    A dwelling unit requires a minimum
    of two 20A small-appliance circuits,
    210.52(B). A multiwire branch circuit
    can be used for this purpose.
                                    SA2
    COPYRIGHT 2002   Lights are not permitted on
    Mike Holt Enterprises, Inc.   the small appliance circuit

Figure 9-3
Required Branch Circuits – Small Appliance

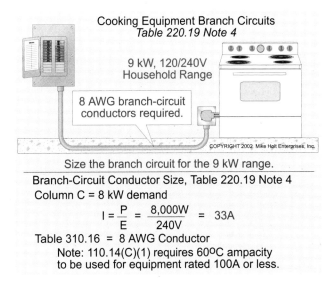

Cooking Equipment Branch Circuits
Table 220.19 Note 4

9 kW, 120/240V
Household Range

8 AWG branch-circuit
conductors required.

COPYRIGHT 2002 Mike Holt Enterprises, Inc.

Size the branch circuit for the 9 kW range.
Branch-Circuit Conductor Size, Table 220.19 Note 4
Column C = 8 kW demand

$$I = \frac{P}{E} = \frac{8,000W}{240V} = 33A$$

Table 310.16 = 8 AWG Conductor
Note: 110.14(C)(1) requires 60ºC ampacity
to be used for equipment rated 100A or less.

Figure 9-4
Cooking Equipment Branch Circuits

❏ **More than 12 kW, Note 1 of Table 220.19**

What is the branch-circuit load for one 14 kW range? Figure 9–5.

(a) 33A   (b) 37A   (c) 50A   (d) 58A

• Answer: (b) 37A

Step 1:   Since the range exceeds 12 kW, follow Note 1 of Table 220.19. The first step is to determine the demand load as listed in Column C of Table 220.19 for one unit = 8 kW.

Step 2:   We must increase the Column C value (8 kW) by 5% for each kW that the average range (in this case 14 kW) exceeds 12 kW. This results is an increase of the Column C value (8 kW) by 10%,

8 kW x 1.1 = 8.8 kW or 8,800W.

Step 3:   Convert the demand load in kW to amperes. $I = \frac{P}{E} = \frac{8,800W}{240V} = 36.7A$

❏ **One Counter-Mounted Cooking Unit**

What is the branch-circuit load for one 6 kW counter-mounted cooking unit? Figure 9–6.

(a) 15A   (b) 20A   (c) 25A   (d) 30A

• Answer: (c) 25A

$$I = \frac{P}{E} = \frac{6,000W}{240V} = 25A$$

ONE COUNTER-MOUNTED COOKING UNIT AND UP TO TWO OVENS [220.19 Note 4]

To calculate the load for one counter-mounted cooking unit and up to two wall-mounted ovens, complete the following steps:

Step 1:   Total Load. Add the nameplate ratings of the cooking appliances and treat this total as one range.

Step 2:   Table 220.19 Demand. Determine the demand load for one unit from Table 220.19, Column C.

Step 3:   Net Computed Load. If the total nameplate rating exceeds 12 kW, increase Column C (8 kW) 5% for each kW, or major fraction (0.5 kW), that the combined rating exceeds 12 kW.

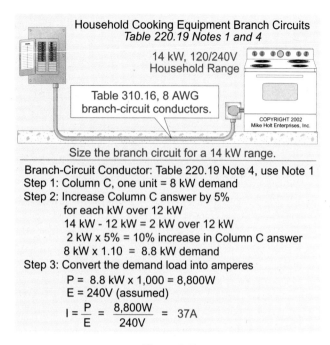

Figure 9-5
Household Cooking Equipment Branch Circuits

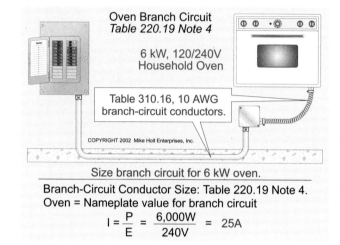

Figure 9-6
Oven Branch Circuit

❏ **Cooktop and One Oven**

What is the branch-circuit load for one 6 kW counter-mounted cooking unit and one 3 kW wall-mounted oven? Figure 9–7.

(a) 25A   (b) 38A   (c) 33A   (d) 42A

• Answer: (c) 33A

Step 1:   Total connected load: 6 kW + 3 kW = 9 kW.

Step 2:   Demand load for one range, Column C of Table 220.19: 8 kW.

Step 3: $I = \dfrac{P}{E} = \dfrac{8,000W}{240V} = 33.3A$

❏ **Cooktop and Two Ovens**

What is the branch-circuit load for one 6 kW counter-mounted cooking unit and two 4 kW wall-mounted ovens? Figure 9–8.

(a) 22A   (b) 27A   (c) 33A   (d) 37A

• Answer: (d) 37A

Step 1:   The total connected load: 6 kW + 4 kW + 4 kW = 14 kW.

Step 2:   Demand load for one range, Column C or Table 220.19: 8 kW.

Step 3:   Since the total (14 kW) exceeds 12 kW, we must increase Column C (8 kW) 5% for each kW that the total exceeds 12 kW.

         14 kW exceeds 12 kW by 2 kW, which results in a 10% increase of the Column C value.

         8 kW × 1.1 = 8.8 kW

Step 4: $I = \dfrac{P}{E} = \dfrac{8,800\ W}{240V} = 36.7A$

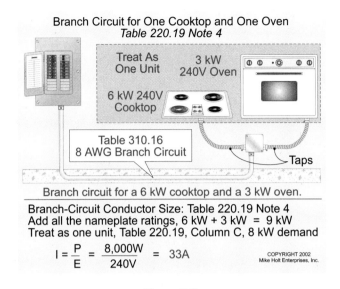

Branch Circuit for One Cooktop and One Oven
Table 220.19 Note 4

Treat As One Unit

3 kW 240V Oven

6 kW 240V Cooktop

Table 310.16
8 AWG Branch Circuit

Taps

Branch circuit for a 6 kW cooktop and a 3 kW oven.

Branch-Circuit Conductor Size: Table 220.19 Note 4
Add all the nameplate ratings, 6 kW + 3 kW = 9 kW
Treat as one unit, Table 220.19, Column C, 8 kW demand

$$I = \frac{P}{E} = \frac{8,000W}{240V} = 33A$$

COPYRIGHT 2002
Mike Holt Enterprises, Inc.

Figure 9-7
Branch Circuit for One Cooktop and One Oven

Branch Circuit for Two Ovens and One Cooktop
Table 220.19 Note 4

Add the NAMEPLATE ratings of all units then size as ONE UNIT using Table 220.19.

6 kW Cooktop

4 kW Oven

4 kW Oven

Taps, see 210.19(A)(3)

Table 310.16
8 AWG Branch Circuit

COPYRIGHT 2002
Mike Holt Enterprises, Inc.

Branch circuit for a 6 kW cooktop and 2- 4 kW ovens.

Branch Circuit Conductor: Table 220.19 Note 4, use Note 1
Step 1: Total nameplate of all 3 units, 6 kW + 4 kW + 4 kW = 14 kW
Step 2: Treat as one 14 kW range. Column C for one unit is 8 kW.
Increase Column C answer (8 kW) by 5% for every kW over 12 kW
14 kW - 12 kW = 2 kW = 10%, 8 kW x 1.10 = 8.8 kW demand
I = P/E = 8,800W/240V = 37A

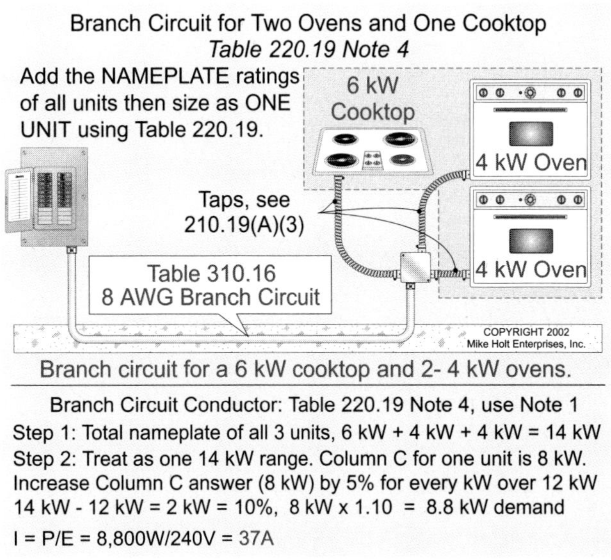

Figure 9-8
Branch Circuit for Two Ovens and One Cooktop

9–6 LAUNDRY RECEPTACLE(S) CIRCUIT [210.11(C)(2)]

One 20A branch circuit for the laundry receptacle outlet (or outlets) is required and this laundry circuit cannot serve any other outlet such as the laundry room lights [210.52(F)]. The *NEC* does not require a separate circuit for the washing machine, but does require a separate circuit for the laundry room receptacle or receptacles. Figure 9–9. If the washing machine receptacle(s) are located in the garage or basement, GFCI protection might be required, see 210.8(A)(2) and (5) and their exceptions.

Feeder and Service

Each dwelling unit shall have a feeder load consisting of 1,500 VA for the 20A laundry circuit [220.16(B)].

Other Related *Code* Rules

A laundry outlet must be within 6 ft of a washing machine [210.50(C)].
Laundry area receptacle outlet required [210.52(F)].

9–7 LIGHTING AND RECEPTACLES

General Lighting Volt-Ampere Load [220.3(A)]

The *NEC* requires a minimum 3 VA per sq ft for the general lighting and general-use receptacles for the purpose of determining branch circuits and feeder/service calculations. The dimensions for determining the area shall be computed from the outside dimensions of the building and shall not include open porches, garages or spaces not adaptable for future use. Figure 9–10.

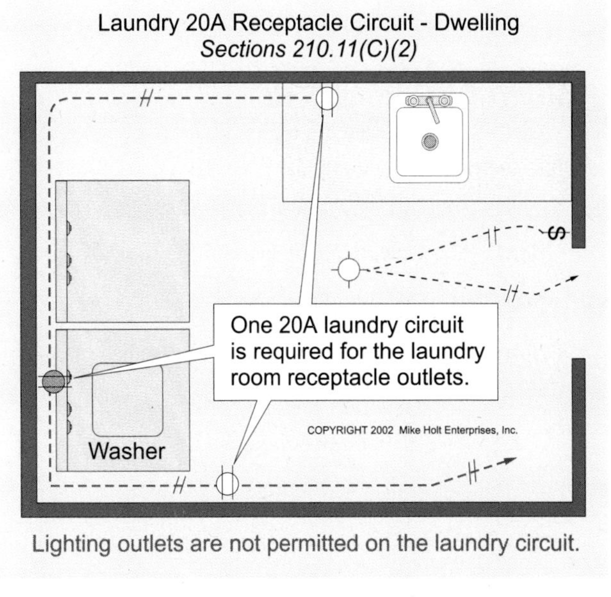

Laundry 20A Receptacle Circuit - Dwelling
Sections 210.11(C)(2)

One 20A laundry circuit is required for the laundry room receptacle outlets.

COPYRIGHT 2002 Mike Holt Enterprises, Inc.

Washer

Lighting outlets are not permitted on the laundry circuit.

Figure 9-9
Required Branch Circuits – Laundry Circuit

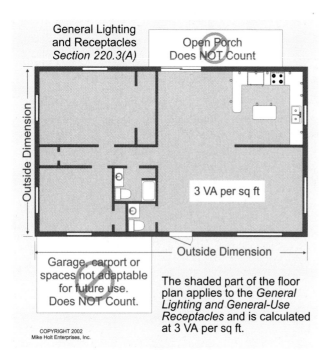

Figure 9-10
General Lighting and Receptacles

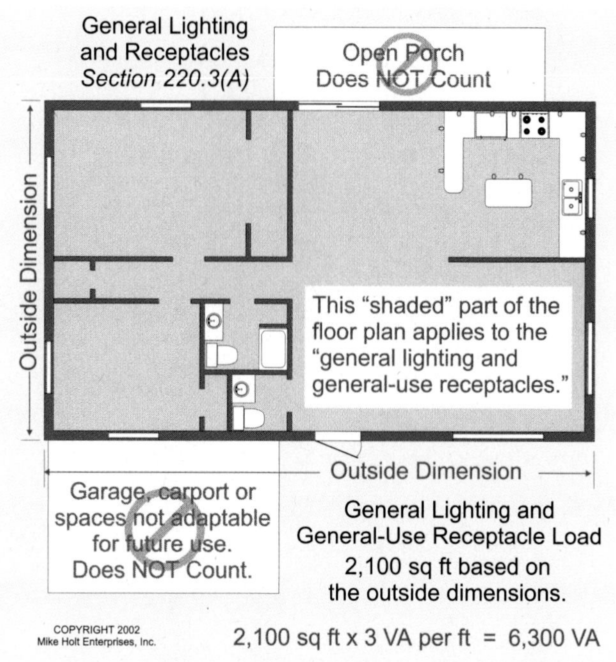

Figure 9-11
General Lighting and Receptacles

Note: The 3 VA each sq ft rule for general lighting includes all 15 and 20A general-use receptacles, but it does not include the appliance or laundry circuit receptacles. See 220.3(10) for details.

❑ **General Lighting Load [Table 220.3(A)]**

What is the general lighting and receptacle load for a 2,100 sq ft dwelling unit that has 34 convenience receptacles and 12 lighting luminaires rated 100W each? Figure 9–11.

(a) 2,100 VA   (b) 4,200 VA   (c) 6,300 VA   (d) 8,400 VA

• Answer: (c) 6,300 VA

2,100 sq ft $\times$ 3 VA = 6,300 VA

Note: No additional load is required for general-use receptacles and lighting outlets, see 220.3(A)(10).

NUMBER OF CIRCUITS REQUIRED [Chapter 9, Example No. D1(a)]

The number of branch circuits required for general lighting and receptacles shall be determined from the general lighting load and the rating of the circuits [210.11(A)].

To determine the number of branch circuits for general lighting and receptacles, follow these steps:

Step 1:   Determine the general lighting VA load: Living area square footage $\times$ 3 VA.

Step 2:   Determine the general lighting ampere load: Amperes = $^{VA}/_{E}$.

Step 3:   Determine the number of branch circuits:

$$\frac{\text{General Light Amperes (Step 2)}}{\text{Circuit Amperes (100\%)}}$$

❑ **Number of 15 Ampere Circuits**

How many 15A circuits are required for a 2,100 sq ft dwelling unit. Figure 9–12.

(a) 2 circuits   (b) 3 circuits   (c) 4 circuits   (d) 5 circuits

• Answer: (c) 4 circuits

Step 1:   General lighting VA:
            2,100 sq ft $\times$ 3 VA = 6,300 VA

Step 2:   General lighting amperes: $I = \dfrac{VA}{E}$

            $I = \dfrac{6,300\ VA}{120V^*} = 53A$

*Use 120V, 1Ø unless specified otherwise.

Step 3:   Determine the number of circuits:

$\dfrac{\text{General Light Amperes}}{\text{Circuit Amperes}} = \dfrac{53A}{15A} = 3.53$ or 4 circuits

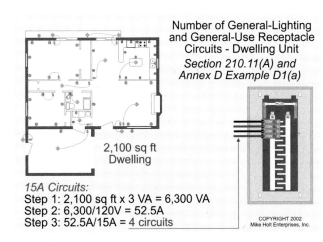

Number of General-Lighting and General-Use Receptacle Circuits - Dwelling Unit
Section 210.11(A) and Annex D Example D1(a)

2,100 sq ft
Dwelling

15A Circuits:
Step 1: 2,100 sq ft x 3 VA = 6,300 VA
Step 2: 6,300/120V = 52.5A
Step 3: 52.5A/15A = 4 circuits

COPYRIGHT 2002
Mike Holt Enterprises, Inc.

❏ **Number of 20A Circuits**

How many 20A circuits are required for a 2,100 sq ft dwelling unit?

(a) 2 circuits   (b) 3 circuits
(c) 4 circuits   (d) 5 circuits
  • Answer: (b) 3

Figure 9-12

Number of Lighting and General-Use Receptacle Circuits

Step 1:   General lighting VA: 2,100 sq ft $\times$ 3 VA = 6,300 VA

Step 2:   General lighting amperes: $I = \dfrac{VA}{E} = \dfrac{6,300\ VA}{120V} = 53A$

Step 3:   Determine the number of circuits: $\dfrac{\text{General Lighting Amperes}}{\text{Circuit Amperes}} = \dfrac{53A}{20A} = 2.65$ or 3 circuits

PART B - STANDARD METHOD – FEEDER/SERVICE LOAD CALCULATIONS

9–8 DWELLING UNIT FEEDER/SERVICE LOAD CALCULATIONS (Part II of Article 220)

The following steps can be used to determine the feeder or service size for a dwelling unit using the standard method contained in Part II of Article 220:

Step 1:   **General Lighting and Receptacles, Small-Appliance and Laundry Circuits [Table 220.11].**

The *NEC* recognizes that the general lighting and receptacle, small-appliance and laundry circuits will not all be on, or loaded, at the same time and permits a demand factor to be applied to the total connected load [220.16]. To determine the feeder demand load for these loads, follow these steps:

Total Connected Load—Determine the total connected load for (1) general lighting and receptacles (3 VA per sq ft), (2) two small-appliance circuits each at 1,500 VA and (3) one laundry circuit at 1,500 VA.

Demand Factor—Apply Table 220.11 demand factors to the total connected load (Step 1).

First 3,000 VA at 100% demand

Remainder VA at 35% demand

Step 2:   **Air-Conditioning versus Heat.**
Because the air-conditioning and heating loads are not on at the same time, it is permissible to omit the smaller of the two loads [220.21]. The air-conditioning demand load is calculated at 100% and the fixed electric heating demand load is calculated at 100% [220.15].

Step 3:   **Appliances [220.17].**

A 75% demand factor is permitted to be applied when four or more appliances fastened in place, such as dishwasher, waste disposal, trash compactor, water heater, etc. are on the same feeder. This does not apply to motors [220.14], space-heating equipment [220.15], clothes dryers [220.18], cooking appliances [220.19], or air-conditioning equipment.

Step 4:   **Clothes Dryer [220.18].**

The feeder or service demand load for electric clothes dryers located in a dwelling unit shall not be less than 5,000W or the nameplate rating if greater than 5,000W. A feeder or service dryer load is not required if the dwelling unit does not contain an electric dryer!

Step 5:   **Cooking Equipment [220.19].**

Household cooking appliances rated over $1^3/_4$ kW can have the feeder and service calculated according to the demand factors of 220.19, Table and Notes 1, 2, and 3.

Step 6:   **Feeder and Service Conductor Size.**

400 Amperes and Less: For 3-wire, 120/240V, 1Ø systems, the feeder or service conductors can be sized to Table 310.15(B)(6). All Others: Table 310.16. The grounded (neutral) conductor is sized to the maximum unbalanced load [220.22] using Table 310.16.

Over 400 Amperes: The ungrounded and grounded (neutral) conductors are sized according to Table 310.16.

9–9 DWELLING UNIT FEEDER/SERVICE CALCULATION EXAMPLES

Step 1: General Lighting, Small Appliance and Laundry Demand [Table 220.11]

The demand factors in Table 220.11 apply to the 3 VA per sq ft for general lighting [Table 220.3(A)], small-appliance circuit [220.16(A)] and laundry circuit [220.16(B)].

General Lighting No. 1

What is the general lighting, small appliance and laundry demand load for a 2,700 sq ft dwelling unit? Figure 9–13.

(a) 8,100 VA   (b) 12,600 VA   (c) 2,700 VA   (d) 6,360 VA

• Answer: (d) 6,360 VA

General Lighting/Receptacles	(2,700 sq ft × 3 VA)	8,100 VA
Small-Appliance Circuits	(1,500 VA × 2)	3,000 VA
Laundry Circuit	(1,500 VA × 1)	1,500 VA
Total Connected Load		12,600 VA
First 3,000 VA at 100%		−3,000 VA × 1.00 = 3,000 VA
Remainder at 35%		9,600 VA × 0.35 = + 3,360 VA
		6,360 VA

General Lighting No. 2

What is the general lighting, small appliance and laundry demand load for a 6,540 sq ft dwelling unit?

(a) 8,100 VA   (b) 12,600 VA   (c) 2,700 VA   (d) 10,392 VA

• Answer: (d) 10,392 VA

General Lighting/Receptacles	(6,540 sq ft × 3 VA)	19,620 VA
Small-Appliance Circuits	(1,500 VA × 2)	3,000 VA
Laundry Circuit	(1,500 VA × 1)	1,500 VA
Total connected load		24,120 VA
First 3,000 VA at 100%		−3,000 VA × 1.00 = 3,000 VA
Remainder at 35%		21,120 VA × 0.35 = + 7,392 VA
		10,392 VA

General Lighting, Small Appliance and Laundry Demand
Table 220.11

Figure 9-13
General Lighting, Small-Appliance and Laundry Demand

Heat Versus Air Conditioning - Service Loads
Section 220.15

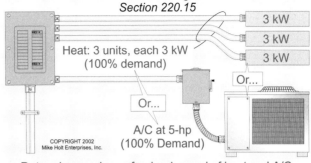

Determine service or feeder demand of heat and A/C.
Heat and A/C are not used at the same time,
see 220.21. Omit smaller.
A/C at 100%   = 230 V x 28A = 6,440 VA (omit A/C)
Heat at 100% = 3,000 VA x 3 units = 9,000 W

Figure 9-14
Heat versus Air-Conditioning – Service Loads

Step 2: Air-Conditioning Versus Heat [220.15]

Compare the A/C load at 100% against the heating load at 100% and omit the smaller of the two loads [220.21].

Air-Conditioning versus Heat No. 1

What is the service demand load for 5-hp, 230V A/C versus three 3 kW baseboard heaters? Figure 9–14.

(a) 6,400 VA   (b) 3,000W   (c) 8,050 VA   (d) 9,000W

• Answer: (d) 9,000W

A/C 230V × 28A*

A/C = 6,440 VA, omit [220.21]

Heat [220.15] 3,000W × 3 = 9,000W

*Table 430.148

Air-Conditioning versus Heat No. 2

What is the service or feeder demand load for A/C (5-hp, 230 V) and an electric space heater (5 kW)?

(a) 5,000 VA   (b) 3,910 VA   (c) 6,440 VA   (d) 10,350 VA

• Answer: (c) 6,440 VA

A/C 5-hp = 230V × 28A = 6,440 VA

Omit heat because it's less than the A/C load [220.21]

Step 3: Appliance Demand Load [220.17]

Appliance No. 1

What is the demand load for a waste disposal (940 VA), a dishwasher (1,250 VA) and a water heater (4,500 VA)?

(a) 5,018 VA   (b) 6,690 VA   (c) 8,363 VA   (d) 6,272 VA

• Answer: (b) 6,690 VA

Waste Disposal	940 VA
Dishwasher	1,250 VA
Water Heater	4,500 VA
	6,690 VA

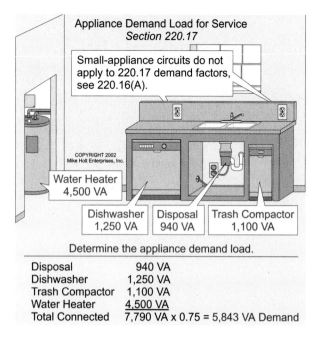

Figure 9-15

Appliance Demand Load for Service

Figure 9-16

Dryer Demand Load for Service

Appliance No. 2

What is the demand load for a waste disposal (940 VA), a dishwasher (1,250 VA), a trash compactor (1,100 VA) and a water heater (4,500 VA)? Figure 9–15.

    (a) 7,790 VA   (b) 5,843 VA   (c) 7,303 VA   (d) 9,738 VA

    • Answer: (b) 5,843 VA

Waste Disposal	940 VA
Dishwasher	1,250 VA
Trash Compactor	1,100 VA
Water Heater	4,500 VA
	7,790 VA × 0.75 = 5,843 VA

Step 4: Dryer Demand Load [220.18]

Dryer No. 1

What is the service and feeder demand load for a 4 kW dryer? Figure 9–16.

    (a) 4,000W   (b) 3,000W   (c) 5,000W   (d) 5,500W

    • Answer: (c) 5,000W

The dryer load must not be less than 5,000 VA.

Dryer No. 2

What is the service and feeder demand load for a 5.5 kW dryer?

    (a) 4,000W   (b) 3,000W   (c) 5,000W   (d) 5,500W

    • Answer: (d) 5,500W

The dryer load must not be less than the nameplate rating if greater than 5,000 VA.

Step 5: Cooking Equipment Demand Load [220.19]

Note 3 – Not over 3¹/₂ kW – Column A

What is the service and feeder demand load for two 3 kW cooking appliances in a dwelling unit?

    (a) 3 kW   (b) 4.8 kW   (c) 4.5 kW   (d) 3.9 kW

    • Answer: (c) 4.5 kW

3 kW × 2 units = 6 kW × 0.75 = 4.5 kW

Note 3 – Not over 8³/₄ – Column B

What is the service and feeder demand load for one 6 kW cooking appliance in a dwelling unit?

(a) 6 kW   (b) 4.8 kW

(c) 4.5 kW   (d) 3.9 kW

• Answer: (b) 4.8 kW

$6 \text{ kW} \times 0.8 = 4.8 \text{ kW}$

Note 3 – Column A and B

What is the service and feeder demand load for two 3 kW ovens and one 6 kW cooktop in a dwelling unit? Figure 9–17.

(a) 6 kW   (b) 4.8 kW

(c) 4.5 kW   (d) 9.3 kW

• Answer: (d) 9.3 kW

Column A demand: $(3 \text{ kW} \times 2) \times 0.75 = 4.5 \text{ kW}$

Column B demand: $6 \text{ kW} \times 0.8 \qquad = \underline{4.8 \text{ kW}}$

Total demand   9.3 kW

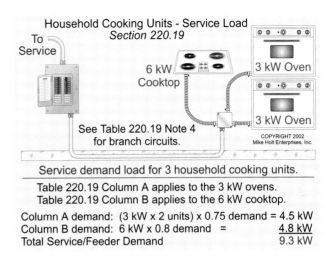

Household Cooking Units - Service Load
Section 220.19

See Table 220.19 Note 4 for branch circuits.

COPYRIGHT 2002
Mike Holt Enterprises, Inc.

Service demand load for 3 household cooking units.

Table 220.19 Column A applies to the 3 kW ovens.
Table 220.19 Column B applies to the 6 kW cooktop.

Column A demand: (3 kW x 2 units) x 0.75 demand = 4.5 kW
Column B demand: 6 kW x 0.8 demand   =   <u>4.8 kW</u>
Total Service/Feeder Demand   9.3 kW

Figure 9-17

Household Cooking Units – Service Load

Column C – Not over 12 kW

What is the service and feeder demand load for an 11.5 kW range in a dwelling unit?

(a) 11.5 kW   (b) 8 kW   (c) 9.2 kW   (d) 6 kW

• Answer: (b) 8 kW

Table 220.19, Column C: 8 kW

Over 12 kW – Note 1

What is the service and feeder demand load for a 13.6 kW range in a dwelling unit?

(a) 8.8 kW   (b) 8 kW   (c) 9.2 kW   (d) 6 kW

• Answer: (a) 8.8 kW

Step 1:   13.6 kW exceeds 12 kW by one kW and one major fraction of a kW. Column C value (8 kW) must be increased 5% for each kW or major fraction of a kW (0.5 kW or larger) over 12 kW.

Step 2: $8 \text{ kW} \times 1.1 = 8.8 \text{ kW}$

Step 6: Service Conductor Size [Table 310.15(B)(6)]

Service Conductor 1 AWG

What size THHN feeder or service conductor (120/240V, 1Ø) is required for a 225A service demand load?

(a) 1/0 AWG   (b) 2/0 AWG   (c) 3/0 AWG   (d) 4/0 AWG

• Answer: (c) 3/0 AWG

Service Conductor 2 AWG

What size THHN feeder or service conductor (208Y/120V, 1Ø) is required for a 225A service demand load?

(a) 1/0 AWG   (b) 2/0 AWG   (c) 3/0 AWG   (d) 4/0 AWG

• Answer: (d) 4/0 AWG

Table 310.15(B)(6) does not apply to 208Y/120V systems because the grounded (neutral) conductor carries current even if the loads are balanced.

PART C - Optional METHOD – FEEDER/SERVICE LOAD CALCULATIONS

9-10 DWELLING UNIT OPTIONAL FEEDER/SERVICE CALCULATIONS [220.30]

Instead of sizing the dwelling unit feeder and/or service conductors according to the standard method (Part II of Article 220), an optional method [220.30] (Part IV) can be used. The following steps can be used to determine the demand load:

Step 1:   **220.30(B) General Loads.**

The calculated load shall not be less than 100% for the first 10 kW, plus 40% of the remainder of the following loads:

(1) Small-Appliance and Laundry Branch Circuits: 1,500 VA for each 20A small-appliance and laundry branch circuit.

(2) General Lighting and Receptacles: 3 VA per sq ft

(3) Appliances: The nameplate VA rating of all appliances and motors that are fastened in place (permanently connected) or located on a specific circuit.

Note: Be sure to use the range and dryer at nameplate rating!

Step 2:   **220.30(C) Heating and Air-Conditioning Load.**

Include the largest of the following:

(1) 100 percent of the nameplate rating of the air-conditioning equipment.

(2) 100 percent of the heat-pump compressors and supplemental heating unless the controller prevents both the compressor and supplemental heating from operating at the same time.

(3) 100 percent of the nameplate ratings of electric thermal storage and other heating systems where the usual load is expected to be continuous at the full nameplate value. Systems qualifying under this selection shall not be figured under any other selection in this table.

(4) 65 percent of the nameplate rating(s) of the central electric space heating, including integral supplemental heating in heat pumps where the controller prevents both the compressor and supplemental heating from operating at the same time.

(5) 65 percent of the nameplate rating(s) of electric space heating, if there are less than four separately controlled units.

(6) 40 percent of the nameplate rating(s) of electric space heating of four or more separately controlled units.

Step 3:   **Size Service/Feeder Conductors [310.15(B)(6)].**

400 Amperes and Less The ungrounded conductors are permitted to be sized to Table 310.15(B)(6) for 120/240V, 1Ø systems up to 400A. The grounded (neutral) conductor must be sized to carry the unbalanced load per Table 310.16.

Over 400 Amperes The ungrounded and grounded (neutral) conductors are sized according to Table 310.16.

9–11 DWELLING UNIT OPTIONAL CALCULATION EXAMPLES

❏ **Optional Load Calculation No. 1**

What size service conductor is required for a 1,500 sq ft dwelling unit containing the following loads?

Dishwasher (1,200 VA)
Disposal (900 VA)
Cooktop (6,000 VA)
A/C (5-hp)

Water Heater (4,500 VA)
Dryer (4,000 VA)
Two Ovens (each 3,000 VA)
Electric Heating (10 kW)

   (a) 6 AWG   (b) 4 AWG   (c) 3 AWG   (d) 2 AWG

   • Answer: (c) 3 AWG

Step 1: General Loads [220.30(B)].

Small Appliance Circuits (1,500 VA × 2 circuits)	3,000 VA
General Lighting (1,500 sq ft × 3 VA)	4,500 VA
Laundry Circuit	1,500 VA
Appliances (nameplate)	
Dishwasher	1,200 VA
Water Heater	4,500 VA
Disposal	900 VA
Dryer	4,000 VA
Cooktop	6,000 VA
Ovens 3,000 VA × 2 units	6,000 VA
Total Connected Load	31,600 VA
First 10,000 at 100%	- 10,000 VA × 1.00 = 10,000 VA
Remainder at 40%	21,600 VA × 0.40 = + 8,640 VA
Demand load	18,640 VA

Step 2: Air-Conditioning Versus Heat [220.30(C)].

Air-conditioning at 100% versus 65% electric space-heating

Air Conditioner: 230V × 28A = 6,440 VA (omit)

Electric heat: 10,000 VA × 0.65 = 6,500 VA

Step 3: Service/Feeder Conductors [310.15(C)(6)].

General Loads	18,640 VA
Heat	+ 6,500 VA
Total Demand Load	25,140 VA

   I = VA/V, I = 25,140 VA/240V = 105A

Note: The feeder/service conductor is sized to 110A, 3 AWG.

❏ Optional Load Calculation No. 2

What size service conductor is required for a dwelling unit with a 300 sq ft open porch and a 150 sq ft carport that contain the following:

Dishwasher (1.5 kVA)	Waste disposal (1 kVA)	Trash compactor (1.5 kVA)
Water Heater (6 kW)	Range (14 kW)	Dryer (4.5 kVA)
A/C (5-hp)	Electric Heat (four, each 2.5 kW)	

(a) 5 AWG　　　　　(b) 4 AWG　　　　　(c) 3 AWG　　　　　(d) 2 AWG

• Answer: (d) 2 AWG

Step 1: General Loads [220.30(B)].

Small Appliance Circuits (1500 VA × 2 circuits)	3,000 VA
General Lighting (2330 sq ft × 3 VA)	6,990 VA
Laundry Circuit	1,500 VA

Appliances (nameplate)

Dishwasher	1,500 VA
Disposal	1,000 VA
Trash Compactor	1,500 VA
Water Heater	6,000 VA
Range	14,000 VA
Dryer (nameplate)	4,500 VA
Total Connected Load	39,900 VA
First 10,000 at 100%	10,000 VA × 1.00 = 10,000 VA
Remainder at 40%	29,000 VA × 0.40 = 11,996 VA
Demand Load	21,996 VA

Step 2: Air-Conditioning Versus Heat [220.30(C)].

Air-conditioning at 100% versus 40% electric heating

Air conditioner: 230V × 28A [Table 430.148] = 6,440 VA

Electric Heat: 10,000 VA × 0.40 = 4,000 VA, omit [220.21]

Step 3: Service/Feeder Conductors [310.15(B)(6)].

General Loads	21,996 VA
A/C	6,440 VA
Total Demand Load	28,436 VA

I = VA/Volts,

I = 28,436 VA/240V

I = 118A, 2 AWG

9–12 NEUTRAL CALCULATIONS – GENERAL [220.22]

The feeder/service neutral load is the maximum unbalanced demand load between the grounded (neutral) conductor and any ungrounded conductor as determined by Article 220 Part II.

Because phase-to-phase loads are not connected to the grounded (neutral) conductor, they are not considered when sizing the grounded conductor.

❏ Neutral not Over 200A

What size 120/240V, 1Ø ungrounded and grounded (neutral) conductor is required for a 375A two-family dwelling unit demand load, of which 175A consist of 240V loads? Figure 9–18.

(a) Two – 400 kcmil and 1 – 350
(b) Two – 350 kcmil and 1 – 350
(c) Two – 500 kcmil and 1 – 500
(d) Two – 400 kcmil and 1 – 3/0 AWG
  • Answer: (d) Two – 400 kcmil and 1 – 3/0 AWG

The ungrounded conductor is sized at 375A (400 kcmil THHN) according to Table 310.15(B)(6). The grounded (neutral) conductor must be sized to carry the maximum unbalanced load of 200A (120V loads only).

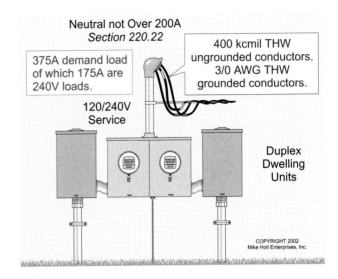

Figure 9-18
Neutral not Over 200A.

Cooking Appliance Feeder/Service Neutral Load [220.22]

The feeder or service cooking appliance neutral load for household cooking appliances such as electric ranges, wall-mounted ovens, or counter-mounted cooking units shall be calculated at 70% of the demand load as determined by 220.19.

❏ Range Neutral

What is the neutral load (in amperes) for one 14 kW range?
(a) 3.85 kW   (b) 5.5 kW   (c) 6.2 kW   (d) 8.8 kW
  • Answer: (c) 6.2 kW

Step 1: Since the range exceeds 12 kW, we must comply with Note 1 of Table 220.19. The first step is to determine the demand load as listed in Column C of Table 220.19 for one unit = 8 kW.

Step 2: We must increase the Column C value (8 kW) by 5% for each kW that the average range (in this case 14 kW) exceeds 12 kW. This results in an increase of Column C value (8 kW) by 10%.

8 kW × 1.1 = 8.8 kW × 1,000 = 8,800W

Neutral load at 70%: 8,800W × 0.70 = 6.2 kW.

Dryer Feeder/Service Neutral Load [220.22]

The feeder and service dryer neutral demand load for electric clothes dryers shall be calculated at 70% of the demand load determined by 220.18.

❏ Dryer Over 5 kW

What is the feeder/service neutral load for one 5.5 kW dryer?
(a) 3.85 kW   (b) 5.5 kW   (c) 4.6 kW   (d) none of these
  • Answer: (a) 3.85 kW
  5.5 kW × 0.7 = 3.85 kW

Unit 9 – One-Family Dwelling Unit Load Calculations Summary Questions

Calculations Questions

Part A – General Requirements

9–2 Voltages

1. Unless other voltages are specified, feeder and service loads shall be computed at a nominal system voltage(s) of _____.
   (a) 120   (b) 120/240   (c) 208Y/120   (d) any of these

9–3 Fraction of an Ampere

2. When a calculation results in _____ of an ampere or larger, such fractions may be rounded up to the next ampere.
   (a) 0.15   (b) 0.50   (c) 0.49   (d) 0.051

9–4 Appliance (Small) Circuits

3. A minimum of _____ 20A small-appliance branch circuits are required for receptacle outlets in the kitchen, dining room, breakfast room, pantry and similar areas.
   (a) one   (b) two   (c) three   (d) none of these

4. When sizing the feeder or service, each dwelling unit shall have a minimum feeder load of _____ for the two small-appliance branch circuits.
   (a) 1,500 VA   (b) 2,000 VA   (c) 3,000 VA   (d) 4,500 VA

9–5 Cooking Equipment – Branch Circuit [Table 220.19, Note 4]

5. What is the branch-circuit demand load (in amperes) for one 11 kW range?
   (a) 21A   (b) 27A   (c) 38A   (d) 33A

6. What is the branch-circuit load (in amperes) for one 13.5 kW range?
   (a) 33A   (b) 37A   (c) 50A   (d) 58A

Wall-Mounted Oven and/or Counter-Mounted Cooking Unit

7. What is the branch-circuit load (in amperes) for one 6 kW wall-mounted oven, 240V, 1Ø?
   (a) 15A   (b) 25A   (c) 30A   (d) 40A

8. • What is the branch-circuit load (in amperes) for one 4.8 kW counter-mounted cooking unit, 240V, 1Ø?
   (a) 15A   (b) 20A   (c) 25A   (d) 30A

9. What is the branch-circuit load (in amperes) for one 6 kW counter-mounted cooking unit and one 3 kW wall-mounted oven, 240V, 1Ø?
   (a) 25A   (b) 38A   (c) 33A   (d) 42A

10. What is the branch-circuit load (in amperes) for one 6 kW counter-mounted cooking unit and two 4 kW wall-mounted ovens, 240V, 1Ø?
    (a) 22A   (b) 27A   (c) 33A   (d) 37A

9–6 Laundry Receptacle(s) Circuit

11. The *NEC* does not require a separate circuit for the washing machine, but requires a separate circuit for the laundry room receptacle or receptacles, of which one can be for the washing machine.
    (a) True   (b) False

12. Each dwelling unit shall have a feeder load consisting of _____ for the 20A laundry circuit.
    (a) 1,500 VA   (b) 2,000 VA   (c) 3,000 VA   (d) 4,500 VA

9–7 Lighting and Receptacles

13. The *NEC* requires a minimum of 3 VA per each sq ft for the required general lighting and receptacles. The dimensions for determining the area shall be computed from the outside of the building and shall not include _____.
    (a) open porches   (b) garages
    (c) spaces not adaptable for future use   (d) all of these

14. The 3 VA each sq ft for general lighting includes all 20A rating or less general-use receptacles, but not the appliance or laundry circuit receptacles.
    (a) True   (b) False

15. What is the general lighting load and receptacle load for a 2,100 sq ft home that has 14 extra convenience receptacles and nine extra recessed luminaires rated at 100W each?
    (a) 2,100 VA   (b) 4,200 VA   (c) 6,300 VA   (d) 8,400 VA

Number of General Lighting Circuits

16. The number of branch circuits required for general lighting and receptacles shall be determined from the general lighting load and the rating of the circuits.
    (a) True   (b) False

17. How many 15A general lighting circuits are required for a 2,340 sq ft home?
    (a) 2   (b) 3   (c) 4   (d) 5

18. How many 20A general lighting circuits are required for a 2,100 sq ft home?
    (a) 2   (b) 3   (c) 4   (d) 5

19. How many 20A circuits are required for the general lighting and receptacles for a 1,500 sq ft dwelling unit?
    (a) 2   (b) 3   (c) 4   (d) 5

Part II – Standard Method – Load Calculation Questions

9–8 Dwelling Unit Feeder/Service Load Calculations (Part II of Article 240)

20. The *NEC* recognizes that the general lighting and receptacles, small-appliance and laundry circuits will not all be on or loaded at the same time and permits a _____ to be applied to the total of these loads.
    (a) demand factor   (b) adjustment factor
    (c) correction factor   (d) correction

21. Because the air-conditioning and heating loads are not on at the same time (simultaneously), it is permissible to omit the smaller of the two loads when determining the air-conditioning versus heat demand load.
    (a) True   (b) False

22. A load demand factor of 75 percent is permitted for _____ or more appliances fastened in place such as dishwashers, waste disposals, trash compactors, water heaters, etc.
    (a) one   (b) two   (c) three   (d) four

23. The feeder or service demand load for electric clothes dryers located in dwelling units shall not be less than _____.
    (a) 5,000W   (b) nameplate rating   (c) a or b   (d) the greater of a or b

24. • A feeder or service dryer load is required even if the dwelling unit does not contain an electric dryer.
    (a) True   (b) False

25. Household cooking appliances rated $1^3/_4$ kW can have the feeder and service loads calculated according to the demand factors of Table 220.19.
    (a) True   (b) False

26. Dwelling unit feeder and service ungrounded (hot) conductors are sized according to Table 310.15(B)(6) for _____.
    (a) 3-wire, 1Ø, 120/240V systems up to 400A
    (b) 3-wire, 1Ø, 120/208V systems up to 400A
    (c) 4-wire, 1Ø, 120/240V systems up to 400A
    (d) none of these

27. What is the general lighting load for a 2,700 sq ft dwelling unit?
    (a) 8,100 VA   (b) 12,600 VA   (c) 2,700 VA   (d) 6,840 VA

28. What is the total connected load for general lighting and receptacles, small-appliance and laundry circuits for a 6,540 sq ft dwelling unit?
    (a) 8,100 VA   (b) 12,600 VA   (c) 2,700 VA   (d) 24,120 VA

29. What is the feeder or service demand load for one air conditioner (5-hp, 230 V) and three baseboard heaters (3 kW)?
    (a) 6,400 VA   (b) 3,000 VA   (c) 8,050 VA   (d) 9,000 VA

30. What is the feeder or service demand load for an air conditioner (5-hp, 230 V) and electric space heater (10 kW)?
    (a) 10,000 VA   (b) 3,910 VA   (c) 6,440 VA   (d) 10,350 VA

31. What is the feeder or service demand load for a waste disposal (940 VA), dishwasher (1,250 VA) and a water heater (4,500 VA)?
    (a) 5,018 VA   (b) 6,690 VA   (c) 8,363 VA   (d) 6,272 VA

32. • What is the feeder or service demand load for a waste disposal (940 VA), dishwasher (1,250 VA), water heater (4,500 VA) and a trash compactor (1,100 VA)?
    (a) 7,790 VA   (b) 5,843 VA   (c) 7,303 VA   (d) 9,738 VA

33. What is the feeder or service demand load for a 4 kW dryer?
    (a) 4 kW   (b) 3 kW   (c) 5 kW   (d) 6 kW

34. What is the feeder or service demand load for a 5.5 kW dryer?
    (a) 4 kW   (b) 3 kW   (c) 5 kW   (d) 6 kW

35. What is the feeder or service demand load for two 3 kW ranges?
    (a) 3 kW   (b) 4.8 kW   (c) 4.5 kW   (d) 3.9 kW

36. What is the feeder or service demand load for one 6 kW cooking appliance?
    (a) 6 kW   (b) 4.8 kW   (c) 4.5 kW   (d) 3.9 kW

37. What is the feeder or service demand load for one 6 kW and two 3 kW cooking appliances?
    (a) 6 kW   (b) 4.8 kW   (c) 4.5 kW   (d) 9.3 kW

38. What is the feeder or service demand load for an 11.5 kW range?
    (a) 11.5 kW   (b) 8 kW   (c) 9.2 kW   (d) 6 kW

39. What is the feeder or service demand load for a 13.6 kW range?
    (a) 8.8 kW   (b) 8 kW   (c) 9.2 kW   (d) 6 kW

40. What size 120/240V, 1Ø service or feeder THHN copper conductors are required for a dwelling unit that has a 190A service demand load?
    (a) 1/0 AWG   (b) 2/0 AWG   (c) 3/0 AWG   (d) 4/0 AWG

41. • What is the net computed demand load for the general lighting, small appliance and laundry circuits for a 1,500 sq ft dwelling unit?
    (a) 4,500 VA   (b) 9,000 VA   (c) 5,100 VA   (d) none of these

42. What is the feeder or service demand load for an air conditioner (3-hp, 230 V) with a blower ($^1/_4$-hp) and electric space heating (4 kW)?

    (a) 3,910 VA   (b) 4,577 VA   (c) 4,000 VA   (d) none of these

43. What is the feeder or service demand load for a dwelling unit that has one water heater (4 kW) and one dishwasher (1.5 kW)?

    (a) 4 kW   (b) 5.5 kW   (c) 4.13 kW   (d) none of these

44. What is the net computed load for a 4.5 kW dryer?

    (a) 4.5 kW   (b) 5 kW   (c) 5.5 kW   (d) none of these

45. What is the total net computed demand load for a 14 kW range?

    (a) 8 kW   (b) 8.4 kW   (c) 8.8 kW   (d) 14 kW

46. If the total demand load is 30 kW for a 120/240V dwelling unit, what is the service and feeder copper conductor size?

    (a) 4 AWG   (b) 3/0 AWG   (c) 2 AWG   (d) 1/0 AWG

47. If the service raceway contains 2 AWG service conductors, what size copper-bonding jumper is required for the service raceway?

    (a) 8 AWG   (b) 6 AWG   (c) 4 AWG   (d) 3 AWG

48. If the service contains 2 AWG conductors, what is the minimum size grounding electrode conductor required?

    (a) 8 AWG   (b) 6 AWG   (c) 4 AWG   (d) 3 AWG

Part C - Optional Method – Load Calculation Questions

9–10 Dwelling Unit Optional Feeder/Service Calculations (220.30)

49. When sizing the feeder or service conductor according to the optional method we must:

    (a) Determine the total connected load of the general lighting and receptacles, small-appliance and laundry branch circuits, and the nameplate VA rating of all appliances and motors.

    (b) Determine the demand load by applying the following demand factor to the total connected load: First 10 kVA at 100%, remainder at 40%.

    (c) Determine the air-conditioning versus heat demand load.

    (d) all of these.

The following information applies to question 50.

A 1,500 sq ft dwelling unit contains the following loads:

Dishwasher (1.5 kVA)	Water heater (5 kW)
Waste disposal (1 kVA)	Dryer (5.5 kW)
Cooktop (6 kW)	Two ovens (each 3 kW)
A/C (3-hp, 230V)	Electric heat (10 kW)

50. • Use the optional method of calculation to determine what size aluminum conductor is required for the 120/240V, 1Ø service.

    (a) 1 AWG   (b) 1/0 AWG   (c) 2/0 AWG   (d) 3/0 AWG

Grounding and Bonding Library

Grounding and bonding problems are at epidemic levels. Surveys repeatedly show 90% of power quality problems are due to poor grounding and bonding. Mike has applied electrical theory to this very difficult to understand Article, making it easier for students to grasp the concepts of Grounding & Bonding. Additionally, Mike has color coded the text so you can easily differentiate between Grounding and Bonding.

Order the entire Grounding and Bonding Library including the textbook, two videos, two DVDs and the CD-Rom

Call Today 1.888.NEC.CODE or visit us online at www.NECcode.com for the latest information and pricing.

The following information applies to question 51.

A 2,330 sq ft residence with a 300 sq ft porch and 150 sq ft carport contains the following:

Dishwasher (1.5 kW)   Waste Disposal (1 kVA)
Trash Compactor (1.5 kVA)   Water Heater (6 kW)
Range (14 kW)   Dryer (4.5 kW)
A/C Unit (5-hp 230 V)   Baseboard Heat (Four @ 2.5 kW each)

51.   Use the optional method to determine what size aluminum conductor is required for the 120/240V, 1Ø service.
      (a) 1 AWG   (b) 1/0 AWG   (c) 2/0 AWG   (d) 3/0 AWG

9–12 Neutral Calculations General

52.   The feeder and service neutral load is the maximum unbalanced demand load between the grounded (neutral) conductor and any one ungrounded (hot) conductor as determined by Article 220, Part II. Since 240V loads cannot be connected to the neutral conductor, 240V loads are not considered for sizing the feeder neutral conductor.
      (a) True   (b) False

53.   • What size 1Ø, 3-wire feeder is required for a 475A demand load of which 275A consists of 240V loads? Size the conductors based on 75°C terminals in accordance with 110.14(C).
      (a) Two 750 kcmil and one 350 kcmil   (b) Two 500 kcmil and one 350 kcmil
      (c) Two 750 kcmil and one 500 kcmil   (d) Two 750 kcmil and one 3/0 AWG

54.   The feeder or service neutral load for household cooking appliances such as electric ranges, wall-mounted ovens or counter-mounted cooking units, shall be calculated at _____ of the demand load as determined by 220.19.
      (a) 50%   (b) 60%   (c) 70%   (d) 80%

55.   What is the dwelling unit service neutral load for one 9 kW range?
      (a) 5.6 kW   (b) 6.5 kW   (c) 3.5 kW   (d) 12 kW

56.   The service neutral demand load for household electric clothes dryers shall be calculated at _____ of the demand load as determined by 220.18.
      (a) 50%   (b) 60%   (c) 70%   (d) 80%

57.   What is the feeder neutral load for one 6 kW household dryer?
      (a) 4.2 kW   (b) 4.7 kW   (c) 5.4 kW   (d) 6.6 kW

☆ Challenge Questions

9–5 Cooking Equipment – Branch Circuit [Table 220.19 Note 4]

58.   The branch-circuit demand load for one 18 kW range is _____.
      (a) 12 kW   (b) 8 kW   (c) 10.4 kW   (d) 18 kW

59.   A dwelling unit kitchen has the following appliances: one 9 kW cooktop and one wall-mounted oven rated 5.3 kW. The branch-circuit demand load for these appliances is _____.
      (a) 14.3 kW   (b) 12 kW   (c) 8.8 kW   (d) 8 kW

9–8 Dwelling Unit Feeder/Service Load Calculations (Part II of Article 240)

60.   An apartment building contains 20 units, each 840 sq ft. What is the general lighting feeder demand load for each dwelling unit? Note: Laundry facilities are provided on the premises for all tenants and no laundry circuit is required in each unit, see 210.52(F) Ex. 1.
      (a) 3,520 VA   (b) 3,882 VA   (c) 4,220 VA   (d) 6,300 VA

61. What is the demand load for the general lighting, receptacles, small appliance circuits and laundry circuits for an 1,800 sq ft dwelling unit?
    (a) 5,400 VA   (b) 7,900 VA   (c) 5,415 VA   (d) 6,600 VA

62. How many 15A, 120V branch circuits are required for general-use receptacle and lighting for a 1,800 sq ft dwelling unit?
    (a) 1 circuit   (b) 2 circuits   (c) 3 circuits   (d) 4 circuits

63. How many 20A, 120V branch circuits are required for general-use receptacle and lighting for a 2,800 sq ft dwelling unit?
    (a) 2 circuits   (b) 6 circuits   (c) 7 circuits   (d) 4 circuits

Appliance Demand Factors [220.17]

64. A dwelling unit contains one of each of the following: washing machine (1.2 kW), water heater (4 kW), dishwasher (1.2 kW), trash compactor (1.5 kW). The appliance demand load added to the service is _____.
    (a) 5.9 kW   (b) 6.7 kW   (c) 7.7 kW   (d) 8.8 kW

65. • A dwelling unit contains one of each of the following: water heater (4 kW), dishwasher ($^1/_2$-hp), dryer, pool pump ($^3/_4$-hp), cooktop (6 kW), oven (6 kW), A/C (4-hp, 230 V), heat (6 kW). What is the appliance demand load for the dwelling unit?
    (a) 6.7 kW   (b) 4 kW   (c) 9 kW   (d) 11 kW

Dryer Calculation [220.18]

66. A dwelling unit contains a 5.5 kW electric clothes dryer. What is the feeder demand load for the dryer?
    (a) 3.38 kW   (b) 4.5 kW   (c) 5 kW   (d) 5.5 kW

Cooking Equipment Demand Load [220.19]

67. The feeder load for twelve 1.75 kW cooktops is _____.
    (a) 21 kW   (b) 10.5 kW   (c) 13.5 kW   (d) 12.5 kW

68. The minimum neutral for a branch circuit to an 8 kW household range is _____. See Article 210.
    (a) 10 AWG   (b) 12 AWG   (c) 8 AWG   (d) 6 AWG

69. What is the maximum dwelling unit feeder or service demand load for fifteen 8 kW cooking units?
    (a) 38.4 kW   (b) 33 kW   (c) 88 kW   (d) 27 kW

Note 2 of Table 220.19

70. What is the feeder demand load for five 10 kW, five 14 kW and five 16 kW household ranges?
    (a) 70 kW   (b) 23 kW   (c) 14 kW   (d) 33 kW

Note 3 to Table 220.19

71. A dwelling unit has one 6 kW cooktop and one 6 kW oven. What is the minimum feeder demand load for the cooking appliances?
    (a) 13 kW   (b) 8.8 kW   (c) 7.8 kW   (d) 8.2 kW

72. The maximum feeder demand load for an 8 kW range is _____.
    (a) 8.5 kW   (b) 8 kW   (c) 6.3 kW   (d) 12 kW

73. The minimum feeder demand load for five 5 kW ranges, two 4 kW ovens and four 7 kW cooking units is _____.
    (a) 61 kW   (b) 22 kW   (c) 19.5 kW   (d) 18 kW

74. The minimum feeder demand load for two 3 kW wall-mounted ovens and one 6 kW cooktop is _____.
    (a) 9.3 kW   (b) 11 kW   (c) 8.4 kW   (d) 9.6 kW

9–9 Dwelling Unit Feeder/Service Calculations Examples

75. • Both units of a duplex apartment require a 100A main, the resulting 200A service would require _____ THHN.
    (a) 1/0 AWG   (b) 2/0 AWG   (c) 3/0 AWG   (d) 4 AWG

76. After all demand factors have been taken into consideration, the demand load for a dwelling unit is 21,560 VA. The minimum service size for this residence would be _____ if the optional method of service calculations were used for a 120/240V, 1Ø system. Be sure to read 220.30 carefully!
    (a) 90A   (b) 100A   (c) 110A   (d) 125A

77. After all demand factors have been taken into consideration, the net computed load for a service is 24,221, 120/240V, 1Ø system. The minimum service size is _____.
    (a) 150A   (b) 175A   (c) 125A   (d) 110A

9–10 Dwelling Unit Optional Feeder/Service Calculations [220.30]

78. Using the optional calculations method, determine the demand load for a 6 kW central space heating and a 4 kW air-conditioning unit.
    (a) 9,000W   (b) 3,900W   (c) 4,000W   (d) 5,000W

79. The total connected load of a dwelling unit is 25 kVA, not including heat or air-conditioning. If the heat is separately controlled in five rooms (10 kW) and the air-conditioning is 6 kW, then the total service demand load is _____ if the optional method is used.
    (a) 22 kVA   (b) 29 kVA   (c) 35 kVA   (d) 36 kVA

80. Using the optional method, determine the service for the following loads: 1,200 sq ft first floor, plus 600 sq ft on the second floor, a 200 sq ft open porch, water heater (4 kW), dishwasher ($^1/_2$-hp), clothes dryer (4 kW), A/C (5-hp), heat (6 kW), pool pump ($^3/_4$-hp), oven (6 kW), cooktop (6 kW). The service source voltage is 120/240V, 1Ø.
    (a) 200A   (b) 175A   (c) 125A   (d) 110A

81. An 1,800 sq ft residence contains the following: a 4 kW water heater, 10 kW heat separated in five rooms, one 1.5 kW dishwasher, one 6 kW range, one 4.5 kW dryer, two 3 kW ovens, one 6 kW air conditioner. The 120/240V, 1Ø service for the loads would be _____. The optional method of service calculations is used.
    (a) 175A   (b) 110A   (c) 125A   (d) 150A

82. A dwelling unit has 1,200 sq ft on the first floor, 600 sq ft upstairs (unfinished but adaptable for future use), a 200 sq ft open porch, and the following loads: pool pump, ($^3/_4$-hp), range (13.9 kW), dishwasher (1.2 kW), water heater (4 kW), dryer (4 kW), A/C (5-hp), heat (6 kW). Using the optional calculation method, what size service and feeder conductor is required for this 120/240V, 1Ø service?
    (a) 4 AWG   (b) 3 AWG   (c) 2 AWG   (d) 1 AWG

NEC Questions – Articles 430-501

Article 430 Motors, Motor Circuits and Controllers

83. Motor controllers and terminals of control circuit devices are required to be connected to copper conductors unless identified as approved for use with a different type of conductor.
    (a) True   (b) False

84. Torque requirements for motor control circuit device terminals shall be a minimum of _____ pound-inches (unless otherwise indicated) for screw-type pressure terminals used for 14 AWG and smaller copper conductors.
    (a) 7   (b) 9   (c) 10   (d) 15

85. When motors are provided with terminal housings, the housings shall be of _____ and of substantial construction.
    (a) steel   (b) iron   (c) metal   (d) copper

86. A motor terminal housing with rigidly mounted motor terminals shall have a minimum of _____ between line terminals for a 230V motor.
    (a) $1/4$ in.   (b) $3/8$ in.   (c) $1/2$ in.   (d) $5/8$ in.

87. Motors shall be located so that adequate _____ is provided and so that maintenance, such as lubrication of bearings and replacing of brushes, can be readily accomplished.
    (a) space   (b) ventilation   (c) protection   (d) all of these

88. Open motors having commutators shall be located or protected so that sparks cannot reach adjacent combustible material, but this shall not prohibit the installation of these motors _____.
    (a) on wooden floors   (b) over combustible fiber
    (c) under combustible material   (d) none of these

89. Branch-circuit conductors supplying a single continuous duty motor shall have an ampacity not less than _____ rating.
    (a) 125 percent of the motor's nameplate current rating
    (b) 125 percent of the motor's full-load current (as listed in the *NEC*)
    (c) 125 percent of the motor's full locked-rotor
    (d) 80 percent of the motor's full-load current

90. Feeder tap conductors supplying motor circuits, with an ampacity of at least one-third that of the feeder, shall not exceed _____ in length.
    (a) 10 ft   (b) 15 ft   (c) 20 ft   (d) 25 ft

91. Overload devices are intended to protect motors, motor control apparatus and motor branch-circuit conductors against _____.
    (a) excessive heating due to motor overloads   (b) excessive heating due to failure to start
    (c) short circuits and ground faults   (d) a and b

92. Motor overload protection is not required where _____.
    (a) conductors are oversized by 125 percent
    (b) conductors are part of a limited-energy circuit
    (c) it might introduce additional hazards
    (d) short-circuit protection is provided

93. An overload device used to protect continuous-duty motors (rated more than 1-hp) shall be selected to trip or rated at no more than _____ percent of the motor nameplate full-load current rating for motors with a marked service factor not less than 1.15.
    (a) 110   (b) 115   (c) 120   (d) 125

94. The ultimate trip current of a thermally protected motor with a full-load current not exceeding 9A shall not exceed _____ percent of the motor full-load current as listed in Tables 430.148, 430.149 and 430.150.
    (a) 140   (b) 156   (c) 170   (d) none of these

95. When an overload relay, selected in accordance with 430.32(A)(1) and (B)(1), is not sufficient to start the motor or carry the load, higher-size sensing elements or incremental settings are permitted to be used as long as the trip current does not exceed _____ percent of the motor full-load current rating when the motor is marked with a service factor of 1.12.
    (a) 100   (b) 110   (c) 120   (d) 130

96. Each continuous-duty motor of _____ or less that is not permanently installed, not automatically started and is within sight of the controller, shall be permitted to be protected against overload by the branch-circuit short-circuit and ground-fault protective device.
    (a) 1-hp   (b) 2-hp   (c) 3-hp   (d) 4-hp

97. Motor overload protection shall not be shunted or cut out during the starting period if the motor is _____.
    (a) not automatically started   (b) automatically started
    (c) manually started   (d) all of these

98. When fuses are used for motor overload protection, a fuse shall be inserted in each ungrounded conductor and also in the grounded conductor if the supply system is _____ with one conductor grounded.
(a) 2-wire, 3Ø dc
(b) 2-wire, 3Ø ac
(c) 3-wire, 3Ø dc
(d) 3-wire, 3Ø ac

99. The minimum number of overload unit(s) required for a 3Ø ac motor shall be _____.
(a) one
(b) two
(c) three
(d) any of these

100. Overload relays and other devices for motor overload protection that are not capable of opening _____ shall be protected by fuses or circuit breakers.
(a) short circuits
(b) clearing overloads
(c) ground faults
(d) a and c

101. A motor _____ device that can restart a motor automatically after overload tripping shall not be installed if automatic restarting of the motor can result in injury to persons.
(a) short-circuit
(b) ground-fault
(c) overcurrent
(d) overload

102. If a(n) _____ shutdown is necessary to reduce hazards to persons, the overload sensing devices shall be permitted to be connected to a supervised alarm instead of causing immediate interruption of the motor circuit.
(a) emergency
(b) normal
(c) orderly
(d) none of these

103. The motor branch-circuit short-circuit and ground-fault protective device shall be capable of carrying the _____ current of the motor.
(a) varying
(b) starting
(c) running
(d) continuous

104. The maximum rating or setting of motor branch-circuit short-circuit and ground-fault protective devices for a 1Ø motor is _____ percent for an inverse-time breaker.
(a) 125
(b) 175
(c) 250
(d) 300

105. • A feeder shall have a protective device with a rating or setting _____ branch-circuit short-circuit and ground-fault protective device for any motor in the group, plus the sum of the full-load currents of the other motors of the group.
(a) not greater than the largest
(b) 125 percent of the largest rating
(c) equal to the largest rating
(d) none of these

106. Overcurrent protection for motor control circuits shall not exceed 400 percent if the conductor does not extend beyond the motor control equipment enclosure.
(a) True
(b) False

107. Motor control circuit conductors that extend beyond the motor control equipment enclosure shall be required to have short-circuit and ground-fault protection sized not greater than _____ percent of the conductor ampacity as listed in Table 430.72(B) for 60°C conductors.
(a) 100
(b) 150
(c) 300
(d) 500

108. Motor control circuit transformers, with a primary current rating at less than 2A, can have the primary protection device set no more than _____ percent of the rated primary current rating.
(a) 150
(b) 200
(c) 400
(d) 500

109. Motor control circuits shall be arranged so that they will be disconnected from all sources of supply when the disconnecting means is in the open position. Where separate devices are used for the motor and control circuit, they shall be located immediately adjacent to each other.
(a) True
(b) False

110. If the control circuit transformer is located in the controller enclosure, the transformer shall be connected to the _____ side of the control circuit disconnect.
(a) line
(b) load
(c) adjacent
(d) none of these

111. The branch-circuit protective device shall be permitted to serve as the controller for a stationary motor rated at _____ or less that is normally left running and cannot be damaged by overload or failure to start.
(a) $1/8$-hp
(b) $1/4$-hp
(c) $3/8$-hp
(d) $1/2$-hp

112. The motor controller shall have horsepower ratings at the application voltage not _____ the horsepower rating of the motor.
    (a) lower than   (b) higher than   (c) equal to   (d) none of these

113. Each motor shall be provided with an individual controller.
    (a) True   (b) False

114. A disconnecting means is required to disconnect the _____ from all ungrounded supply conductors.
    (a) motor   (b) motor or controller   (c) controller   (d) motor and controller

115. • A _____ shall be located in sight from the motor location and the driven machinery location.
    (a) controller   (b) protection device
    (c) disconnecting means   (d) all of these

116. The motor disconnecting means is not required to be in sight from the motor and the driven machinery location provided _____.
    (a) the controller disconnecting means is capable of being individually locked in the open position.
    (b) the provisions for locking are permanently installed on, or at the switch or circuit breaker used as the controller disconnecting means.
    (c) locating the motor disconnecting means within sight of the motor is impractical or introduces additional or increased hazards to people or property.
    (d) all of these

117. The disconnecting means for the controller and motor must open all ungrounded supply conductors and shall be designed so that no pole can be operated independently.
    (a) True   (b) False

118. Where more than one motor disconnecting means is provided in the same motor branch circuit, only one of the disconnecting means is required to be readily accessible.
    (a) True   (b) False

119. The motor disconnecting means can be a _____.
    (a) circuit breaker   (b) motor-circuit switch rated in horsepower
    (c) molded case switch   (d) any of these

120. If the motor disconnecting means is a motor-circuit switch, it shall be rated in _____.
    (a) horsepower   (b) watts   (c) amperes   (d) locked-rotor current

121. A branch-circuit overcurrent protection device such as a plug fuse may serve as the disconnecting means for a stationary motor of $^1/_8$-hp or less.
    (a) True   (b) False

122. The disconnecting means for a torque motor shall have an ampere rating of at least _____ percent of the motor nameplate current.
    (a) 100   (b) 115   (c) 125   (d) 175

123. The disconnecting means for a 50-hp, 460V, 3Ø induction motor (FLC 65A) shall have an ampere rating of not less than _____.
    (a) 126A   (b) 75A   (c) 91A   (d) 63A

124. An oil switch used for both a controller and disconnect is permitted on a circuit whose rating _____ or 100A.
    (a) does not exceed 300V   (b) exceeds 600V
    (c) does not exceed 600V   (d) none of these

125. • Exposed live parts of motors and controllers operating at _____ or more between terminals shall be guarded against accidental contact by enclosure or by location.
    (a) 24V   (b) 50V   (c) 150V   (d) 60V

Article 440 Air-Conditioning and Refrigerating Equipment

126.   Article 440 applies to electric motor-driven air-conditioning and refrigerating equipment that has a hermetic refrigerant motor-compressor.
(a) True   (b) False

127.   A disconnecting means that serves a hermetic refrigerant motor-compressor, and is selected on the basis of the nameplate rated-load current or branch-circuit selection current, whichever is greater, shall have an ampere rating of _____ percent of the nameplate rated-load current or branch-circuit selection current, whichever is greater.
(a) 125   (b) 80   (c) 100   (d) 115

128.   The disconnecting means for air-conditioning and refrigerating equipment shall be _____ from the air-conditioning or refrigerating equipment.
(a) readily accessible   (b) within sight   (c) a or b   (d) a and b

129.   Disconnecting means shall be located within sight from and readily accessible from the air-conditioning or refrigerating equipment. The disconnecting means can be installed _____ the air-conditioning or refrigerating equipment, but not on panels that are designed to allow access to the air-conditioning or refrigeration equipment.
(a) on   (b) within   (c) a or b   (d) none of these

130.   The rating of the branch-circuit short-circuit and ground-fault protection device for an individual motor-compressor shall not exceed _____ percent of the rated-load current or branch-circuit selection current, whichever is greater, if the protection device will carry the starting current of the motor.
(a) 100   (b) 125   (c) 175   (d) none of these

131.   A hermetic motor-compressor controller shall have a _____ current rating not less than the respective rating(s) on the compressor.
(a) continuous-duty full-load   (b) locked-rotor
(c) a or b   (d) a and b

132.   The attachment plug and receptacle shall not exceed _____ at 250V for a cord-and-plug-connected air-conditioning motor-compressor.
(a) 15A   (b) 20A   (c) 30A   (d) 40A

133.   The total rating of a plug-connected room air conditioner, connected to the same branch circuit where lighting units or other appliances are also supplied, shall not exceed _____ percent of the branch circuit rating.
(a) 80   (b) 70   (c) 50   (d) 40

134.   When supplying a 120V rated (nominal) room air conditioner, the length of the flexible supply cord shall not exceed _____
(a) 4 ft.   (b) 6 ft.   (c) 8 ft.   (d) 10 ft.

Article 445 Generators

135.   Each generator shall be provided with a _____ giving the manufacturer's name, the rated frequency, power factor, number of phases (if of alternating current) and the rating in kilowatts or kilovolt-amperes.
(a) list   (b) faceplate   (c) nameplate   (d) sticker

136.   The ampacity of ungrounded (phase) conductors from the generator terminals to the first overcurrent protection devices shall not be less than _____ percent of the nameplate rating of the generator.
(a) 75   (b) 115   (c) 125   (d) 140

137.   Unless two restrictive conditions exist, a generator shall be equipped with a disconnecting means to disconnect the generator, its protective devices and all control apparatus entirely from the circuits supplied by the generator.
(a) True   (b) False

Article 450 Transformers and Transformer Vaults (Including Secondary Ties)

138. According to Article 450, a transformer rated 600V, nominal, or less, and whose primary current rating is 9A or more, is protected against overcurrent only when _____.
    (a) an individual overcurrent device on the primary side is set at not more than 125 percent of the rated primary current of the transformer.
    (b) a secondary overcurrent device is set at not more than 125 percent of the rated secondary current of the transformer, and a primary overcurrent device is set at not more than 250 percent of the rated primary current of the transformer.
    (c) a or b
    (d) none of these

139. For a transformer rated 600V, nominal, or less, if the primary overcurrent protection device is sized at 250 percent of the primary current, what size secondary overcurrent protection device is required if the secondary current is 42A?
    (a) 40A   (b) 70A   (c) 60A   (d) 90A

140. What size "primary only" overcurrent protection is required for a 45 kVA transformer, rated 600V nominal, or less, that has a primary current rating of 54A?
    (a) 70   (b) 80   (c) 90   (d) 100

141. A secondary tie of a transformer is a circuit operating at _____ nominal, or less, between phases that connects two power sources or power-supply points.
    (a) 600V   (b) 1,000V   (c) 12,000V   (d) 35,000V

142. Each transformer shall be provided with a nameplate giving the name of the manufacturer, rated kilovolt-amperes, frequency, primary and secondary voltage, impedance of transformers _____ and larger, required clearances for transformers with ventilating openings, and the amount and kind of insulating liquid where used.
    (a) 112 kVA   (b) 25 kVA   (c) 33 kVA   (d) 50 kVA

143. All transformers and transformer vaults shall be readily accessible to qualified personnel for inspection and maintenance, or _____
    (a) dry-type transformers 600V, nominal, or less, located in the open on walls, columns, or structures, shall not be required to be readily accessible.
    (b) dry-type transformers rated not more than 50 kVA and not over 600V can be installed in hollow spaces of buildings.
    (c) a or b
    (d) none of these

144. Transformers of more than _____ rating shall be installed in a transformer room of fire-resistant construction.
    (a) 35,000 kVA   (b) 87$^1/_2$ kVA   (c) 112$^1/_2$ kVA   (d) 75 kVA

145. Dry-type transformers installed indoors rated over _____ shall be installed in a vault.
    (a) 1,000V   (b) 112$^1/_2$ kVA   (c) 50,000V   (d) 35,000V

146. Transformers, with nonflammable dielectric fluid rated over 35,000V and installed in a vault, shall be furnished with a _____ when installed indoors.
    (a) liquid confinement area
    (b) pressure-relief vent
    (c) means for absorbing or venting any gases generated by arcing
    (d) all of these

147. Transformer vaults shall be located where they can be ventilated to the outside air without using flues or ducts wherever such an arrangement is _____.
    (a) permitted   (b) practicable   (c) required   (d) all of these

148. The walls and roofs of transformer vaults shall be constructed of materials that have adequate structural strength for the conditions with a minimum fire resistance of _____ hour(s).
    (a) 1   (b) 2   (c) 3   (d) 4

149. Each doorway leading into a transformer vault from the building interior shall be provided with a tight-fitting door having a minimum fire rating of _____ hours.
   (a) 2   (b) 4   (c) 5   (d) 3

150. Personnel doors for transformer vaults shall _____ and be equipped with panic bars, pressure plates or other devices that are normally latched but open under simple pressure.
   (a) be clearly identified   (b) swing out
   (c) a and b   (d) a or b

151. Where practicable, transformer vaults containing more than _____ kVA transformer capacity shall be provided with a drain or other means that will carry off any accumulation of oil or water in the vault unless local conditions make this impracticable.
   (a) 100   (b) 150   (c) 200   (d) 250

Article 455 Phase Converters

152. For phase converters serving variable loads, the ampacity of the 1Ø supply conductors shall not be less than _____ percent of the 1Ø input full-load amperes listed on the phase converter nameplate.
   (a) 75   (b) 100   (c) 125   (d) 150

153. Means shall be provided to disconnect simultaneously all _____ supply conductors to the phase converter.
   (a) ungrounded   (b) grounded   (c) grounding   (d) all of these

154. The phase converter disconnecting means shall be _____ and located in sight from the phase converter.
   (a) protected from physical damage   (b) readily accessible
   (c) easily visible   (d) clearly identified

Article 460 Capacitors

155. Capacitors shall be _____ so that persons cannot come into accidental contact or bring conducting materials into accidental contact with exposed energized parts, terminals or buses associated with them.
   (a) enclosed   (b) located   (c) guarded   (d) any of these

156. The residual voltage of a capacitor, rated not over 600V, shall be reduced to 50V, nominal, or less within _____ after the capacitor is disconnected from the source of supply.
   (a) 15 seconds   (b) 45 seconds   (c) 1 minute   (d) 2 minutes

157. The ampacity of capacitor circuit conductors shall not be less than _____ percent of the rated current of the capacitor.
   (a) 100   (b) 115   (c) 125   (d) 135

157. A capacitor operating at over 600V, nominal, shall be provided with means to reduce the residual voltage to 50V or less within _____ after it is disconnected from the source of supply.
   (a) 15 seconds   (b) 45 seconds   (c) 1 minute   (d) 5 minutes

Article 470 Resistors and Reactors

159. For installations of resistors and reactors, a thermal barrier shall be required if the space between them and any combustible material is less than _____
   (a) 2 in.   (b) 3 in.   (c) 6 in.   (d) 12 in.

160. Resistors and reactors over 600V, nominal, shall be isolated by _____ to protect personnel from accidental contact with energized parts.
   (a) an enclosure   (b) elevation   (c) a or b   (d) a and b

Article 480 Storage Batteries

161. Nominal Battery Voltage: The voltage computed on the basis of _____ per cell for the lead-acid type and _____ per cell for the alkali type.
    (a) 6V, 12V   (b) 12V, 24V   (c) 2V, 1.2V   (d) 1.2V, 2V

162. A vented alkaline-type battery operating at less than 250V shall be installed with not more than _____ cells in the series circuit in any one tray.
    (a) 10   (b) 12   (c) 18   (d) 20

163. Provisions shall be made for sufficient diffusion and ventilation of the gases from the storage battery to prevent the accumulation of a(n) _____ mixture.
    (a) corrosive   (b) explosive   (c) toxic   (d) all of these

164. Each vented cell of a battery shall be equipped with _____ that is designed to prevent destruction of the cell due to ignition of gases within the cell by an external spark or flame under normal operating conditions.
    (a) pressure relief   (b) a flame arrester   (c) fluid level indicators   (d) none of these

Chapter 5 Special Occupancies

Article 500 Hazardous (Classified) Locations, Classes I, II and III, Divisions 1 and 2

165. All areas designated as hazardous shall be properly _____ and the documentation shall be available to those authorized to design, install, inspect, maintain or operate electrical equipment at these locations.
    (a) cleaned   (b) documented   (c) maintained   (d) all of these

166. Hazardous locations are classified based on the properties of the _____ that may be present and the likelihood that a flammable or combustible concentration or quantity is present.
    (a) flammable vapors   (b) flammable gases or liquids
    (c) combustible dusts or fibers   (d) all of these

167. Through the exercise of ingenuity in the layout of the electrical installation for hazardous (classified) locations, it is frequently possible to locate much of the equipment in less hazardous or in nonhazardous locations and thus to reduce the amount of special equipment required.
    (a) True   (b) False

168. Class I, Division 1 locations are those in which ignitable concentrations of _____ can exist under normal operating conditions.
    (a) combustible dust   (b) easily ignitable fibers or flyings
    (c) flammable gases or vapors   (d) flammable liquids or gases

169. A Class I, Division 1 location is a location in which _____.
    (a) ignitable concentrations of flammable gases or vapors can exist under normal operating conditions
    (b) ignitable concentrations of such gases or vapors may exist frequently because of repair or maintenance operations or because of leakage
    (c) breakdown or faulty operation of equipment or processes might release ignitable concentrations of flammable gases or vapors, and might also cause simultaneous failure of electrical equipment.
    (d) all of these

170. A Class I, Division 2 location usually includes locations where volatile flammable liquids or flammable gases or vapors are used but that, in the judgment of the authority having jurisdiction, would become hazardous only in case of an accident or of some unusual operating condition.
    (a) True   (b) False

171. When determining a Class I, Division 2 location, _____ is a factor that merits consideration in determining the classification and extent of each location.
      (a) the quantity of flammable material that might escape in case of an accident
      (b) the adequacy of ventilating equipment
      (c) the record of the industry or business with respect to explosions or fires
      (d) all of these

172. Class II locations are those that are hazardous because of the presence of _____.
      (a) combustible dust   (b) easily ignitable fibers or flyings
      (c) flammable gases or vapors   (d) flammable liquids or gases

173. Locations in which combustible dust is in the air under normal operating conditions in quantities sufficient to produce explosive or ignitible mixtures are classified as _____.
      (a) Class I, Division 2   (b) Class II, Division 1
      (c) Class II, Division 2   (d) Class III, Division 1

174. Class III locations are those that are hazardous because of the presence of _____.
      (a) combustible dust   (b) easily ignitable fibers or flyings
      (c) flammable gases or vapors   (d) flammable liquids or gases

175. Class III, Division _____ location(s) include areas where easily ignitable fibers or materials producing combustible flyings are handled, manufactured or used.
      (a) 1   (b) 2   (c) 3   (d) all of these

176. Locations in which easily ignitable combustible fibers are stored or handled other than in the process of manufacture are designated as _____.
      (a) Class II, Division 2   (b) Class III, Division 1
      (c) Class III, Division 2   (d) nonhazardous

177. • A Class III, Division_____ location is where easily ignitable fibers or combustible flying material are stored or handled but not manufactured.
      (a) 1   (b) 2   (c) 3   (d) all of these

178. In Class _____ locations for Groups A, B, C and D, the classification involves determinations of maximum explosion pressure, maximum safe clearance between parts of a clamped joint in an enclosure and the minimum ignition temperature of the atmospheric mixture.
      (a) I   (b) II   (c) III   (d) all of these

179. Equipment in an area containing acetylene gas shall be rated for a Class I, Group A location.
      (a) True   (b) False

180. Electrical equipment installed in hazardous (classified) locations shall be constructed for the class, division and group. A Group C atmosphere contains _____.
      (a) hydrogen   (b) ethylene   (c) propylene oxide   (d) all of these

181. An atmosphere listed as Group E contains combustible metal dusts.
      (a) True   (b) False

182. An atmosphere listed as Group F contains carbon black, charcoal, coal or coke dusts that have been sensitized by other materials so that they present an explosive hazard.
      (a) True   (b) False

183. An atmosphere listed as Group G contains combustible dusts such as flour, grain, wood, plastic and other chemicals.
      (a) True   (b) False

184. For Class _____ locations, the classification of Groups E, F and G involves the tightness of the joints of assembly and shaft openings to prevent entrance of dust in the dust-ignition proof enclosure. The blanketing effect of layers of dust on the equipment may cause overheating, electrical conductivity of the dust and the ignition temperature of the dust.
      (a) I   (b) II   (c) III   (d) all of these

185. Equipment is required to be identified not only for the class of location but also for the explosive, combustible or ignitible properties of the specific _____ that will be present.
(a) gas or vapor   (b) dust   (c) fiber or flyings   (d) all of these

186. Suitability of equipment for a specific purpose, environment or application may be determined by:
(a) Equipment listing or labeling.
(b) Evidence of equipment evaluation from a qualified testing laboratory or inspection agency concerned with product evaluation.
(c) Other evidence acceptable to the authority having jurisdiction, such as a manufacturer's self evaluation or an owner's engineering judgment.
(d) any of the above

187. Equipment installed in hazardous locations shall be approved and shall be marked to show the _____.
(a) class
(b) group
(c) operating temperature, or temperature class reference to 40°C ambient
(d) all of these

188. All conduits referred to in hazardous locations shall be threaded with a _____ taper per foot.
(a) $\frac{1}{2}$ in.   (b) $\frac{1}{4}$ in.   (c) 1 in.   (d) all of these

Article 501 Class I Locations

189. Meters, instruments and relays including kilowatt-hour meters, instrument transformers, resistors, rectifiers and thermionic tubes that are installed in Class I, Division 1 locations shall be installed in explosionproof enclosures or purged and pressurized enclosures.
(a) True   (b) False

190. Meters, instruments and relays installed in Class I, Division 2 locations can have switches, circuit breakers and make-and-break contacts of push buttons, relays, alarm bells and horns installed in enclosures that are approved for general purpose if current-interrupting contacts are _____.
(a) immersed in oil
(b) enclosed within a hermetically-sealed chamber
(c) a or b
(d) a and b

191. Which of the following wiring methods are permitted in a Class I, Division 1 location?
(a) Threaded rigid metal conduit   (b) Threaded IMC
(c) MI cable   (d) all of these

192. Wiring methods permitted in Class I, Division 1 locations include _____.
(a) threaded rigid metal or threaded steel intermediate metal conduit
(b) flexible fittings listed for Class I, Division 1 locations
(c) boxes approved for Class I, Division 1 locations
(d) all of these

193. ITC-HL cables, listed for use in Class I, Division 1 locations, with a gas/vaportight, continuous-corrugated, metallic sheath, an overall jacket of suitable polymeric material and provided with termination fittings listed for the application can be installed in Class I, Division 1 _____ establishments with restricted public access.
(a) commercial   (b) industrial   (c) institutional   (d) all of these

194. When provisions shall be made in a Class I, Division 2 location for limited flexibility, such as motor termination, flexible metal conduit with listed fittings may be used.
(a) True   (b) False

195.   Boxes, enclosures, fittings and joints are not required to be explosionproof in a Class I, Division 2 location. However, if arcs or sparks (such as from make-or-break contacts) can result from equipment being utilized, that equipment shall be installed in an explosionproof enclosure meeting the requirements for Class I, Division 1 locations.
       (a) True   (b) False

196.   Sealing compound is employed with MI cable terminal fittings in Class I locations for the purpose of _____.
       (a) preventing the passage of gas or vapor
       (b) excluding moisture and other fluids from the cable insulation
       (c) limiting a possible explosion
       (d) preventing the escape of powder

197.   Conduits of $1^1/_2$ in. or smaller entering an explosionproof enclosure that houses switches intended to interrupt current in the normal performance of the function, shall not be required to be sealed if the current-interrupting contacts are within a chamber hermetically sealed against the entrance of gases and vapors.
       (a) True   (b) False

198.   Each conduit leaving a Class I, Division 1 location requires a seal to be located on either side of the hazardous location boundary. Unions, couplings, boxes or fittings are permitted between the seal and the point where the conduit leaves the Division 1 location.
       (a) True   (b) False

199.   A sealing fitting shall be installed within _____ of either side of the boundary where a conduit leaves a Class I, Division 1 location. The seal fitting shall be designed and installed so as to minimize the amount of gas or vapor within the Division 1 portion of the conduit being communicated beyond the seal.
       (a) 5 ft   (b) 6 ft   (c) 8 ft   (d) 10 ft

200.   Where the Class I, Division 1 boundary is beneath the ground, the sealing fitting shall be installed _____. Except for listed explosionproof reducers at the conduit seal, there shall be no union, coupling, box or fitting between the conduit seal and the point at which the conduit leaves the ground.
       (a) after the conduit leaves the ground   (b) before the conduit leaves the ground
       (c) within 10 ft   (d) none of these

Unit 9 NEC Exam – NEC Code Order 430.6 – 501.3

1. For general motor applications, the motor branch-circuit short-circuit and ground-fault protection device must be sized based on the _____ amperes.
    (a) motor nameplate   (b) NEMA standards
    (c) *NEC* Table   (d) Factory Mutual

2. The motor _____ current as listed in Tables 430.147 through 430.150 must be used for sizing motor branch-circuit conductors and short-circuit ground-fault protection devices.
    (a) nameplate   (b) full-load   (c) load   (d) none of these

3. When motors are provided with terminal housings, the housings shall be of _____ and of substantial construction.
    (a) steel   (b) iron   (c) metal   (d) copper

4. When an overload relay, selected in accordance with 430.32(A)(1) and (B)(1), is not sufficient to start the motor or carry the load, higher-size sensing elements or incremental settings are permitted to be used as long as the trip current does not exceed _____ percent of the motor full-load current rating when the motor is marked with a service factor of 1.12.
    (a) 100   (b) 110   (c) 120   (d) 130

5. When fuses are used for motor overload protection, a fuse shall be inserted in each ungrounded conductor and also in the grounded conductor if the supply system is _____ with one conductor grounded.
    (a) 2-wire, 3Ø dc   (b) 2-wire, 3Ø ac   (c) 3-wire, 3Ø dc   (d) 3-wire, 3Ø ac

6. The minimum number of overload unit(s) required for a 3Ø ac motor shall be _____.
    (a) one   (b) two   (c) three   (d) any of these

7. Overload relays and other devices for motor overload protection that are not capable of opening _____ shall be protected by fuses or circuit breakers.
    (a) short circuits   (b) clearing overloads
    (c) ground faults   (d) a and c

8. If a(n) _____ shutdown is necessary to reduce hazards to persons, the overload sensing devices shall be permitted to be connected to a supervised alarm instead of causing immediate interruption of the motor circuit.
    (a) emergency   (b) normal   (c) orderly   (d) none of these

9. A feeder shall have a protective device with a rating or setting _____ branch-circuit short-circuit and ground-fault protective device for any motor in the group, plus the sum of the full-load currents of the other motors of the group.
    (a) not greater than the largest   (b) 125 percent of the largest rating
    (c) equal to the largest rating   (d) none of these

10. A disconnecting means is required to disconnect the _____ from all ungrounded supply conductors.
    (a) motor   (b) motor or controller
    (c) controller   (d) motor and controller

11. A _____ shall be located in sight from the motor location and the driven machinery location.
    (a) controller   (b) protection device
    (c) disconnecting means   (d) all of these

12. If the motor disconnecting means is a motor-circuit switch, it shall be rated in _____.
    (a) horsepower   (b) watts
    (c) amperes   (d) locked-rotor current

13. Exposed live parts of motors and controllers operating at _____ or more between terminals shall be guarded against accidental contact by enclosure or by location.
    (a) 24V   (b) 50V   (c) 150V   (d) 60V

14. The rating of the branch-circuit short-circuit and ground-fault protection device for an individual motor-compressor shall not exceed _____ percent of the rated-load current or branch-circuit selection current, whichever is greater, if the protection device will carry the starting current of the motor.
    (a) 100   (b) 125   (c) 175   (d) none of these

15. The attachment plug and receptacle shall not exceed _____ at 250V for a cord-and-plug-connected air-conditioning motor-compressor.
    (a) 15A   (b) 20A   (c) 30A   (d) 40A

16. The total rating of a plug-connected room air conditioner, connected to the same branch circuit where lighting units or other appliances are also supplied, shall not exceed _____ percent of the branch circuit rating.
    (a) 80   (b) 70   (c) 50   (d) 40

17. When supplying a 120V rated (nominal) room air conditioner, the length of the flexible supply cord shall not exceed _____.
    (a) 4 ft   (b) 6 ft   (c) 8 ft   (d) 10 ft

18. The ampacity of ungrounded (phase) conductors from the generator terminals to the first overcurrent protection devices shall not be less than _____ percent of the nameplate rating of the generator.
    (a) 75   (b) 115   (c) 125   (d) 140

19. Transformers, with nonflammable dielectric fluid rated over 35,000V and installed in a vault, shall be furnished with a _____ when installed indoors.
    (a) liquid confinement area
    (b) pressure-relief vent
    (c) means for absorbing or venting any gases generated by arcing
    (d) all of these

20. Transformer vaults shall be located where they can be ventilated to the outside air without using flues or ducts wherever such an arrangement is _____.
    (a) permitted   (b) practicable   (c) required   (d) all of these

21. The walls and roofs of transformer vaults shall be constructed of materials that have adequate structural strength for the conditions with a minimum fire resistance of _____ hour(s).
    (a) 1   (b) 2   (c) 3   (d) 4

22. Personnel doors for transformer vaults shall _____ and be equipped with panic bars, pressure plates or other devices that are normally latched but open under simple pressure.
    (a) be clearly identified   (b) swing out
    (c) a and b   (d) a or b

23. Locations in which combustible dust is in the air under normal operating conditions in quantities sufficient to produce explosive or ignitible mixtures are classified as _____.
    (a) Class I, Division 2   (b) Class II, Division 1
    (c) Class II, Division 2   (d) Class III, Division 1

24. A Class III, Division_____ location is where easily ignitable fibers or combustible flying material are stored or handled but not manufactured.
    (a) 1   (b) 2   (c) 3   (d) all of these

25. Meters, instruments and relays installed in Class I, Division 2 locations can have switches, circuit breakers and make-and-break contacts of push buttons, relays, alarm bells and horns installed in enclosures that are approved for general purpose if current-interrupting contacts are _____.
    (a) immersed in oil
    (b) enclosed within a hermetically-sealed chamber
    (c) a or b
    (d) a and b

Unit 9 NEC Exam – Random Order 110.26 – 501.4

1. A branch-circuit overcurrent protection device such as a plug fuse may serve as the disconnecting means for a stationary motor of $^1/_8$-hp or less.
   (a) True   (b) False

2. A Class I, Division 1 location is a location in which _____.
   (a) ignitable concentrations of flammable gases or vapors can exist under normal operating conditions
   (b) ignitable concentrations of such gases or vapors may exist frequently because of repair or maintenance operations or because of leakage
   (c) breakdown or faulty operation of equipment or processes might release ignitable concentrations of flammable gases or vapors, and might also cause simultaneous failure of electrical equipment.
   (d) all of these

3. A Class I, Division 2 location usually includes locations where volatile flammable liquids or flammable gases or vapors are used but that, in the judgment of the authority having jurisdiction, would become hazardous only in case of an accident or of some unusual operating condition.
   (a) True   (b) False

4. A hermetic motor-compressor controller shall have a _____ current rating not less than the respective rating(s) on the compressor.
   (a) continuous-duty full-load   (b) locked-rotor
   (c) a or b   (d) a and b

5. All conduits referred to in hazardous locations shall be threaded with a _____ inch taper per foot.
   (a) $^1/_2$ (b) $^3/_4$ (c) 1   (d) all of these

6. An atmosphere listed as Group G contains combustible dusts such as flour, grain, wood, plastic, and other chemicals.
   (a) True   (b) False

7. An oil switch used for both a controller and disconnect is permitted on a circuit whose rating _____ or 100A.
   (a) does not exceed 300V   (b) exceeds 600V
   (c) does not exceed 600V   (d) none of these

8. Boxes, enclosures, fittings, and joints are not required to be explosionproof in a Class I, Division 2 location. However, if arcs or sparks (such as from make-or-break contacts) can result from equipment being utilized, that equipment must be installed in an explosionproof enclosure meeting the requirements for Class I, Division 1 locations.
   (a) True   (b) False

9. Dry-type transformers installed indoors rated over _____ shall be installed in a vault.
   (a) 1,000V   (b) $112^1/_2$ kVA   (c) 50,000V   (d) 35,000V

10. Each continuous-duty motor of _____ or less that is not permanently installed, not automatically started, and is within sight of the controller, shall be permitted to be protected against overload by the branch-circuit short-circuit and ground-fault protective device.
    (a) 1-hp   (b) 2-hp   (c) 3-hp   (d) 4-hp

11. Feeder tap conductors supplying motor circuits, with an ampacity of at least one-third that of the feeder, shall not exceed _____ in length.
    (a) 10 ft   (b) 15 ft   (c) 20 ft   (d) 25 ft

12. For a transformer rated 600V, nominal, or less, if the primary overcurrent protection device is sized at 250 percent of the primary current, what size secondary overcurrent protection device is required if the secondary current is 42A?
    (a) 40A   (b) 70A   (c) 60A   (d) 90A

13. If the control circuit transformer is located in the controller enclosure, the transformer must be connected to the _____ side of the control circuit disconnect.
    (a) line   (b) load   (c) adjacent   (d) none of these

14. Locations in which easily ignitable combustible fibers are stored or handled other than in the process of manufacture are designated as _____.
(a) Class II, Division 2   (b) Class III, Division 1
(c) Class III, Division 2   (d) nonhazardous

15. Motor overload protection is not required where _____.
(a) conductors are oversized by 125 percent
(b) conductors are part of a limited-energy circuit
(c) it might introduce additional hazards
(d) short-circuit protection is provided

16. Motor overload protection shall not be shunted or cut out during the starting period if the motor is _____.
(a) not automatically started   (b) automatically started
(c) manually started   (d) all of these

17. Overload devices are intended to protect motors, motor control apparatus and motor branch-circuit conductors against _____.
(a) excessive heating due to motor overloads
(b) excessive heating due to failure to start
(c) short circuits and ground faults
(d) a and b

18. A motor _____ device that can restart a motor automatically after overload tripping shall not be installed if automatic restarting of the motor can result in injury to persons.
(a) short-circuit   (b) ground-fault
(c) overcurrent   (d) overload

19. According to Article 450, a transformer rated 600V, nominal, or less, and whose primary current rating is 9A or more, is protected against overcurrent only when _____.
(a) an individual overcurrent device on the primary side is set at not more than 125 percent of the rated primary current of the transformer.
(b) a secondary overcurrent device is set at not more than 125 percent of the rated secondary current of the transformer, and a primary overcurrent device is set at not more than 250 percent of the rated primary current of the transformer.
(c) a or b
(d) none of these

20. All areas designated as hazardous shall be properly _____ and the documentation shall be available to those authorized to design, install, inspect, maintain or operate electrical equipment at these locations.
(a) cleaned   (b) documented
(c) maintained   (d) all of these

21. All transformers and transformer vaults shall be readily accessible to qualified personnel for inspection and maintenance, or _____
(a) dry-type transformers 600V, nominal, or less, located in the open on walls, columns or structures, shall not be required to be readily accessible.
(b) dry-type transformers rated not more than 50 kVA and not over 600V can be installed in hollow spaces of buildings.
(c) a or b
(d) none of these

22. An overload device used to protect continuous-duty motors (rated more than 1-hp) shall be selected to trip or rated at no more than _____ percent of the motor nameplate full-load current rating for motors with a marked service factor not less than 1.15.
(a) 110   (b) 115   (c) 120   (d) 125

23. Branch-circuit conductors supplying a single continuous duty motor shall have an ampacity not less than _____ rating.
    (a) 125 percent of the motor's nameplate current rating
    (b) 125 percent of the motor's full-load current as listed in the *NEC*
    (c) 125 percent of the motor's full locked-rotor
    (d) 80 percent of the motor's full-load current

24. Class III, Division _____ location(s) include areas where easily ignitable fibers or materials producing combustible flyings are handled, manufactured or used.
    (a) 1   (b) 2   (c) 3   (d) all of these

25. Disconnecting means must be located within sight from and readily accessible from the air-conditioning or refrigerating equipment. The disconnecting means can be installed _____ the air-conditioning or refrigerating equipment but not on panels that are designed to allow access to the air-conditioning or refrigeration equipment.
    (a) on   (b) within   (c) a or b   (d) none of these

26. Electrical equipment installed in hazardous (classified) locations must be constructed for the class, division and group. A Group C atmosphere contains _____.
    (a) hydrogen   (b) ethylene   (c) propylene oxide (d) all of these

27. Equipment installed in hazardous locations must be approved and shall be marked to show the _____.
    (a) class
    (b) group
    (c) operating temperature, or temperature class reference to 40°C ambient
    (d) all of these

28. Equipment is required to be identified not only for the class of location but also for the explosive, combustible or ignitible properties of the specific _____ that will be present.
    (a) gas or vapor   (b) dust
    (c) fiber or flyings   (d) all of these

29. ITC-HL cables, listed for use in Class I, Division 1 locations, with a gas/vaportight, continuous-corrugated, metallic sheath, an overall jacket of suitable polymeric material and provided with termination fittings listed for the application can be installed in Class I, Division 1 _____ establishments with restricted public access.
    (a) commercial   (b) industrial   (c) institutional   (d) all of these

30. Meters, instruments and relays including kilowatt-hour meters, instrument transformers, resistors, rectifiers and thermionic tubes that are installed in Class I, Division 1 locations must be installed in explosionproof enclosures or purged and pressurized enclosures.
    (a) True   (b) False

31. Motor control circuit conductors that extend beyond the motor control equipment enclosure shall be required to have short-circuit and ground-fault protection sized not greater than _____ percent of the conductor ampacity as listed in Table 430.72(B) for 60°C conductors.
    (a) 100   (b) 150   (c) 300   (d) 500

32. Overcurrent protection for motor control circuits shall not exceed 400 percent if the conductor does not extend beyond the motor control equipment enclosure.
    (a) True   (b) False

33. Suitability of equipment for a specific purpose, environment or application may be determined by:
    (a) Equipment listing or labeling.
    (b) Evidence of equipment evaluation from a qualified testing laboratory or inspection agency concerned with product evaluation.
    (c) Other evidence acceptable to the authority having jurisdiction, such as a manufacturer's self evaluation or an owner's engineering judgment.
    (d) any of the above

34. The maximum rating or setting of motor branch-circuit short-circuit and ground-fault protective devices for a 1Ø motor is _____ percent for an inverse-time breaker.
    (a) 125   (b) 175   (c) 250   (d) 300

35. The motor disconnecting means is not required to be in sight from the motor and the driven machinery location provided _____.
    (a) the controller disconnecting means is capable of being individually locked in the open position
    (b) the provisions for locking are permanently installed on, or at the switch or circuit breaker used as the controller disconnecting means
    (c) locating the motor disconnecting within sight of the motor is impractical or introduces additional or increased hazards to people or property
    (d) all of these

36. Unless two restrictive conditions exist, a generator shall be equipped with a disconnecting means to disconnect the generator, its protective devices and all control apparatus entirely from the circuits supplied by the generator.
    (a) True   (b) False

37. What size "primary only" overcurrent protection is required for a 45 kVA transformer, rated 600V nominal, or less, that has a primary current rating of 54A?
    (a) 70   (b) 80   (c) 90   (d) 100

38. When considering lighting outlets in dwelling units, a vehicle door in a garage is considered as an outdoor entrance.
    (a) True   (b) False

39. When considering whether equipment is effectively grounded, the structural metal frame of a building shall be permitted to be used as the required equipment grounding conductor for ac equipment.
    (a) True   (b) False

40. When LFMC is used as a fixed raceway, it must be secured within _____ in. on each side of the box and shall be at intervals not exceeding _____ ft.
    (a) 12, 4$^{1}/_{2}$ (b) 18, 3   (c) 12, 3   (d) 18, 4

41. When normally enclosed live parts are exposed for inspection or servicing, the working space, if in a passageway or general open space, shall be suitably _____.
    (a) accessible   (b) guarded   (c) open   (d) enclosed

42. When sizing a feeder, the appliance loads in dwelling units can apply a demand factor of 75 percent of nameplate ratings for _____ or more appliances fastened in place on the same feeder.
    (a) two   (b) three   (c) four   (d) five

43. When the service disconnecting means consists of more than one switch or circuit breaker, the combined ratings of all the switches or circuit breakers used _____ than the rating required by 230.79.
    (a) can be more   (b) shall not be less
    (c) must be more   (d) none of these

44. Where a branch circuit supplies continuous loads or any combination of continuous and noncontinuous loads, the rating of the overcurrent device shall not be less than the noncontinuous load plus 125 percent of the continuous load.
    (a) True   (b) False

45. Where a grounded (neutral) conductor is installed and the neutral-to-case bond is not at the source of the separately derived system, the grounded (neutral) conductor shall be routed with the derived phase conductors and shall not be smaller than the required grounding electrode conductor specified in Table 250.66, but shall not be required to be larger than the largest ungrounded derived phase conductor.
    (a) True   (b) False

46. Where a lighting outlet(s) is installed for interior stairways, there must be a wall switch at each floor landing that includes an entryway where the stairway between floor levels has four risers or more.
    (a) True   (b) False

47. Where a portion of the dwelling unit basement is finished into one or more habitable rooms, each separate unfinished portion shall have a receptacle outlet installed.
    (a) True   (b) False

48. Where conductors are run in parallel in multiple raceways or cables, the equipment grounding conductor, where used, shall be run in parallel in each raceway or cable.
    (a) True   (b) False

49. Where corrosion protection is necessary and the conduit is threaded in the field, the threads must be coated with a(n) _____ electrically conductive, corrosion-resistance compound.
    (a) marked   (b) listed   (c) labeled   (d) approved

50. Exposed structural steel that is interconnected to form a steel building frame, is not intentionally grounded, and may become energized must be grounded to:
    (a) The service equipment enclosure
    (b) The grounded (neutral) conductor at the service
    (c) The grounding electrode conductor where of sufficient size
    (d) any of these

CHAPTER 3
Advanced NEC Calculations And Code Questions

Scope of Chapter 3

Unit 10

Multifamily Dwelling Unit Load Calculations

OBJECTIVES

After reading this unit, the student should be able to explain the following concepts:

Air-conditioning versus heat
Appliance (small) branch circuits
Appliance demand load
Clothes dryer demand load

Cooking equipment calculations
Dwelling unit calculations – optional method

Dwelling unit calculations – standard method
Laundry circuit
Lighting and receptacles

After reading this unit, the student should be able to explain the following terms:

General lighting
General-use receptacles
Optional method

Rounding
Standard method
Unbalanced demand load

Voltages

10–1 MULTIFAMILY DWELLING UNIT CALCULATIONS – STANDARD METHOD

When determining the ungrounded and grounded (neutral) conductor for multifamily dwelling units (Figure 10–1), apply the following steps:

Step 1: **General Lighting and Receptacles, Small-Appliance and Laundry Circuits [Table 220.11]**

The *NEC* recognizes that the general lighting and receptacles, small-appliance and laundry circuits will not all be on, or loaded, at the same time. The *NEC* permits the following demand factors to be applied to these loads [220.16]:

(a) Total Connected Load – Determine the total connected general lighting and receptacles (3 VA per sq ft), small appliance (3,000 VA) and the laundry (1,500 VA) circuit load of all dwelling units. The laundry load (1,500 VA) can be omitted if laundry facilities which are available to all building occupants are provided on the premises [210.52(F) Ex. 1].

(b) Demand Factor – Apply Table 220.11 demand factors to the total connected load (Step 1a).
First 3,000 VA at 100% demand
Next 117,000 VA at 35% demand
Remainder at 25% demand

Step 2: **Air-Conditioning versus Heat [220.15, 220.21]**

Because the air-conditioning and heating loads are not on at the same time (simultaneously), it is permissible to omit the smaller of the two loads.

(a) Air-conditioning. The air-conditioning demand load shall be calculated at 100%.

(b) Heat. Electric space-heating loads shall be computed at 100% of the total connected load [220.15].

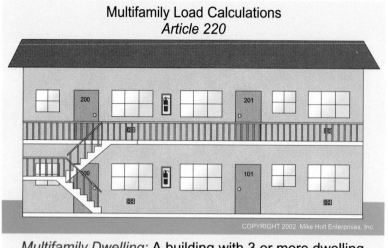

Multifamily Dwelling: A building with 3 or more dwelling units. Examples: apartment buildings, condominiums, some hotels and motels.

Figure 10-1
Multifamily Load Calculations

Step 3:　**Appliances [220.17].**

A demand factor of 75% is permitted for four or more appliances fastened in place such as a dishwasher, waste disposal, trash compactor, water heater, etc. This does not apply to space-heating equipment [220.15], clothes dryers [220.18], cooking appliances [220.19] or air-conditioning equipment.

Step 4:　**Clothes Dryers [220.18].**

The feeder or service demand load for electric clothes dryers located in dwelling units shall not be less than 5,000W, or the nameplate rating (whichever is greater) adjusted according to the demand factors listed in Table 220.18.

Note: A dryer load is not required if the dwelling unit does not contain an electric dryer. Laundry room dryers shall not have their loads calculated according to this method. This is covered in Unit 11.

Step 5:　**Cooking Equipment [220.19].**

Household cooking appliances rated over $1\frac{3}{4}$ kW can have their feeder and service loads calculated according to the demand factors of 220.19, Table and Notes.

Step 6:　**Feeder and Service Conductor Size.**
400A and less. The ungrounded conductors are sized according to Table 310.15(B)(6) for 120/240V, 1Ø systems.
Over 400A. The ungrounded conductors are sized according to Table 310.16 to the calculated unbalanced demand load.

10–2 MULTIFAMILY DWELLING UNIT CALCULATION EXAMPLES – STANDARD METHOD

Step 1. General Lighting, Small-Appliance and Laundry Demand [Table 220.11]

General Lighting Load No. 1
What is the demand load for an apartment building that contains 20 units? Each apartment is 840 sq ft.
Note: Laundry facilities are provided on the premises for all tenants [210.52(F)].
(a) 5,200 VA　　　　　(b) 40,590 VA　　　　　(c) 110,400 VA　　　　　(d) none of these
　• Answer: (b) 40,590 VA

General Lighting	(840 sq ft × 3 VA)	2,520 VA
Small-Appliance Circuits		3,000 VA
Laundry Circuit		0 VA
Total connected load for one unit		5,520 VA

Demand Factor [Table 220.11] (5,520 VA × 20 units)	110,400 VA	
First 3,000 VA at 100%	− 3,000 VA × 1.00 =	3,000 VA
Next 117,000 VA at 35%	107,400 VA × 0.35 =	37,590 VA
Total Demand Load		40,590 VA

General Lighting Load No. 2

What is the general lighting net computed demand load for a 20-unit apartment building? Each unit is 990 sq ft,

(a) 74,700 VA   (b) 149,400 VA   (c) 51,300 VA   (d) 105,600 VA

• Answer: (c) 51,300 VA

General Lighting	(990 sq ft × 3 VA)	2,970 VA
Small-Appliance Circuits		3,000 VA
Laundry Circuit		1,500 VA
Total connected load for one unit		7,470 VA

Demand Factor [Table 220.11] (7,470 VA × 20 units)	149,400 VA	
First 3,000 VA at 100%	− 3,000 VA × 1.00 =	3,000 VA
	146,400 VA	
Next 117,000 VA at 35%	−117,000 VA × 0.35 =	40,950 VA
Remainder VA at 25%	29,400 VA × 0.25 =	7,350 VA
Total Demand Load		51,300 VA

Step 2. Air-Conditioning Versus Heat [220.15]

Air-Conditioning versus Heat No. 1

What is the net computed air-conditioning versus heat load for a 40-unit multifamily building that has an A/C (3-hp, 230 volt) and two baseboard heaters (3 kW) in each unit?

(a) 160 kW   (b) 240 kW   (c) 60 kW   (d) 50 kW

• Answer: (b) 240 kW

Air-conditioning 230V × 17A = 3,910 VA

3,910 VA × 40 = 156,400 VA, Omit smaller than heat [220.21]

Heat [220.15] 3 kW × 2 units = 6 kW × 40 units = 240 kW

Air-Conditioning versus Heat No. 2

What is the air-conditioning versus heat demand load for a 25-unit multifamily building that has air-conditioning (3-hp, 230 V) and electric heat (5 kW)?

(a) 160 kVA   (b) 125 kVA   (c) 6 kVA   (d) 5 kVA

• Answer: (b) 125 kVA

Air-Conditioning 230V × 17A* = 3,910 VA

3,910 VA × 25 units = 97,750 VA, Omit

Heat [220.15] 5 kW × 25 units = 125 kW

* Table 430.148

Step 3. Appliance Demand Load [220.17]

Appliance Load No. 1

What is the appliance demand load for a 20-unit multifamily building that contains a 940 VA waste disposal, a 1,250 VA dishwasher, and a 4,500 VA water heater?

(a) 100 kVA   (b) 134 kVA   (c) 7 kVA   (d) 5 kVA

• Answer: (a) 100 kVA

Waste Disposal	940 VA
Dishwasher	1,250 VA
Water Heater	4,500 VA
Total Demand Load	6,690 VA × 20 units × 0.75 = 100,350 VA

Appliance Load No. 2

What is the appliance demand load for a 35-unit multifamily building that contains a 900 VA waste disposal, a 1,200 VA dishwasher and a 5,000 VA water heater.

(a) 71 kVA   (b) 142 kVA   (c) 107 kVA   (d) 186 kVA

• Answer: (d) 186 kVA

Waste Disposal	900 VA
Dishwasher	1,200 VA
Water Heater	5,000 VA
Total Demand Load	7,100 VA × 35 units × 0.75 = 186,375 VA/1,000 = 186 kVA

Step 4. Dryer Demand Load [220.18]

Dryer Load No. 1

A 10-unit multifamily dwelling building contains a 4.5 kW electric clothes dryer in each unit. What is the feeder and service dryer demand load for the building?

(a) 5 kW   (b) 25 kW   (c) 60 kW   (d) none of these

• Answer: (b) 25 kW

Demand Load = 5 kW × 10 units = 50 kVA × 50% = 25 kW

Dryer Load No. 2

A 20-unit multifamily dwelling building contains a 5.25 kVA electric clothes dryer in each unit. What is the feeder and service dryer demand load for the building?

(a) 5 kVA   (b) 27 kVA   (c) 60 kVA   (d) 37.8 kVA

• Answer: (d) 37.8 kVA

20 dryers × 5.25 kVA = 105 kVA × [47 - (number of dryers - 11)]

Demand = 105 kVA × [36%] = 37.8 kVA

Step 5. Cooking Equipment Demand Load [220.19]

❏ Table 220.19 Column C – not Over 12 kW

What is the feeder and service demand load for five 9 kW ranges?

(a) 9 kW   (b) 45 kW   (c) 20 kW   (d) none of these

• Answer: (c) 20 kW

❏ Table 220.19 Note 1 – Over 12 kW

What is the feeder and service demand load for three ranges rated 15.6 kW each?

(a) 15 kW   (b) 14 kW   (c) 17 kW   (d) 21 kW

• Answer: (c) 17 kW (closest answer)

Step 1:   "Column C" demand load: 14 kW (3 units).

Step 2:   The average range (15.6 kW) exceeds 12 kW by 3.6 kW. Increase "Column C" demand load (14 kW) by 20%: 14 kW × 1.2 = 16.8 kW.

❏ Table 220.19 Note 2 – Unequal Ratings Over 12 kW

What is the feeder and service demand load for three ranges rated 9 kW and three ranges rated 14 kW?

(a) 36 kW   (b) 42 kW   (c) 78 kW   (d) 22 kW

• Answer: (d) 22 kW

Step 1: Determine the total connected load.

9 kW (minimum 12 kW)	3 ranges × 12 kW =	36 kW
14 kW	3 ranges × 14 kW =	42 kW
Total connected load		78 kW

Step 2: Determine the average range rating, 78 kW/6 units = 13 kW average rating.

Step 3: Demand load Table 220.19 "Column C": 6 ranges = 21 kW.

Step 4: The average range (13 kW) exceeds 12 kW by 1 kW. Increase "Column C" demand load (21 kW) by 5%: 21 kW × 1.05 = 22.05 kW.

❏ **Table 220.19 Note 3 – Less than 3$^1/_2$ kW – Column A**
What is the feeder and service demand load for ten 3 kW ovens?

(a) 10 kW   (b) 30 kW   (c) 15 kW   (d) 20 kW

• Answer: (c) 15 kW closest answer

3 kW × 10 units = 30 kW × 0.49 = 14.70 kW

❏ **Table 220.19 Note 3 – Less than 8$^3/_4$ kW – Column B**
What is the feeder and service demand load for eight 6 kW cooktops?

(a) 10 kW   (b) 17 kW   (c) 14.7 kW   (d) 48 kW

• Answer: (b) 17 kW

6 kW × 8 units = 48 kW × 0.36 = 17.28 kW

Step 6. Service Conductor Size [Table 310.15(B)(6)]

Service Conductor Size No. 1
What size aluminum service conductors are required for a 120/240V, 1Ø multifamily building that has a total demand load of 93 kVA? Figure 10–2.

(a) 300 kcmil   (b) 350 kcmil   (c) 500 kcmil   (d) 600 kcmil

• Answer: (d) 600 kcmil aluminum

I = VA/E

I = 93,000 VA/240V = 388A,

600 kcmil aluminum, rated 400A

Service Conductor Size No. 2
What size service conductors are required for a multifamily building that has a total demand load of 270 kVA for a 120/208V, 3Ø system?

Note: Service conductors are run in parallel.

(a) 2 – 300 kcmil per phase   (b) 2 – 350 kcmil per phase
(c) 2 – 500 kcmil per phase   (d) 2 – 600 kcmil per phase

• Answer: (c) 2 – 500 kcmil per phase

I = VA/(E × 1.732)

I = 270,000 VA/(208V × 1.732) = 750A

Amperes per parallel set:

750A/2 raceways = 375A per phase.

500 kcmil conductor has an ampacity of 380A [Table 310.16 at 75ºC].

Two sets of 500 kcmil conductors (380A × 2) can be protected by an 800A protection device [240.4(B)].

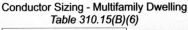

Conductor Sizing - Multifamily Dwelling
Table 310.15(B)(6)

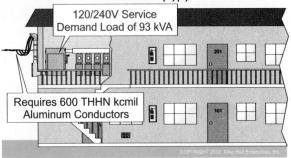

Determine the aluminum service conductor size.

Step 1: Convert VA into amperes, I = VA/E
VA = 93 kVA x 1,000 = 93,000 VA
E = 240V single phase (given)

$$I = \frac{VA}{E} = \frac{93,000\ VA}{240V} = 388A$$

Step 2: Table 310.15(B)(6) = 600 kcmil aluminum
Note: Table 310.15(B)(6) can be used on dwelling units for services up to 400A 120/240V single phase.

Figure 10-2
Conductor Sizing – Multifamily Dwelling

10–3 MULTIFAMILY DWELLING UNIT CALCULATIONS SAMPLE – STANDARD METHOD

What is the demand load for a 25 unit apartment building? Each apartment is 1,000 sq ft and contains the following: air-conditioning (28A, 240 V), heat (7.5 kW), dishwasher (1.2 kVA), waste disposal (1.5 kVA), water heater (4.5 kW), dryer (4.5 kW), and range (15.5 kW). System voltage is 120/240V, 1Ø.

Step 1. General Lighting, Small Appliance, and Laundry Demand [Table 220.11]

Each unit contains:

General Lighting	(1,000 sq ft × 3 VA)	3,000 VA
Small Appliance Circuits		3,000 VA
Laundry Circuit		1,500 VA
		7,500 VA

Total demand load (7,500 VA × 25 units)		187,500 VA
First 3,000 VA at 100% =	– 3,000 VA × 1.00 =	3,000 VA
	184,500 VA	
Next 117,000 VA at 35%	– 117,000 VA × 0.35 =	40,950 VA
Remainder at 25%	67,500 VA × 0.25 =	16,875 VA
		60,825 VA

Step 2. Air-Conditioning Versus Heat [220.15]

Air-conditioning 230V × 28A = 6,440 VA

6,440 VA × 25 units = 161,000 VA, omit [220.21]

Heat [220.15] 7,500 VA × 25 units = 187,500 VA

Step 3. Appliance Demand Load [220.17]

Waste Disposal	1,200 VA
Dishwasher	1,500 VA
Water Heater	4,500 VA
Total Connected Load	7,200 VA × 25 units × 0.75 = 135,000 VA

Step 4. Dryer Demand Load [220.18]

Total Connected Load = 5 kW × 25 units = 125 kW

Total Demand Load = 125 kW × Demand Factor

Demand Factor = [35% – [0.5 – (number of dryers – 23)]

Demand Factor = 35% – [0.5 – (25 – 23)]

Demand Factor = 35% – [0.5 – (2)]

Demand Factor = 35% – [1]

Demand Load = 125 kW × 34% = 42,500 W

Step 5. Cooking Equipment Demand Load [220.19]

Step 1: Column "A" demand load for 25 units = 40 kW.

Step 2: The average range (15.5 kW) exceeds 12 kW by 3.5 kW. Increase Column "A" demand load (40 kW) by 20%. Range demand load: 40 kW × 1.2 = 48 kW

Step 6. Service Conductor Size

Step 1:	Total general lighting, small-appliance, laundry demand load	60,825 VA
Step 2:	Total heat demand load [220.15] (7,500 VA × 25 units)	187,500 VA
Step 3:	Total appliance demand load	135,000 VA
Step 4:	Total demand dryer load	42,500 VA
Step 5:	Range demand load	48,000 VA
	Total Demand Load	473,825 VA

Service conductor amperes = VA/E = 473,825 VA/240V = 1,974A

10–4 MULTIFAMILY DWELLING UNIT CALCULATIONS [220.32] – OPTIONAL METHOD

Instead of sizing the ungrounded conductors according to the standard method (Part II of Article 220), the optional method can be used for feeders and service conductors in multifamily dwelling units. Follow these steps for determining the demand load:

Step 1:   **Total Connected Load.**

Add up the following loads:

(a) General Lighting. 3 VA per sq ft.

(b) Small-Appliance and Laundry Branch Circuit. 1,500 VA for each 20A small-appliance and laundry branch circuit.

(c) Appliances. The *nameplate* VA rating of all appliances and motors fastened in place (permanently connected), but not the air-conditioning or heating load. Be sure to use the range and dryer nameplate ratings!

(d) Air-Conditioning versus Heat. Determine which load is larger, air-conditioning or heat.

   (1) Air-conditioning or heat-pump compressors at 100%.

   (2) Heat at 100%.

Step 2:   **Demand Load.**

The net computed demand load is determined by applying the demand factor from Table 220.32 to the total connected load (Step 1). The net computed demand load (kVA) can be converted to amperes by:

$$I\,(1\varnothing) = \frac{VA}{E} \qquad\qquad I\,(3\varnothing) = \frac{VA}{(E \times 1.732)}$$

Step 3:   **Feeder and Service Conductor Size.**

400A and less. The ungrounded conductors are sized according to Table 310.15(B)(6) for 120/240V, 1Ø systems up to 400A.

Over 400A. The ungrounded conductors are sized according to Table 310.16 based on the calculated demand load.

When do you use the standard method verses the optional method? For the purpose of exam preparation, always use the standard load calculation unless the question specifies the optional method. In the field, you will probably want to use the optional method because it results in a smaller service.

10–5 MULTIFAMILY – OPTIONAL METHOD EXAMPLE [220.32]

A 120/240V, 1Ø system supplies a 12-unit multifamily building. Each 1,500 sq ft unit contains:

Dryer (4.5 kVA)	Washing Machine (1.2 kVA)
Range (14.4 kW)	Dishwasher (1.5 kVA)
Water Heater (4 kW)	Heat (5 kW)
A/C (3-hp with 1/8-hp compressor fan)	

General Lighting Load

Using the optional calculations, what is the demand load for the building general lighting and general-use receptacles, small appliance and laundry circuits?

(a) 44 kVA   (b) 90 kVA   (c) 108 kVA   (d) 60 kVA

• Answer: (a) 44 kVA
(a) General Lighting (1,500 sq ft × 3 VA)	4,500 VA
(b) Small Appliance Circuits	3,000 VA
(c) Laundry Circuit	1,500 VA
	9,000 VA × 12 units × 0.41 = 44,280 VA

Note: The washing machine is calculated as part of the 1,500 VA laundry circuit.

Air-Conditioning Versus Heat Load

Using the optional calculations, what is the demand load for the building air-conditioning versus heat?

(a) 52 kVA   (b) 60 kVA   (c) 30 kVA   (d) 25 kVA

• Answer: (d) 25 kVA [220.32]

A/C (3-hp) 230V × 17A [Tbl 430.148]	3,910 VA
Compressor ($^1/_8$-hp) 230V × 1.45A* [Table 430.148]	334 VA
*One-half the value of a $^1/_4$-hp motor	4,244 VA × 12 units × 0.41 = 20,880 VA, omit

Heat [220.15]: 5,000W × 12 units = 60,000 VA × 0.41 = 24,600W.

Appliance Demand Load

Using the optional calculations, what is the demand load for the appliances?

(a) 41 kVA   (b) 53 kVA   (c) 33 kVA   (d) 27 kVA

• Answer: (d) 27 kVA [220.32]

Water Heater	4,000 VA
Dishwasher	1,500 VA
	5,500 VA × 12 units = 66,000 VA × 0.41 = 27,060 VA

Note: The washing machine is calculated as part of the laundry circuit.

Dryer Demand Load

Using the optional calculations, what is the demand load for the 4.5 kVA dryer?

(a) 60 kW   (b) 25 kW   (c) 55 kW   (d) 22 kW

• Answer: (d) 22 kW

4.5 kW* × 12 units × 0.41 = 22.14 kW

* Nameplate rating.

Range Demand Load

Using the optional calculations, what is the demand load for the 14.4 kW ranges?

(a) 71 kW   (b) 42 kW
(c) 78 kW   (d) 67 kW

• Answer: (a) 71 kW

14.4 kW × 12 units × 0.41 = 70.85 kW

Service Conductor Size

If the total demand load equals 189 kVA, what is the service conductor size? Service is 120/240V, 1Ø. Figure 10–3.

(a) 600A   (b) 800A
(c) 1,000A   (d) 1,200A

• Answer: (b) 800A

I = VA/E
I = 189,000 VA/240V
I = 788A

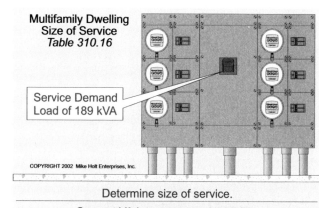

Multifamily Dwelling
Size of Service
Table 310.16

Service Demand
Load of 189 kVA

COPYRIGHT 2002 Mike Holt Enterprises, Inc.

Determine size of service.

Convert VA into amperes, I = VA/E
VA = 189 kVA x 1,000 = 189,000 VA
E = 240V single phase (given)

$$I = \frac{VA}{E} = \frac{189{,}000 \text{ VA}}{240V} = 788A$$

Table 310.15(B)(6) does not apply to services over 400A.

Figure 10-3
Multifamily Dwelling – Size of Service

Unit 10 – Multifamily Dwelling Unit Load Calculations Summary Questions

Calculations Questions

10–1 Multifamily Dwelling Unit Calculations – Standard Method

1. • Each unit of a 20-unit apartment building is 840 sq ft. What is the general lighting feeder demand load for the building? Note: Laundry facilities are provided on the premises for all tenants.
    (a) 5,200 VA   (b) 40,590 VA   (c) 110,400 VA   (d) none of these

2. Each unit of a 20-unit apartment building is 990 sq ft. What is the general lighting net computed demand load for the building?
    (a) 74,700 VA   (b) 149,400 VA   (c) 51,300 VA   (d) 105,600 VA

3. A 40-unit multifamily building has an air conditioner (3-hp 230V) and two baseboard heaters (3 kW) in each unit. What is the net computed demand load for the air-conditioning and heat?
    (a) 160 kW   (b) 240 kW   (c) 60 kW   (d) 50 kW

4. A 25-unit multifamily building has air-conditioning (3-hp 230 V) and electric heat (5 kW). What is the air-conditioning and heat net computed demand load?
    (a) 160 kW   (b) 125 kW   (c) 6 kW   (d) 5 kW

5. In a 16-unit multifamily building, each unit contains a waste disposal (940 VA), a dishwasher (1,250 VA) and a water heater (4,500 VA). What is the service demand load for these appliances?
    (a) 100 kVA   (b) 134 kVA   (c) 80 kVA   (d) 5 kVA

6. Each unit in a 28-unit apartment building contains a waste disposal (900 VA), dishwasher (1,200 VA) and a water heater (5,000 VA). What is the feeder and service demand load for the appliances?
    (a) 149 kVA   (b) 142 kVA   (c) 107 kVA   (d) 186 kVA

7. A multifamily dwelling (40 unit) contains a 4.5 kW electric clothes dryer in each unit. What is the feeder and service demand load for all the dryers?
    (a) 53 kW   (b) 27 kW   (c) 60 kW   (d) None of these

8. What is the demand load for ten 5.25 kW dryers installed in dwelling units of a multifamily building?
    (a) 26 kW   (b) 37 kW   (c) 60 kW   (d) 37 kW

9. What is the demand load for twelve 3.25 kW ovens?
    (a) 10 kW   (b) 18 kW   (c) 15 kW   (d) 20 kW

10. What is the demand load for eight 7 kW cooktops?
    (a) 20 kW   (b) 17 kW   (c) 14.7 kW   (d) 48 kW

11. • What is the demand load for five 12.4 kW ranges?
    (a) 9 kW   (b) 45 kW   (c) 20 kW   (d) None of these

12. What is the demand load for three ranges rated 15.5 kW?
    (a) 15 kW   (b) 14 kW   (c) 17 kW   (d) 21 kW

13. What is the feeder and service demand load for three ranges rated 11 kW and three ranges rated 14 kW?
    (a) 36 kW   (b) 42 kW   (c) 78 kW   (d) 22 kW

14. • What size aluminum service conductors are required for a 120/240V, 1Ø multifamily building that has a total demand load of 90 kW?

    (a) 500 kcmil   (b) 600 kcmil   (c) 700 kcmil   (d) 800 kcmil

15. What size copper service conductors are required for a multifamily building that has a total demand load of 260 kW for a 120/208V wye, 3Ø system?

    (a) 2 - 300 kcmil   (b) 2 - 350 kcmil   (c) 2 - 500 kcmil   (d) 2 - 600 kcmil

---

The following information applies to the next five questions.

A multifamily building has 12 units. Each is 1,500 sq ft and contains the following:
System voltage 120/240V, 1Ø.

Dryer (4.5 kW)   Washing machine (1.2 kVA)
Range (14.45 kW)   Dishwasher (1.5 kVA)
Water heater (4 kW)   Heat (5 kW)
A/C (3-hp with $1/8$-hp compressor fan)

16. • What is the building's net computed load for the general lighting, small appliance and laundry in VA?

    (a) 40 kVA   (b) 108 kVA   (c) 105 kVA   (d) 90 kVA

17. What is the building's demand load in kVA for an air conditioner (3-hp with a $1/8$-hp compressor fan) versus electric heat (5 kW)?

    (a) 52 kW   (b) 60 kW   (c) 30 kW   (d) 105 kW

18. What is the building demand load for the appliances in VA?

    (a) 40 kVA   (b) 55 kVA   (c) 43 kVA   (d) 50 kVA

19. What is the building demand load for the 4.5 kW dryers?

    (a) 60.4 kW   (b) 27.6 kW   (c) 55.3 kW   (d) 45.7 kW

20. What is the building demand load for the 14.45 kW range?

    (a) 27 kW   (b) 35 kW   (c) 168 kW   (d) 30 kW

21. If the total demand load of a multifamily dwelling unit is 206 kVA, what is the service and feeder conductor size? The system voltage is 120/240, 1Ø.

    (a) 600A   (b) 800A   (c) 1,000A   (d) 1200A

22. • If a service consists of parallel 400 kcmil conductors, routed in separate raceways, what size bonding jumper would be required for each service raceway?

    (a) 1 AWG   (b) 1/0 AWG   (c) 2/0 AWG   (d) 3/0 AWG

23. • If 400 kcmil copper service conductors are in parallel in three raceways, what is the minimum size grounding electrode conductor required?

    (a) 1/0 AWG   (b) 2/0 AWG   (c) 3/0 AWG   (d) 4/0 AWG

10–3 Multifamily Dwelling Unit Calculations [220.32] – Optional Method

24. • When determining the service (using optional calculations) for a multifamily dwelling, the total connected load shall have the demand factors of Table 220.32 applied. When determining the total connected load, _____ shall be used.

    (a) 125% of the air-conditioning load plus 100% of the space-heating load
    (b) the larger of the air-conditioning load or the space-heating load
    (c) the sum of the air-conditioning load and the space-heating load
    (d) 125% of the sum of the air-conditioning load and the space-heating load

25. A multifamily building has 60 units. Each unit is 1,500 sq ft. Using the optional dwelling unit calculations, what is the net computed load for the building general lighting and general-use receptacles, small-appliance and laundry circuits?

    (a) 145 kVA   (b) 190 kVA   (c) 108 kVA   (d) 130 kVA

26. A 60 unit multifamily dwelling has an air conditioner (3-hp with a $^1/_8$-hp blower) and electric heat (5 kW) in each unit. Using the optional method, what is the air-conditioning versus heat demand load?
    (a) 50 kVA   (b) 61 kVA   (c) 30 kVA   (d) 72 kVA

27. Using the optional method for dwelling unit calculations, what is a 60 unit multifamily building net demand load if each unit has a water heater (4 kW) and a dishwasher (1.5 kW)?
    (a) 80 kW   (b) 50 kW   (c) 30 kW   (d) 60 kW

28. Each unit of a 60 unit apartment building has a 4 kW dryer. Using the optional method, the demand load that would be added to the service is _____.
    (a) 75 kW   (b) 240 kW   (c) 72 kW   (d) 58 kW

29. Using the optional method for dwelling unit calculations, what is the 60 unit multifamily building net computed load for a 4.5 kW dryer?
    (a) 65 kW   (b) 25 kW   (c) 55 kW   (d) 75 kW

30. Using the optional method for dwelling unit calculations, what is a 60 unit multifamily building demand load if each apartment has a 14 kW range?
    (a) 150 kW   (b) 50 kW   (c) 100 kW   (d) 200 kW

31. If the total demand load is 270 kVA, what is the service and feeder conductor size? The service is 208/120V wye, 3Ø.
    (a) 600A   (b) 800A   (c) 1,000A   (d) 1200A

32. Each unit of a 20 unit multifamily dwelling has 900 sq ft of living space and contains one of each of the following: air-conditioning (5-hp), heat (5 kW), water heater (5 kW), range (14 kW). The service for this apartment building is approximately _____ if the optional method calculation is used. (120/240 V)
    (a) 200 kVA   (b) 250 kVA   (c) 280 kVA   (d) 320 kVA

33. • If the service conductors are in parallel in two raceways (500 kcmil), what size bonding jumper is required for each service raceway?
    (a) 2 AWG   (b) 1 AWG   (c) 1/0 AWG   (d) 2/0 AWG

34. • If the service conductors are in parallel in two raceways (500 kcmil), what is the minimum size copper grounding electrode conductor required?
    (a) 1/0 AWG   (b) 2/0 AWG   (c) 3/0 AWG   (d) 4/0 AWG

☆ Challenge Questions

General Lighting and Receptacle Calculations [Table 220.11]

35. Each dwelling unit of a 20 unit multifamily building has 900 square feet of living space. What is the general lighting load, before demand factors have been applied, for the multifamily dwelling unit apartment building?
    (a) 45 kVA   (b) 60 kVA   (c) 37 kVA   (d) 54 kVA

36. A multifamily apartment building contains 20 units and each unit has 840 sq ft of living space. What is the general lighting feeder demand load for the multifamily building if laundry facilities are provided on the premises for all tenants and no laundry circuit is installed in each unit?
    (a) 35 kVA   (b) 41 kVA   (c) 45 kVA   (d) 63 kVA

Appliance Demand Factors [220.17].

37. A multifamily apartment building contains 20 units. Each unit contains a waste disposal (900 VA), a dishwasher (1,200 VA) and a water heater (5,000 VA). What is the feeder demand load for the appliances in this building?
    (a) 106,500 VA   (b) 117,100 VA   (c) 137,000 VA   (d) 60,000

Dryer Calculation [220.18]

38. The nameplate rating for each household dryer in a 10 unit apartment building is 4 kW. This would add _____ to the service size.
     (a) 20 kW　　　　　　(b) 25 kW　　　　　　(c) 40 kW　　　　　　(d) 50 kW

Ranges – Note 1 of Table 220.19

39. • The demand load for thirty 15.8 kW household ranges is _____.
     (a) 31 kW　　　　　　(b) 47 kW　　　　　　(c) 54 kW　　　　　　(d) 33 kW

Ranges – Note 2 of Table 220.19

40. What is the feeder demand load for five 10 kW, five 14 kW and five 16 kW household ranges?
     (a) 210 kW　　　　　　(b) 30 kW　　　　　　(c) 14 kW　　　　　　(d) 33 kW

41. What kW would be added to service loads for ten 12 kW, eight 14 kW and two 9 kW household ranges?
     (a) 33 kW　　　　　　(b) 35 kW　　　　　　(c) 36.75 kW　　　　　　(d) 29.35 kW

Ranges – Note 3 to Table 220.19

42. What is the minimum demand load for five 5 kW cooktops, two 4 kW ovens and four 7 kW ranges?
     (a) 15.5 kW　　　　　　(b) 8.8 kW　　　　　　(c) 19.5 kW　　　　　　(d) 18.2 kW

43. • What is the maximum dwelling unit feeder or service demand load for fifteen 8 kW cooking units?
     (a) 38.4 kW　　　　　　(b) 30 kW　　　　　　(c) 120 kW　　　　　　(d) none of these

Ranges – Note 5 of Table 220.19

44. A school has twenty 10 kW ranges installed in the home economics class. The minimum load this would add to the service is _____.
     (a) 35 kW　　　　　　(b) 44.8 kW　　　　　　(c) 56 kW　　　　　　(d) 160 kW

Neutral Calculation [220.22]

45. The service neutral demand load for household electric clothes dryers shall be calculated at _____ of the demand load as determined by 220.18.
     (a) 50%　　　　　　(b) 60%　　　　　　(c) 70%　　　　　　(d) 80%

46. The feeder neutral demand for fifteen 9 kW cooking units would be _____.
     (a) 38.4 kW　　　　　　(b) 30 kW　　　　　　(c) 120 kW　　　　　　(d) 21 kW

47. What is the feeder neutral load in amperes for ten 5 kW household dryers?
     (a) 17.5 kW　　　　　　(b) 21.5 kW　　　　　　(c) 27.5 kW　　　　　　(d) 32.5 kW

48. An 11 unit multifamily dwelling contains a 4 kW electric clothes dryer in each unit. What is the feeder or service neutral demand load (in amperes)?
     (a) 18.9 kW　　　　　　(b) 22.3 kW　　　　　　(c) 29.3 kW　　　　　　(d) 32.9 kW

49. • The feeder and service neutral demand load can be reduced 70 percent for that portion of the unbalanced load over 200A. This applies to _____ systems.
     (a) 3-wire, 1Ø, 120/240V　　　　　　　　　　(b) 3-wire, 1Ø, 120/208V
     (c) 4-wire, 3Ø, 120/208V　　　　　　　　　　(d) a and c

50. • What is the dwelling unit service neutral load (in amperes) for ten 9 kW ranges?
     (a) 12.4 kW　　　　　　(b) 13.5 kW　　　　　　(c) 15.5 kW　　　　　　(d) 17.5 kW

NEC Questions – Articles 501-600

Article 501 Class I Locations

51. A seal fitting is required for each conduit run passing from a Class I, Division 2 location into an unclassified location and shall be located no more than _____ from the boundary.
(a) 3 ft   (b) 6 ft   (c) 10 ft   (d) 20 ft

52. No seal is required if a conduit (with no unions, couplings, boxes or fittings) passes through a Class I, Division 2 location if the termination points of the unbroken conduit are in at least _____ within the unclassified location.
(a) 6 in.   (b) 12 in.   (c) 18 in.   (d) 24 in.

53. When seals are required for Class I locations, they shall comply with the following rule(s):
(a) They shall be listed for Class I locations and shall be accessible.
(b) The minimum thickness of the sealing compound shall not be less than the trade size of the sealing fitting and, in no case, less than $^5/_8$ in.
(c) Splices and taps shall not be made in the conduit seal.
(d) all of these

54. The minimum thickness of sealing compound in Class I, Division 1 and 2 locations shall not be less than the trade size of the conduit or sealing fitting and, in no case, less than _____
(a) $^1/_8$ in.   (b) $^1/_4$ in.   (c) $^3/_8$ in.   (d) $^5/_8$ in.

55. The cross-sectional area of the conductors permitted in a seal shall not exceed _____ percent of the cross-sectional area of rigid metal conduit unless the seal is specifically listed for a higher percentage of conductor fill.
(a) 25   (b) 50   (c) 100   (d) 125

56. MC-HL cable containing shielded cables and/or twisted pair cables in a Class I, Division 1 location shall not require the removal of the shielding material or the separation of the twisted pairs, provided the termination is accomplished by a(n) _____ means to minimize the entrance of gases or vapors and to prevent propagation of flame into the cable core.
(a) approved   (b) listed   (c) acceptable   (d) none of these

57. In Class I, Division 2 locations, shielded cables and twisted pair cables shall not require the removal of the shielding material or separation of the twisted pairs, provided the termination is by an approved means to minimize the entrance of _____ and prevent propagation of flame into the cable core.
(a) gases   (b) vapors   (c) dust   (d) a or b

58. Switches, circuit breakers, motor controllers and fuses including pushbuttons, relays and similar devices that are installed in Class I, Division 1 locations shall be provided with approved explosionproof enclosures and they shall be identified as a complete assembly for use in Class I locations.
(a) True   (b) False

59. A standard circuit breaker mounted in a Class I, Division 2 location with make-and-break contacts and not hermetically sealed or oil-immersed shall be installed in a Class I, Division 1 rated enclosure.
(a) True   (b) False

60. In Class I, Division 2 locations, fused or unfused disconnect and isolating switches for transformers or capacitor banks that are not intended to interrupt current in the normal performance of the function for which they are installed shall be permitted to be installed in general-purpose enclosures.
(a) True   (b) False

61. When flexible metal conduit or LFMC is used as permitted in Class I, Division 2 locations, it shall be installed with an _____ bonding jumper that is installed in parallel with each conduit in compliance with 250.102.
(a) internal   (b) external   (c) a or b   (d) a and b

Article 502 Class II Locations

62. In Class II locations, where electrically conductive dust is present, flexible connections can be made with _____.
    (a) flexible metal conduit
    (b) AC armored cable
    (c) hard-usage cord
    (d) liquidtight flexible metal conduit with listed fittings

63. In a Class II, Division 1 location where dust from magnesium, aluminum, aluminum bronze powders or other metals of similarly hazardous characteristics may be present, fuses, switches, motor controllers and circuit breakers shall have enclosures specifically approved for such locations.
    (a) True   (b) False

64. In Class II, Division 1 and 2 locations, an approved method of bonding is the use of _____.
    (a) bonding jumpers with approved fittings
    (b) double locknut types of contacts
    (c) locknut-bushing types of contacts
    (d) any of the above are approved methods of bonding

Article 503 Class III Locations

65. Luminaires in a Class III location that may be exposed to physical damage shall be protected by a _____ guard.
    (a) plastic   (b) metal   (c) suitable   (d) explosionproof

66. The power supply to contact conductors of a crane in a Class III location shall be _____.
    (a) isolated from all other systems
    (b) equipped with an acceptable ground detector
    (c) have an alarm in the case of a ground fault
    (d) all of these

Article 505 Class I, Zone 0, 1, and 2 Locations

67. A multiwire branch circuit is not permitted in a Class I, Zone 1, location unless all conductors of the circuit can be opened simultaneously.
    (a) True   (b) False

Article 511 Commercial Garages, Repair and Storage

68. Article _____ contains the requirements for the wiring of occupancy locations used for service and repair operations in connection with self-propelled vehicles (including passenger automobiles, buses, trucks, tractors, etc.) in which volatile flammable gases are used for fuel or power.
    (a) 500   (b) 501   (c) 511   (d) 514

69. The requirements in Article 511 apply to locations used for service and repair operations in connection with self-propelled vehicles such as _____, in which volatile flammable liquids or flammable gases are used for fuel or power.
    (a) buses   (b) trucks   (c) tractors   (d) all of these

70. Parking garages used for parking or storage and where no repair work is done except for exchange of parts and routine maintenance requiring no use of electrical equipment, open flame, welding or the use of volatile flammable liquids, are not classified as hazardous locations.
    (a) True   (b) False

71. The _____ of alcohol-based windshield washer fluid shall not cause the areas used for service and repair operations in connection with self-propelled vehicles to be classified as hazardous.
    (a) storage   (b) handling
    (c) dispensing into motor vehicles   (d) any of these

72. For each floor area inside a commercial garage, the entire area up to a level of _____ above the floor shall be considered to be a Class I, Division 2 location.
     (a) 6 in.   (b) 12 in.   (c) 18 in.   (d) 24 in.

73. The floor area shall not be classified if the enforcing agency determines that there is mechanical ventilation that provides a minimum of four air changes per hour or _____ cu ft per minute of exchanged air for each square foot of floor area.
     (a) 1   (b) 2   (c) 3   (d) 4

74. Any pit or depression below a garage floor level shall be considered to be a Class I, Division _____ location up to floor level.
     (a) 1   (b) 2   (c) 3   (d) not classified

75. • Any pit or depression in a commercial garage in which six air changes per hour are exhausted at the floor level of the pit shall be permitted to be judged by the enforcing agency to be a _____ location.
     (a) Class I, Division 2   (b) Class II, Division 2
     (c) Class II, Division 1   (d) Class I, Division 1

76. Areas adjacent to defined locations in commercial garages where flammable vapors are not likely to be released shall not be classified when mechanically ventilated at a rate of _____ or more air changes per hour, or when effectively cut off by walls or partitions.
     (a) two   (b) four   (c) six   (d) none of these

77. Raceways embedded in a masonry wall or buried beneath a floor shall be considered _____.
     (a) outside the hazardous area if any extensions pass through such areas
     (b) within the hazardous area if any connections or extensions lead into or through such areas
     (c) hazardous if beneath a hazardous area
     (d) outside any hazardous location

78. A permanently mounted luminaire (fixture) in a commercial garage and located over lanes on which vehicles are commonly driven shall be located not less than _____ above floor level.
     (a) 10 ft   (b) 12 ft   (c) 14 ft   (d) none of these

79. In a commercial garage, over a Class I location, equipment less than _____ above the floor level that may produce arcs, sparks or particles of hot metal, shall be of the totally enclosed type or constructed so as to prevent the escape of sparks or hot metal particles.
     (a) 6 ft   (b) 10 ft   (c) 12 ft   (d) 18 ft

Article 513 Aircraft Hangars

80. Stock rooms and similar areas adjacent to aircraft hangars, but effectively isolated and adequately ventilated, shall be designated as _____ locations.
     (a) Class I, Division 2   (b) Class II, Division 1
     (c) Class II, Division 2   (d) nonhazardous

81. Which of the following areas of an aircraft hangar are not classified as a Class I, Division 1 or 2 location?
     (a) Any pit or depression below the level of the hangar floor.
     (b) Areas adjacent to and not suitably cut off from the hangar.
     (c) Areas within the vicinity of aircraft.
     (d) Adjacent areas where adequately ventilated and where effectively cut off from the hangar.

Article 514 Motor Fuel Dispensing Facilities

82. Underground gasoline dispenser wiring shall be installed in _____.
    (a) threaded rigid metal conduit
    (b) threaded intermediate metal conduit
    (c) rigid nonmetallic conduit when buried under not less than 2 ft of cover
    (d) any of these

83. A listed seal shall be _____.
    (a) provided in each conduit run entering a dispenser
    (b) provided in each conduit run leaving a dispenser
    (c) the first fitting after the conduit emerges from the earth or concrete
    (d) all of these

84. Each circuit leading to or through dispensing equipment, including equipment for remote pumping systems, shall be provided with a switch or other acceptable means to disconnect _____ from the source of supply all conductors of the circuit, including the grounded conductor, if any.
    (a) automatically   (b) simultaneously   (c) manually   (d) individually

85. Each circuit leading to or through a dispensing pump shall be provided with a switch or other acceptable means to disconnect simultaneously from the source of supply all conductors of the circuit, including the _____ conductor, if any.
    (a) grounding   (b) grounded   (c) bonding   (d) all of these

86. Single-pole breakers utilizing approved handle ties cannot be used for the circuit disconnect for gasoline dispensing equipment.
    (a) True   (b) False

87. Each circuit leading to dispensing equipment shall be provided with a clearly identified and readily accessible switch or other acceptable means to disconnect all conductors of the circuit.
    (a) True   (b) False

88. The emergency controls for attended self-service stations shall be located no more than _____ from the gasoline dispensers.
    (a) 20 ft   (b) 50 ft   (c) 75 ft   (d) 100 ft

89. The emergency controls for unattended self-service stations shall be located not less than _____ or more than _____ from the gasoline dispensers.
    (a) 10 ft, 25 ft   (b) 20 ft, 50 ft   (c) 20 ft, 100 ft   (d) 50 ft, 100 ft

90. Each dispensing device shall be provided with a means to remove all external voltage sources, including feedback, during periods of maintenance and service of the dispensing equipment. The disconnecting means shall be either inside or adjacent to the dispensing device.
    (a) True   (b) False

Article 515 Bulk Storage Plants

91. An aboveground tank in a bulk storage plant is a Class I, Division 1 location within _____ from the open end of a vent, extending in all directions.
    (a) 12 ft   (b) 10 ft   (c) 6 ft   (d) 5 ft

92. Aboveground bulk (dust) storage tanks shall be classified as _____ for the space between 5 ft and 10 ft from the open end of a vent, extending in all directions.
    (a) Class I, Division 1   (b) Class I, Division 2
    (c) Class II, Division 1   (d) Class II, Division 2

Article 516 Spray Application, Dipping, and Coating Processes

93. The authority having jurisdiction may judge a location utilized for _____ as a nonhazardous location providing the conditions of the *Code* are met.
    (a) drying or curing   (b) dipping and coating   (c) spraying operations   (d) none of these

94. Locations where flammable paints are dried, with the ventilating equipment interlocked with the electrical equipment, may be designated as a(n) _____ location.
    (a) Class I, Division 2   (b) Unclassified
    (c) Class II, Division 2   (d) Class II, Division 1

Article 517 Health Care Facilities

95. The requirements of Article 517 (Health Care Facility) apply to buildings or portions of buildings in which medical, _____ or surgical care is provided.
    (a) psychiatric   (b) nursing   (c) obstetrical   (d) any of these

96. The patient care area is any portion of a health care facility, including the business offices, corridors, lounges, day rooms, dining rooms or similar areas.
    (a) True   (b) False

97. The patient care area is any portion of a health care facility where patients are intended to be _____.
    (a) examined   (b) treated   (c) registered   (d) a or b

98. All branch circuits serving patient care areas shall be installed in a metal raceway or cable that is listed in 250.118 as an acceptable grounding return path, such as EMT or AC cable.
    (a) True   (b) False

99. The outer metal sheath of interlocked MC cable is not listed as an acceptable grounding return path; however, if it contains an insulated 12 AWG or larger equipment grounding conductor, it can be used to supply branch circuits in patient care areas of health care facilities.
    (a) True   (b) False

100. In areas used for patient care, the grounding terminals of all receptacles and all non-current-carrying conductive surfaces of fixed electric equipment _____ shall be grounded by an insulated copper equipment grounding conductor.
    (a) operating at over 100V   (b) likely to become energized
    (c) subject to personal contact   (d) all of these

101. In health care facilities, patient bed receptacles located in general care areas shall be supplied by at least two branch circuits. The branch circuits for these receptacles can originate from two separate transfer switches on the emergency system.
    (a) True   (b) False

102. Each general care area patient bed location shall be provided with _____ receptacle(s).
    (a) 1 single or 1 duplex   (b) 6 single or 3 duplex
    (c) 2 single or 1 duplex   (d) 4 single or 2 duplex

103. Receptacles located within the patient care areas of pediatric wards, rooms or areas shall be listed _____.
    (a) tamper resistant   (b) isolated
    (c) GFCI-protected   (d) specification grade

104. In critical care areas of Health Care Centers, each patient bed location shall be provided with a minimum of _____ receptacles.
    (a) 10   (b) 6   (c) 3   (d) 4

105. A(n) _____ system must supply major electric equipment necessary for patient care and basic hospital operation.
    (a) emergency   (b) equipment   (c) life safety   (d) none of these

106. The cover plates for receptacles or receptacles supplied from the emergency system for essential electrical systems in hospitals must have a distinctive color or marking so as to be readily identifiable.
      (a) True   (b) False

107. Which one of the following functions shall not be connected to the life safety branch circuits in a hospital?
      (a) Exit signs.   (b) Elevators.
      (c) Administrative office lighting.   (d) Communications systems.

108. The essential electrical systems in a health care facility shall have sources of power from _____.
      (a) a normal source generally supplying the entire electrical system
      (b) one or more alternate sources for use when the normal source is interrupted
      (c) a or b
      (d) a and b

109. The receptacles or the cover plates for the receptacles supplied from the emergency system for essential electrical systems in nursing homes must have a distinctive color or marking so as to be readily identifiable.
      (a) True   (b) False

110. In a location where flammable anesthetics are employed, the entire area shall be classified as Class I, Division 1 that shall extend _____.
      (a) upward to the structural ceiling   (b) upward to a level 8 ft above the floor
      (c) upward to a level 5 ft above the floor (d) 10 ft in all directions

111. In a health care facility, receptacles and attachment plugs in a hazardous (classified) location within an anesthetizing area shall be listed for use in Class I, Group _____ locations.
      (a) A   (b) B   (c) C   (d) D

112. In anesthetizing locations, low-voltage equipment that is frequently in contact with the bodies of persons or has exposed current-carrying elements shall _____.
      (a) operate on an electrical potential of 10V or less   (b) be moisture resistant
      (c) be intrinsically safe or double-insulated   (d) all of these

113. The maximum internal current that can flow through the line isolation monitor when any point of the isolated system is grounded shall be _____ when used in a health care facility.
      (a) 15A or less   (b) no more than 1A   (c) 1 mA   (d) 10 mA

Article 518 Places of Assembly

114. A place of assembly is a building, portion of a building, or structure designed or intended for the assembly of _____ or more persons.
      (a) 50   (b) 100   (c) 150   (d) 200

115. Examples of places of assembly could include, but are not limited to _____.
      (a) restaurants   (b) conference rooms   (c) pool rooms   (d) all of these

116. Which of the following wiring methods can be installed in a place of assembly?
      (a) Metal raceways.
      (b) MC cable.
      (c) AC cable containing an insulated equipment grounding conductor.
      (d) all of these

117. In places of assembly, nonmetallic raceways encased in not less than _____ of concrete shall be permitted.
      (a) 1 in.   (b) 2 in.   (c) 3 in.   (d) none of these

Article 520 Theaters, Audience Areas of Motion Picture and Television Studios, Performance Areas, and Similar Locations

118. • Article 520 applies to the performance area, which includes the stage and audience seating area associated with a _____ stage structure, whether indoors or outdoors, which is used for the presentation of theatrical or musical productions or public presentations.
    (a) temporary   (b) permanent   (c) a or b   (d) a and b

119. The wiring methods permitted in theaters, audience areas of motion picture and television studios, performance areas and similar locations are _____.
    (a) any metal raceway
    (b) nonmetallic raceways encased in 2 in. of concrete
    (c) MC or AC cable with an insulated equipment grounding conductor
    (d) any of these

120. The wiring methods in theaters, audience areas of motion picture and television studios, performance areas and similar locations for control, signal and communications can be _____.
    (a) Communications circuits - CM cable listed for the location
    (b) Class 2 remote-control and signaling circuits - CL2 cable listed for the location
    (c) optical fiber raceways listed for the location
    (d) a or b, but not c

121. All theater fixed stage switchboards that are not completely enclosed, dead-front and dead-rear or recessed into a wall, shall be provided with a metal hood extending the full length of the board to protect all equipment from falling objects.
    (a) True   (b) False

122. On fixed stage equipment, portable or strip luminaires and connector strips shall be wired with conductors having insulation rated suitable for the temperature but not less than _____.
    (a) 75°C   (b) 90°C   (c) 125°C   (d) 200°C

123. The pilot light provided within a portable stage switchboard enclosure shall have overcurrent protection rated or set at not more than _____.
    (a) 10A   (b) 15A   (c) 20A   (d) 30A

124. Flexible conductors, including cable extensions, used to supply portable stage equipment shall be _____ cords or cables.
    (a) listed   (b) extra-hard usage   (c) hard usage   (d) a and b

125. All lights and receptacles installed in theater dressing rooms adjacent to the mirrors and above the dressing table counter(s) shall be controlled by wall switches in the dressing rooms.
    (a) True   (b) False

Article 525 Carnivals, Circuses, Fairs and Similar Events

126. Electrical wiring in and around water attractions such as bumper boats for carnivals, circuses and fairs must comply with the requirements of Article 680 - Pools and Fountains.
    (a) True   (b) False

127. Carnival and circus overhead wiring outside of tents and concession areas where the voltage-to-ground does not exceed 150 shall maintain a vertical clearance of _____ above platforms, projections or surfaces from which they might be reached.

    (a) 3 ft   (b) 6 ft   (c) 8 ft   (d) 10 ft

128. When installed indoors for carnivals, circuses and fairs, flexible cords and cables shall be listed for wet locations and shall be sunlight resistant.

    (a) True   (b) False

129. Cord connectors for carnivals, circuses and fairs can be laid on the ground when the connectors are _____ for a wet location but they must not be placed in audience traffic paths or within areas accessible to the public unless guarded.

    (a) listed   (b) labeled   (c) approved   (d) all of these

130. Wiring for an amusement ride, attraction, tent or similar structure shall not be supported by any other ride or structure unless specifically designed for the purpose.

    (a) True   (b) False

131. Wiring for temporary lighting located inside tents and concession areas at carnivals, circuses and fairs shall be securely installed, and where subject to physical damage, shall be provided with mechanical protection.

    (a) True   (b) False

132. The continuity of the grounding conductor system used to reduce electrical shock hazards at receptacles for cord-and-plug-connected equipment shall be verified each time the portable electrical equipment is connected.

    (a) True   (b) False

Article 527 Temporary Installations

133. There is no time limit for temporary electrical power and lighting except that it shall be removed upon completion for which of the following?

    (a) construction or remodeling   (b) maintenance or repair
    (c) demolition of buildings   (d) all of these

134. Temporary electrical power and lighting installations are permitted for a period not to exceed 90 days for _____ decorative lighting and similar purposes.

    (a) Christmas   (b) New Year's   (c) July 4th   (d) all of these

135. Temporary electrical power and lighting is permitted for emergencies and _____.

    (a) tests   (b) experiments   (c) developmental work   (d) all of these

136. Temporary wiring shall be _____ immediately upon the completion of construction or purpose for which it was installed.

    (a) disconnected   (b) removed   (c) de-energized   (d) any of these

137. NM and NMC cables can be used for temporary wiring in structures of a height of _____.

    (a) 18 ft   (b) 6 ft   (c) 12 ft   (d) any of these

138. It is permitted to use individual open conductors for a period not to exceed 90 days for holiday decorative lighting and similar purposes when the circuit voltage-to-ground does not exceed _____.

    (a) 50V   (b) 125V   (c) 150V   (d) 277V

139. All receptacles for temporary branch circuits are required to be electrically connected to the _____ conductor.

    (a) grounded   (b) grounding   (c) equipment grounding   (d) grounding electrode

140. Receptacles for construction sites shall not be installed on the _____ or connected to _____ that supply temporary lighting.

    (a) same branch circuit, feeders
    (b) feeders, multiwire branch circuits
    (c) same branch circuit, multiwire branch circuits
    (d) all of these

141. Multiwire branch circuits for temporary wiring shall be provided with a means to disconnect simultaneously all _____ conductors at the power outlet or panelboard where the branch circuit originated.
(a) underground   (b) overhead   (c) ungrounded   (d) grounded

142. At construction sites, boxes are not required for temporary wiring splices of _____.
(a) multiconductor cords   (b) multiconductor cables
(c) a an b   (d) none of these

143. Flexible cords and cables used for temporary wiring shall be protected _____.
(a) from accidental damage   (b) where passing through doorways
(c) from sharp corners and projections   (d) all of these

144. For temporary installations, cable assemblies, as well as flexible cords and cables, shall be supported at intervals that ensure protection from physical damage. Support shall be in the form of _____ or similar type fittings designed to not damage the cable or cord assembly.
(a) staples   (b) cable ties   (c) straps   (d) all of these

145. Vegetation cannot be used for support of overhead spans of _____.
(a) branch circuits   (b) feeders   (c) service conductors   (d) a or b

146. All _____, 125V receptacle outlets that are not a part of the permanent wiring of the building or structure and are in use by personnel for temporary power shall have ground-fault circuit-interrupter protection for personnel.
(a) 15A   (b) 20A   (c) 30A   (d) all of these

147. All 125V, 1Ø, 15 , 20 and 30A receptacle outlets used by personnel for temporary power shall have ground-fault circuit-interrupter protection for personnel. GFCI protection can be provided by a _____ or other devices such as GFCI adapters incorporating listed GFCI protection.
(a) circuit breaker   (b) receptacle   (c) cord set   (d) all of these

148. Ground-fault protection for personnel is required for receptacles rated more than 30A or other than 125V. The ground-fault protection shall be supplied by a(n) _____.
(a) GFCI protection device
(b) Assured Equipment Grounding Conductor Program
(c) AFCI protection device
(d) a or b

149. For temporary wiring over 600V, nominal, suitable _____ shall be provided to prevent access of other than authorized and qualified personnel.
(a) fencing   (b) barriers   (c) signs   (d) a or b

Article 530 Motion Picture and Television Studios and Similar Locations

150. Each receptacle of dc plugging boxes shall be rated at not _____ when used on a stage or set of a motion picture studio.
(a) more than 30A   (b) less than 20A   (c) less than 30A   (d) more than 20A

151. A professional-type projector uses _____film.
(a) 35 mm   (b) 70 mm   (c) a or b   (d) none of these

152. A switch for the control of parking lights in a theater may be installed inside the projection booth.
(a) True   (b) False

Article 540 Motion Picture Projection Rooms

153. Conductors supplying outlets for arc and xenon projectors of the professional type shall not be smaller than _____.
(a) 12 AWG   (b) 10 AWG   (c) 8 AWG   (d) 6 AWG

Article 545 Manufactured Buildings

154. The *NEC* specifies wiring methods for prefabricated buildings (manufactured buildings).
(a) True   (b) False

155. The plans, specifications and other building details for construction of manufactured buildings are included in the _____ details.
(a) manufactured building   (b) building component
(c) building structure   (d) building system

156. Service-entrance conductors for a manufactured building shall be installed _____.
(a) after erection at the building site
(b) before erection where the point of attachment is known prior to manufacture
(c) before erection at the building site
(d) a or b

Article 547 Agricultural Buildings

157. The distribution point is an electrical supply point from which _____ to agricultural buildings, associated farm dwelling(s) and associated buildings under single management are supplied.
(a) service drops or service laterals   (b) feeders or branch circuits
(c) a or b   (d) none of these

158. The distribution point is known as the _____.
(a) center yard pole   (b) meter pole
(c) common distribution point   (d) all of these

159. An equipotential plane is an area where wire mesh or other conductive elements are embedded in or placed under concrete bonded to _____.
(a) all metal structures
(b) fixed nonelectrical equipment that may become energized
(c) the grounding electrode system
(d) all of these

160. The purpose of the equipotential plane is to prevent a difference in voltage within the agricultural building area.
(a) True   (b) False

161. Enclosures and fittings installed in areas of agricultural buildings where excessive dust may be present shall be designed to minimize the entrance of dust and must have no openings through which dust could enter the enclosure. Only dust ignitionproof enclosures and fittings can be used for this purpose.
(a) True   (b) False

162. In damp or wet locations of agricultural buildings, equipment enclosures and fittings shall be located or equipped to prevent moisture from _____ within the enclosure, box, conduit body or fitting.
(a) entering   (b) accumulating   (c) a or b   (d) none of these

163. In agricultural building locations where surfaces are periodically washed or sprayed with water, enclosures and fittings shall be listed for use in wet locations and the enclosures shall be _____.
(a) weatherproof   (b) watertight   (c) rainproof   (d) any of these

164. Where _____ may be present in an agricultural building, enclosures and fittings must have corrosion-resistance properties suitable for the conditions.
(a) wet dust   (b) corrosive gases or vapors
(c) other corrosive conditions   (d) any of these

165. All 125V, 1Ø, 15 and 20A general-purpose receptacles installed _____ of agricultural buildings must have ground-fault circuit-interrupter protection for personnel.
(a) in areas having an equipotential plane   (b) outdoors
(c) in wet areas   (d) any of these

166. General-purpose receptacles rated 125V, 1Ø, _____ shall be GFCI-protected if they are located in an agricultural livestock building that has an equipotential plane.
    (a) 15A   (b) 20A   (c) 30A   (d) a and b

167. Where livestock is housed, that portion of the equipment grounding conductor run underground to the remote building disconnecting means shall be insulated or covered _____.
    (a) aluminum   (b) copper   (c) copper-clad aluminum (d) none of these

168. An equipotential plane shall be installed in all concrete floor confinement areas of livestock buildings that contain metallic equipment that is accessible to animals and likely to become energized.
    (a) True   (b) False

169. Outdoor concrete confinement areas, such as feedlots, must have equipotential planes installed around metallic equipment that is accessible to animals and likely to become energized. The equipotential plane must encompass the area around the equipment where the animal stands while accessing the equipment.
    (a) True   (b) False

170. An equipotential plane is not required in dirt confinement areas containing metallic equipment that is accessible to animals and likely to become energized, provided GFCI protection is provided for electrical equipment accessible to animals in the confinement areas.
    (a) True   (b) False

Article 550 Mobile Homes, Manufactured Homes and Mobile Home Parks

171. A _____ is a factory-assembled structure transportable in one or more sections that is built on a permanent chassis and designed to be used as a dwelling, with or without a permanent foundation.
    (a) manufactured home   (b) mobile home   (c) dwelling unit   (d) all of these

172. In reference to mobile/manufactured homes, examples of portable appliances could be _____, but only if these appliances are not built in.
    (a) refrigerators   (b) range equipment   (c) clothes washers   (d) all of these

173. The power supply to the mobile home shall be a _____.
    (a) One 50A mobile home power-supply cord with attachment plug
    (b) Permanently installed feeder
    (c) a or b
    (d) none of these

174. A mobile home that is factory-equipped with gas or oil-fired central heating equipment and cooking appliances shall be permitted to be supplied with a listed mobile home power-supply cord rated _____.
    (a) 30A   (b) 35A   (c) 40A   (d) 50A

175. Ground-fault circuit-interrupter (GFCI) protection in a mobile home is required for _____.
    (a) receptacle outlets installed outdoors
    (b) receptacles within 6 ft of a wet bar sink
    (c) all receptacles in bathrooms including receptacles in luminaires (light fixtures)
    (d) all of these

176. The maximum spacing of receptacle outlets over, or adjacent to, countertops in the kitchen in a mobile home is _____.
    (a) 6 ft   (b) 12 ft   (c) 3 ft   (d) none of these

177. The receptacle outlet for mobile and manufactured home heat tape that is used to protect cold water inlet piping shall be _____ protected and it shall be connected to an interior branch circuit where all of the receptacles of the circuit are _____ protected.
    (a) AFCI   (b) GFCI   (c) a or b   (d) none of these

178.   Aluminum conductors, aluminum-alloy conductors and copper-clad aluminum conductors are permitted only for branch-circuit wiring.
        (a) True   (b) False

179.   All branch circuits that supply 15 and 20A, 125V outlets in bedrooms of _____ shall be protected by arc-fault circuit interrupter(s).
        (a) mobile homes   (b) manufactured homes   (c) a and b   (d) none of these

180.   • What is the total park electrical wiring system load, after applying the demand factors permitted in Article 550, for a small mobile home park having six mobile homes?
        (a) 4,640 VA   (b) 27,840 VA   (c) 96,000 VA   (d) none of these

181.   Mobile home service equipment shall be located adjacent to the mobile home and not mounted in or on the mobile home. The service equipment shall be located in sight from but not more than _____ from the exterior wall of the mobile home it serves.
        (a) 15 ft   (b) 20 ft   (c) 30 ft   (d) none of these

182.   Service equipment for a manufactured home can be installed in or on a manufactured home provided that which of the following conditions are met?
        (a) The manufacturer must include in its written installation instructions information indicating that the home must be secured in place by an anchoring system or installed on and secured to a permanent foundation.
        (b) The manufacturer must include in its written installation instructions one method of grounding the service equipment at the installation site. The instructions must clearly state that other methods of grounding are found in Article 250.
        (c) A red warning label must be mounted on or adjacent to the service equipment.
        (d) all of these

183.   An outdoor disconnecting means for a mobile home shall be installed so the bottom of the enclosure is not less than _____ above the finished grade or working platform.
        (a) 1 ft   (b) 2 ft   (c) 3 ft   (d) 6 ft

Article 551 Recreational Vehicles and Recreational Vehicle Parks

184.   The working space clearance for a distribution panelboard located in a recreational vehicle shall be no less than _____.
        (a) 24 in. wide   (b) 30 in. deep   (c) 30 in. wide   (d) a and b

185.   A minimum of 70 percent of all recreational vehicle sites with electrical supply shall each be equipped with a _____ receptacle.
        (a) 15A   (b) 20A   (c) 30A   (d) 50A

186.   • Electrical service and feeders of a recreational vehicle park shall be calculated at a minimum of _____ per site equipped with only 20A supply facilities (not including tent sites).
        (a) 1,200 VA   (b) 2,400 VA   (c) 3,600 VA   (d) 9,600 VA

Article 553 Floating Buildings

187.   Feeders to floating buildings can be installed in _____ with approved fittings where flexible connections are required for services.
        (a) rigid nonmetallic conduit   (b) liquidtight metallic conduit
        (c) liquidtight nonmetallic conduit   (d) portable power cables

Article 555 Marinas and Boatyards

188.   Private, noncommercial docking facilities _____ for the use of the owner or residents of the associated single-family dwelling are not covered by this article.
        (a) constructed   (b) occupied   (c) a or b   (d) a and d

189.   The electrical datum plane (land areas subject to tidal fluctuation) is a horizontal plane _____ above the highest high tide under normal circumstances.
        (a) 1 ft   (b) 2 ft   (c) 3 ft   (d) none of these

190.   A _____ is an enclosed assembly that can include receptacles, circuit breakers, fused switches, fuses, watthour meter(s) and monitoring means approved for marine use. All such enclosures shall be arranged with a weep hole to discharge condensation.
        (a) marine power receptacle   (b) marine outlet
        (c) marine power outlet   (d) any of these

191.   Service equipment for floating docks or marinas shall be located _____ the floating structure.
        (a) next to   (b) on   (c) 100 ft from   (d) 20 ft from

192.   The feeder for six 20A receptacles supplying shore power for boats shall be calculated at _____ percent of the sum of the rating of the receptacles.
        (a) 70   (b) 80   (c) 90   (d) 100

193.   The demand percentage used to calculate 45 receptacles in a boatyard feeder is _____ percent.
        (a) 90   (b) 80   (c) 70   (d) 50

194.   Where shore power accommodations provide two receptacles specifically for an individual boat slip, and these receptacles have different voltages, only the receptacle with the _____ kW demand shall be required to be calculated.
        (a) smaller   (b) larger   (c) all receptacles   (d) none of these

195.   In marinas or boatyards, the *NEC* requires a(n) _____ disconnecting means which allows individual boats to be isolated from their supply circuit.
        (a) accessible   (b) readily accessible   (c) remote   (d) any of these

196.   The disconnecting means for a boat shall be readily accessible, not more than _____ from the receptacle it controls and shall be in the supply circuit ahead of the receptacle.
        (a) 12 in.   (b) 24 in.   (c) 30 in.   (d) none of these

197.   Receptacles that provide shore power for boats shall be rated not less than _____ and shall be single outlet type.
        (a) a 20A duplex receptacle
        (b) a 20A single receptacle of the locking and grounding type
        (c) a 30A
        (d) none of these

Chapter 6 Special Equipment

Article 600 Electric Signs and Outline Lighting

198.   Each commercial building and each commercial occupancy with ground floor access for pedestrians shall have at least one outside sign outlet in an accessible location at each entrance. The outlet(s) shall be supplied by a branch circuit rated at least _____ that supplies no other load.
        (a) 15A   (b) 20A   (c) a or b   (d) none of these

199.   Branch circuits that supply signs and outline lighting systems containing incandescent and fluorescent forms of illumination shall be rated not to exceed _____.
        (a) 20A   (b) 30A   (c) 40A   (d) 50A

200.   Each sign and outline lighting system, or feeder/branch circuit supplying a sign or outline lighting system, shall be controlled by an externally operable switch or circuit breaker that opens all _____ conductors.
        (a) ungrounded   (b) grounded   (c) grounding   (d) all of these

Unit 10 NEC Exam – NEC Code Order 503.9 – 555.19

1. Luminaires in a Class III location that may be exposed to physical damage shall be protected by a(n) _____ guard.
    (a) plastic　　　　　(b) metal　　　　　(c) suitable　　　　　(d) explosionproof

2. Any pit or depression in a commercial garage in which six air changes per hour are exhausted at the floor level of the pit shall be permitted to be judged by the enforcing agency to be a _____ location.
    (a) Class I, Division 2　　　　　　　　(b) Class II, Division 2
    (c) Class II, Division 1　　　　　　　　(d) Class I, Division 1

3. Raceways embedded in a masonry wall or buried beneath a floor shall be considered _____.
    (a) outside the hazardous area if any extensions pass through such areas
    (b) within the hazardous area if any connections or extensions lead into or through such areas
    (c) hazardous if beneath a hazardous area
    (d) outside any hazardous location

4. Underground gasoline dispenser wiring shall be installed in _____.
    (a) threaded rigid metal conduit
    (b) threaded intermediate metal conduit
    (c) rigid nonmetallic conduit when buried under not less than 2 ft of cover
    (d) any of these

5. Locations where flammable paints are dried, with the ventilating equipment interlocked with the electrical equipment, may be designated as a(n) _____ location.
    (a) Class I, Division 2　　　　　　　　(b) Unclassified
    (c) Class II, Division 2　　　　　　　　(d) Class II, Division 1

6. The patient care area is any portion of a health care facility, including the business offices, corridors, lounges, day rooms, dining rooms or similar areas.
    (a) True　　　　　　(b) False

7. Receptacles located within the patient care areas of pediatric wards, rooms or areas shall be listed _____.
    (a) tamper resistant　　　　　　　　(b) isolated
    (c) GFCI-protected　　　　　　　　(d) specification grade

8. In critical care areas of Health Care Centers, each patient bed location shall be provided with a minimum of _____ receptacles.
    (a) 10　　　　　(b) 6　　　　　(c) 3　　　　　(d) 4.

9. The essential electrical systems in a health care facility shall have sources of power from _____.
    (a) a normal source generally supplying the entire electrical system
    (b) one or more alternate sources for use when the normal source is interrupted
    (c) a or b
    (d) a and b

10. In places of assembly, nonmetallic raceways encased in not less than _____ of concrete shall be permitted.
    (a) 1 in.　　　　　(b) 2 in.　　　　　(c) 3 in.　　　　　(d) none of these

11. On fixed stage equipment, portable or strip luminaires and connector strips shall be wired with conductors having insulation rated suitable for the temperature but not less than _____.
    (a) 75 ℃　　　　　(b) 90 ℃　　　　　(c) 125 ℃　　　　　(d) 200 ℃

12. All receptacles for temporary branch circuits are required to be electrically connected to the _____ conductor.
    (a) grounded　　　　　　　　(b) grounding
    (c) equipment grounding　　　　　　　　(d) grounding electrode

13. Multiwire branch circuits for temporary wiring shall be provided with a means to disconnect simultaneously all _____ conductors at the power outlet or panelboard where the branch circuit originated.
    (a) underground　　(b) overhead　　(c) ungrounded　　(d) grounded

14. The *NEC* specifies wiring methods for prefabricated buildings (manufactured buildings).
    (a) True   (b) False

15. The plans, specifications and other building details for construction of manufactured buildings are included in the _____ details.
    (a) manufactured building   (b) building component
    (c) building structure   (d) building system

16. Service-entrance conductors for a manufactured building shall be installed _____.
    (a) after erection at the building site
    (b) before erection where the point of attachment is known prior to manufacture
    (c) before erection at the building site
    (d) a or b

17. The power supply to the mobile home shall be _____.
    (a) one 50A mobile home power-supply cord with attachment plug
    (b) permanently installed feeder
    (c) a or b
    (d) none of these

18. A mobile home that is factory-equipped with gas or oil-fired central heating equipment and cooking appliances shall be permitted to be supplied with a listed mobile home power-supply cord rated _____.
    (a) 30A   (b) 35A   (c) 40A   (d) 50A

19. The maximum spacing of receptacle outlets over or adjacent to countertops in the kitchen in a mobile home is _____.
    (a) 6 ft   (b) 12 ft   (c) 3 ft   (d) none of these

20. What is the total park electrical wiring system load, after applying the demand factors permitted in Article 550, for a small mobile home park having six mobile homes?
    (a) 4,640 VA   (b) 27,840 VA   (c) 96,000 VA   (d) none of these

21. The working space clearance for a distribution panelboard located in a recreational vehicle shall be no less than _____.
    (a) 24 in. wide   (b) 30 in. deep   (c) 30 in. wide   (d) a and b

22. A minimum of 70 percent of all recreational vehicle sites with electrical supply shall each be equipped with a _____ receptacle.
    (a) 15A   (b) 20A   (c) 30A   (d) 50A

23. Electrical service and feeders of a recreational vehicle park shall be calculated at a minimum of _____ volt-ampere per site equipped with only 20A supply facilities (not including tent sites).
    (a) 1,400   (b) 2,400   (c) 3,300   (d) 9,200

24. The feeder for six 20A receptacles supplying shore power for boats shall be calculated at _____ percent of the sum of the rating of the receptacles.
    (a) 70   (b) 80   (c) 90   (d) 100

25. Receptacles that provide shore power for boats shall be rated not less than _____ and shall be single outlet type.
    (a) a 20A duplex receptacle
    (b) a 20A single receptacle of the locking and grounding type
    (c) 30A
    (d) none of these

Unit 10 NEC Exam – Random Order 110.27 – 547.9

1. The service disconnecting means shall be installed at a(n) _____location.
   (a) dry   (b) readily accessible
   (c) outdoor   (d) indoor

2. The small-appliance branch circuits can supply the _____ as well as the kitchen.
   (a) dining room   (b) refrigerator
   (c) breakfast room   (d) all of these

3. The smallest size fixture wire permitted in the *NEC* is _____ AWG.
   (a) 22   (b) 20   (c) 18   (d) 16

4. There shall be no reduction in the size of the grounded conductor on _____ type lighting loads.
   (a) dwelling unit   (b) hospital   (c) nonlinear   (d) motel

5. Threadless couplings and connectors used with RMC and installed in wet locations shall be of the _____ type.
   (a) raintight   (b) wet and damp location
   (c) nonabsorbent   (d) weatherproof

6. Threadless couplings approved for use with IMC in wet locations shall be of the _____ type.
   (a) rainproof   (b) raintight
   (c) moistureproof   (d) concrete-tight

7. Through the exercise of ingenuity in the layout of the electrical installation for hazardous (classified) locations, it is frequently possible to locate much of the equipment in less hazardous or in nonhazardous locations and thus reduce the amount of special equipment required.
   (a) True   (b) False

8. Torque requirements for motor control circuit device terminals shall be a minimum of _____ pound-inches (unless otherwise indicated) for screw-type pressure terminals used for 14 AWG and smaller copper conductors.
   (a) 7   (b) 9   (c) 10   (d) 15

9. TPT and TST cords shall be permitted in lengths not exceeding _____ when attached directly, or by means of a special type of plug, to a portable appliance rated 50W or less.
   (a) 8 ft   (b) 10 ft   (c) 15 ft   (d) can't be used at all

10. Two conductor NM cables cannot be stapled on edge.
    (a) True   (b) False

11. Two or more grounding electrodes that are effectively bonded together shall be considered as a single grounding electrode system.
    (a) True   (b) False

12. Type _____ fuse adapters shall be designed so that once inserted in a fuseholder they cannot be easily removed.
    (a) A   (b) E   (c) S   (d) P

13. Type SE cable shall be permitted to be formed in a _____ and taped with a self-sealing weather-resistant thermoplastic.
    (a) loop   (b) circle   (c) gooseneck   (d) none of these

14. Types of pipeline resistive heaters are heating _____.
    (a) blankets   (b) tape   (c) coils   (d) a and b

15. UF cable shall not be used _____.
    (a) in any hazardous (classified) location
    (b) embedded in poured cement, concrete or aggregate
    (c) where exposed to direct rays of the sun, unless identified as sunlight-resistant
    (d) all of these

16. UF cable used with a 24V landscape lighting system is permitted to have a minimum cover of _____
     (a) 6 in.   (b) 12 in.   (c) 18 in.   (d) 24 in.

17. Under the optional method for calculating a single-family dwelling, general loads beyond the initial 10 kW are to be assessed at a _____ percent demand factor.
     (a) 40   (b) 50   (c) 60   (d) 75

18. Underfloor raceways shall be laid so that a straight line from the center of one _____ to the center of the next _____ will coincide with the centerline of the raceway system.
     (a) termination point   (b) junction box
     (c) receptacle   (d) panelboard

19. Underground-copper service conductors shall not be smaller than _____ copper.
     (a) 3 AWG   (b) 4 AWG   (c) 6 AWG   (d) 8 AWG

20. Unless an individual switch is provided for each luminaire located over combustible material, lampholders shall be located at least _____ above the floor or shall be located or guarded so that the lamps cannot be readily removed or damaged.
     (a) 3 ft   (b) 6 ft   (c) 8 ft   (d) 10 ft

21. Unless specifically permitted in 400.7, flexible cords and cables shall not be used where _____.
     (a) run through holes in walls, ceilings or floors
     (b) run through doorways, windows or similar openings
     (c) attached to building surfaces
     (d) all of these

22. Unless specified otherwise, live parts of electrical equipment operating at _____ or more shall be guarded.
     (a) 12V   (b) 15V   (c) 50V   (d) 24V

23. Use of FCC cable systems shall be permitted on wall surfaces in _____.
     (a) surface metal raceways   (b) cable trays
     (c) busways   (d) any of these

24. Using standard load calculations, the feeder demand factor for five household clothes dryers is _____ percent.
     (a) 70   (b) 85   (c) 50   (d) 100

25. Water heaters having a capacity of _____ gallons or less shall have a branch-circuit rating not less than 125 percent of the nameplate rating of the water heater.
     (a) 60   (b) 75   (c) 90   (d) 120

26. What is the maximum cord-and-plug-connected load permitted on a 15A multioutlet receptacle that is supplied by a 20A circuit?
     (a) 12   (b) 16   (c) 20   (d) 24

27. When an electric-discharge luminaire is mounted directly over an outlet box, the luminaire must provide access to the conductor wiring within the outlet box.
     (a) True   (b) False

28. When an underground metal water-piping system is used as a grounding electrode, effective bonding shall be provided around insulated joints and sections around any equipment that is likely to be disconnected for repairs or replacement. Bonding conductors shall be made of _____ to permit removal of such equipment while retaining the integrity of the bond.
     (a) stranded wire   (b) flexible conduit
     (c) sufficient length   (d) none of these

29. When bonding enclosures, metal raceways, frames, fittings and other metal noncurrent-carrying parts, any nonconductive paint, enamel, or similar coating shall be removed at _____.
     (a) contact surfaces   (b) threads
     (c) contact points   (d) all of these

30. When connected to a branch circuit supplying _____ or more receptacles or outlets, a receptacle shall not supply a total cord-and-plug-connected load in excess of the maximum specified in Table 210.21(B)(2).
    (a) two   (b) three   (c) four   (d) five

31. When ENT is installed concealed in walls, floors and ceilings of buildings exceeding three floors above grade, a thermal barrier must be provided having a minimum _____ minute finish rating as listed for fire-rated assemblies.
    (a) 5   (b) 10   (c) 15   (d) 30.

32. When equipment grounding conductor(s) are installed in a metal box, an electrical connection is required between the equipment grounding conductor and the metal box enclosure by means of a _____.
    (a) grounding screw   (b) soldered connection
    (c) listed grounding device   (d) a or c

33. When multiple ground rods are used for a grounding electrode, they shall be separated not less than _____ apart.
    (a) 6 ft   (b) 8 ft   (c) 20 ft   (d) 12 ft

34. When nails are used to mount knobs for the support of open wiring on insulators, they shall not be smaller than _____-penny.
    (a) six   (b) eight   (c) ten   (d) none of these

35. When provisions must be made in a Class I, Division 2 location for limited flexibility, such as motor termination, flexible metal conduit with listed fittings may be used.
    (a) True   (b) False

36. When screws are used to mount knobs for the support of open wiring on insulators, they shall be of a length sufficient to penetrate the wood to a depth equal to at least _____ the height of the knob and the full thickness of the cleat.
    (a) one-eighth   (b) one-quarter   (c) one-third   (d) one-half

37. When service-entrance conductors exceed 1100 kcmil for copper, the required grounded conductor for the service must be sized not less than _____ percent of the area of the largest ungrounded service-entrance (phase) conductor.
    (a) 15   (b) 19   (c) $12\frac{1}{2}$ (d) 25

38. When switches, cutouts or other equipment operating at 600V, nominal, or less are installed in a room or enclosure where there are exposed live parts or exposed wiring operating at over 600V, nominal, the high-voltage equipment shall be effectively separated from the space occupied by _____ by a suitable partition, fence or screen.
    (a) the access area   (b) the low-voltage equipment
    (c) unauthorized persons   (d) motor control equipment

39. When two to six service disconnecting means in separate enclosures are grouped at one location and supply separate loads from one service drop or lateral, _____ set(s) of service-entrance conductors shall be permitted to supply each or several such service equipment enclosures.
    (a) one   (b) two   (c) three   (d) four

40. Where installed in raceways, conductors _____ and larger shall be stranded.
    (a) 10 AWG   (b) 6 AWG   (c) 8 AWG   (d) 4 AWG

41. Where it is unlikely that two or more loads will be in use simultaneously, it shall be permissible to use only the _____ loads on at any given time in computing the total load to a feeder.
    (a) smaller   (b) larger
    (c) difference between the   (d) none of these

42. Where livestock is housed, that portion of the equipment grounding conductor run underground to the remote building disconnecting means shall be insulated or covered _____.
    (a) aluminum   (b) copper
    (c) copper-clad aluminum   (d) none of these

43. Where NMC cable is run at angles with joists in unfinished basements, it shall be permissible to secure cables not smaller than _____ conductors directly to the lower edges of the joist.
    (a) two, 6 AWG   (b) three, 8 AWG
    (c) three, 10 AWG   (d) a or b

44. Where practicable, transformer vaults containing more than _____ transformer capacity shall be provided with a drain or other means that will carry off any accumulation of oil or water in the vault unless local conditions make this impracticable.
    (a) 100 kVA   (b) 150 kVA   (c) 200 kVA   (d) 250 kVA

45. Which of the following conductor types are required to be used when FMC is installed in a wet location?
    (a) THWN   (b) XHHW   (c) THW   (d) any of these

46. Which of the following metal parts shall be protected from corrosion both inside and out?
    (a) Ferrous raceways.   (b) Metal elbows.
    (c) Boxes.   (d) all of these

47. Which of the following switches must indicate whether they are in the (open) OFF or (closed) ON position?
    (a) General-use switches.   (b) Motor-circuit switches.
    (c) Circuit breakers.   (d) all of these

48. Which of the following wiring methods are permitted in a Class I, Division 1 location?
    (a) Threaded rigid metal conduit   (b) Threaded IMC
    (c) MI cable   (d) all of these

49. Wiring methods and equipment installed behind panels designed to permit access (such as drop ceilings) shall be so arranged and secured as to permit the removal of panels to give access to electrical equipment.
    (a) True   (b) False

50. Wiring methods installed behind panels that allow access, such as the space above a dropped ceiling, are required to be _____ according to their applicable articles.
    (a) supported   (b) painted
    (c) in a metal raceway   (d) all of these

Unit 11

Commercial Load Calculations

PART A – GENERAL

11-1 GENERAL REQUIREMENTS

Article 220 provides the requirement for branch circuits, feeders and services. In addition to this article, other articles are applicable such as Branch Circuits – 210, Feeders – 215, Services – 230, Overcurrent Protection – 240, Wiring Methods – 300, Conductors – 310, Appliances – 422, Electric Space-Heating Equipment – 424, Motors – 430 and Air-Conditioning – 440.

11-2 CONDUCTOR AMPACITY [ARTICLE 100]

The ampacity of a conductor is the rating, in amperes, that a conductor can carry continuously without exceeding its insulation temperature rating [*NEC* Definition – Article 100]. The allowable ampacities listed in Table 310.16 are affected by ambient temperature, conductor insulation and conductor bundling [310.10]. Figure 11–1.

Continuous Loads

Conductors are sized at 125% of the continuous load before any adjustment factor and the overcurrent protection devices are sized at 125% of the continuous load [210.19(A), 215.2(A)(1), and 230.42].

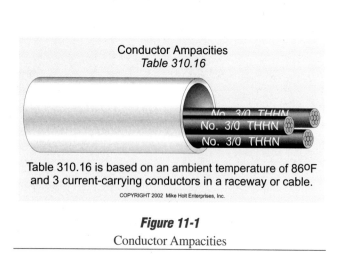

Conductor Ampacities
Table 310.16

Table 310.16 is based on an ambient temperature of 86°F
and 3 current-carrying conductors in a raceway or cable.

COPYRIGHT 2002 Mike Holt Enterprises, Inc.

Figure 11-1
Conductor Ampacities

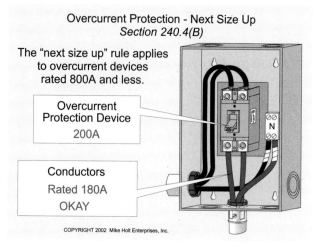

Overcurrent Protection - Next Size Up
Section 240.4(B)

The "next size up" rule applies
to overcurrent devices
rated 800A and less.

Overcurrent
Protection Device
200A

Conductors
Rated 180A
OKAY

COPYRIGHT 2002 Mike Holt Enterprises, Inc.

Figure 11-2
Overcurrent Protection – Next Size Up

11–3 CONDUCTOR OVERCURRENT PROTECTION [240.4]

The purpose of overcurrent protection devices is to protect conductors and equipment against excessive or dangerous temperatures [240.1 FPN]. There are many rules in the *NEC* for conductor protection, and there are many different installation applications where the general rule of protecting the conductor at its ampacity does not apply. Examples would be motor circuits, air-conditioning, tap conductors, etc. See 240.4 for specific rules on conductor overcurrent protection.

Next Size Up Okay [240.4(B)] If the ampacity of a conductor does not correspond with the standard ampere rating of a fuse or circuit breaker as listed in 240.6(A), the next size up protection device is permitted. This only applies if the conductors do not supply multioutlet receptacles and if the next size up overcurrent protection device does not exceed 800A. Figure 11–2.

Standard Size Overcurrent Devices [240.6(A)] The following is a list of some of the standard ampere ratings for fuses and inverse-time circuit breakers: 15, 20, 25, 30, 35, 40, 45, 50, 60, 70, 80, 90, 100, 110, 125, 150, 175, 200, 225, 250, 300, 350, 400, 500, 600, 800, 1000, 1200 and 1600A.

11–4 VOLTAGES [220.2(A)]

Unless other voltages are specified, branch-circuit, feeder and service loads shall be computed at a nominal system voltage of 120, 120/240, 208Y/120, 240, 480Y/277 or 480. Figure 11–3.

11–5 ROUNDING AN AMPERE [220.2(B)]

The rules for rounding an ampere require that when the calculation results in a fraction of an ampere of 0.50 and more, we "round up" to the next ampere. If the calculation is 0.49 of an ampere or less, we "round down" to the next lower ampere.

PART B – LOADS

11–6 AIR-CONDITIONING

Air-Conditioning Branch Circuit Branch-circuit conductors and protection for air-conditioning equipment are marked on the equipment nameplate [110.3(B)]. The nameplate values are determined by the use of 440.32 for conductor sizing and 440.22(A) for short-circuit protection. 440.32 specifies that branch-circuit conductors must be sized no less than 125% of the air conditioner rating, and 440.22(A) specifies that the short-circuit protection device must be sized between 175% and up to 225% of the air-conditioning rating.

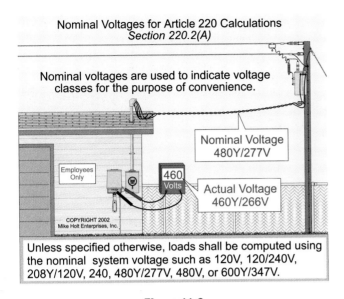

Figure 11-3

Nominal Voltages for Article 220 Calculations

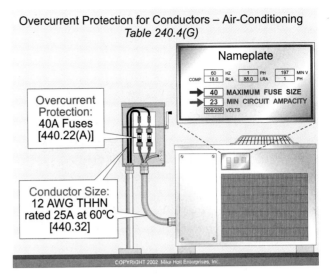

Figure 11-4

Overcurrent Protection for Conductors – Air Conditioning

❑ Air-Conditioning Branch Circuit [240.4(G)]

An air conditioner has a motor-compressor rated-load amperage of 18A. The nameplate indicates the minimum circuit ampacity of 23A and the maximum fuse size of 40A. What size conductor and protection are required? Figure 11–4.

(a) 12 AWG with a 40A breaker

(b) 12 AWG with a 40A fuse

(c) 14 AWG with a 30A breaker

(d) 10 AWG with a 40A breaker

• Answer: (b) 12 AWG with a 40A fuse

Conductor [440.32]. The conductor must be sized at no less than 125% of the motor-compressor rated-load current. 18A × 1.25 = 22.5A, 12 AWG THHN rated at 25A at 60°C

Short-Circuit Protection [440.22(A)]. The short-circuit protection must be a fuse [110.3(B)] because the nameplate specified "fuse." However, the protection device shall not be greater than 225% of the motor-compressor current. 18A × 2.25 = 40.5A, next size down, 40A [240.6(A)]

Note: I know many feel the conductor should be larger or the short-circuit protection smaller, but this is not the *NEC* requirement.

Air-Conditioning Feeder/Service Conductors. Feeder circuit conductors that supply air-conditioning equipment must be sized no less than 100%.

VA Rating – The VA rating of an air conditioner is determined by multiplying the voltage rating of the unit by its ampere rating. For the purpose of this book, we will determine the unit ampere rating by using motor FLC ratings as listed in Article 430, Table 430.148 and Table 430.150.

❑ Air-Conditioning Feeder/Service Conductor

What is the demand load required for the A/C equipment of a 12 unit office building where each unit contains a 5-hp, 230V A/C?

(a) 77 kVA   (b) 73 kVA   (c) 45 kVA   (d) none of these

• Answer: (a) 77 kVA

A/C 5-hp, 230V × 28A = 6,440 VA.

6,440 VA × 12 units = 77,280 VA/1,000 = 77.28 kVA.

11–7 DRYERS

The branch-circuit conductors and overcurrent protection device for commercial dryers are sized to the appliance nameplate rating. The feeder demand load for dryers is calculated at 100% of the appliance rating. Table 220.18 demand factors do not apply to commercial dryers.

❑ Dryer Branch Circuit

What size branch-circuit conductor and overcurrent protection is required for a 7 kW dryer rated 240V when the dryer is located in the laundry room of a multifamily dwelling? Figure 11–5.

(a) 12 AWG with a 20A breaker
(b) 10 AWG with a 20A breaker
(c) 12 AWG with a 30A breaker
(d) 10 AWG with a 30A breaker

• Answer: (d) 10 AWG with a 30A breaker

$I = P/E = 7,000W/240V = 29A$

The ampacity of the conductor and overcurrent device must not be less than 29A [240.4].

Table 310.16, 10 AWG conductor at 60°C is rated 30A.

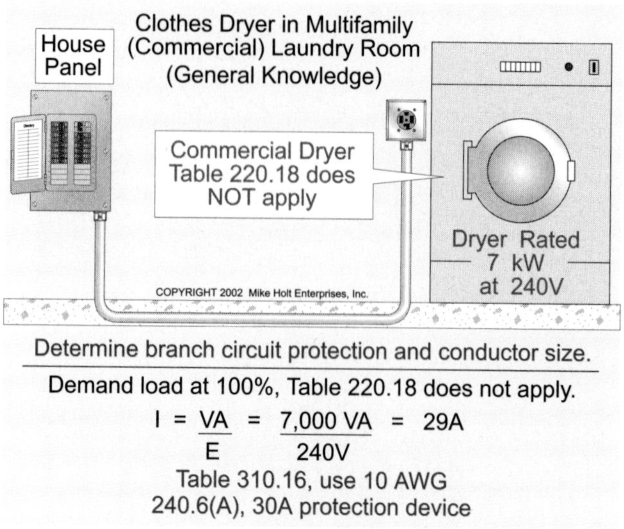

❑ Dryer Feeder

What is the service demand load for ten 7 kW dryers located in a laundry room?

(a) 70 kW   (b) 52.5 kW
(c) 35 kW   (d) none of these

• Answer: (a) 70 kW

The *NEC* does not permit a demand factor for commercial dryers, therefore the dryer demand load must be calculated at 100%: 7 kW × 10 units = 70 kW.

Note: If the dryers are on continuously, the conductor and protection device must be sized at 125% of the load [210.19(A), 215.3, and 230.42].

Clothes Dryer in Multifamily (Commercial) Laundry Room (General Knowledge)

House Panel

Commercial Dryer Table 220.18 does NOT apply

Dryer Rated 7 kW at 240V

COPYRIGHT 2002 Mike Holt Enterprises, Inc.

Determine branch circuit protection and conductor size.

Demand load at 100%, Table 220.18 does not apply.

$$I = \frac{VA}{E} = \frac{7,000\ VA}{240V} = 29A$$

Table 310.16, use 10 AWG
240.6(A), 30A protection device

Figure 11-5

Clothes Dryer in Multifamily Laundry Room

11–8 ELECTRIC HEAT

Electric Heat Branch Circuit [424.3(B)]

Branch-circuit conductors and the overcurrent protection device for electric heating shall be sized to not be less than 125% of the total heating load including blower motors.

❑ Electric Heat

What size conductors and protection are required for a 240V, 15 kW, 3Ø heat strip with a 5.4A blower motor? Figure 11–6.

(a) 10 AWG with 30A protection   (b) 6 AWG with 50A protection
(c) 8 AWG with 40A protection   (d) 6 AWG with 60A protection

• Answer: (d) 6 AWG with 60A protection [424.3(B)]

Conductors and protection must not be less than 125% of the total load.

$I = VA/(E \times \sqrt{3}) = 15,000\ VA/(240V \times 1.732) = 36A$

$36A + 5.4A = 41.4A \times 1.25 = 51.75A$

Conductor sized to the 60°C column ampacities of Table 310.16 [110.14(C)(1)(a)], 6 AWG rated 55A.

Overcurrent protection device sized no less than 52A, 60A protection device [240.6(A)].

Electric Heat Feeder/Service Demand Load [220.15]

The feeder/service demand load for electric heating equipment is calculated at 100% of the total heating load.

❑ Electric Heat Feeder/Service Demand Load

What is the feeder/service demand load for a building that has seven 208V, 10 kW, 3Ø heat strips with a 5.4A blower motor (1,945 VA) for each unit?

(a) 84 kVA   (b) 53 kVA   (c) 129 kVA   (d) 154 kVA

• Answer: (a) 84 kVA

10,000 VA + 1,945 VA = 11,945 VA × 7 units = 83,615 VA/1,000 = 83.6 kVA

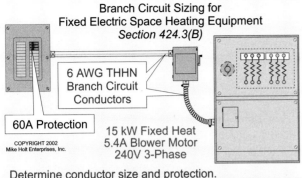

Branch Circuit Sizing for
Fixed Electric Space Heating Equipment
Section 424.3(B)

6 AWG THHN
Branch Circuit
Conductors

60A Protection

COPYRIGHT 2002
Mike Holt Enterprises, Inc.

15 kW Fixed Heat
5.4A Blower Motor
240V 3-Phase

Determine conductor size and protection.

The overcurrent protection device and the branch
circuit conductors are sized at 125% of the total load
of the motors and the heaters.

Combined Load of Motors and Heaters

$$I\ 3\text{-ph} = \frac{VA}{(E \times \sqrt{3})} \quad \frac{15{,}000W}{(240V \times 1.732)} = 36A$$

Total Load = 5.4A + 36A = 41.4A
41.4A x 1.25 = 51.75A

Table 310.16 = 6 AWG wire   240.6(A) = 60A device

Figure 11-6

Branch-Circuit Sizing for Fixed Electric Space-Heating
Equipment

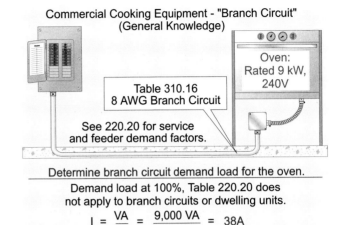

Commercial Cooking Equipment - "Branch Circuit"
(General Knowledge)

Oven:
Rated 9 kW,
240V

Table 310.16
8 AWG Branch Circuit

See 220.20 for service
and feeder demand factors.

Determine branch circuit demand load for the oven.
Demand load at 100%, Table 220.20 does
not apply to branch circuits or dwelling units.

$$I = \frac{VA}{E} = \frac{9{,}000\ VA}{240V} = 38A$$

COPYRIGHT 2002
Mike Holt Enterprises, Inc.

Figure 11-7

Commercial Cooking Equipment – "Branch Circuit"
(General Knowledge)

11–9 KITCHEN EQUIPMENT

Kitchen Equipment Branch Circuit

Branch-circuit conductors and overcurrent protection for commercial kitchen equipment are sized according to the appliance nameplate rating.

❏ Kitchen Equipment Branch Circuit No. 1

What is the branch-circuit demand load (in amperes) for one 9 kW oven rated 240V? Figure 11–7.

(a) 38A   (b) 27A   (c) 32A   (d) 33A

• Answer: (a) 38A

I = P/E

I = 9,000W/240V

I = 38A

❏ Kitchen Equipment Branch Circuit No. 2

What is the branch-circuit load for one 208V, 14.47 kW, 3Ø range?

(a) 60A   (b) 40A   (c) 50A   (d) 30A

• Answer: (b) 40A

I = P/(E × 1.732)

I = 14,470W/(208V × 1.732)

I = 40A

Kitchen Equipment Feeder/Service Demand Load [220.20]

The service demand load for thermostatic control or intermittent use commercial kitchen equipment is determined by applying the demand factors from Table 220.20 to the total connected kitchen equipment load. *The feeder or service demand load cannot be less than the two largest appliances!*

Note: The demand factors of Table 220.20 do not apply to space-heating, ventilating or air-conditioning equipment.

❏ **Kitchen Equipment Feeder/Service No. 1**

What is the demand load for the following kitchen equipment loads? Figure 11–8.

Water Heater	5 kW	Booster Heater	7.5 kW
Mixer	3 kW	Oven	5 kW
Dishwasher	1.5 kW	Waste Disposal	1 kW

(a) 15 kW   (b) 23 kW   (c) 12.5 kW   (d) none of these

• Answer: (a) 15 kW

Water Heater	5 kW
Booster Heater	7.5 kW
Mixer	3 kW
Oven	5 kW
Dishwasher	1.5 kW
Waste disposal	1 kW
Total connected	23 kW × 0.65 = 14.95 kW.

❏ **Kitchen Equipment Feeder/Service No. 2**

What is the demand load for the following kitchen equipment loads?

Water Heater	10 kW*	Booster Heater	15 kW*
Mixer	4 kW	Oven	6 kW
Dishwasher	1.5 kW	Waste Disposal	1 kW

(a) 24.4 kW   (b) 38.2 kW   (c) 25.0 kW   (d) 18.9

• Answer: (c) 25 kW*

Water Heater	10 kW
Booster Heater	15 kW
Mixer	4 kW
Oven	6 kW
Dishwasher	1.5 kW
Waste Disposal	1 kW
Total connected	37.5 kW × 0.65 = 24.4 kW

* The demand load cannot be less than the two largest appliances 10 kW + 15 kW = 25 kW.

11–10 LAUNDRY EQUIPMENT

Laundry equipment circuits are sized to the appliance nameplate rating. For exam purposes, it is generally accepted that a laundry circuit is not considered a continuous load and assume all commercial laundry circuits to be rated 1,500 VA unless noted otherwise in the question.

❏ **Laundry Equipment**

What is the demand load for 10 washing machines located in a laundry room? Figure 11–9.

(a) 1,500 VA   (b) 15,000 VA

(c) 1,125 VA   (d) none of these

• Answer: (b) 15,000 VA

1,500 VA × 10 units = 15,000 VA

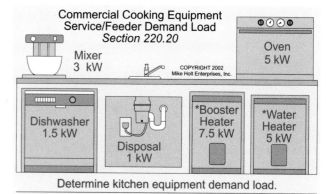

Commercial Cooking Equipment
Service/Feeder Demand Load
Section 220.20

Determine kitchen equipment demand load.

Table 220.20, 6 units, 65% of connected load.

Water heater	*5.00 kW	*The demand load cannot
Booster heater	*7.50 kW	be less than the sum of
Mixer	3.00 kW	two largest appliances.
Oven	5.00 kW	
Dishwasher	1.50 kW	Water heater 5.00 kW
Disposal	1.00 kW	Booster heater 7.50 kW
Total Connected	23.00 kW	12.50 kW
Table 220 demand	x .65 df	
Demand	14.95 kW	Feeder/Service Demand
		14.95 kW

Figure 11-8
Commercial Cooking Equipment
Services/Feeder Demand Load

Commercial Laundry Circuits

Washers
75 Cents

Do Not
Use Dye

10 Washing Machines
Assume 1,500 VA Each

COPYRIGHT 2002
Mike Holt Enterprises, Inc.

Determine feeder/service demand for 10 washers.

Demand load at 100%, there are no demand factors for commercial laundry circuits. Use nameplate value if given. Use 1,500 VA per washer if not given.

1,500 VA x 10 units = 15,000 VA Demand Load

Figure 11-9
Commercial Laundry Circuits

General Lighting Demand Load - Hotel/Motel
Tables 220.3(A) and 220.11

Hotel - 40 Units
600 sq ft per unit

COPYRIGHT 2002
Mike Holt Enterprises, Inc.

Determine general lighting demand load.

Table 220.3(A), lighting load is 2 VA per sq ft.
600 sq ft x 40 units x 2 VA per sq ft. = 48,000 VA
600 sq ft x 40 x 2 VA per sq ft. = 48,000 VA
Tbl 220.11, 1st 20,000 VA, 50% - 20,000 VA x 0.5 = 10,000 VA
          Next 80,000 VA, 40% 28,000 VA x 0.4 = 11,200 VA
Demand Load   = 21,200 VA

Figure 11-10
General Lighting Demand Load – Hotel/Motel

11–11 LIGHTING – DEMAND FACTORS [Table 220.3(A) and 220.11]

The *NEC* requires a minimum load per sq ft for general lighting depending on the type of occupancy [Table 220.3(A)]. For the guest rooms of hotels and motels, hospitals and storage warehouses, the general lighting demand factors of Table 220.11 can be applied to the general lighting load.

Hotel or Motel Guest Rooms – General Lighting [Table 220.3(A) and Table 220.11]

The general lighting demand load of 2 VA each sq ft [Table 220.3(A)] for the guest rooms of hotels and motels is permitted to be reduced according to the demand factors listed in Table 220.11.

General Lighting Demand Factors

First 20,000 VA at 50% demand factor
Next 80,000 VA at 40% demand factor
Remainder VA at 30% demand factor

❏ Hotel General Lighting Demand

What is the general lighting demand load for a 40-room hotel? Each unit contains 600 sq ft of living area. Figure 11–10.

(a) 48 kVA   (b) 24 kVA   (c) 20 kVA   (d) 21 kVA

• Answer: (d) 21 kVA

40 units × 600 sq ft (24,000 sq ft × 2 VA)	48,000 VA
First 20,000 VA at 50 %	– 20,000 VA × 0.5 = 10,000 VA
Next 80,000 VA at 40 %	28,000 VA × 0.4 = 11,200 VA
	21,200 VA

11–12 LIGHTING WITHOUT DEMAND FACTORS [Table 220.3(A), 215.2(A)(1) and 230.42].

The general lighting load for commercial occupancies other than guest rooms of motels and hotels, hospitals, and storage warehouses is assumed continuous and shall be calculated at 125% of the general lighting load as listed in Table 220.3(A).

❏ Store Lighting

What is the general lighting load for a 21,000 sq ft store? Figure 11–11.

(a) 40 kVA   (b) 63 kVA   (c) 79 kVA   (d) 81 kVA

• Answer: (c) 79 kVA

21,000 sq ft × 3 VA × 1.25 = 78,750 VA

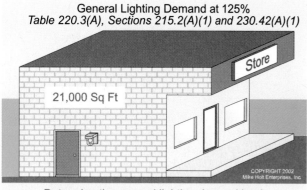

General Lighting Demand at 125%
Table 220.3(A), Sections 215.2(A)(1) and 230.42(A)(1)

Determine the general lighting demand load.

Table 220.3(A), store lighting is 3 VA per sq ft.
21,000 sq ft x 3 VA per sq ft = 63,000 VA lighting load
Store lighting is a continuous load at 125%
63,000 VA lighting load x 1.25 = 78,750 VA demand load

Figure 11-11
General Lighting Demand at 125%

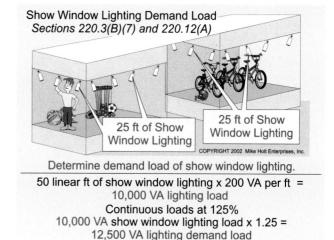

Show Window Lighting Demand Load
Sections 220.3(B)(7) and 220.12(A)

Determine demand load of show window lighting.

50 linear ft of show window lighting x 200 VA per ft =
10,000 VA lighting load
Continuous loads at 125%
10,000 VA show window lighting load x 1.25 =
12,500 VA lighting demand load

Figure 11-12
Show-Window Lighting Demand Load

❏ **Club Lighting**

What is the general lighting load for a 4,700 sq ft dance club?

(a) 4,700 VA   (b) 9,400 VA   (c) 11,750 VA   (d) 250 kVA

• Answer: (c) 11,750 VA

4,700 sq ft × 2 VA × 1.25 = 11,750 VA

❏ **School Lighting**

What is the general lighting load for a 125,000 sq ft school?

(a) 125 kVA   (b) 375 kVA   (c) 469 kVA   (d) 550 kVA

• Answer: (c) 469 kVA

125,000 sq ft × 3 VA × 1.25 = 468,750 VA

11–13 LIGHTING – MISCELLANEOUS

Show-Window Lighting [220.3(B)(7) and 220.12(A)]

The demand load for each linear foot of show-window lighting shall be calculated at 200 VA per ft. Show-window lighting is assumed to be a continuous load; see Example D3 in Annex D for the requirements for show-window branch circuits.

❏ **Show-Window Load**

What is the demand load in kVA for 50 ft of show-window lighting? Figure 11–12.

(a) 6 kVA   (b) 7.5 kVA   (c) 9 kVA   (d) 12.5 kVA

• Answer: (d) 12.5 kVA

50 ft × 200 VA per ft = 10,000 VA × 1.25 = 12,500 VA

11–14 MULTIOUTLET RECEPTACLE ASSEMBLY [220.3(B)(8)]

Each 5 ft or fraction of a foot, of multioutlet receptacle assembly shall be considered to be 180 VA for service calculations. When a multioutlet receptacle assembly is expected to have a number of appliances used simultaneously, each ft or fraction of a ft shall be considered as 180 VA for service calculations. A multioutlet receptacle assembly is not generally considered to be a continuous load.

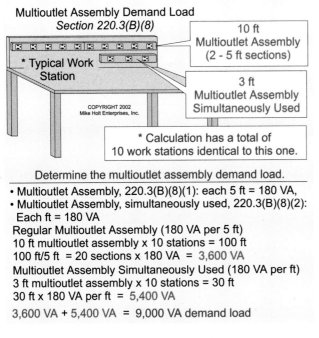

Multioutlet Assembly Demand Load
Section 220.3(B)(8)

* Typical Work Station

COPYRIGHT 2002
Mike Holt Enterprises, Inc.

10 ft
Multioutlet Assembly
(2 - 5 ft sections)

3 ft
Multioutlet Assembly
Simultaneously Used

* Calculation has a total of
10 work stations identical to this one.

Determine the multioutlet assembly demand load.
• Multioutlet Assembly, 220.3(B)(8)(1): each 5 ft = 180 VA,
• Multioutlet Assembly, simultaneously used, 220.3(B)(8)(2):
  Each ft = 180 VA
Regular Multioutlet Assembly (180 VA per 5 ft)
10 ft multioutlet assembly x 10 stations = 100 ft
100 ft/5 ft = 20 sections x 180 VA = 3,600 VA
Multioutlet Assembly Simultaneously Used (180 VA per ft)
3 ft multioutlet assembly x 10 stations = 30 ft
30 ft x 180 VA per ft = 5,400 VA

3,600 VA + 5,400 VA = 9,000 VA demand load

Figure 11-13
Multioutlet Assembly Demand Load

Receptacle Outlets for Article 220 Calculations
Section 220.3(B)(9)

ONE Outlet

180 VA
One Strap

180 VA
One Strap

COPYRIGHT 2002 Mike Holt Enterprises, Inc.

TWO Outlets

360 VA
Two Straps

360 VA
Two Straps

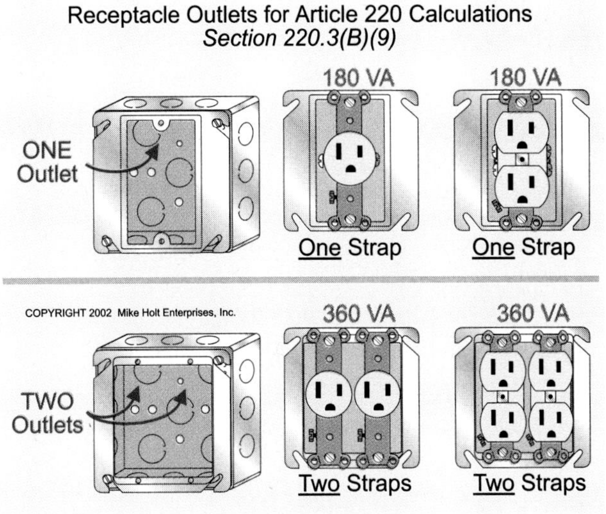

Figure 11-14
Receptacle Outlet Load Calculations

❏ Multioutlet Receptacle Assembly

What is the demand load for 10 work stations that have 10 ft of multioutlet receptacle assembly and 3 ft of multioutlet receptacle assembly simultaneously used? Figure 11–13.

(a) 5 kVA   (b) 6 kVA   (c) 7 kVA   (d) 9 kVA

• Answer: (d) 9 kVA

100 ft/5 ft per section (20 sections × 180 VA)   3,600 VA

30 ft/1 ft per section (30 sections × 180 VA)   5,400 VA

9,000 VA/1,000 = 9 kVA

11–15 RECEPTACLES VA LOAD [220.13]

Receptacle VA Load [220.3(B)(9)]

The minimum load for each commercial or industrial general-use receptacle outlet shall be 180 VA per strap [220.3(B)(9)]. Receptacles are generally not considered to be a continuous load. Figure 11–14.

Number of Receptacles Permitted on a Circuit

The maximum number of receptacle outlets permitted on a commercial or industrial circuit is dependent on the circuit ampacity. The number of receptacles per circuit is calculated by dividing the VA rating of the circuit by 180 VA for each receptacle strap.

❏ Receptacles per 15A Circuit [220.3(B)(9)]

How many receptacle outlets are permitted on a 15A, 120V circuit? Figure 11–15.

(a) 10   (b) 13   (c) 15   (d) 20

• Answer: (a) 10

The total circuit VA load for a 15A circuit is 120V × 15A = 1,800 VA.

The number of receptacle outlets per circuit = 1,800 VA/180 VA = 10 receptacles.

Note: 15A circuits are permitted for commercial and industrial occupancies according to the *NEC*, but some local codes require a minimum 20A rating for commercial and industrial circuits [310.5].

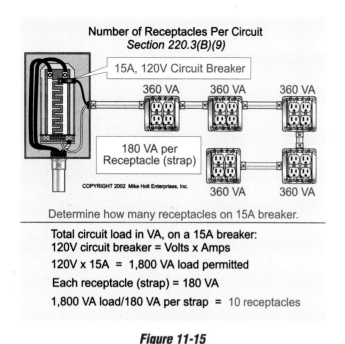

Number of Receptacles Per Circuit
Section 220.3(B)(9)

15A, 120V Circuit Breaker

360 VA   360 VA   360 VA

180 VA per Receptacle (strap)

360 VA   360 VA

COPYRIGHT 2002 Mike Holt Enterprises, Inc.

Determine how many receptacles on 15A breaker.

Total circuit load in VA, on a 15A breaker:
120V circuit breaker = Volts x Amps
120V x 15A = 1,800 VA load permitted
Each receptacle (strap) = 180 VA
1,800 VA load/180 VA per strap = 10 receptacles

Figure 11-15
Number of Receptacles per Circuit

Receptacle Demand Loads
Section 220.13

Building contains 150 receptacles.

COPYRIGHT 2002
Mike Holt Enterprises, Inc.

Shipping and Receiving

Determine the demand load for 150 receptacles.

220.3(B)(9), each receptacle = 180 VA
Table 220.13: First 10 kVA at 100%, remainder at 50%

150 receptacles x 180 VA = 27,000 VA
1st 10,000 VA at 100%   = -10,000 VA x 1.00 = 10,000 VA
Remainder at 50% =   17,000 VA x 0.50 = +8,500 VA
Receptacle Demand Load =   18,500 VA

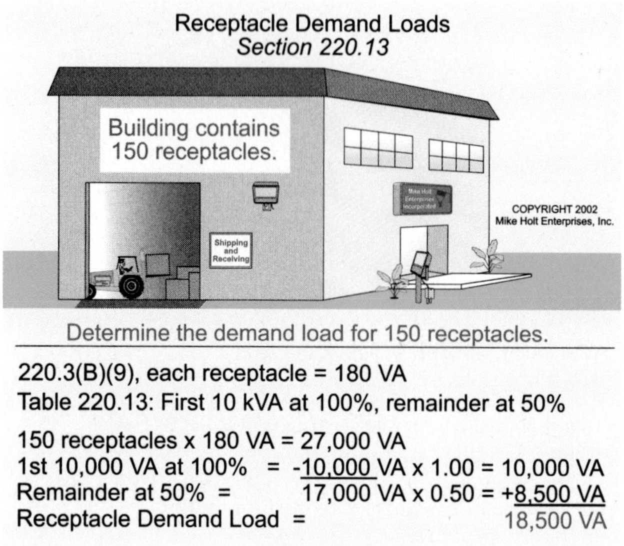

Figure 11-16
Receptacle Demand Loads

❑ **Receptacles Per 20A Circuit [220.3(B)(9)]**
How many receptacle outlets are permitted on a 20A, 120V circuit?
(a) 10   (b) 13   (c) 15   (d) 20
• Answer: (a) 10
The total circuit VA load for a 20A circuit is 120V × 20A = 2,400 VA.
The number of receptacle outlets per circuit = 2,400 VA/180 VA = 13 receptacles.

Receptacle Service Demand Load [220.13]
In other than dwelling units, receptacle loads computed at not more than 180 VA per outlet in accordance with 220.3(B)(9) and fixed multioutlet assemblies computed in accordance with 220.3(B)(8) shall be permitted to be added to the lighting loads and made subject to the demand factors given in Table 220.11, or they shall be permitted to be made subject to the demand factors given in Table 220.13. Receptacle loads are generally not considered to be a continuous load.

Table 220.13 Receptacle Demand Factors:
First 10 kVA at 100% demand factor
Remainder kVA at 50% demand factor

❑ **Receptacle Service Demand Load**
What is the service demand load for 150-20A, 120V general-use receptacles in a commercial building? Figure 11–16.
(a) 27 kVA   (b) 14 kVA   (c) 4 kVA   (d) 19 kVA
• Answer: (d) 19 kVA
Total receptacle load   (150 receptacles × 180 VA)   27,000 VA
First 10 kVA at 100%   – 10,000 VA × 1.00 = 10,000 VA
Remainder at 50%   17,000 VA × 0.50 =   8,500 VA
Total receptacle demand load   18,500 VA/1,000 = 18.5 kVA

11–16 BANKS AND OFFICES – GENERAL LIGHTING AND RECEPTACLES

Receptacle Demand [Table 220.13] The receptacle demand load is calculated at 180 VA for each receptacle strap [220.3(B)(9)] if the number of receptacles are known, or 1 VA for each sq ft if the number of receptacles are unknown [Table 220.3(A) Note b].

Bank - General Lighting and Receptacle Demand Load
Section 220.3(A) Note[b]

Bank - 18,000 sq ft
Number of Receptacles Unknown

Determine demand load for lighting and receptacles.

Table 220.3(A), Lighting load for a bank is 3 1/2 VA per sq ft

18,000 sq ft x 3 1/2 VA per sq ft = 63,000 VA lighting load

63,000 VA x 1.25 = 78,750 VA

18,000 sq ft x 1 VA per ft = 18,000 VA receptacle load

78,750 VA lighting + 18,000 VA receptacle = 96,750 VA demand load

Figure 11-17
Bank – General Lighting and Receptacle Demand Load

Signs - Commercial Buildings, Pedestrian Access
Section 220.3(B)(6)

Commercial Building
With Pedestrian Access
[600.5(A)]

Minimum load for exterior
outline lighting is 1,200 VA.

Determine service demand load for exterior sign.

Continuous loads on feeders are calculated at 125%.

1,200 VA sign outlet x 1.25 = 1,500 VA demand load

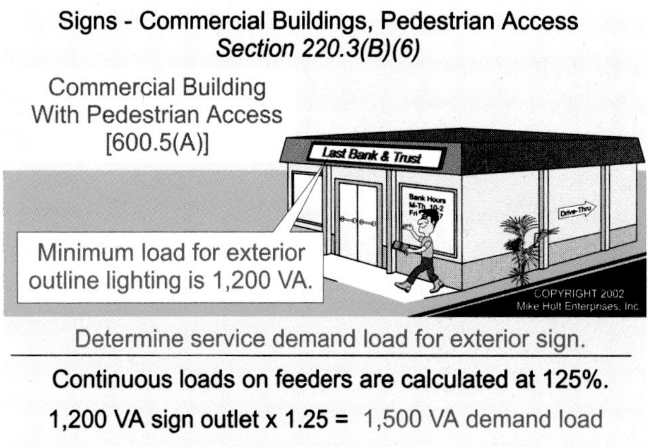

Figure 11-18
Signs – Commercial Buildings, Pedestrian Access

❑ Bank General Lighting and Receptacle

What is the general lighting demand load (including receptacles) for an 18,000 sq ft bank? The number of receptacles is unknown. Figure 11–17.

(a) 68 kVA   (b) 110 kVA   (c) 84 kVA   (d) 97 kVA

• Answer: (d) 97 kVA

(a) General Lighting (18,000 sq ft $\times$ 3.5 VA $\times$ 1.25)   78,750 VA

(b) Receptacle Load (18,000 sq ft $\times$ 1 VA)   <u>18,000 VA</u>

Total demand load   96,750 VA

❑ Office General Lighting and Receptacle

What is the general lighting demand load for a 28,000 sq ft office building that has 160 receptacles?

(a) 111 kVA   (b) 110 kVA   (c) 128 kVA   (d) 142 kVA

• Answer: (d) 142 kVA

(a) General Lighting (28,000 sq ft 3.5 VA $\times$ 1.25)   122,500 VA

(b) Receptacle Load (160 receptacles $\times$ 180 VA)   28,800 VA

First 10,000 VA at 100%   <u>10,000 VA</u> $\times$ 1.00 = 10,000 VA

Remainder at 50%   18,800 VA $\times$ 0.50 = <u>9,400 VA</u>

141,900 VA/1,000 = 142 kVA

11–17 SIGNS [220.3(B)(6) and 600.5]

The *NEC* requires each commercial occupancy that is accessible to pedestrians to be provided with at least one 20A branch circuit for a sign [600.5(A)]. The load for the required exterior signs or outline lighting shall be a minimum of 1,200 VA [220.3(B)(6)]. A sign outlet is considered to be a continuous load and the feeder load must be sized at 125% of the continuous load [215.2(A)(1), and 230.42].

❑ Sign Demand Load

What is the demand load for one electric sign? Figure 11–18.

(a) 1,200 VA   (b) 1,500 VA   (c) 1,920 VA   (d) 2,400 VA

• Answer: (b) 1,500 VA

1,200 VA $\times$ 1.25 = 1,500 VA

11–18 NEUTRAL CALCULATIONS [220.22]

The neutral load is considered the maximum unbalanced demand load between the grounded (neutral) conductor and any one ungrounded (hot) conductor as determined by the calculations in Article 220, Part B. This means that line-to-line loads are not considered when sizing the grounded (neutral) conductor.

Reduction Over 200A

The feeder/service net computed load for 3-wire 1Ø or 4-wire, 3Ø systems supplying linear loads can be reduced for that portion of the unbalanced load over 200A by a multiplier of 70 percent.

❑ **Reduction Over 200A**

What is the neutral demand load for a balanced 400A, 3-wire, 120/240V feeder?

(a) 400A   (b) 340A   (c) 300A   (d) 500A

• Answer: (b) 340A

Total Neutral Load	400A
First 200A at 100%	200A × 1.00 = 200A
Remainder at 70%	200A × 0.70 = 140A
	340A

Reduction not Permitted

The neutral demand load shall not be permitted to be reduced for 3-wire, 1Ø, 208Y/120V or 480Y/277V circuits consisting of two line wires and the common conductor (neutral) of a 4-wire, 3Ø wye system. This is because the common (neutral) conductor of a 3-wire circuit connected to a 4-wire, 3Ø wye system carries approximately the same current as the phase conductors. See 310.15(B)(4)(b). This can be proven with the following formula:

$$I_{Neutral} = \sqrt{(I_{Line\ 1}^2 + I_{Line\ 2}^2) - (I_{Line\ 1} \times I_{Line\ 2})}$$

$Line_1$ = Neutral Current of Line 1
$Line_2$ = Neutral Current of Line 2

❑ **Three-Wire Wye Neutral Current**

What is the neutral demand load for a balanced 300A 3-wire 208Y/120V feeder supplied from a 4-wire, 3Ø wye system? Figure 11–19.

(a) 100A   (b) 200A   (c) 300A   (d) none of these

• Answer: (c) 300A

$$I_{Neutral} = \sqrt{(I_{Line\ 1}^2 + I_{Line\ 2}^2) - (I_{Line\ 1} \times I_{Line\ 2})}$$

$$I_{Neutral} = \sqrt{(300A^2 + 300A^2) - 300A \times 300A} = \sqrt{180,000 - 90,000} = \sqrt{90,000} = 300A$$

Nonlinear Loads

The neutral demand load cannot be reduced for nonlinear loads supplied from a 3Ø, 4-wire, wye-connected system because they produce *triplen harmonic currents* that add on the neutral conductor, which can require the neutral conductor to be larger than the ungrounded conductor load; see 220.22 FPN 2.

PART C – LOAD CALCULATION EXAMPLES

MARINA [555.12]

The *NEC* permits a demand factor to apply to the receptacle outlets for boat slips at a marina. The demand factors of 555.6 are based on the number of receptacles on the feeder. The receptacles must also be balanced between the lines to determine the number of receptacles on any given line.

❑ **Marina Receptacle Outlet Demand [555.12]**

What size 120/240V, 1Ø, service is required for a marina that has twenty 120V, 20A receptacles and twenty 240V, 30A receptacles? Figure 11–20.

(a) 200A   (b) 420A   (c) 560A   (d) 720A

• Answer: (c) 560A

	Line 1	Line 2
Ten 20A 120V	200A	200A
Twenty 30A 240V	600A	600A
	800A	800A × 0.7 = 560A per line

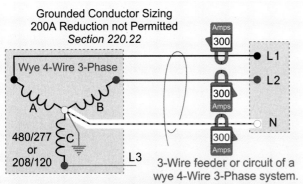

Grounded Conductor Sizing
200A Reduction not Permitted
Section 220.22

Wye 4-Wire 3-Phase

A   B   C

480/277 or 208/120   L3

3-Wire feeder or circuit of a wye 4-Wire 3-Phase system.

The 70% reduction of the unbalanced load over 200A does NOT apply to wye 3-wire 1-phase circuits of a wye 4-wire 3-phase system because the neutral carries about the same current as the phase conductors.

Proof: Determine neutral current of 300A 3-wire feeder.

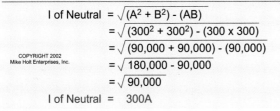

$$\text{I of Neutral} = \sqrt{(A^2 + B^2) - (AB)}$$
$$= \sqrt{(300^2 + 300^2) - (300 \times 300)}$$
$$= \sqrt{(90{,}000 + 90{,}000) - (90{,}000)}$$
$$= \sqrt{180{,}000 - 90{,}000}$$
$$= \sqrt{90{,}000}$$
$$\text{I of Neutral} = 300A$$

COPYRIGHT 2002
Mike Holt Enterprises, Inc.

Figure 11-19
Grounded Conductor Sizing

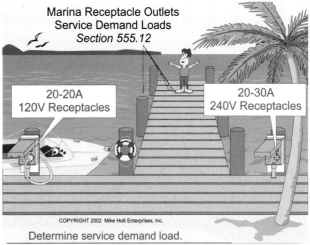

Marina Receptacle Outlets
Service Demand Loads
Section 555.12

20-20A 120V Receptacles

20-30A 240V Receptacles

COPYRIGHT 2002 Mike Holt Enterprises, Inc.

Determine service demand load.

555.12 - Demand load based on number of receptacles. Balance receptacles to determine maximum on any line.

	Line 1	Line 2
20- 30A 240V receptacles	20 x 30A----------	20 x 30A
20- 20A 120V receptacles	10 x 20A	10 x 20A
	30- 800A	30- 800A

555.12, demand load for 30 receptacle is 70%.
800A x 0.7 demand factor = 560A demand load per line

Figure 11-20
Marina Receptacle Outlet - Service Demand Loads

MOBILE/MANUFACTURED HOME PARK [550.31]

The service demand load for a mobile/manufactured home park is sized according to the demand factors of Table 550.31 to the larger of 16,000 VA for each mobile/manufactured home lot or the calculated load for each mobile/manufactured home site according to 550.18.

❏ **Mobile/Manufactured Home Park**

What is the demand load for a mobile/manufactured home park that has facilities for 35 sites? The system is 120/240V, 1Ø. Figure 11–21.

(a) 400A   (b) 560A   (c) 800A   (d) 1,000A

• Answer: (b) 560A

$16{,}000 \text{ VA} \times 35 \text{ sites} \times 0.24 = 134{,}400 \text{ VA}$

$I = VA/E = 134{,}400 \text{ VA}/240V = 560A$

RECREATIONAL VEHICLE PARK [551.73]

Recreational vehicle parks are calculated according to the demand factor of Table 551.73. The total calculated load is based on:

• 2,400 VA for each 20A supply facilities site,
• 3,600 VA for each 20A and 30A supply facilities site, and
• 9,600 VA for each 50A, 120/240V supply facilities site.

❏ **Recreational Vehicle Park**

What is the demand load for a recreational vehicle park that has ten 20A supply facilities, fifteen 20 and 30A supply facilities and twenty 50A supply facilities? The system is 120/240V, 1Ø. Figure 11–22.

(a) 420A   (b) 460A

(c) 520A   (d) 600A

• Answer: (b) 460A

Mobile/Manufactured Home Park Site Demand Load
Section 550.31

COPYRIGHT 2002 Mike Holt Enterprises, Inc.

Park Contains 35 Sites

Determine service demand load for 35 sites.

550.31, minimum site size is 16,000 VA.
Table 550.31 demand factor for 35 sites is 24%.
16,000 VA x 35 sites x 0.24 = 134,400 VA demand load

$$I = \frac{VA}{E} = \frac{134{,}400 \text{ VA}}{240V} = 560A$$

Figure 11-21
Mobile Home Park – Service Demand Loads

Recreational Vehicle Park Site Demand Load
Section 551.73

Park Contains:
10 - 20A supply facilities
15 - 30/20A supply facilities
20 - 50A supply facilities

Determine service demand load for 45 sites.

Determine VA of all sites, apply Tbl 551.73 demand factor

20A site   = 2,400 VA per site x 10 sites =   24,000 VA
30/20A site = 3,600 VA per site x 15 sites =   54,000 VA
50A site   = 9,600 VA per site x 20 sites = 192,000 VA
Total connected site VA   45 sites = 270,000 VA

Table 551.73 demand for 45 sites = 41%
270,000 VA x 0.41 = 110,700 VA demand load

$$I = \frac{VA}{E} = \frac{110{,}700\ VA}{240V} = 461A$$

Figure 11-22
Recreational Vehicle Park Site Demand Load

The service for the recreational vehicle park is sized according to the demand factors of Table 551.73

Ten 20A sites (2,400 VA × 10 sites)	24,000 VA
Fifteen 20/30A sites (3,600 VA × 15 sites)	54,000 VA
Twenty 50A sites (9,600 VA × 20 sites)	192,000 VA
Demand Factor [Table 551.73]	270,000 VA × 0.41 = 110,700 VA

I = VA/E = 110,700 VA/240V = 461A

RESTAURANT – OPTIONAL METHOD [220.36]

An optional method of calculating the demand service load for a restaurant is permitted. The following steps can be used to determine the service size:

Step 1: Determine the total connected load. Add the nameplate rating of all loads at 100% and include both the air conditioner and heat load.

Step 2: Apply the demand factors from from Table 220.36 to the total connected load [Step 1].

❑ **All-Electric Restaurant**

What is the demand load for an all-electric restaurant (208Y/120, 3Ø) that has a total connected load of 300 kVA?

(a) 420A   (b) 472A   (c) 520A   (d) 600A

  • Answer: (b) 472A

Total Connected Load	300 kVA
First 200 kVA at 80%	200 kVA × .80 = 160 kVA
201-325 kVA at 10%	100 kVA × .10 = 10 kVA
Total Demand Load	170 kVA

I = 170,000/(208 × 1.732) = 472A

❑ **Not All-Electric Restaurant**

What is the demand load for a not all-electric restaurant (208Y/120, 3Ø) that has a total connected load of 300 kVA?

(a) 420A   (b) 460A   (c) 520A   (d) 695A

  • Answer: (d) 695A

Total Connected Load   300 kVA
First 200 kVA at 100%   200 kVA ×1.00 = 200 kVA
201-325 kVA at 50%   100 kVA × 0.50 = 50 kVA
Total Demand Load   250 kVA
I = 250,000/(208 × 1.732) = 695A

❏ Restaurant Optional Method

Using the optional method, what is the service size for the loads listed in Figure 11–23?

Step 1: Determine the total connected load.

(1) Air-conditioning demand [Table 430.150]
    VA = 208V× 59.4A × 1.732   21,399 VA

(2) Kitchen equipment
    Ovens 10 kVA × 2 units   20,000 VA
    Mixers 3 kVA × 2 units   6,000 VA
    Water Heaters 7.5 kVA × 1 unit   7,500 VA

(3) Lighting [Table 220.3(A)]
    (8,400 sq ft × 2 VA)   16,800 VA

(4) Multioutlet assembly [220.3(B)(8)]
    147 ft/5 ft = thirty 5 ft sections
    (180 VA × 30)   5,400 VA
    Simultaneously used (180 VA × 20) 3,600 VA

(5) Receptacles [220.3(B)(9)]
    (180 VA × 40 receptacles)   7,200 VA

(6) Separate circuits
    (120V × 20A × 10 circuits)   24,000 VA

(7) Sign [220.3(B)(6), and 600.5]   1,200 VA

Totals
(1) Air-conditioning   21,399 VA
(2) Kitchen equipment   33,500 VA
(3) Lighting   16,800 VA
(4) Multioutlet assembly   9,000 VA
(5) Receptacles   7,200 VA
(6) Separate circuits   24,000 VA
(7) Sign   1,200 VA
Total connected load   113,099 VA

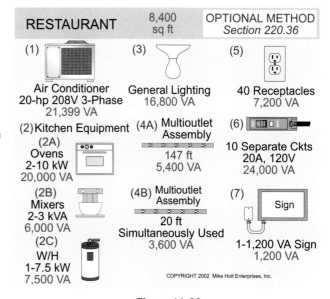

RESTAURANT   8,400 sq ft   OPTIONAL METHOD Section 220.36

(1) Air Conditioner 20-hp 208V 3-Phase 21,399 VA
(3) General Lighting 16,800 VA
(5) 40 Receptacles 7,200 VA
(2) Kitchen Equipment
(2A) Ovens 2-10 kW 20,000 VA
(2B) Mixers 2-3 kVA 6,000 VA
(2C) W/H 1-7.5 kW 7,500 VA
(4A) Multioutlet Assembly 147 ft 5,400 VA
(4B) Multioutlet Assembly 20 ft Simultaneously Used 3,600 VA
(6) 10 Separate Ckts 20A, 120V 24,000 VA
(7) Sign 1-1,200 VA Sign 1,200 VA
COPYRIGHT 2002 Mike Holt Enterprises, Inc.

Figure 11-23
Restaurant Demand Load Optional Method

Step 2:   Apply the demand factors from Table 220.36 to the total connected load.

<u>All-Electric</u> If the restaurant is all-electric, a demand factor of 80% is permitted to apply to the first 250 kVA.

113,099 VA × 0.8 = 90,479 VA

I = VA/(E × √3) = 90,479 VA/(208V × 1.732) = 251A

<u>Not All-Electric</u> If the restaurant is not all-electric, then a demand factor of 100% applies to the first 250 kVA.

113,099 VA at 100% = 113,099 VA

I = VA/(E × √3) = 113,099 VA/(208V × 1.732) = 314A

SCHOOL – OPTIONAL METHOD [220.34]

An optional method of calculating the demand service load for a school is permitted. The following steps can be used to determine the service size:

Step 1:   Add the nameplate rating of all loads at 100% and select the larger of the A/C versus heat load.

Step 2:   Determine the average VA per sq ft by dividing the total connected load (Step 1) by the sq ft of the building.

Step 3:   Determine the demand VA per sq ft by applying Table 220.34 demand factors to the average VA per sq ft.

Step 4:   Determine the net VA by multiplying the demand VA per sq ft (Step 3) by the sq ft of the school building.

❏ **School Optional Method**

What is the service size for the following loads? The system voltage is 208Y/120V, 3Ø. Figure 11–24.

(1) Air-conditioning	50,000 VA
(2) Cooking equipment	40,000 VA
(3) Lighting	100,000 VA
(4) Multioutlet assembly	10,000 VA
(5) Receptacles	40,000 VA
(6) Separate circuits	40,000 VA
	280,000 VA

(a) 300A   (b) 400A   (c) 500A   (d) 600A

• Answer: (c) 500A

Step 1:   Determine the average VA per sq ft (total connected load/sq ft area):
280,000 VA/10,000 sq ft = 28 VA per sq ft

Step 2:   Apply the demand factors from Table 220.34 to the VA per sq ft

Average VA per sq ft	28 VA
First 3 VA at 100%	– 3 VA × 1.00 = 3.00 VA
	25 VA
Next 17 VA at 75%	– 17 VA × 0.75 = 12.75 VA
Remainder at 25%	8 VA × 0.25 = 2.00 VA
	17.75 VA

Step 3:   Net VA per sq ft = 17.75 VA × 10,000 sq ft = 177,500 VA
Service Size – I = VA/(E × $\sqrt{3}$) = 177,500 VA/(208V × 1.732) = 493A

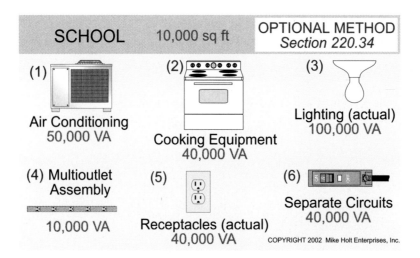

Figure 11-24
School – Optional Service Demand Load

SERVICE DEMAND LOAD USING THE STANDARD METHOD

For exam preparation purposes, you are not expected to calculate the total demand load for a commercial building. However, you are expected to know how to determine the demand load for the individual loads (Steps 1 through 10 below). For your own personal knowledge, you can use the following steps to determine the total demand load for a commercial building for the purpose of sizing the service.

Determine the Demand Load for Each Type of Load

Step 1: Determine the general lighting load: Table 220.3(A) and Table 220.11.
Step 2: Determine the receptacle demand load [220.13].
Step 3: Determine the appliance demand load at 100%.
Step 4: Determine the demand load for show-windows at 125%, [220.12].
Step 5: Determine the demand load for multioutlet assembly, [220.3(B)(8)].
Step 6: Determine the larger of:
　　　　　Air-conditioning at 100%.
　　　　　Heat at 100%, [220.15].
Step 7: Determine the demand load for educational cooking equipment, [220.19 Note 5].
Step 8: Determine the kitchen equipment demand load, [Table 220.20].
Step 9: All other noncontinuous loads at 100%.
Step 10: All other continuous loads at 125%, [215.2(A)(1), and 230.42].

Determine the Total Demand Load

Add up the individual demand loads of Steps 1 through 10.

Determine the Service Size

Divide the total demand load (Part B) by the system voltage.

$1\emptyset = I = VA/E$　　　　　　　$3\emptyset = I = VA/(E \times 1.732)$

PART D – LOAD CALCULATION EXAMPLES

BANK (120/240V, 1Ø). Figure 11–25.

(1) Air-conditioning [Table 430.148]
　　10-hp, 1Ø (230V × 50A)　　　　　　　　　　　　　　　　11,500 VA

(2) Heat [220.15 and 220.21] 10 kW omit

(3) Lighting [Table 220.3(A)]
　　General Lighting (30,000 sq ft × 3.5 VA = 105,000 VA* × 1.25)　　131,250 VA*
　　Actual Lighting (200 units × 1.65A × 120V × 1.25 = 49,500 VA, omit)

(4) Motor [Table 430.148] 5-hp, 1Ø (230V × 28A)　　　　　　　　6,440 VA
　　(Some exams use 240V × 28A = 6,720 VA)

(5) Receptacles [220.13] (400 receptacles × 180 VA)　　72,000 VA
　　　　　　　　　　　　　　　　　　　　　　　−10,000 VA　× 1.00 =　10,000 VA*
　　　　　　　　　　　　　　　　　　　　　　　 62,000 VA　× 0.50 =　31,000 VA*

(6) Separate Circuits (30 circuits × 20A × 120 V)　　　　　　　　72,000 VA*
(7) Sign [220.3(B)(6)] (1,200 VA* × 1.25)　　　　　　　　　　　　1,500 VA*
*Neutral load

SUMMARY	Overcurrent Protection	Conductor	Neutral
(1) Air conditioner	11,500 VA	11,500 VA	0 VA
(3) Lighting	131,250 VA	131,250 VA	131,250 VA
(4) Motor	6,440 VA	6,440 VA	0 VA
(5) Receptacles	41,000 VA	41,000 VA	41,000 VA
(6) Separate Circuits	72,000 VA	72,000 VA	72,000 VA
(7) Sign	1,500 VA	1,500 VA	1,500 VA
	263,690 VA	263,690 VA	245,750 VA

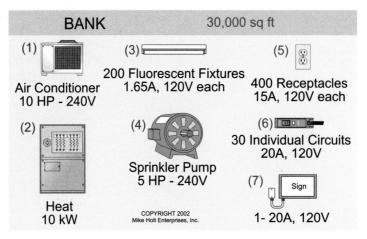

Figure 11-25
Bank

Feeder/service size I = VA/E = 263,690 VA /240V = 1,099A

Feeder and service conductor I = VA/E = 286,690 VA/240V = 1,099A

Neutral Amperes The neutral conductor is permitted to be reduced according to the requirements of 220.22 for that portion of the unbalanced load that exceeds 200A. There shall be no reduction of the neutral capacity for that portion of the load that consists of nonlinear loads supplied from a 4-wire, wye-connected, 3Ø system. Since this system is a 120/240V, 1Ø system, we are permitted to reduce the neutral by 70% for that portion that exceeds 200A.

I = VA/E I = 245,750/240V   1,024A
First 200A − 200A × 1.00 =   200A
Remainder   824A × 0.7 =   <u>577A</u>
                         777A

❏ **Summary Questions**

The service overcurrent protection device must be sized no less than 1,099A. The service conductors must have an ampacity no less than the overcurrent protection device rating [240.4(C)]. The grounded (neutral) conductor must not be less than 777A.

Overcurrent Protection: What size service overcurrent protection device is required?
(a) 800A   (b) 1,000A   (c) 1,200A   (d) 1,600A
    • Answer: (c) 1,200A [240.6(A)].

Service Conductor Size: What size service THHN conductors are required in each raceway if the service is parallel in four raceways?
(a) 250 kcmil   (b) 300 kcmil   (c) 350 kcmil   (d) 400 kcmil

    • Answer: (c) 350 kcmil
    1,200A/4 raceways = 300A per raceway, sized based on 75°C terminal rating [110.14(C)(2)]
    350 kcmil rated 310A × 4 = 1,240A [240.4(C) and Table 310.16]

Note: 300 kcmil THHN has an ampacity of 320A at 90°C, but we must size the conductors at 75°C, not 90°C!

Service Neutral Conductor Size: What size service grounded (neutral) conductor is required in each of the four raceways?
(a) 1/0 AWG   (b) 2/0 AWG   (c) 250 kcmil   (d) 350 kcmil

    • Answer: (b) 2/0 AWG
    The grounded (neutral) conductor must be sized no less than:
    (1) 12$\frac{1}{2}$% of the area of the line conductor, 350,000 × 4 = 1,400,000 × 0.125 = 175,000 [250.24(B)]
       175,000/4 = 43,750 circular mil, 3 AWG per raceway [Chapter 9, Table 8].
    (2) When paralleling conductors, no grounded (neutral) conductor can be smaller than 1/0 AWG [310.4].
    (3) The service neutral conductor must have an ampacity of at least 777A.
       777A/4 = 194A, Table 310.16, 3/0 AWG is required at 75°C [110.14(C)]

Grounding Electrode Conductor Size [250.66(B)]: What size grounding electrode conductor is required to a concrete encased electrode?

(a) 4 AWG   (b) 1/0 AWG   (c) 2/0 AWG   (d) 3/0 AWG

• Answer: (a) 4 AWG

OFFICE BUILDING (480Y/277V, 3Ø). Figure 11–26.

(1) Air-conditioning (Table 430.150)
460V × 7.6A × 1.732 = (6,055 VA × 15 units)   90,825 VA

(2) Lighting (*NEC*) [Table 220.3(A) and 220.10(B)]
28,000 sq ft × 3.5 VA × 1.25 = 122,500 VA (omit)
Lighting Electric Discharge (Actual) [215.2(A)(1) and 230.42]
500 lights × 0.75A × 277V = 103,875 VA × 1.25   129,844 VA*

(3) Lighting, Track (Actual) [220.12(B)] 150 VA per 2 ft
200 ft/2 ft = 100 sections × 150 VA = 15,000 VA × 1.25   18,750 VA*

(4) Receptacles [220.13](Actual) (200 receptacles × 180 VA)   36,000 VA
                                                            −10,000 VA × 100% =   10,000 VA*
                                                             26,000 VA × 50% =   13,000 VA*

(5) Multioutlet Assembly [220.3(B)(8)]
147 ft/5 ft = thirty 5 ft sections (30 sections × 180 VA)   5,400 VA*
Simultaneously used (20 ft × 180 VA)   3,600 VA*

(6) Separate Circuits (noncontinuous loads) (120V × 20A × 30 circuits)   72,000 VA*

(7) Sign [220.3(B)(6)] (1,200 VA × 1.25)   1,500 VA*

*Neutral load

SUMMARY	Overcurrent Protection	Conductor	Neutral
(1) Air-conditioning	90,825 VA	90,825 VA	0 VA
(2) Lighting (actual)	129,844 VA	129,844 VA	129,844 VA
(3) Lighting (track)	18,750 VA	18,750 VA	18,750 VA
(4) Receptacles (actual)	23,000 VA	23,000 VA	23,000 VA
(5) Multioutlet assembly	9,000 VA	9,000 VA	9,000 VA
(6) Separate circuits	72,000 VA	72,000 VA	72,000 VA
(7) Sign	1,500 VA	1,500 VA	1,500 VA
	344,919 VA	344,919 VA	254,094 VA

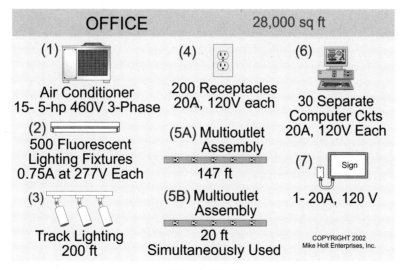

Figure 11-26
Office

Feeder/service protection device I = VA/(E × √3), I = 344,919 VA/(480 × 1.732) = 415A

Feeder and service conductor amperes I = VA/(E × √3), I = 344,919 VA/(480 × 1.732) = 415A

Neutral conductor I = VA/(E × √3), I = 254,094 VA/(480V × 1.732) = 306A

❏ **Summary**

The service overcurrent protection device and conductor must be sized no less than 415A and the grounded (neutral) conductor must not be less than 306A.

Overcurrent Protection: What size service overcurrent protection device is required?

(a) 300A   (b) 350A   (c) 450A   (d) 500A

• Answer: (c) 450A [240.6(A)]

Since the total demand load for the overcurrent device is 415A, the minimum service permitted is 450A.

Service Conductor Size: What size service conductors are required if total calculated demand load is 415A?

(a) 300 kcmil   (b) 400 kcmil   (c) 500 kcmil   (d) 600 kcmil

• Answer: (d) 600 kcmil rated 420A [240.4(B) and Table 310.16]

The service conductors must have an ampacity of at least 415A protected by a 450A overcurrent protection device [240.4(B)] and the conductors must be selected based on 75°C insulation rating [110.14(C)(2)].

Service Neutral Conductor Size: What size service grounded (neutral) conductor is required for this service?

(a) 1/0 AWG   (b) 3/0   (c) 250 kcmil   (d) 300 kcmil

• Answer: (d) 300 kcmil

The grounded (neutral) conductor must be sized no less than:

(1) The required grounding electrode conductor [250.24(B) and Table 250.66], 1/0 AWG.

(2) An ampacity of at least 270A, Table 310.16, 300 kcmil is rated 280A at 75°C [110.14(C)].

Note: Because there is 125A of electric-discharge lighting [103,875 VA/(480 × 1.732)], the neutral conductor is not permitted to be reduced for that portion of the load in excess of 200A [220.22].

RESTAURANT - Standard Load Calculation (208Y/120 Volt, 3-Phase). Figure 11–27.

(1) Air-conditioning [Table 430.150] 208V × 59.4A =   21,399VA

(2) Kitchen equipment [220.20]

Ovens (10 kW × 2 units)   20.0 kW

Mixer (3 kW × 2 units)   6.0 kW

Water Heaters (7.5 kW × 1 unit)   7.5 kW

Total connected   33.5 kW × 0.7 × 1,000 = 23,450 kW

(3) Lighting *NEC* [Table 220.3(A) and 220.10(B)]

8,400 sq ft × 2 VA × 1.25 = 21,000 VA, omit

(4) Lighting, Track 150 VA per two ft [220.12(B)]

50 ft/2 = 25 sections × 150 VA × 1.25   4,688 VA

100 lights × 1.65A × 120V × 1.25   24,750 VA

(5) Multioutlet Assembly [220.3(B)(8)]

50/5 = ten 5 ft sections (10 sections × 180 VA)   1,800 VA

Simultaneous used (20 ft × 180 VA)   3,600 VA

(6) Receptacles (Actual) [220.13] (40 receptacles × 180 VA)   7,200 VA

(7) Separate Circuits [215.2(A)(1) and 230.42] (120V × 20A × 10 circuits)   24,000 VA

(8) Sign [220.3(B)(6)] (1,200 VA × 1.25)   1,500 VA

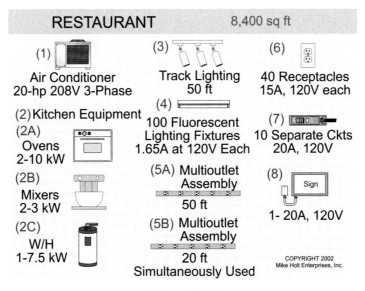

Figure 11-27
Restaurant

SUMMARY	Overcurrent Protection	Conductor
(1) Air-conditioning	21,399 VA	21,399 VA
(2) Kitchen equipment	23,450 VA	23,450 VA
(3) Lighting (track)	4,688 VA	4,688 VA
(4) Lighting (actual)	24,750 VA	24,750 VA
(5) Multioutlet assembly	5,400 VA	5,400 VA
(6) Receptacles (actual)	7,200 VA	7,200 VA
(7) Separate circuits	24,000 VA	24,000 VA
(8) Sign	1,500 VA	1,500 VA
	112,387 VA	112,387 VA

Feeder and service protection device $I = VA/(E \times \sqrt{3}) = 112,387 \text{ VA}/(208 \times 1.732) = 312A$
Feeder and service conductor $I = VA/(E \times \sqrt{3}) = 112,387 \text{ VA}/(208 \times 1.732) = 312A$

❏ **Summary Questions**

The service overcurrent protection device must be sized no less than 312A. The service conductors must be no less than 312A.

Overcurrent Protection: What size service overcurrent protection device is required?
(a) 300A   (b) 350A   (c) 450A   (d) 500A
   • Answer: (b) 350A [240.6(A)]
   Since the total demand load for the overcurrent device is 312A, the minimum service permitted is 350A [240.6(A)].

Service Conductors Size: What size service conductors are required?
(a) 4/0 AWG   (b) 250 kcmil   (c) 300 kcmil   (d) 400 kcmil
   • Answer: (d) 400 kcmil
   The service conductors must have an ampacity of at least 312A and be protected by a 350A protection device [240.4(B)]. Next size up is permitted and the conductors must be selected according to Table 310.16, 400 kcmil is based on 75°C insulation rating [110.14(C)(2)] and is rated 335A.

Unit 11 – Commercial Load Calculations Summary Questions

Calculations Questions

Part A – General

11–2 Conductor Ampacity

1. The _____ of a conductor is the rating in amperes a conductor can carry continuously without exceeding its insulation temperature rating. The allowable ampacities as listed in Table 310.16 are affected by ambient temperature, current flow, conductor insulation and conductor bundling [310.10].
   (a) load rating   (b) ampacity   (c) demand load   (d) continuous factor

11–3 Conductor Overcurrent Protection

2. The purpose of conductor _____ is to protect the conductors against excessive or dangerous temperatures. If the ampacity of a conductor does not correspond with the standard ampere rating of a fuse or circuit breaker, the next size up protection device is permitted. This applies only if the conductors do not supply multioutlet receptacles and if the next size overcurrent protection device does not exceed 800A.
   (a) short-circuit protection   (b) ground-fault protection
   (c) overload protection   (d) overcurrent protection

3. The following is a partial list of some of the standard ampere ratings for overcurrent protection devices (fuses and inverse-time circuit breakers): 15, 25, 35, 45, 80, 90, 110, 175, 250 and 350.
   (a) True   (b) False

11–4 Voltages

4. Unless other voltages are specified, branch-circuit, feeder and service loads shall be computed at a nominal system voltage of _____.
   (a) 600Y/347   (b) 240/120   (c) 208Y/120   (d) any of these

11–5 Rounding and Amperes [220.2(B)]

5. There are no *NEC* rules for rounding when a calculation results in a fraction of an ampere, but it does contain the following note: except where the computations result in a _____ of an ampere or larger, such fractions may be dropped.
   (a) 0.05   (b) 0.5   (c) 0.49   (d) 0.51

Part B – Loads

11–6 Air-Conditioning

6. What is the feeder or service demand load required for the air-conditioning of a 6 unit office building? Each unit contains one air conditioner rated 3-hp, 230V, 1Ø.
   (a) 13.2 kVA   (b) 23.4 kVA   (c) 45.5 kVA   (d) 52.1 kVA

11–7 Dryers

7. What size branch-circuit conductor and overcurrent protection is required for a 5 kW dryer rated 240V located in the laundry room of a multifamily dwelling?
   (a) 12 AWG with a 20A overcurrent current protection device
   (b) 10 AWG with a 20A overcurrent current protection device
   (c) 12 AWG with a 30A overcurrent current protection device
   (d) 10 AWG with a 30A overcurrent current protection device

8. What is the feeder and service demand for eight 6.75 kW dryers in a commercial laundromat?
   (a) 70 kW   (b) 54 kW   (c) 35 kW   (d) 27 kW

11–8 Electric Heat

9. • What size conductor and protection is required for a 480V, 15 kW, 1Ø heat strip that has a 1.6A blower motor?
   (a) 10 AWG THHN with 30A protection   (b) 6 AWG THHN with 45A protection
   (c) 10 AWG THHN with 40A protection   (d) 4 AWG THHN with 70A protection

10. What is the feeder and service demand load in VA for a building that has four 230V, 20 kW, 1Ø heat strips that have a 5.4A blower motor for each unit?
    (a) 100 kW   (b) 50 kW   (c) 125 kW   (d) 85 kW

11–9 Kitchen Equipment

11. What is the branch-circuit demand load in amperes for one 11.4 kW oven rated 240V?
    (a) 60A   (b) 27A   (c) 48A   (d) 33A

12. • What is the feeder and service demand load for the following?

    Water Heater   9 kW
    Booster heater   12 kW
    Mixer   3 kW
    Oven   2 kW
    Dishwasher   1.5 kW
    Waste disposal   1 kW
    (a) 19 kW   (b) 21 kW   (c) 29 kW   (d) 12 kW

13. What is the feeder and service demand load for the following?

    Water heater   14 kW
    Booster heater   11 kW
    Mixer   7 kW
    Oven   9 kW
    Dishwasher   1.5 kW
    Waste disposal   3 kW
    (a) 25 kW   (b) 30 kW   (c) 20 kW   (d) 45 kW

11–10 Laundry Equipment

14. What is the feeder and service demand load for seven commercial washing machines at 1,500 VA each?
    (a) 10,500 VA   (b) 15,000 VA   (c) 1,125 VA   (d) none of these

11–11 Lighting – Demand Factors

15. What is the feeder demand load for 50 ft of multioutlet assembly and 10 ft of multioutlet assembly simultaneously used?
    (a) 3,600 VA   (b) 12,000 VA   (c) 7,200 VA   (d) 5,500 VA

16. What is the general lighting feeder or service demand load for a 24-room motel, 685 sq ft each?
    (a) 10 kVA   (b) 15 kVA   (c) 20 kVA   (d) 25 kVA

11–12 Lighting Without Demand Factors

17. What is the general lighting and receptacle feeder or service demand load for a 250,000 sq ft storage warehouse that has 200 receptacles?
    (a) 25 kVA   (b) 55 kVA   (c) 95 kVA   (d) 105 kVA

18. What is the feeder and service (overcurrent protection device) general lighting demand load for a 3,200 sq ft dance club?
    (a) 3,200 VA   (b) 6,400 VA   (c) 8,000 VA   (d) 12,000 VA

19. What is the feeder and service general lighting load for a 90,000 sq ft school?
    (a) 238 kVA   (b) 338 kVA   (c) 90 kVA   (d) 270 kVA

11–13 Lighting – Miscellaneous

20. How many 2 × 4 fluorescent luminaires, each rated 277V, 0.8A, can be connected to a 20A circuit? The four lamps are rated 40W each and the luminaire is to be on for more than three hours.
    (a) 5   (b) 7   (c) 9   (d) 20

21. How many 360W, 120V incandescent luminaires that will operate continuously can be connected to a 20A, 120V circuit?
    (a) 5   (b) 7   (c) 9   (d) 12

22. • What is the VA demand feeder and service load for 130 ft of show-window lighting?
    (a) 26 kVA   (b) 33 kVA   (c) 9 kVA   (d) 11 kVA

11–14 Multioutlet Receptacle Assembly

23. How many receptacle outlets are permitted on a 20A, 120V circuit in a commercial occupancy?
    (a) 10   (b) 13   (c) 15   (d) 20

11–15 Receptacle VA Load

24. What is the service demand load for 110 receptacles (15 or 20A, 125V) in a commercial building?
    (a) 5 kVA   (b) 10 kVA   (c) 15 kVA   (d) 20 kVA

25. What is the service demand load for the general lighting and receptacles for a 30,000 sq ft bank?
    (a) 111 kVA   (b) 151 kVA   (c) 123 kVA   (d) 175 kVA

26. • What is the service general lighting demand and receptacle load for a 10,000 sq ft office building with 75 receptacles?
    (a) 25 kVA   (b) 35 kVA   (c) 45 kVA   (d) 55 kVA

11–17 Signs

27. What is the feeder demand load for sizing the overcurrent protection device for one electric sign?
    (a) 1,400 VA   (b) 1,500 VA   (c) 1,920 VA   (d) 2,400 VA

11–18 Neutral Calculations

28. What is the neutral current for a balanced 150A, 3-wire, 208Y/120V feeder?
    (a) 0A   (b) 150A   (c) 250A   (d) 300A

Part C – Load Calculations

Marina

29. A marina has the following: 24 slips (20A, 240V receptacles) and 30 slips designed for boats over 20 ft (30A, 240V receptacles). What size service is required for the marina?
    (a) 400A   (b) 550A   (c) 900A   (d) 1,000A

30. What size conductor is required for the service if the total demand load per phase is 570A and the service is paralleled in two raceways?
    (a) 4/0 AWG   (b) 250 kcmil   (c) 300 kcmil   (d) 350 kcmil

Mobile Home Park

31. What is the feeder and service load for a mobile home park that has the facilities for 42 sites? The system is 120/240V, 1Ø.

     (a) 644A   (b) 512A   (c) 732A   (d) 452A

Recreational Vehicle Park

32. A recreational vehicle park has 17 sites with only 20A, 240V receptacles, 35 sites with 20 and 30A, 240V receptacles and 10 sites with 50A, 240V receptacles. The feeder and service demand load is approximately _____.

     (a) 355A   (b) 450A   (c) 789A   (d) 1,114A

Restaurant – Optional Method

33. A new restaurant has a total connected load of 400 kVA and is all electric. What is the demand load for the service?

     (a) 210 kVA   (b) 325 kVA   (c) 300 kVA   (d) 275 kVA

34. A new restaurant has a total connected load of 400 kVA and is not all electric. What is the demand load for the service?

     (a) 296 kVA   (b) 325 kVA   (c) 300 kVA   (d) 275 kVA

School – Optional Method

35. A 28,000 sq ft school has a total connected load of 590,000 VA. What is the demand load?

     (a) 350 kVA   (b) 400 kVA   (c) 450 kVA   (d) 500 kVA

☆ Challenge Questions

General Commercial Calculations

36. • If the service high-leg conductor only supplies 3Ø loads, what size service conductor is required for the high-leg? 3Ø loads: A/C (10-hp 230V, 3Ø) and Heat (10 kW 230V, 3Ø).

     (a) 10 AWG TW   (b) 12 AWG THW   (c) 10 AWG THHN   (d) 8 AWG THW

37. • A commercial office building has forty-six 250W, 120V lights installed on a 208Y/120V, 3Ø system. If the luminaires were on continuously, how many 20A, 120V circuits would be required?

     (a) 9 circuits   (b) 7 circuits   (c) 5 circuits   (d) 4 circuits

Conductor Sizing

38. • Three parallel service raceways are installed. Each raceway contains three 500 kcmil THHN conductors. What size bonding jumper is required for each service raceway?

     (a) 250,000   (b) 2/0 AWG   (c) 1/0 AWG   (d) 4/0 AWG

39. • If each service raceway contains 300 kcmil conductors, what size bonding jumper is required for the raceway?

     (a) 1 AWG   (b) 2 AWG   (c) 1/0 AWG   (d) 4 AWG

11–9 Kitchen Equipment

40. What is the kitchen equipment demand load for one 14 kW range, one 1.5 kW water heater, one 3/4 kW mixer, one 2 kW dishwasher, one 3 kW booster heater and one 3 kW coffee machine?

     (a) 17 kW   (b) 14 kW   (c) 15 kW   (d) 24 kW

11–11 Lighting – Demand Factors

41. • Each unit of a 100 unit hotel is 12 × 15 ft. In addition, there is an office that is 60 × 20 ft and there are hallways of 120 sq ft. What is the general lighting demand load?

     (a) 29 kVA   (b) 37 kVA   (c) 23 kVA   (d) 25 kVA

11–15 Receptacle VA Load

42. If in the hall and other areas of a motel there are 100 receptacle outlets (not in the motel rooms), the demand load added to the service for these receptacles would be _____.
    (a) 0 VA   (b) 10,000 VA   (c) 14,000 VA   (d) 20,000 VA

43. What is the receptacle demand load for a 20,000 sq ft office building?
    (a) 10,000 VA   (b) 40,000 VA   (c) 30,000 VA   (d) 15,000 VA

11–16 Banks and Offices – General Lighting and Receptacle

44. What is the general lighting and general-use receptacle load for a 30,000 sq ft bank?
    (a) 162 kVA   (b) 123 kVA   (c) 173 kVA   (d) 151 kVA

11–18 Neutral Calculations [220.22]

45. • A 208Y/120V, 3Ø service has a total connected load of 1,450A:
    (1) 600A of these are phase-to-phase loads
    (2) 300A are balanced 120V fluorescent lighting
    (3) 550A of other 120V loads
    The neutral demand for this service is _____.
    (a) 1,150A   (b) 650A   (c) 745A   (d) 420A

Part C – Load Calculations

Marina Calculations [555]

46. A marina shore power facility has twenty 20A, 240V receptacles, seventeen 30A, 240V receptacles and seven 50A, 240V receptacles. After applying demand factors, the service demand load for the shore power boxes is _____.
    (a) 1,260A   (b) 630A   (c) 1,160A   (d) 625A

Mobile/Manufactured Home Parks [550]

47. A 75-site mobile home park is designed for mobile homes that have a 14,000 VA load per site. The service demand load for the park is _____.
    (a) 1,200 kVA   (b) 264 kVA   (c) 222 kVA   (d) 201 kVA

Motel

48. A 40 unit motel (300 sq ft in each unit) has 3 kVA of air-conditioning and 4 kW of heat in each unit, and every two units share one 1.5 kW water heater. What is the demand load for the motel?
    (a) 281 kVA   (b) 202 kVA   (c) 226 kVA   (d) 251 kVA

Recreational Vehicle Park [551]

49. A recreational vehicle park has 42 sites: 3 sites rated 50A, 240/120V, 30 sites rated 30/20A and 9 sites rated 20A. The minimum feeder demand load for these sites would be _____.
    (a) 158 kVA   (b) 101 kVA   (c) 139 kVA   (d) 65 kVA

Restaurant Calculations

50. • The branch-circuit rating required for a 12 kW range in a restaurant would be _____.
    (a) 50A   (b) 45A   (c) 35A   (d) 30A

51. • A new restaurant has a total connected lighting load of 30 kVA. The kitchen equipment includes two gas stoves, one gas grill, three gas ovens, one 75-gallon gas water heater, one 5 kW dishwasher, two 2 kW coffee makers, five 2 kW kitchen appliances on their own circuit and ten 1.5 kVA small-appliance circuits. Using the optional method, the service demand load would be closest to _____.
     (a) 64 kVA   (b) 50 kVA   (c) 45 kVA   (d) 38 kVA

School – Optional Method [220.34]

52. • Using the optional method, what is the demand load (VA per sq ft) for a 10,000 sq ft school that has a total connected load of 320 kVA?
     (a) 15.75 VA per sq ft   (b) 18.75 VA per sq ft   (c) 29.00 VA per sq ft   (d) 12.75 VA per sq ft

53. Using the optional method, what is the demand load (kVA) for a 20,000 sq ft school that has a total connected load of 160 kVA?
     (a) 135 kVA   (b) 106 kVA   (c) 120 kVA   (d) 112 kVA

NEC Questions – Articles 600-725

Article 600 Electric Signs and Outline Lighting

54. The disconnecting means for each circuit leading to a sign located within a fountain shall be located in accordance with _____.
     (a) 430.102   (b) 440.14   (c) 680.12   (d) all or these

55. Signs or outline lighting systems operated by electronic or electromechanical controllers located external to the sign or outline lighting system shall be permitted to have a disconnecting means located _____.
     (a) within sight of the controller   (b) in the same enclosure with the controller
     (c) a or b   (d) none of these

56. Sign and outline lighting enclosures for live parts other than lamps and neon tubing shall _____.
     (a) have ample structural strength and rigidity   (b) be constructed of metal or shall be listed
     (c) if of sheet steel, be at least 0.016 in. thick   (d) all of these

57. The bottom of sign and outline lighting enclosures shall not be less than _____ above areas accessible to vehicles.
     (a) 12 ft   (b) 14 ft   (c) 16 ft   (d) 18 ft

58. An outdoor portable electric sign shall have a ground-fault circuit interrupter _____.
     (a) located on the sign
     (b) located in the power supply cord within 12 in. of the attachment plug
     (c) as an integral part of the attachment plug of supply cord
     (d) b or c

59. Ballasts, transformers and electronic power supplies for electric signs shall be located where accessible and shall be securely fastened in place. A working space at least _____ shall be provided.
     (a) 3 ft high, 3 ft wide by 3 ft deep
     (b) 4 ft high, 3 ft wide by 3 ft deep
     (c) 6 ft high, 3 ft wide by 3 ft deep
     (d) none of these

60. Ballasts, transformers and electronic power supplies for signs installed in suspended ceilings can be connected to the branch circuit by a _____.
     (a) fixed wiring method   (b) flexible wiring method
     (c) flexible cord   (d) a or b

Article 604 Manufactured Wiring Systems

61. Manufactured wiring systems shall be constructed with _____.
    (a) listed AC or MC cable   (b) 10 or 12 AWG copper-insulated conductors
    (c) conductors suitable for nominal 600V   (d) all of these

62. Each section of a manufactured wiring system shall be marked to identify _____.
    (a) its location   (b) the type of cable, flexible cord, or conduit
    (c) the size of the wires installed   (d) suitability for wet or damp locations

63. Manufactured wiring systems constructed with MC cable shall be supported and secured at intervals not exceeding

    _____.
    (a) 3 ft   (b) 4$\frac{1}{2}$ ft   (c) 6 ft   (d) none of these

Article 610 Cranes and Hoists

64. All exposed metal parts of cranes, hoists, and accessories shall _____ into a continuous electrical conductor.
    (a) be bonded with 6 AWG or larger conductors to be made
    (b) be metallically joined together
    (c) be grounded
    (d) not be grounded or made

65. The conductors to the hoistway door interlocks from the hoistway riser shall be flame retardant and suitable for a
    temperature of not less than _____, and the conductors shall be SF or equivalent.
    (a) 200°C   (b) 60°C   (c) 90°C   (d) 110°C

Article 620 Elevators, Dumbwaiters, Escalators, Moving Walks, Wheelchair Lifts, and Stairway Chair Lifts

66. • The minimum size parallel conductors permitted for elevator lighting is _____ provided the combined ampacity is
    equivalent to at least that of a 14 AWG wire.
    (a) 14 AWG   (b) 20 AWG   (c) 16 AWG   (d) 1/0 AWG

67. The minimum size conductor for operating control and signaling circuits in an elevator is _____ .
    (a) 20 AWG   (b) 16 AWG   (c) 14 AWG   (d) 12 AWG

68. A separate _____ is required for the elevator car lights, receptacle(s), auxiliary lighting power source, and ventilation on
    each elevator car.
    (a) branch circuit   (b) disconnect   (c) connection   (d) none of these

69. Where multiple driving machines are connected to a single elevator, escalator, moving walk, or pumping unit, there
    shall be one disconnecting means to disconnect the _____.
    (a) motor(s)   (b) control valve operating magnets   (c) a and b   (d) none of these

70. Duty on escalator and moving walk driving machine motors shall be rated as _____.
    (a) full time   (b) continuous   (c) various   (d) long term

71. All 125V, 1Ø, 15 and 20A receptacles installed in machine rooms and spaces for elevators, escalators, moving walks
    and lifts shall have ground-fault circuit-interrupter protection by a _____.
    (a) GFCI receptacle
    (b) GFCI circuit breaker
    (c) neither a or b since GFCI protection is not required
    (d) a or b

72. 125V, 1Ø, 15 and 20A receptacles installed in pits, in hoistways, on elevator car tops, and in escalator and moving walk
    wellways shall be _____.
    (a) on a GFCI-protected circuit   (b) of the GFCI type
    (c) a or b   (d) none of these

73. Each elevator shall have a single means for disconnecting all ungrounded main power-supply conductors for each unit:
    (a) Excluding the emergency power system.
    (b) Including the emergency or standby power system.
    (c) Excluding the emergency power system if it is automatic.
    (d) No elevator shall operate on an emergency power system.

Article 625 Electric Vehicle Charging System

74. Electric vehicle supply equipment must have a listed system of protection against electric shock of personnel, set at no more than 5 mA.
    (a) True   (b) False

Article 630 Electric Welders

75. Each arc welder shall have overcurrent protection rated or set at not more than _____ percent of the rated primary current of the welder.
    (a) 100   (b) 125   (c) 150   (d) 200

76. The ampacity for the supply conductors for a resistance welder with a duty cycle of 15 percent and a primary current of 21A is _____.
    (a) 9.45A   (b) 8.19A   (c) 6.72A   (d) 5.67A

Article 640 Audio Signal Processing, Amplification, and Reproduction Equipment

77. Installed audio distribution cable that is not terminated at equipment and not identified for future use with a tag is considered abandoned.
    (a) True   (b) False

78. All abandoned audio distribution cables shall be removed.
    (a) True   (b) False

79. Audio cables installed exposed on the surface of ceiling and sidewalls shall be supported by the structural components of the building in such a manner that the cable will not be damaged by normal building use. Such cables shall be secured to structural components by _____ designed and installed so as not to damage the cable.
    (a) straps   (b) staples   (c) hangers   (d) any of these

Article 645 Information Technology Equipment

80. Under a raised floor, liquidtight flexible conduit shall be permitted to enclose branch-circuit conductors for information technology communication equipment.
    (a) True   (b) False

81. Nonmetallic raceways of all types can be installed within the raised floor area in an information technology equipment room.
    (a) True   (b) False

82. Interconnecting cables under raised floors that support information technology equipment must have a cable that is listed as Type _____ cable having adequate fire-resistant characteristics suitable for use under raised floors of an information technology equipment room.
    (a) RF   (b) UF   (c) LS   (d) DP

83. _____ cables such as CL2, CM, MP or CATV can be installed within the raised floor area in an information technology equipment room.
    (a) Control   (b) Signal   (c) Communications   (d) all of these

84. _____ conductor cables 4 AWG and larger marked for use in cable trays or for CT use, can be within a raised floor of an information technology equipment room.
    (a) Green
    (b) Insulated
    (c) Single
    (d) all of these

85. Abandoned cables under an information technology room raised floor shall be removed, unless the cables are contained within a metal raceway.
    (a) True
    (b) False

86. In information technology equipment rooms, a single disconnecting means is permitted to control _____.
    (a) only the HVAC systems to the room
    (b) only the power to electronic computer/data-processing equipment
    (c) the electronic computer/data-processing equipment and the building supply
    (d) the HVAC systems to the room and power to electronic computer/data-processing equipment

87. Where a button is used as a means to disconnect power to electronic equipment in the information technology equipment room, pushing the button "in" must disconnect the power.
    (a) True
    (b) False

88. Each unit of an information technology system supplied by a branch circuit shall have a manufacturer's nameplate that includes the _____.
    (a) rating in volts
    (b) operating frequency
    (c) total load in amperes
    (d) all of these

90. Power for sensitive electronic equipment, called "Technical Power" is a separately derived 1Ø, 3-wire system with _____ volts to a grounded neutral conductor on each of two ungrounded conductors. The line-to-line voltage is _____.
    (a) 30/60
    (b) 60/120
    (c) 120/120
    (d) none of these

Article 647 Sensitive Electronic Equipment

90. The purpose of a 60/120V power system is to reduce objectionable noise in sensitive electronic equipment locations. Its use is restricted to _____ occupancies that are under close supervision by qualified personnel.
    (a) commercial
    (b) industrial
    (c) a and b
    (d) no restrictions

Article 650 Pipe Organs

91. Electric pipe organ circuits shall be arranged so that all conductors shall be protected from overcurrent by an overcurrent protection device rated at not more than _____.
    (a) 20A
    (b) 15A
    (c) 6A
    (d) none of these

Article 660 X-Ray Equipment

92. X-ray equipment mounted on a permanent base equipped with wheels and/or casters for moving while completely assembled is defined as _____.
    (a) portable
    (b) mobile
    (c) movable
    (d) room

93. A disconnecting means for X-ray equipment shall have adequate capacity for at least _____.
    (a) 50 percent of the input required for the momentary rating of equipment
    (b) 100 percent of the input required for the momentary rating of equipment
    (c) 50 percent of the input requirement for the long-time rating
    (d) 125 percent of the input requirement

94. The ampacity requirements for a disconnecting means for X-ray equipment shall be based on the greater of _____ percent of the input required for the momentary rating or 100 percent of the input required for the long-time rating.
    (a) 125
    (b) 100
    (c) 50
    (d) none of these

95. The ampacity of supply branch-circuit conductors and the overcurrent protection devices for X-ray equipment shall not be less than _____.
(a) 50 percent of the momentary rating   (b) 100 percent of the long-time rating
(c) the larger of a or b   (d) the smaller of a or b

96. Size 18 or 16 AWG fixture wires can be used for the control and operating circuits of X-ray and auxiliary equipment when protected by an overcurrent protection device not larger than _____.
(a) 15A   (b) 20A   (c) 25A   (d) 30A

Article 668 Electrolytic Cells

97. An assembly of electrically interconnected electrolytic cells supplied by a source of dc power is called a _____.
(a) battery pack   (b) cell line   (c) electrolytic cell bank   (d) battery storage bank

Article 670 Industrial Machinery

98. Where overcurrent protection is provided as part of the industrial machine, the machine shall be marked to read _____.
(a) overcurrent protection provided at machine supply terminals
(b) this unit contains overcurrent protection
(c) fuses or circuit breaker enclosed
(d) with the overcurrent protection category and type

Article 675 Electrically Driven or Controlled Irrigation Machines

99. A center pivot irrigation machine may have hand-portable motors.
(a) True   (b) False

100. An electrically driven or controlled machine with one or more motors that are not hand-portable, and used primarily to transport and distribute water for agricultural purposes, is called a(n) _____.
(a) irrigation machine   (b) electric water distribution system
(c) center pivot irrigation machine   (d) automatic water distribution system

Article 680 Swimming Pools, Fountains and Similar Installations

101. The scope of Article 680 includes _____.
(a) wading and decorative pools   (b) fountains
(c) hydromassage bathtubs   (d) all of these

102. • A spa or hot tub is a hydromassage pool or tub for recreational or therapeutic use designed for the immersion of users. They are not generally designed or intended to have the contents drained or discharged after each use.
(a) True   (b) False

103. A pool capable of holding water to a maximum depth of _____ is a storable pool.
(a) 18 in.   (b) 36 in.   (c) 42 in.   (d) none of these

104. A wet-niche luminaire (lighting fixture) is intended to be installed in a _____.
(a) transformer   (b) forming shell
(c) hydromassage bathtub   (d) all of these

105. Ground-fault circuit interrupters (GFCI) protecting a 120V wet-niche light on a pool or spa shall only be of the circuit-breaker type.
(a) True   (b) False

106. Overhead utility service conductors that operate at not over 600V must maintain a _____ clearance in any direction to the water level, edge of water surface, base of diving platform or permanently anchored raft.
(a) 14 ft   (b) 16 ft   (c) 20 ft   (d) $22^{1}/_{2}$ ft

107. Cables for communications systems such as _____ shall be located no less than 10 ft from the water's edge of swimming and wading pools, diving structures, observation stands, towers or platforms.
      (a) telephone   (b) radio   (c) CATV   (d) all of these

108. Overhead network-powered broadband communications systems conductors shall be located no less than _____ from the water's edge of swimming and wading pools or the base of diving structures.
      (a) 10 ft   (b) 12 ft   (c) 18 ft   (d) none of these

109. The maximum load permitted on a heating element in a pool heater shall not exceed _____.
      (a) 20A   (b) 35A   (c) 48A   (d) 60A

110. Underground rigid nonmetallic wiring that is located less than 5 ft from the inside wall of the pool or spa shall be buried not less than _____
      (a) 6 in.   (b) 10 in.   (c) 12 in.   (d) 18 in.

111. An underground rigid nonmetallic raceway shall be not less than _____ from the inside wall of the pool or spa, unless space limitations prevent otherwise.
      (a) 8 ft   (b) 10 ft   (c) 5 ft   (d) 25 ft

112. Underground outdoor pool or spa equipment rooms or pits must have adequate drainage to prevent water accumulation during normal operation or filter maintenance.
      (a) True   (b) False

113. Electric equipment shall not be installed in rooms or pits that do not have adequate drainage to prevent water accumulation only during abnormal operation.
      (a) True   (b) False

114. The "maintenance" disconnect for pool equipment applies to all utilization equipment, including lighting.
      (a) True   (b) False

115. Pool-associated motors shall be grounded using a minimum size 12 AWG insulated copper conductor and this grounding conductor shall be installed in _____.
      (a) rigid nonmetallic conduit
      (b) electrical metallic tubing where installed on or within buildings
      (c) flexible metal conduit
      (d) a or b

116. Single, locking and grounding-type receptacles for water-pump motors or other loads directly related to the circulation and sanitation system of a permanently installed pool or fountain can be located _____ from the inside walls of the pool or fountain, if the receptacle is GFCI-protected.
      (a) 3 to 6 ft   (b) 5 to 10 ft   (c) 10 to 15 ft   (d) 10 to 20 ft

117. Receptacles that provide power for water-pump motors or for other loads directly related to the circulation and sanitation system shall be located at least _____ from the inside walls of the pool.
      (a) 3 ft   (b) 5 ft   (c) 10 ft   (d) 12 ft

118. In dwelling units, a 125V receptacle is required to be installed a minimum of 10 ft and a maximum of 20 ft from the inside wall of the pool.
      (a) True   (b) False

119. One 125V, 1Ø, 15 or 20A receptacle can be installed not less than _____ measured horizontally from the inside wall of the pool at a dwelling unit if the dimensions of the lot do not allow the required receptacle outlet to be 10 ft from the water.
      (a) 3 ft   (b) 5 ft   (c) 6 ft   (d) none of these

120. Receptacles located within _____ of the inside walls of a pool or fountain shall be protected by a ground-fault circuit interrupter.
      (a) 8 ft   (b) 10 ft   (c) 15 ft   (d) 20 ft

121. All outdoor 120V through 240V, 1Ø, 15 and 20A receptacles for pool, spa and hot tub pump motors shall be _____.
      (a) AFCI-protected   (b) GFCI-protected   (c) approved   (d) listed

122. In outdoor pool areas, ceiling-suspended (paddle) fans shall be installed above the pool or the area extending _____ horizontally from the inside walls of the pool, and at a height not less than 12 ft above the maximum water level of the pool.
      (a) 3 ft   (b) 5 ft   (c) 10 ft   (d) 12 ft

123. Luminaires or ceiling fans mounted less than 12 ft above the water level cannot be installed within _____ of an outdoor pool, fountain or spa.
      (a) 3 ft   (b) 5 ft   (c) 10 ft   (d) 8 ft

124. Switching devices shall be at least 5 ft horizontally from the inside walls of a pool unless the switch is listed as being acceptable for use within 5 ft. An example of a switch that meets this requirement would be a pneumatic switch listed for this purpose.
      (a) True   (b) False

125. A pool transformer used for the supply of underwater luminaires, together with the transformer enclosure, is required to _____.
      (a) be of the isolated winding type
      (b) have a grounded metal barrier between the primary and secondary windings
      (c) be listed for the purpose
      (d) all of these

126. A ground-fault circuit interrupter (GFCI) shall be installed in the branch circuit supplying swimming pool luminaires that operate at more than _____ to eliminate shock hazard during relamping.
      (a) 6V   (b) 14V   (c) 15V   (d) 18V

127. Forming shells for wet-niche luminaires shall be installed with the top level of the fixture lens not less than _____ below the normal water level of the pool or spa.
      (a) 6 in.   (b) 12 in.   (c) 18 in.   (d) 24 in.

128. When rigid nonmetallic conduit extends from the pool light forming shell to a suitable junction box, an 8 AWG _____ conductor shall be installed in the raceway.
      (a) solid bare   (b) solid insulated   (c) stranded insulated   (d) b or c

129. Wet niche luminaires shall be connected to an equipment grounding conductor that is not smaller than _____ AWG.
      (a) 10   (b) 6   (c) 8   (d) 12

130. A pool light junction box that is connected to a conduit that extends directly to a forming shell or mounting bracket of a no-niche luminaire (fixture) shall be _____ for this use.
      (a) listed   (b) labeled   (c) marked   (d) a and b

131. In swimming pools, the junction box that extends to the forming shell of the luminaire (fixture) shall be listed as a pool light junction box. In addition, the junction box shall be located not less than _____ above the ground level or pool deck or not less than _____ above the maximum water level.
      (a) 8 in., 4 in.   (b) 4 in., 8 in.   (c) 6 in., 12 in.   (d) 12 in., 6 in.

132. Junction boxes for pool lighting shall not be located less than _____ from the inside wall of a pool unless separated by a fence or wall.
      (a) 3 ft   (b) 4 ft   (c) 6 ft   (d) 8 ft

133. A pool light junction box that has a raceway that extends directly to underwater pool light forming shells shall be located not less than _____ from the outdoor pool or spa.
      (a) 2 ft   (b) 3 ft   (c) 4 ft   (d) 6 ft

134. The enclosure for a transformer or ground-fault circuit interrupter that is connected to a conduit that extends directly to a pool light forming shell shall be _____ for this purpose.
      (a) labeled   (b) listed   (c) identified   (d) a and b

135.   A forming shell shall be provided with a number of grounding terminals that shall be _____ the number of conduit entries.

(a) one more than   (b) two more than   (c) the same as   (d) none of these

136.   The feeder to a swimming pool panelboard at a separate building or structure is permitted to be supplied with any Chapter 3 wiring method provided the feeder has a separate insulated copper equipment grounding conductor.

(a) True   (b) False

137.   An 8 AWG or larger solid copper bonding conductor shall be extended or attached to any remote panelboard or service equipment enclosure to eliminate voltage gradients in the pool area.

(a) True   (b) False

138.   • When bonding pool and spa equipment, a solid 8 AWG copper conductor shall be run back to the service equipment. This conductor shall be unbroken.

(a) True   (b) False

139.   The pool structure, including the reinforcing metal of the pool shell and deck, shall be bonded together.

(a) True   (b) False

140.   • When bonding together pool reinforcing steel and welded wire fabric (wire-mesh) with tie-wire, the tie-wires shall be _____.

(a) stainless steel   (b) accessible   (c) made tight   (d) none of these

141.   Metal conduit and metal piping within _____ of the inside walls of the pool that are not separated from the pool by a permanent barrier are required to be bonded.

(a) 4 ft   (b) 5 ft   (c) 8 ft   (d) 10 ft

142.   • Which of the following shall be bonded?

(a) electric equipment within 5 ft of the inside wall of the pool

(b) pool structural steel

(c) metal fittings within or attached to the pool

(d) all of these

143.   Stainless steel, copper or copper alloy clamps approved for direct burial use can be used to connect the bonding conductor to the common bonding grid of a swimming pool.

(a) True   (b) False

144.   The components that are required to be connected to the common bonding grid of a swimming pool shall be connected using a minimum size of 8 AWG _____ conductor.

(a) solid insulated   (b) solid bare   (c) solid covered   (d) any of these

145.   The _____ pool bonding conductor shall be terminated either by exothermic welding or by pressure connectors that are labeled as being suitable for the purpose.

(a) 8 AWG   (b) insulated or bare   (c) copper   (d) all of these

146.   Radiant heating cables embedded in or below the pool deck shall _____.

(a) not be installed within 5 ft horizontally from the inside walls of the pool

(b) be mounted at least 12 ft vertically above the pool deck

(c) not be permitted

(d) none of these

147.   All electric equipment, including power-supply cords, used with storable pools shall be protected by _____.

(a) fuses   (b) circuit breakers   (c) double-insulation   (d) GFCI

148.   In spas or hot tubs, a clearly labeled emergency shutoff or control switch for the purpose of stopping the motors(s) that provide power to the recirculation system and jet system shall be installed. The emergency shutoff control switch shall be _____ to the users and located not less than 5 ft away, and within sight of, the spa or hot tub. This requirement does not apply to single-family dwelling units.

(a) accessible   (b) readily accessible   (c) available   (d) none of these

149. The wiring for spas and hot tubs installed outdoors, such as receptacles, switches, lighting locations, grounding and bonding must comply with the same requirements as permanently installed pools.
(a) True   (b) False

150. A luminaire (fixture) that is located within 5 ft of the inside wall of an indoor spa or hot tub shall be _____.
(a) a minimum of 7 ft 6 in. above maximum water level
(b) protected by a ground-fault circuit interrupter
(c) a or b
(d) a and b

151. Receptacle outlet(s) for a _____ shall be GFCI-protected.
(a) self-contained spa or hot tub
(b) packaged spa or hot tub equipment assembly
(c) field assembled spa or hot tub with a heater load of 50A or less
(d) all of these

152. The maximum length of exposed cord in a fountain shall be limited to _____ .
(a) 3 ft   (b) 4 ft   (c) 6 ft   (d) 10 ft

153. A portable electric sign cannot be placed in or within _____ from the inside walls of a fountain.
(a) 3 ft   (b) 5 ft   (c) 10 ft   (d) none of these

154. GFCI protection is required for all 125V, 1Ø receptacles located within 5 ft measured _____ from the inside walls of a hydromassage bathtub.
(a) vertically   (b) horizontally   (c) across   (d) none of these

155. Hydromassage bathtub equipment shall be _____ without damaging the building structure or building finish.
(a) readily accessible   (b) accessible
(c) within sight   (d) none of these

Article 690 Solar Photovoltaic Systems

156. The photovoltaic disconnecting means must _____.
(a) be installed at a readily accessible location either outside of a building or structure or inside nearest the point of entrance of the system conductors.
(b) be suitable for the prevailing conditions.
(c) consist of not more than six switches or six circuit breakers.
(d) all of these

157. Any structure with a photovoltaic power system that is not connected to a utility service source must have a permanent plaque or directory on the exterior of the structure that identifies the location of system disconnecting means and that the structure contains a stand-alone electrical power system.
(a) True   (b) False

158. Structures with a utility service and a photovoltaic system must have a permanent plaque or directory that identifies the location of the service and the photovoltaic system disconnecting means, if they are not at the same location.
(a) True   (b) False

Article 692 Fuel Cell Systems

159. A fuel cell is an electrochemical system that consumes fuel to produce an electrical current. The main chemical reaction used in a fuel cell for producing electrical power is not combustion.
(a) True   (b) False

160. A fuel cell system typically consists of a reformer, stack, power inverter and auxiliary equipment.
(a) True   (b) False

161. The fuel cell system shall be evaluated and _____ for its intended application prior to installation.
    (a) approved   (b) identified   (c) listed   (d) marked

Article 695 Fire Pumps

162. Feeder conductors supplying fire-pump motors and accessory equipment shall be sized no less than _____ percent of the sum of the motor full-load currents as listed in Table 430.148 and/or 430.150, plus 100 percent of the ampere rating of the fire-pump accessory equipment.
    (a) 100   (b) 125   (c) 250   (d) 600

163. The voltage at the line terminals of a fire-pump motor controller (locked-rotor current) shall not drop more than _____ percent below normal (controller-rated voltage) under motor-starting conditions.
    (a) 5   (b) 10   (c) 15   (d) any of these

164. When a fire-pump motor is operating at 115 percent of the full-load current rating of the motor, the voltage at the motor terminals shall not drop more than _____ percent below the voltage rating of the motor.
    (a) 5   (b) 10   (c) 15   (d) any of these

Chapter 7 Special Conditions

Article 700 Emergency Systems

165. Transfer equipment, including automatic transfer switches, shall be _____.
    (a) automatic   (b) identified for emergency use
    (c) approved by the authority having jurisdiction   (d) all of these

166. An emergency transfer switch can supply _____.
    (a) emergency loads   (b) computer equipment   (c) UPS-type equipment   (d) all of these

167. Audible and visual signal devices for an emergency system shall be provided, when practicable, for the purpose(s) of indicating _____.
    (a) that the battery is carrying load   (b) derangement of the emergency source
    (c) that the battery charger is not functioning   (d) all of these

168. Wiring from emergency source or emergency source distribution overcurrent protection to emergency loads shall be kept entirely independent of all other wiring and equipment except in _____.
    (a) transfer equipment enclosures
    (b) exit or emergency luminaires supplied from two sources
    (c) a common junction box attached to exit or emergency luminaires supplied from two sources
    (d) all of these

169. Emergency circuit wiring shall be designed and located in such a manner so as to minimize the hazards that might cause failure because of _____.
    (a) flooding   (b) fire   (c) icing   (d) all of these

170. Emergency lighting and/or emergency power in a building or group of buildings will be available within the time period required for the application, but not to exceed _____ seconds.
    (a) 5   (b) 10   (c) 30   (d) 60

171. A storage battery supplying emergency lighting and power shall maintain not less than $87\frac{1}{2}$ percent of full voltage at total load for a period of at least _____ hour(s).
    (a) 1   (b) $1\frac{1}{2}$ (c) 2   (d) $2\frac{1}{2}$

172. Where an internal combustion engine is used as the prime-mover for an emergency system, an on-site fuel supply shall be provided with an on-premise fuel supply for not less than _____ hours full-demand operation of the system.
    (a) 2   (b) 3   (c) 4   (d) 5

173. Where an outdoor-housed generator for emergency circuits is equipped with a readily accessible disconnecting means located within sight of the building or structure supplied, an additional disconnecting means is not required on or at the building or structure for the generator feeder conductors.
     (a) True   (b) False

174. Unit equipment (battery pack) shall be on the same branch circuit as that serving the normal lighting in the area and connected _____ any local switches.
     (a) with   (b) ahead of   (c) after   (d) none of these

175. The alternate source for emergency systems _____ be required to have ground-fault protection of equipment.
     (a) shall   (b) shall not

Article 701 Legally Required Standby Systems

176. A legally required standby system is intended to automatically supply power to _____.
     (a) those systems classed as emergency systems   (b) selected loads
     (c) a and b   (d) none of these

177. A generator set for a required standby system shall _____.
     (a) have means for automatically starting the prime movers
     (b) have two hours full-demand onsite fuel supply if an internal combustion engine
     (c) not be solely dependent on public utility gas system
     (d) all of these

178. Where an outdoor-housed generator for legally required circuits is equipped with a readily accessible disconnecting means located within sight of the building or structure supplied, an additional disconnecting means is not required on or at the building or structure for the generator feeder conductors.
     (a) True   (b) False

179. Where acceptable to the authority having jurisdiction (AHJ), connections ahead of and not within the same cabinet, enclosure or vertical switchboard section as the service disconnecting means are permitted for _____ standby service.
     (a) emergency   (b) legally required   (c) optional   (d) all of these

Article 702 Optional Standby Systems

180. Optional standby systems are typically installed to provide an alternate source of power for _____.
     (a) data-processing and communication systems
     (b) emergency systems for health care facilities
     (c) emergency systems for hospitals
     (d) none of these

181. Optional standby systems shall have adequate capacity and rating for the supply of _____.
     (a) all emergency lighting and power loads
     (b) all equipment to be operated at one time
     (c) 100 percent of the appliance loads and 50 percent of the lighting loads
     (d) 100 percent of the lighting loads and 75 percent of the appliance loads

182. A transfer switch is required for all fixed or portable optional standby power systems for buildings or structures for which an electric-utility supply is either the normal or standby source.
     (a) True   (b) False

183. If the transfer switch for a portable generator switches the _____ conductor, then the portable generator shall be grounded in accordance with 250.30.
     (a) phase   (b) equipment grounding
     (c) grounded (neutral)   (d) all of these

184. If the transfer switch for a portable generator does not switch the _____ conductor, then the equipment grounding conductor shall be bonded to the system grounding electrode.
    (a) phase   (b) equipment grounding
    (c) grounded (neutral)   (d) all of these

185. Conductors for an appliance circuit supplying more than one appliance or appliance receptacle in an installation operating at less than 50V shall not be smaller than _____ copper or equivalent.
    (a) 18 AWG   (b) 14 AWG   (c) 12 AWG   (d) 10 AWG

Article 720 Circuits and Equipment Operating at Less Than 50 Volts

186. Circuits and equipment operating at less than 50V shall use receptacles that are rated at not less than _____.
    (a) 10A   (b) 15A   (c) 20A   (d) 30A

Article 725 Class 1, Class 2, and Class 3 Remote-Control, Signaling and Power-Limited Circuits

187. Class 2, Class 3 and PLTC cable that is not terminated at equipment and not identified for future use with a tag will be considered as abandoned.
    (a) True   (b) False

188. Since Class 3 control circuits permit higher allowable levels of voltage and current, additional _____ are specified to provide protection against the electric shock hazard that could be encountered.
    (a) circuits   (b) safeguards   (c) conditions   (d) requirements

189. All accessible portions of abandoned Class 2, Class 3 and PLTC cables shall be removed.
    (a) True   (b) False

190. Access to electrical equipment shall not be denied by an accumulation of cables that prevents removal of panels, including suspended ceiling panels.
    (a) True   (b) False

191. Cables installed _____ on the surface of ceilings and sidewalls shall be supported by the structural components of the building in such a manner that the cable will not be damaged by normal building use.
    (a) exposed   (b) concealed   (c) hidden   (d) a and b

192. Exposed cables shall be secured to structural components by straps, staples, hangers or similar fittings designed and installed so as not to damage the cable.
    (a) True   (b) False

193. Cables installed _____ to framing members shall be protected against physical damage from penetration by screws or nails by $1^1/_4$ in. separation from the framing member or by a suitable metal plate in accordance with 300.4(D).
    (a) exposed   (b) concealed   (c) parallel   (d) all of these

194. Remote-control circuits to safety-control equipment shall be classified as _____ if the failure of the equipment to operate introduces a direct fire or life hazard.
    (a) Class 1   (b) Class 2   (c) Class 3   (d) Class 1, Division 1

195. Class 1 power-limited circuits shall be supplied from a source having a rated output of not more than 30V. If the voltage rating were 25V, the maximum VA of the circuit would be _____.
    (a) 750 VA   (b) 700 VA   (c) 1,000 VA   (d) 1,200 VA

196. A Class 1 signaling circuit shall not exceed _____.
    (a) 130V   (b) 150V   (c) 220V   (d) 600V

197.   Power-supply and Class 1 circuit conductors shall be permitted to occupy the same cable, enclosure or raceway only where the _____.
(a) equipment powered is functionally associated
(b) circuits involved are not a mixture of ac and dc
(c) Class 1 circuits are never permitted with the power-supply conductors
(d) none of these

198.   Class 1 control circuits using 18 AWG conductors shall use insulation types _____.
(a) RFH-2, RFHH-2 or RFHH-3
(b) TF, TFF, TFN or TFFN
(c) RHH, RHW, THWN or THHN
(d) a and b

199.   When required, the maximum overcurrent protection in amperes for a not inherently limited Class 2 remote-control circuit is _____.
(a) 1.0A   (b) 2.0A   (c) 5.0A   (d) 7.5A

200.   The power source for a Class 2 circuit shall be _____.
(a) a listed Class 2 or 3 transformer
(b) a listed Class 2 or 3 power supply
(c) other listed equipment marked to identify the Class 2 or Class 3 power source
(d) any of these

Unit 11 NEC Exam – NEC Code Order 600.6 – 720.4

1. Branch circuits that supply signs and outline lighting systems containing incandescent and fluorescent forms of illumination shall be rated not to exceed _____.
   (a) 20A   (b) 30A   (c) 40A   (d) 50A

2. Sign and outline lighting enclosures for live parts other than lamps and neon tubing shall _____.
   (a) have ample structural strength and rigidity
   (b) be constructed of metal or shall be listed
   (c) if of sheet steel, be at least 0.016 in. thick
   (d) all of these

3. Ballasts, transformers and electronic power supplies for electric signs shall be located where accessible and shall be securely fastened in place. A working space at least _____ shall be provided.
   (a) 3 ft high, 3 ft wide by 3 ft deep   (b) 4 ft high, 3 ft wide by 3 ft deep
   (c) 6 ft high, 3 ft wide by 3 ft deep   (d) none of these

4. Manufactured wiring systems shall be constructed with _____.
   (a) listed AC or MC cable   (b) 10 or 12 AWG copper-insulated conductors
   (c) conductors suitable for nominal 600V   (d) all of these

5. The conductors to the hoistway door interlocks from the hoistway riser shall be flame retardant and suitable for a temperature of not less than _____and the conductors shall be SF or equivalent.
   (a) 200 °C   (b) 60 °C   (c) 90 °C   (d) 110 °C

6. The minimum size parallel conductors permitted for elevator lighting is _____ provided the combined ampacity is equivalent to at least that of a 14 AWG wire.
   (a) 14 AWG   (b) 20 AWG   (c) 16 AWG   (d) 1/0 AWG

7. Where multiple driving machines are connected to a single elevator, escalator, moving walk or pumping unit, there shall be one disconnecting means to disconnect the _____.
   (a) motor(s)   (b) control valve operating magnets
   (c) a and b   (d) none of these

8. Each elevator shall have a single means for disconnecting all ungrounded main power-supply conductors for each unit _____.
   (a) excluding the emergency power system.
   (b) including the emergency or standby power system.
   (c) excluding the emergency power system if it is automatic.
   (d) no elevator shall operate on an emergency power system.

9. Under a raised floor, liquidtight flexible conduit shall be permitted to enclose branch-circuit conductors for information technology communication equipment.
   (a) True   (b) False

10. In information technology equipment rooms, a single disconnecting means is permitted to control _____.
    (a) only the HVAC systems to the room
    (b) only the power to electronic computer/data-processing equipment
    (c) the electronic computer/data-processing equipment and the building supply
    (d) the HVAC systems to the room and power to electronic computer/data-processing equipment

11. The ampacity requirements for a disconnecting means for X-ray equipment shall be based on the greater of _____ percent of the input required for the momentary rating or 100 percent of the input required for the long-time rating.
    (a) 125   (b) 100   (c) 50   (d) none of these

12. The ampacity of supply branch-circuit conductors and the overcurrent protection devices for X-ray equipment shall not be less than _____.
    (a) 50 percent of the momentary rating   (b) 100 percent of the long-time rating
    (c) the larger of a or b   (d) the smaller of a or b

13. A spa or hot tub is a hydromassage pool or tub for recreational or therapeutic use designed for the immersion of users. They are not generally designed or intended to have the contents drained or discharged after each use.
    (a) True   (b) False

14. An underground rigid nonmetallic raceway must be not less than _____ from the inside wall of the pool or spa unless space limitations prevent otherwise.
    (a) 8 ft   (b) 10 ft   (c) 5 ft   (d) 25 ft

15. Underground outdoor pool or spa equipment rooms or pits must have adequate drainage to prevent water accumulation during normal operation or filter maintenance.
    (a) True   (b) False

16. Electric equipment shall not be installed in rooms or pits that do not have adequate drainage to prevent water accumulation only during abnormal operation.
    (a) True   (b) False

17. Pool-associated motors must be grounded using a minimum size 12 AWG insulated copper conductor and this grounding conductor must be installed in _____.
    (a) rigid nonmetallic conduit
    (b) electrical metallic tubing where installed on or within buildings
    (c) flexible metal conduit
    (d) a or b

18. In dwelling units, a 125V receptacle is required to be installed a minimum of 10 ft and a maximum of 20 ft from the inside wall of the pool.
    (a) True   (b) False

19. When bonding pool and spa equipment, a solid 8 AWG copper conductor must be run back to the service equipment. This conductor must be unbroken.
    (a) True   (b) False

20. When bonding together pool reinforcing steel and welded wire fabric (wire-mesh) with tie-wire, the tie-wires must be _____.
    (a) stainless steel   (b) accessible   (c) made tight   (d) none of these

21. Which of the following must be bonded?
    (a) Electric equipment within 5 ft of the inside wall of the pool
    (b) Pool structural steel
    (c) Metal fittings within or attached to the pool
    (d) all of these

22. The wiring for spas and hot tubs installed outdoors, such as receptacles, switches, lighting locations, grounding and bonding must comply with the same requirements as permanently installed pools.
    (a) True   (b) False

23. Emergency lighting and/or emergency power in a building or group of buildings will be available within the time period required for the application but not to exceed _____ seconds.
    (a) 5   (b) 10   (c) 30   (d) 60

24. Unit equipment (battery pack) shall be on the same branch circuit as that serving the normal lighting in the area and connected _____ any local switches.
    (a) with   (b) ahead of   (c) after   (d) none of these

25. Conductors for an appliance circuit supplying more than one appliance or appliance receptacle in an installation operating at less than 50V shall not be smaller than _____ copper or equivalent.
    (a) 18 AWG   (b) 14 AWG   (c) 12 AWG   (d) 10 AWG

Unit 11 NEC Exam – Random Order 100 – 700.12

1. The branch-circuit conductor and overcurrent protection device for fixed electric space-heating equipment loads shall not be smaller than _____ percent of the total load.
    (a) 80   (b) 100   (c) 125   (d) 150

2. The branch-circuit protective device shall be permitted to serve as the controller for a stationary motor rated at _____-hp or less that is normally left running and cannot be damaged by overload or failure to start.
    (a) $1/8$ (b) $1/4$ (c) $3/8$ (d) $1/2$

3. The building disconnecting means shall be installed at a(n) _____location.
    (a) accessible   (b) readily accessible   (c) outdoor   (d) indoor

4. The *Code* defines a _____ as all circuit conductors between the service equipment, the source of a separately derived system or other power supply source and the final branch-circuit overcurrent device.
    (a) feeder   (b) branch circuit   (c) service   (d) all of these

5. The conductor insulation of MI cable shall be a highly-compressed refractory mineral that will provide proper _____ for all conductors.
    (a) covering   (b) spacing   (c) resistance   (d) none of these

6. The conductors between the final overcurrent protection device and the outlet(s) are known as the _____ conductors.
    (a) feeder   (b) branch-circuit   (c) home run   (d) none of these

7. The conductors, including splices and taps, in a metal surface raceway shall not fill the raceway to more than _____ percent of its cross-sectional area at that point.
    (a) 75   (b) 40   (c) 38   (d) 53

8. The connected load to which the demand factors of Table 220.32 apply shall include the _____ rating of all appliances that are fastened in place, permanently connected or located to be on a specific circuit, such as ranges, wall-mounted ovens, counter-mounted cooking units, clothes dryers, water heaters and space heaters.
    (a) calculated   (b) nameplate   (c) circuit   (d) overcurrent protection

9. The connection of the grounding electrode conductor to a buried grounding electrode (driven ground rod) shall be made with a listed terminal device that is accessible .
    (a) True   (b) False

10. The demand factors of Table 220.20 apply to space heating, ventilating or air-conditioning equipment.
    (a) True   (b) False

11. The demand percentage used to calculate 45 receptacles in a boatyard feeder is _____ percent.
    (a) 90   (b) 80   (c) 70   (d) 50

12. The disconnecting means for a torque motor shall have an ampere rating of at least _____ percent of the motor nameplate current.
    (a) 100   (b) 115   (c) 125   (d) 175

13. The disconnecting means for air-conditioning and refrigerating equipment must be _____ from the air-conditioning or refrigerating equipment.
    (a) readily accessible   (b) within sight
    (c) a or b   (d) a and b

14. The disconnecting means for the controller and motor must open all ungrounded supply conductors and shall be designed so that no pole can be operated independently.
    (a) True   (b) False

15. The distance between a cable or conductor entry and its exit from the box shall be not less than _____ times the outside diameter, over sheath, of that cable or conductor on a 1,000V system.
    (a) 16   (b) 18   (c) 36   (d) 40

16. The emergency controls for unattended self-service stations must be located not less than _____ or more than _____ from the gasoline dispensers.
    (a) 10 ft, 25 ft   (b) 20 ft, 50 ft   (c) 20 ft, 100 ft   (d) 50 ft, 100 ft

17. The equipment grounding conductor shall be identified by _____.
    (a) a continuous outer-green finish
    (b) being bare
    (c) a continuous outer-green finish with one or more yellow stripes
    (d) any of these

18. The ground-fault protection system shall be _____ when first installed on-site.
    (a) inspected   (b) identified
    (c) turned on   (d) performance tested

19. The grounding conductor connection to the grounding electrode shall be made by _____.
    (a) listed lugs   (b) exothermic welding
    (c) listed pressure connectors   (d) any of these

20. The grounding electrode conductor at the service is permitted to terminate on an equipment grounding terminal bar if a (main) bonding jumper is installed between the grounded conductor bus and the equipment grounding terminal.
    (a) True   (b) False

21. The grounding of electrical systems, circuit conductors, surge arresters and conductive non-current-carrying materials and equipment shall be installed and arranged in a manner that will prevent objectionable current over the grounding conductors or grounding paths.
    (a) True   (b) False

22. The handhole of metal luminaire poles can be omitted for metal poles _____ or less in height above finish grade. This is only permitted if the pole is provided with a hinged base and the grounding terminal is accessible within the hinged base.
    (a) 8 ft   (b) 18 ft   (c) 20 ft   (d) none of these

23. The highest current at rated voltage that a device is intended to interrupt under standard test conditions is the _____.
    (a) interrupting rating   (b) manufacturer's rating
    (c) interrupting capacity   (d) GFCI rating

24. The individual conductors in a cablebus shall be supported at intervals not greater than _____ for vertical runs.
    (a) $^1/_2$ ft   (b) 1 ft   (c) $1^1/_2$ ft   (d) 2 ft

25. The maximum allowable-hp rating of a permanently connected appliance when the branch-circuit overcurrent protection device is used as the appliance disconnecting means is _____ or 300 VA.
    (a) $^1/_8$-hp   (b) $^1/_4$-hp   (c) $^1/_2$-hp   (d) $^3/_4$ -hp

26. The maximum internal current that can flow through the line isolation monitor when any point of the isolated system is grounded shall be _____ when used in a health care facility.
    (a) 15A or less   (b) no more than 1A
    (c) 1 mA   (d) 10 mA

27. The maximum weight of a luminaire that may be mounted by the screw-shell of a brass socket is _____
    (a) 2 lbs.   (b) 6 lbs.   (c) 3 lbs.   (d) 50 lbs.

28. The minimum and maximum size of EMT is _____, except for special installations.
    (a) $^5/_{16}$ to 3 in.   (b) $^3/_8$ to 4 in.   (c) $^1/_2$ to 3 in.   (d) $^1/_2$ to 4 in.

29. The motor controller shall have horsepower ratings at the application voltage not _____ than the horsepower rating of the motor.
    (a) lower   (b) higher   (c) equal to   (d) none of these

30. The motor disconnecting means can be a _____.
    (a) circuit breaker
    (b) motor-circuit switch rated in horsepower
    (c) molded case switch
    (d) any of these

31. The *NEC* term to define wiring methods that are not concealed is _____.
    (a) open   (b) uncovered   (c) exposed   (d) bare

32. The noncurrent-carrying metal parts of equipment, such as _____, shall be effectively bonded together.
    (a) service raceways, cable trays or service cable armor
    (b) service equipment enclosures containing service conductors, including meter fittings, boxes or the like, interposed in the service raceway or armor
    (c) the metallic raceway or armor enclosing a grounding electrode conductor
    (d) all of these

33. The number of conductors permitted in a raceway shall be limited to _____.
    (a) permit heat to dissipate
    (b) prevent damage to insulation during installation
    (c) prevent damage to insulation during removal of conductors
    (d) all of these

34. The overall covering of UF cable shall be _____.
    (a) flame retardant
    (b) moisture, fungus and corrosion resistant
    (c) suitable for direct burial in the earth
    (d) all of these

35. The patient care area is any portion of a health care facility where patients are intended to be _____.
    (a) examined   (b) treated   (c) registered   (d) a or b

36. The phase converter disconnecting means shall be _____ and located in sight from the phase converter.
    (a) protected from physical damage   (b) readily accessible
    (c) easily visible   (d) clearly identified

37. The pilot light provided within a portable stage switchboard enclosure shall have overcurrent protection rated or set at not more than _____.
    (a) 10A   (b) 15A   (c) 20A   (d) 30A

38. The raceway or cable for tap conductors to recessed luminaires shall have a minimum length of _____
    (a) 6 in.   (b) 12 in.   (c) 18 in.   (d) 24 in.

39. The radius of the curve of the inner edge of any bend shall not be less than _____ for AC cable.
    (a) five times the largest conductor within the cable
    (b) three times the diameter of the cable
    (c) five times the diameter of the cable
    (d) six times the outside diameter of the conductors

40. The radius of the inner edge of any bend in MI cable shall not be less than five times the external diameter of the metallic sheath for cable not more than _____ in external diameter.
    (a) $1/2$ in.   (b) $3^1/_4$ in.   (c) $5/_8$ in.   (d) $1^1/_2$ in.

41. The requirement for maintaining a 3 ft vertical clearance from the edge of the roof shall not apply to the final conductor span where the conductors are attached to _____.
    (a) a building pole   (b) the side of a building
    (c) an antenna   (d) the base of a building

42. The requirement for maintaining a 3 ft vertical clearance from the edge of the roof shall not apply to the final conductor span where the service drop is attached to _____.
    (a) a service pole   (b) the side of a building
    (c) an antenna   (d) the base of a building

43. The residual voltage of a capacitor, rated not over 600V, shall be reduced to 50V, nominal, or less within _____ after the capacitor is disconnected from the source of supply.
    (a) 15 seconds   (b) 45 seconds
    (c) 1 minute   (d) 2 minutes

44. The secondary circuits of current and potential instrument transformers shall be grounded where the primary windings are connected to circuits of _____ or more to ground and, where on switchboards, shall be grounded irrespective of voltage.
    (a) 300V   (b) 600V   (c) 1,000V   (d) 150V

45. The service conductors between the terminals of the service equipment and a point usually outside the building, clear of building walls, where joined by tap or splice to the service drop is called _____ service-entrance conductors.
    (a) underground   (b) complete   (c) overhead   (d) grounded

46. The service disconnecting means shall be installed at a(n) _____location.
    (a) dry   (b) readily accessible
    (c) outdoor   (d) indoor

47. The small-appliance branch circuits can supply the _____ as well as the kitchen.
    (a) dining room   (b) refrigerator
    (c) breakfast room   (d) all of these

48. Underground rigid nonmetallic wiring that is located less than 5 ft from the inside wall of the pool or spa must be buried not less than _____
    (a) 6 in.   (b) 10 in.   (c) 12 in.   (d) 18 in.

49. Where an internal combustion engine is used as the prime-mover for an emergency system, an on-site fuel supply shall be provided with an on-premise fuel supply for not less than _____ hours full-demand operation of the system.
    (a) 2   (b) 3   (c) 4   (d) 5

50. X-ray equipment mounted on a permanent base equipped with wheels and/or casters for moving while completely assembled is defined as _____.
    (a) portable   (b) mobile   (c) movable   (d) room

Unit 12

Delta-Delta and Delta-Wye Transformer Calculations

OBJECTIVES

After reading this unit, the student should be able to briefly explain the following concepts:

Current flow	Delta transformer sizing	Phase current
Delta transformer voltage	Delta panel schedule in kVA	Wye phase versus line Wye
Delta high leg	Delta panelboard and conductor sizing	transformer loading and balancing
Delta primary and secondary	Delta neutral current	Wye transformer sizing
Line currents	Delta maximum unbalanced load	Wye panel schedule in kVA
Delta primary or secondary phase	Delta/delta example	Wye panelboard and conductor sizing
Currents	Wye transformer voltage	Wye neutral current
Delta phase versus line	Wye voltage triangle	Wye maximum unbalanced load
Delta current triangle	Wye transformer current	Delta/wye example
Delta transformer balancing	Line current	Delta versus wye

After reading this unit, the student should be able to briefly explain the following terms:

Delta connected	Phase load – delta	Wye secondary
kVA rating	Phase load – wye	Winding
Line	Phase voltage	Wye connected
Line current	Ratio	Primary/secondary voltage-ampere
Line voltage	Balanced load	Three-phase load
Phase	Unbalanced load	Single-phase load
Phase current	Delta secondary	

INTRODUCTION

This unit deals with Delta/Delta and Delta/Wye configured transformers. The system voltages used in this unit were selected because of their common industry configuration. Only a few of the concepts contained in this unit are ever on an electrician's exam.

DEFINITIONS

Delta-Connected Transformer

Delta-connected transformers have the windings of three 1Ø transformers connected in series with each other to form a closed circuit. The line conductors are connected to the point where two 1Ø transformers meet. This system is called a Delta System because when it's drawn out it looks like a triangle (greek symbol Δ for the letter delta). Many call it a high-leg system because the voltage from Line 2 to ground is 208V ($120V \times \sqrt{3}$). Figure 12–1.

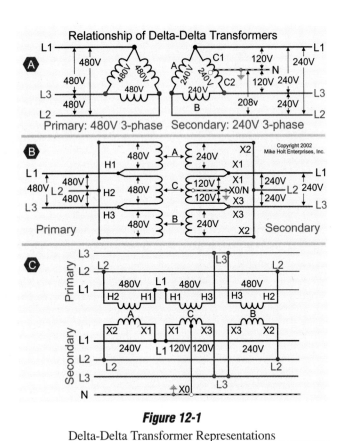

Figure 12-1

Delta-Delta Transformer Representations

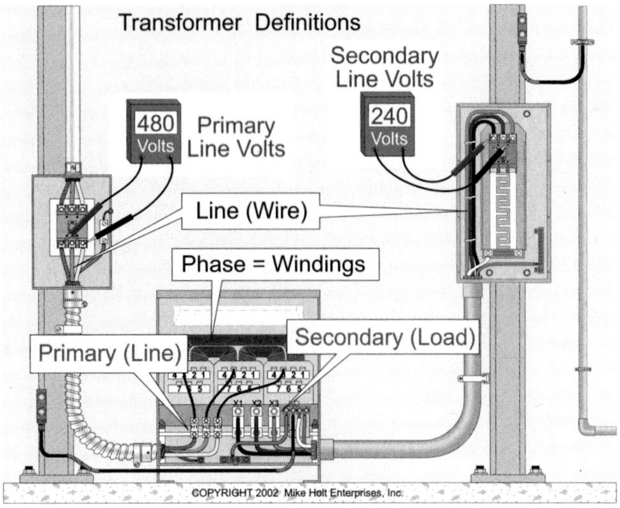

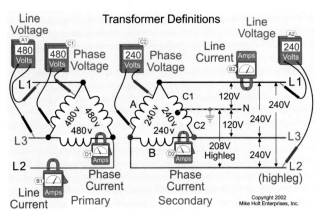

Figure 12-2

Transformer Definitions

Figure 12-3

Transformer Definitions

kVA Rating

Transformers are rated in kilovolt-amperes (kVA) and are sized to the VA rating of the loads they supply.

Line

The ungrounded (hot) conductors. Figure 12–2 and Figure 12–3.

Line Current

The current on the ungrounded conductors. Figure 12–3, B1 and B2:

$$\text{Line Current (1Ø)} = \frac{\text{Line Power}}{\text{Line Voltage, } I_{Line}} = \frac{VA_{Line}}{E_{Line}}$$

$$\text{Line Current (3Ø)} = \frac{\text{Line Power}}{(\text{Line Voltage} \times \sqrt{3}), I_{Line}} = \frac{VA_{Line}}{(E_{Line} \times \sqrt{3})}$$

Delta configured system, the line current is greater than the phase current by $\sqrt{3}$ (1.732).

Line = 100A   Phase = 57.75A

Wye configured system, the line current is equal to the phase current.

Line = 100A   Phase = 100A

Line Voltage

The voltage between any two line (ungrounded) conductors. Figure 12–3, A1 and A2.

Delta configured system, the line voltage equals the phase voltage.

   Example: Line = 240V Phase 240V

Wye configured system, the line voltage is greater than the phase voltage by $\sqrt{3}$ (1.732).

   Example: Line = 208V Phase = 120V

Phase (Winding)

The coil-shaped conductors that serve as the primary or secondary of a transformer. A phase is also called the winding.

Phase Current

The current flowing through the transformer winding. Figure 12–3, D1 and D2.

Delta configured system, the phase current is less than the line current by $\sqrt{3}$ (1.732).

   Line = 100A   Phase = 57.75A

Wye configured system, the phase current is equal to the line current.

   Line = 100A   Phase = 100A

Phase Load

The load on the transformer winding.

Phase Load Delta (Winding)

The phase load is the load on the transformer winding. Figure 12–4.

Phase load of a 3Ø 240V load = line load/3.

   Line = 15 kVA   Phase = 5 kVA

Phase load of a 1Ø 240V load = line load.

   Line = 10 kVA   Phase = 10 kVA

Phase load of a 1Ø 120V load = line load.

   Line 3 kVA   Phase = 3 kVA

Phase Load Wye (Winding)

The phase load is the load on the transformer winding. Figure 12–5.

Phase load of a 3Ø 208V load = line load/3

   Line = 15 kVA   Phase = 5 kVA

Phase load of a 1Ø 208V load = line load/2

   Line = 10 kVA   Phase = 5 kVA

Phase load of a 1Ø 120V load = line load

   Line 3 kVA   Phase = 3 kVA

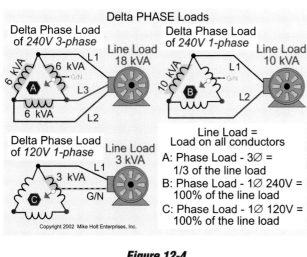

Figure 12-4
Delta Phase Loads

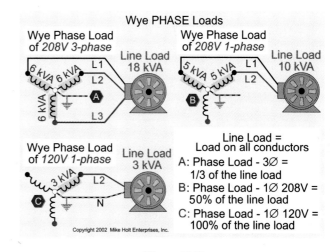

Figure 12-5
Wye Phase Loads

Phase Voltage (Winding Voltage)

The phase voltage is the internal transformer voltage generated across any one winding of a transformer. For a delta secondary, the phase voltage is equal to the line voltage. For wye secondaries, the phase voltage is less than the line voltage. Figure 12–6.

The winding voltage

Delta configured system, the phase voltage is equal to the line voltage.
Line = 240V   Phase = 240V

Wye configured system, the phase voltage is less than the line voltage by $\sqrt{3}$ (1.732).
Line = 208V   Phase = 120V

Ratio

The relationship between the number of primary winding turns as compared to the number of secondary winding turns.

Ratio Phase Voltage

The ratio is the relationship between the number of primary winding turns as compared to the number of secondary winding turns. The ratio is a comparison between the primary phase voltage and the secondary phase voltage. For typical delta/delta systems, the ratio is 2:1. For typical wye systems, the ratio is 4:1. Figure 12–7.

The comparison between the primary phase voltage and the secondary phase voltage.

Delta/Delta configured system, the ratio is 2:1.
Primary = 480V   Secondary = 240V

Delta/Wye configured system, the ratio is 4:1.
Primary = 480V   Secondary = 120V

Unbalanced Load (Neutral Current)

The unbalanced load is the load on the secondary grounded (neutral) conductors.

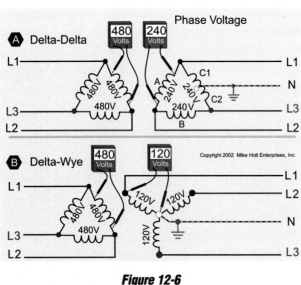

Figure 12-6

Phase Voltage

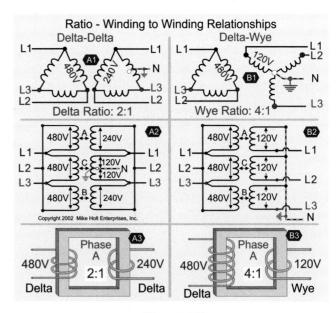

Figure 12-7

Ratio – Winding to Winding Relationships

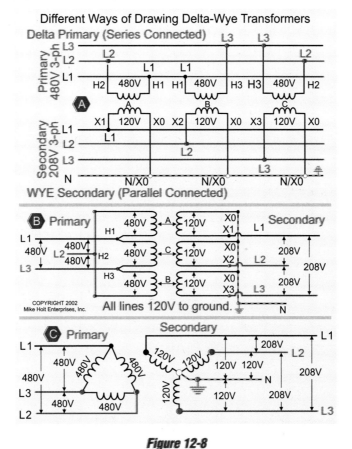

Figure 12-8

Different Ways of Drawing Delta-Wye Transformers

Unbalanced Neutral Load – Delta

The neutral current (unbalanced load) is calculated by subtracting Line 3 current from Line 1 current.

$I_{Neutral} = I_{Line1} - I_{Line3}$

No neutral loads are connected to Line 2 because the voltage from this line to ground is 208V.

Note: Line 2 (high-leg) and its voltage-to-ground is approximately 208V, therefore no neutral loads are connected to this line.

Unbalanced Neutral Load – Wye

The neutral current (unbalanced load) is calculated by using the following formula:

$$I_{Neutral} = \sqrt{(I_{Line\ 1}^2 + I_{Line\ 2}^2 + I_{Line\ 3}^2) - [(I_{Line\ 1} \times I_{Line\ 2}) + (I_{Line\ 2} \times I_{Line\ 3}) + (I_{Line\ 1} \times I_{Line\ 3})]}$$

Winding

The coil-shaped conductors that serve as the primary or secondary of a transformer. A winding is also called the phase.

Wye-Connected Transformers

Wye-connected transformers have one lead from each of three 1Ø transformers connected to a common point (neutral). The other lead from each of the 1Ø transformers is connected to the line conductors. Figure 12–8.

12–1 CURRENT FLOW

When a load is connected to the secondary of a transformer, current will flow through the secondary conductor windings. The current flow in the secondary creates an electromagnetic field that opposes the primary electromagnetic field. The secondary flux lines effectively reduce the strength of the primary flux lines. As a result, less *counter-electromotive force (CEMF)* is generated in the primary winding conductors. With less CEMF to oppose the primary applied voltage, the primary current automatically increases in direct proportion to the secondary current. Figure 12–9.

Note: The primary and secondary line currents are inversely proportional to the voltage ratio of the transformer. This means that the winding with the most number of turns will have a higher voltage and lower current as compared to the winding with the least number of turns, which will have a lower voltage and higher current. Figure 12–10.

The following Tables show the current relationship between kVA and voltage for common size transformers:

Single-Phase Transformers I = VA/E			
kVA Rating	Current at 208Volts	Current at 240 Volts	Current at 480 Volts
7.5	36 A	31 A	16 A
10	48 A	42 A	21 A
15	72 A	63 A	31 A
25	120 A	104 A	52 A
37.5	180 A	156 A	78 A

Three-Phase Transformers I = VA/(E × √3)			
kVA Rating	Current at 208 Volts	Current at 240 Volts	Current at 480 Volts
15	42 A	36 A	18 A
22.5	63 A	54 A	27 A
30	83 A	72 A	36 A
37.5	104 A	90 A	45 A
45	125 A	108 A	54 A
50	139 A	120 A	60 A
75	208 A	180 A	90 A
112.5	313 A	271 A	135 A

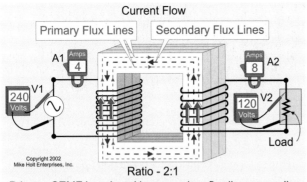

Copyright 2002
Mike Holt Enterprises, Inc.

Ratio - 2:1
Primary CEMF is reduced by secondary flux lines canceling primary flux lines resulting an in increase of primary current.

Figure 12-9
Current Flow

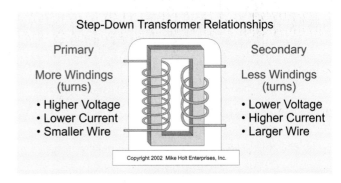

Figure 12-10
Step-Down Transformer Relationships

PART A – DELTA/DELTA TRANSFORMERS

12–2 DELTA TRANSFORMER VOLTAGE

In a delta configured transformer, the line voltage equals the phase voltage ($E_{Line} = E_{Phase}$). Figure 12–11.

Primary Delta Voltage

LINE Voltage	PHASE Voltage
L_1 to L_2 = 480V	Phase A winding = 480V
L_2 to L_3 = 480V	Phase B winding = 480V
L_3 to L_1 = 480V	Phase C winding = 480V

Secondary Delta Voltage

LINE Voltage	PHASE Voltage	NEUTRAL Voltage
L_1 to L_2 = 240V	Phase A winding = 240V	Neutral to L_1 = 120V
L_2 to L_3 = 240V	Phase B winding = 240V	Neutral to L_2 = 208V
L_3 to L_1 = 240V	Phase C winding = 240V	Neutral to L_3 = 120V

12–3 DELTA HIGH-LEG

The term high-leg, wild leg or bastard leg is used to identify the conductor of a delta configured system that has a voltage rating of 208V to ground. The high-leg voltage is the vector sum of the voltage of transformers A and C_1, or transformers B and C_2, which equal 120V × 1.732 = 208V for a 120/240V secondary.

Note: The actual voltage is often less than the nominal system voltage because of voltage drop. Figure 12–12.

❏ **High-leg Voltage**

What is the actual high-leg voltage if the delta configured secondary is 115/230V, 3Ø?

(a) 115V   (b) 230V   (c) 199V   (d) 240V

• Answer: (c) 199V

115V × 1.732 = 199.18 V

12–4 DELTA LINE CURRENTS

In a delta configured transformer, the line current does not equal the phase current.

The line current of a 3Ø transformer can be calculated by the formula:

$$I_{Line} = \frac{VA_{Line}}{E_{Line} \times \sqrt{3}}$$

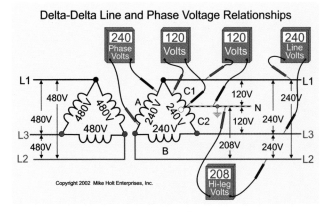

Figure 12-11

Delta-Delta Line and Phase Voltage Relationships

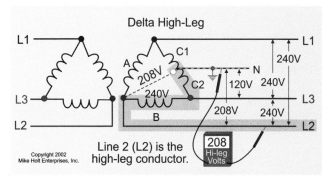

Figure 12-12

Delta High-Leg

❏ **Primary Line Current**

What is the primary line current for a 480 to 240/120V, 150 kVA, 3Ø transformer? Figure 12–13, Part A.

(a) 416A (b) 360A
(c) 180A (d) 144A

• Answer: (c) 180A

$I_{Line} = VA_{Line}/(E_{Line} \times \sqrt{3})$

$I_{Line} = 150,000 \text{ VA}/(480V \times 1.732) = 180A$

❏ **Secondary Line Current**

What is the secondary line current for a 480 to 240V, 150 kVA, 3Ø transformer? Figure 12–13, Part B.

(a) 416A (b) 360A
(c) 180A (d) 144A

• Answer: (b) 360A

$I_{Line} = VA_{Line}/(E_{Line} \times \sqrt{3})$

$I_{Line} = 150,000 \text{ VA}/(240V \times 1.732) = 360A$

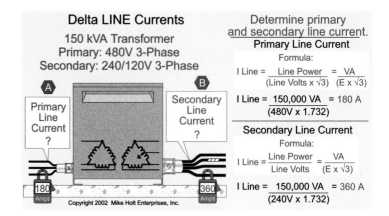

Figure 12-13
Delta Line Currents

12–5 DELTA PHASE CURRENTS

The phase current of a transformer winding is calculated by dividing the phase VA by the phase volts:

$I_{Phase} = VA_{Phase}/E_{Phase}$

The phase load of a 3Ø, 240V load = line load/3.
The phase load of a 1Ø, 240V load = line load.
The phase load of a 1Ø, 120V load = line load.

❏ **Primary Phase Current**

What is the primary phase current for a 480 to 240/120V, 150 kVA, 3Ø transformer? Figure 12–14, Part A.

(a) 416A (b) 360A
(c) 180A (d) 104A

• Answer: (d) 104A

$$I_{Phase} = \frac{VA_{Phase}}{E_{Phase}}$$

$$VA_{Phase} = \frac{150,000 \text{ VA}}{3 \text{ Phases}} = 50,000 \text{ VA}$$

$I_{Phase} = 50,000 \text{ VA}/480V = 104A$

❏ **Secondary Phase Current**

What is the secondary phase current for a 480 to 240/120V, 150 kVA, 3Ø transformer? Figure 12–14, Part B.

(a) 416A (b) 360A
(c) 208A (d) 104A

• Answer: (c) 208A

Phase power = 150,000 VA/3 = 50,000 VA

$$I_{Phase} = \frac{VA_{Phase}}{E_{Phase}} = 50,000 \text{ VA}/240V$$

$I_{Phase} = 208A$

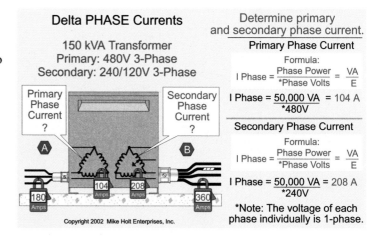

Figure 12-14
Delta Phase Currents

12–6 DELTA PHASE VERSUS LINE CURRENT

Since each line conductor from a delta transformer is actually connected to two transformer windings (phases), the effects of loading on the line (conductors) can be different than on the phase (winding).

❏ **Phase VA (3Ø)**

Because each line from a delta configured transformer is connected to two transformer phases, the line current from a 3Ø load will be greater than the phase current by $\sqrt{3}$.

$I_{Line} = VA_{Line}/(E_{Line} \times \sqrt{3})$
$I_{Phase} = I_{line}/\sqrt{3}$
$I_{Phase} = VA_{Phase}/E_{phase}$

A 240V, 36 kVA, 3Ø load has the following effect on a delta system. Figure 12–15.

LINE: Total line power = 36 kVA
$I_{Line} = VA_{Line}/(E_{Line} \times \sqrt{3})$
$I_{Line} = 36,000\ VA/(240V \times \sqrt{3})$
$I_{Line} = 87A$, or

$I_{Line} = I_{Phase} \times \sqrt{3}$
$I_{Line} = 50A \times 1.732 = 87A$

PHASE: Phase power = 12 kVA (winding)
$I_{Phase} = VA_{Phase}/E_{Phase}$
$I_{Phase} = 12,000\ VA/240V = 50A$, or

$I_{Phase} = I_{Line}/\sqrt{3}$
$I_{Phase} = 87A/1.732 = 50A$

❏ **Phase VA (1Ø) 240V**

The 1Ø line current is equal to the phase current.
$I_{Line} = VA_{Line}/E_{line}$
$I_{Phase} = VA_{Phase}/E_{Phase}$

A 240V, 10 kVA, 1Ø load has the following effect on a delta/delta system. Figure 12–16.

LINE: Total line power = 10 kVA
$I_{Line} = VA_{Line}/E_{Line}$
$I_{Line} = 10,000\ VA/240V = 42A$

PHASE: Phase power = 10 kVA (winding)
$I_{Phase} = VA_{Phase}/E_{Phase}$
$I_{Phase} = 10,000\ VA/240V = 42A$

❏ **Phase VA (1Ø) 120V**

A 120V, 3 kVA, 1Ø load has the following effect on the system. Figure 12–17.

LINE: Line power = 3 kVA
$I_{Line} = VA_{Line}/E_{Line}$
$I_{Line} = 3,000\ VA/120V = 25A$

PHASE: Phase power = 3 kVA
(C_1 or C_2 winding)
$I_{Phase} = VA_{Phase}/E_{Phase}$
$I_{Phase} = 3,000\ VA/120V = 25A$

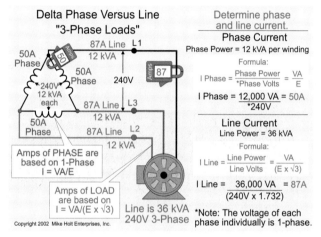

Figure 12-15
Delta Phase Versus Line Loads (3-Phase)

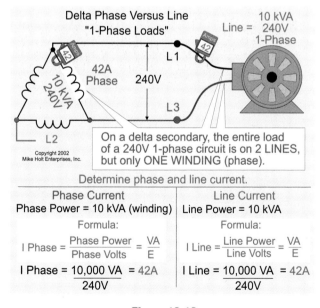

Figure 12-16
Delta Phase Versus Line Load (Single-Phase)

12–7 DELTA CURRENT TRIANGLE

The 3Ø line and phase current of a delta system are not equal; the difference is the square root of 3 ($\sqrt{3}$).

$$I_{Line} = I_{Phase} \times \sqrt{3} \qquad I_{Phase} = I_{Line} / \sqrt{3}$$

The delta triangle can be used to calculate delta 3Ø line and phase currents. Place your finger over the desired item and the remaining items show the formula to use. Figure 12–18.

12–8 DELTA TRANSFORMER BALANCING

To properly size a Delta/Delta transformer, the transformer phases (windings) must be balanced. The following steps are used to balance the transformer:

Step 1:   Determine the VA rating of all loads.

Step 2:   Balance the loads on the transformer windings as follows:
3Ø Loads: one-third of the load on Phase A, one-third of the load on Phase B, and one-third of the load on Phase C.
240V, 1Ø Loads: 100% of the load on Phase A or B. It is permissible to place some 240V, 1Ø load on Phase C when necessary for balance.
120V loads: 100% of the load on C1 or C2.

❏ **Delta Transformer Balancing**

Balance and size a 480 to 240/120V, 3Ø transformer for the following loads: one 208V, 36 kVA, 3Ø heat strip; two 240V, 10 kVA, 3Ø loads; three 120V, 3 kVA loads. Figure 12–19.

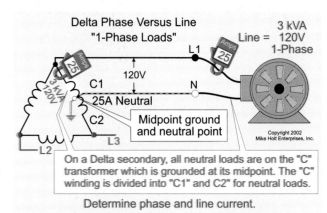

Determine phase and line current.

Phase Current		Line Current	
Phase Power = 3 kVA (C1 or C2 winding)		Line Power = 3 kVA	
Formula:		Formula:	
$I\,Phase = \dfrac{Phase\,Power}{*Phase\,Volts} = \dfrac{VA}{E}$		$I\,Line = \dfrac{Line\,Power}{Line\,Volts} = \dfrac{VA}{E}$	
$I\,Phase = \dfrac{3{,}000\,VA}{120V} = 25A$		$I\,Line = \dfrac{3{,}000\,VA}{120V} = 25A$	

Figure 12-17

Delta Phase Versus Line Load (Single-Phase)

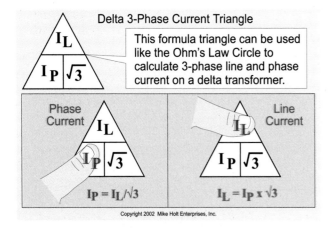

Figure 12-18

Delta 3-Phase Current Triangle

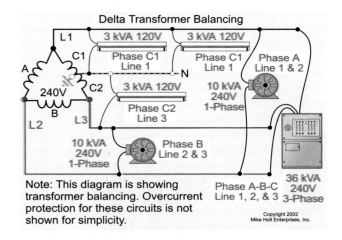

Figure 12-19

Delta Transformer Balancing

	Phase A (L₁ and L₂)	Phase B (L₂ and L₃)	C₁ (L₁)	C₂ (L₃)	Line Total
One 240V, 36 kVA, 3Ø	12 kVA	12 kVA	6 kVA	6 kVA	36,000 VA
Two 240V, 10 kVA, 1Ø	10 kVA	10 kVA			20,000 VA
Three 120V, 3 kVA, 1Ø			6 kVA *	3 kVA *	9,000 VA
	22 kVA	22 kVA	12 kVA	9 kVA	65,000 VA

*Indicates neutral load.

12–9 DELTA TRANSFORMER SIZING

Once you balance the transformer, size the transformer according to the load of each phase. The "C" transformer must be sized using two times the highest of "C₁" or "C₂." The "C" transformer is actually a single unit. If one side has a larger load, that side determines the transformer size.

❏ **Delta Transformer Sizing**

What size 480 to 240/120V transformer is required for the following loads: one 208V, 36 kVA, 3Ø heat strip; two 240V, 10 kVA, 3Ø loads; three 120V, 3 kVA loads?

(a) three 1Ø, 25 kVA transformers   (b) one 3Ø, 75 kVA transformer

(c) a or b   (d) none of these

• Answer: (c) a or b

Phase winding A = 22 kVA

Phase winding B = 22 kVA

Phase winding C = (12 kVA of C₁ × 2) = 24 kVA

12–10 DELTA PANEL BALANCING

When balancing a panelboard in VA, be sure that 3Ø loads are split one-third on each line, 240V, 1Ø loads are split one-half on each line, and 120V loads are placed on line 1 or line 3.

❏ **Delta Panel Example**

Balance a 240/120V, 3Ø panelboard for the following loads: one 240V, 36 kVA, 3Ø heat strip; two 240V, 10 kVA, 3Ø loads; three 120V, 3 kVA loads.

	Line 1	Line 2	Line 3	Line Total
240V, 36 kVA, 3Ø	12 kVA	12 kVA	12 kVA	36,000 VA
240V, 10 kVA, 1Ø	5 kVA	5 kVA		10,000 VA
240V, 10 kVA, 1Ø		5 kVA	5 kVA	10,000 VA
Three 120V, 3 kVA	6 kVA *		3 kVA *	9,000 VA
	23 kVA	22 kVA	20 kVA	65,000 VA

*Indicates neutral load.

12–11 DELTA PANELBOARD AND CONDUCTOR SIZING

To size the panelboard and its conductors, you must balance the loads in amperes.

❏ **Delta Conductors Sizing**

Balance a 240/120V, 3Ø panelboard for the following loads: one 36 kVA, 240V, 3Ø heat strip; two 10 kVA, 240V, 1Ø loads; and three 3 kVA, 120V loads.

	Line 1	Line 2	Line 3	Line Amperes
240V, 36 kVA, 3Ø	87A	87A	87A	36,000 VA/(240V × 1.732)
240V, 10 kVA, 1Ø	42A	42A		10,000 VA/240V
240V, 10 kVA, 1Ø		42A	42A	10,000 VA/240V
Three 120V, 3 kVA	50A *		25A *	3,000 VA/120V
	179A	171A	154A	

*Indicates neutral load.

Why balance the panel in amperes? Why not take the VA per phase and divide by phase voltage?

• Answer: Line current of a 3Ø load is calculated by the formula

$I_L = VA/(E_{Line} \times \sqrt{3})$

$I_L = 36{,}000 \text{ VA}/(240V \times 1.732) = 87A$ per line.

Note: If you took the per line power of 12,000 VA and divided by one line voltage of 120V, you came up with an incorrect line current of 12,000 VA/120V = 100A.

12–12 DELTA NEUTRAL CURRENT

The neutral current is calculated by subtracting Line 3 current from Line 1 current.

$I_{Neutral} = I_{Line1} - I_{Line3}$

No neutral loads are connected to Line 2 because the voltage from this line to ground is 208V.

❏ Delta Neutral Current

What is the neutral current for the following loads: one 240V, 36 kVA, 3Ø heat strip; two 240V, 10 kVA, 1Ø loads; three 3 kVA, 120V loads?

In our continuous example, Line 1 neutral current = 50A and Line 3 neutral current = 25A.

(a) 0A   (b) 25A   (c) 50A   (d) 100A

• Answer: (b) 25A

Neutral current = 50A less 25A

Neutral current = 25A

	Line 1	Line 2	Line 3	Ampere Calculation
240V, 36 kVA, 3Ø	87A	87A	87A	36,000 VA/(240V × 1.732)
240V, 10 kVA, 240V, 1Ø	42A	42A		10,000 VA/240V
240V, 10 kVA, 240V, 1Ø		42A	42A	10,000 VA/240V
Three 120V, 3 kVA	50A *		25A *	3,000 VA/120V

* Indicates neutral (120V) loads

12–13 DELTA MAXIMUM UNBALANCED LOAD

The maximum unbalanced load (neutral) is the actual current on the grounded (neutral) conductor.

❏ Maximum Unbalanced Load

What is the maximum unbalanced load for the following loads: one 240V, 36 kVA, 3Ø heat strip; two 240V, 10 kVA, 1Ø loads; three 3 kVA, 120V loads?

(a) 0 A   (b) 25A   (c) 50 A   (d) none of these

• Answer: (c) 50 A

Maximum unbalanced current equals the line with the largest neutral current.

	Line 1	Line 2	Line 3	Ampere Calculation
240V, 36 kVA, 3Ø	87A	87A	87A	36,000 VA/(240V × 1.732)
240V, 10 kVA, 1Ø	42A	42A		10,000 VA/240V
240V, 10 kVA, 1Ø		42A	42A	10,000 VA/240V
Three 120V, 3 kVA, 1Ø	50A*		25A*	3000 VA/120V

*Indicates neutral (120V) loads

12–14 DELTA/DELTA EXAMPLE

240V, 18 kW 3Ø heater

10-hp, 240V, 3Ø A/C, VA = 230V × 28A × 1.732 = 11,154 VA

2 – 240V, 14 kW, 1Ø ranges

240V, 10 kW, 1Ø water heater

2 – 3-hp 240V, 1Ø motor, VA = 230V × 17A = 3,910 VA

120V, 4.5 kW dishwasher

8 – Lighting circuits 120V, 1.5 kW

Motor VA

VA (1Ø) = Table Volts × Table Amperes, Table 430-148

3-hp VA = 230V × 17A = 3,910 VA, Table 430-150

VA (3Ø) = Table Volts × Table Amperes × $\sqrt{3}$, Table 430-150

10-hp VA = 230V × 28A × 1.732 = 11,154

Note. Some exam testing agencies use 240V instead of the table V.

	Phase A (L₁ and L₂)	Phase B (L₂ and L₃)	C₁ (L₁)	C₂ (L₃)	Line Total
Heat 18 kW (omit A/C)	6,000 VA	6,000 VA	3,000 VA	3,000 VA	18,000 VA
Ranges 240V, 14 kW, 1Ø	14,000 VA	14,000 VA			28,000 VA
Water Heater 240V, 10 kW, 1Ø	10,000 VA				10,000 VA
3-hp, 240V, 1Ø		3,910 VA			3,910 VA
3-hp, 240V, 1Ø		3,910 VA			3,910 VA
Dishwasher 120V, 4.5 kW			4,500 VA *		4,500 VA
Lighting (8 – 1.5 kW), 120V	_____	_____	4,500 VA *	7,500 VA*	12,000 VA
	30,000 VA	27,820 VA	12,000 VA	10,500 VA	80,320 VA

*Indicates neutral (120V) loads.

Note: Phase totals (30,000 VA, 27,820 VA, 22,500 VA) should add up to the Line total (80,320 VA). This is done as a check to make sure all items have been accounted for and added correctly.

10-hp VA = Table Volts × Table Amperes × 1.732

VA = 230V × 28A × 1.732 = 11,154 VA (omit)

Transformer Size

What size transformers are required?

(a) three 30 kVA, 1Ø transformers

(b) one 90 kVA, 3Ø transformer

(c) a or b

(d) none of these

• Answer: (c) a or b

Phase A = 30 kVA

Phase B = 28 kVA

Phase C = 24 kVA (12 kVA × 2)

High-Leg Voltage

What is the high-leg voltage to the ground?

(a) 120V   (b) 208V   (c) 230V   (d) 240V

• Answer: (b) 208V

The vector sum of "A" and "C1" or transformers "B" and "C2," = 120V × 1.732 = 208V.

Neutral kVA

What is the maximum kVA on the neutral?

(a) 3 kVA   (b) 6 kVA   (c) 7.5 kVA   (d) 9 kVA

• Answer: (d) 9 kVA

If Phase C2 loads are not on, then phase C1 neutral loads would impose a 9 kVA load to neutral. (4.5 kW + 4.5 kW). This does not include the 6 kVA on transformer "C" from the heat load since there are no neutrals involved.

Maximum Unbalanced Current

What is the maximum unbalanced load on the neutral?

(a) 25A   (b) 50A   (c) 75A   (d) 100A

• Answer: (c) 75A

I = P/E

I = 9,000 VA/120V = 75A

Note: The neutral must be sized to the carry maximum unbalanced neutral current, which in this case is 75A.

Neutral Current

What is the current on the neutral?

(a) 0A   (b) 13A   (c) 25A   (d) 50A

- Answer: (b) 13A

L1 = 9,000 VA/120V = 75A

L2 = 7,500 VA/120V = 62.5A

Neutral current = 75A less 62.5A = 12.5A

Phase Load (1Ø)

What is the phase load of each 3-hp, 240V, 1Ø motor?

(a) 1,855 VA   (b) 3,910 VA   (c) 978 VA   (d) none of these

- Answer: (b) 3,910 VA

VA = 230V × 17A = 3,910 VA

On a delta system, a 240V, 1Ø load is 100% on one transformer.

Phase Load (3Ø)

What is the phase load for the 10-hp, 3Ø motor?

(a) 11,154 VA   (b) 3,718 VA   (c) 5,577 VA   (d) 6,440 VA

- Answer: (b) 3,718 VA

VA = Volts × Amperes × 1.732

VA = 230V × 28A × 1.732, VA = 11,154 VA

11,154 VA/3 phases = 3,718 VA per phase

High-leg Conductor Size

If only the 3Ø, 18 kW load is on the high-leg, what is the current on the high-leg conductor?

(a) 25A   (b) 43A   (c) 73A   (d) 97A

- Answer: (b) 43A

To calculate current on any conductor:

$I = P/(E \times \sqrt{3})$

$I = 18,000 \text{ VA}/(240V \times 1.732) = 43A$

Voltage Ratio

What is the phase voltage ratio of the transformer?

(a) 4:1   (b) 1:4   (c) 1:2   (d) 2:1

- Answer: (d) 2:1

480 primary phase V to 240 secondary phase V

❏ Panel Loading – VA

Balance the loads on the panelboard in VA.

Line 1	Line 2		Line 3	Line Total
Heat 18 kW, 240V, 3Ø	6,000 VA	6,000 VA	6,000 VA	18,000 VA
Ranges 14 kW, 240V, 1Ø	7,000 VA	7,000 VA		14,000 VA
Ranges 14 kW 240V, 1Ø		7,000 VA	7,000 VA	14,000 VA
Water Heater 240V, 1Ø	5,000 VA	5,000 VA		10,000 VA
3-hp, 240V, 1Ø Motor		1,955 VA	1,955 VA	3,910 VA
3-hp, 240V, 1Ø Motor		1,955 VA	1,955 VA	3,910 VA
Dishwasher 120V, 1Ø	4,500 VA*			4,500 VA
Lighting (8 circuits) 120V, 1Ø	4,500 VA*	_____	7,500 VA *	4,500 VA
	27,000 VA	28,910 VA	24,410 VA	80,320 VA

* Indicates neutral (120V) loads.

❏ **Panel Sizing – Amperes**

Balance the loads from the previous example on the panelboard in amperes.

	Line 1	Line 2	Line 3	Ampere Calculation
Heat 18 kW, 240V, 3Ø	43A	43A	43A	18,000 VA/(240V × 1.732)
Ranges 14 kW, 240V, 1Ø	58A	58A		14,000 VA/240V
Ranges 14 kW, 240V, 1Ø		58A	58A	14,000 VA/240V
Water Heater 240V, 1Ø	42A	42A		10,000 VA/240V
3-hp, 240V, 1Ø Motor		17A	17A	FLC
3-hp, 240V, 1Ø Motor		17A	17A	FLC
Dishwasher 120V, 1Ø	38A*			4,500 VA/120V
Lighting (8 circuits) 120 V, 1Ø	38A*	_____	63A*	7,500 VA/120V
	219A	235A	198A	

*Indicates neutral (120V) loads.

PART B – DELTA/WYE TRANSFORMERS

12–15 WYE TRANSFORMER VOLTAGE

In a wye configured transformer, the line voltage is greater than the phase voltage by $\sqrt{3}$. Figure 12–20.

Reminder: In a delta configured transformer, line current does not equal phase current.

Delta Primary Voltage

LINE Voltage	**PHASE Voltage**
L_1 to L_2 = 480V	Phase Winding A = 480V
L_2 to L_3 = 480V	Phase Winding B = 480V
L_3 to L_1 = 480V	Phase Winding C = 480V

Wye Secondary Voltage

LINE Voltage	**PHASE Voltage**	**NEUTRAL Voltage**
L_1 to L_2 = 208V	Phase Winding A = 120V	Neutral to L_1 = 120V
L_2 to L_3 = 208V	Phase Winding B = 120V	Neutral to L_2 = 120V
L_3 to L_1 = 208V	Phase Winding C = 120V	Neutral to L_3 = 120V

12–16 WYE VOLTAGE TRIANGLE

The line voltage and phase voltage of a wye system are not the same. They differ by a factor of $\sqrt{3}$ (1.732).

Line Voltage $I_{Line} = E_{Phase} \times \sqrt{3}$ **Phase Voltage $I_{Phase} = E_{Line}/\sqrt{3}$**

The wye voltage triangle (Figure 12–21) can be used to calculate wye, 3Ø line and phase voltage. Place your finger over the desired item, and the remaining items show the formula to use. These formulas can be used when either one of the phase or line voltages is known.

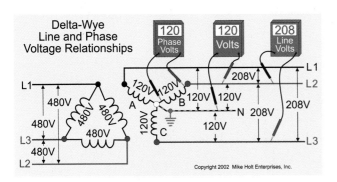

Figure 12-20

Delta-Wye Line and Phase Voltage Relationships

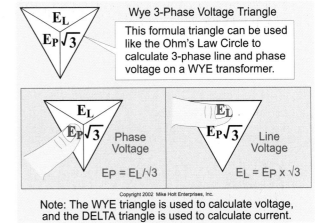

Note: The WYE triangle is used to calculate voltage, and the DELTA triangle is used to calculate current.

Figure 12-21

Wye 3-Phase Voltage Triangle

12–17 WYE TRANSFORMERS CURRENT

In a wye configured transformer, the 3Ø and 1Ø, 120V line current equals the phase current ($I_{Phase} = I_{Line}$). Figure 12–22.

12–18 WYE LINE CURRENT

The line current of both delta and wye transformers can be calculated by the following formula:

$$I_{Line} = VA_{Line}/(E_{Line} \times \sqrt{3})$$

❏ **Primary Line Current Question**

What is the primary line current for a 150 kVA 480 to 208Y/120V, 3Ø transformer? Figure 12–23, Part A.

(a) 416A   (b) 360A   (c) 180A   (d) 144A

• Answer: (c) 180A

$I_{Line} = VA_{Line}/(E_{Line} \times \sqrt{3}) = 150,000$ VA/(480V $\times$ 1.732) = 180A

❏ **Secondary Line Current Question**

What is the secondary line current for a 150 kVA, 480 to 208Y/120V, 3Ø transformer? Figure 12–23, Part B.

(a) 416A   (b) 360A   (c) 180A   (d) 144A

• Answer: (a) 416A

$I_{Line} = VA_{Line}/(E_{Line} \times \sqrt{3}) = 150,000$ VA/(208V $\times$ 1.732) = 416A

12–19 WYE PHASE CURRENT

The phase current of a wye configured transformer winding is the same as the line current. The phase current of a delta configured transformer winding is less than the line current by $\sqrt{3}$ (1.732) .

$$I_{Line} = VA_{Line}/(E_{Line} \times \sqrt{3})$$

❏ **Primary Phase Current Question**

What is the primary phase current for a 150 kVA, 480 to 208Y/120V, 3Ø transformer? Figure 12–24, Part A.

(a) 416A   (b) 360A   (c) 180A   (d) 104A

• Answer: (d) 104A

$$I_{Line} = \frac{VA_{Line}}{E_{Line} \times \sqrt{3}} = 150,000 \text{ VA}/(480\text{V} \times 1.732) = 180\text{A}$$

$$I_{Phase} = \frac{VA_{Phase}}{E_{Phase}} = 50,000 \text{ VA}/480\text{V} = 104\text{A, or } I_{Phase} = I_{Line}/\sqrt{3} = 180\text{A}/1.732 = 104\text{A}$$

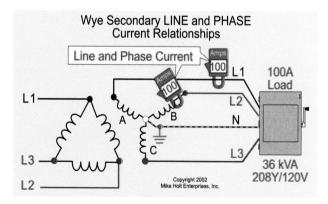

Figure 12-22

Wye Secondary Line and Phase Current Relationships

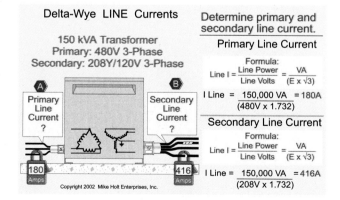

Figure 12-23

Delta-Wye Line Currents

❏ **Secondary Phase Current Question**

What is the secondary phase current for a 150 kVA, 480 to 208Y/120V, 3∅ transformer? Figure 13-24, Part B.

(a) 416A   (b) 360A   (c) 208A   (d) 104A

• Answer: (a) 416A

$$I_{Line} = \frac{VA_{Line}}{E_{Line} \times \sqrt{3}} = 150,000 \text{ VA}/(208V \times 1.732) = 416A, \text{ or}$$

$$I_{Phase} = \frac{VA_{Phase}}{E_{Phase}} = 50,000 \text{ VA}/120V = 416A$$

12–20 WYE PHASE VERSUS LINE CURRENT

Since each line conductor from a wye transformer is connected to a different transformer winding (phase), the effects of 3∅ loading on the line are the same as the phase. Figure 12–25.

$$I_{Line} = VA_{Line}/(E_{Line} \times \sqrt{3})$$

❏ **Line Versus Phase, 208V, 3∅**

A 36 kVA, 208V, 3∅ load has the following effect on the system:

LINE: Line power = 36 kVA

$I_{Line} = VA_{Line}/(E_{Line} \times \sqrt{3})$

$I_{Line} = 36,000 \text{ VA}/(208V \times \sqrt{3}) = 100A$

PHASE: Phase power = 12 kVA (any winding)

$I_{Phase} = VA_{Phase}/E_{Phase}$

$I_{Phase} = 12,000 \text{ VA}/120V = 100A$

❏ **Line Versus Phase, 208V, 1∅**

A 10 kVA, 208V, 1∅ load has the following effect on the system. Figure 12–26:

LINE: Line power = 10 kVA

$I_{Line} = VA_{Line}/E_{Line}$

$I_{Line} = 10,000 \text{ VA}/208V = 48A$

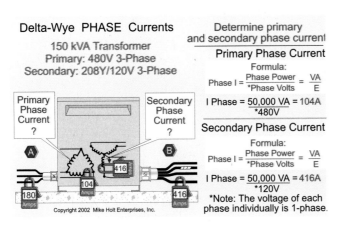

Figure 12-24
Delta-Wye Phase Currents

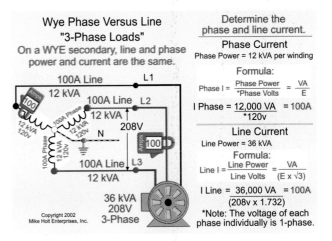

Figure 12-25
Wye Phase Versus Line Load (3-Phase)

❏ **Line Versus Phase, 120V, 1Ø**

    PHASE: Phase power = 10 kVA (winding)

    $I_{Phase} = VA_{Phase}/E_{Phase}$

    $I_{Phase} = 5,000\ VA/120V = 42A$

❏ **Phase Load (1Ø) 120V**

    A 3 kVA, 120V, 1Ø load has the following effect on the system. Figure 12–27:

    LINE: Line power = 3 kVA

    $I_{Line} = VA_{Line}/E_{Line}$

    $I_{Line} = 3,000\ VA/120V = 25A$

    PHASE: Phase power = 3 kVA (any winding)

    $I_{Phase} = VA_{Phase}/E_{Phase}$

    $I_{Phase} = 3,000\ VA/120V = 25A$

12–21 WYE TRANSFORMER BALANCING

To properly size a delta-wye transformer, the secondary transformer phases (windings) or the line conductors must be balanced. Balancing the panel (line conductors) is identical to balancing the transformer for wye configured transformers! The following steps should be helpful to balance the transformer:

Step 1:   Determine the VA rating of all loads.

Step 2:   Split 3Ø loads: one-third on Phase A, one-third on Phase B and one-third on Phase C.

Step 3:   Split 1Ø, 208V loads (largest to smallest): one-half on each phase (A to B, B to C, or A to C).

Step 4:   Place 120V loads (largest to smallest): 100% on any phase.

❏ **Transformer Balancing Example**

Balance and size a 480 to 208Y/120V, 3Ø transformer for the following loads: 36 kVA, 208V, 3Ø heat strip; two 10 kVA, 208V, 1Ø loads; three 3 kVA, 120V loads. Figure 12–28.

	Phase A (L₁)	Phase B (L₂)	Phase C (L₃)	Line Total
36 kVA, 208V, 3Ø	12 kVA	12 kVA	12 kVA	36 kVA
10 kVA, 208V, 1Ø	5 kVA	5 kVA		10 kVA
10 kVA, 208V, 1Ø		5 kVA	5 kVA	10 kVA
Three 3 kVA, 120V	6 kVA *		3 kVA *	9 kVA
	23 kVA	22 kVA	20 kVA	65 kVA

* Indicates neutral (120V) loads.

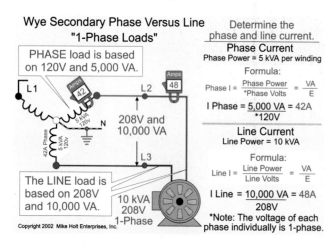

Figure 12-26

Wye Phase Versus Line Load (Single-Phase)

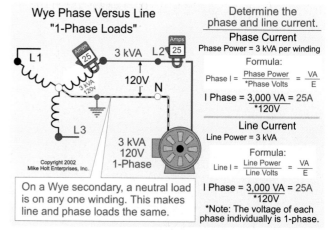

Figure 12-27

Wye Phase Versus Line Load (Single-Phase)

12–22 WYE TRANSFORMER SIZING

Once you balance the transformer, you can size the transformer according to the load on each phase.

❏ **Transformer Sizing Example**

What size 480 to 208Y/120V, 3Ø phase transformer is required for the following loads: 208V, 36 kVA, 3Ø heat strip; two 208V, 10 kVA, 1Ø loads; three 120V, 3 kVA loads?

(a) three 1Ø, 25 kVA transformers

(b) one 3Ø, 75 kVA transformer

(c) a or b

(d) none of these

• Answer: (c) a or b

Phase A = 23 kVA

Phase B = 22 kVA

Phase C = 20 kVA

	Phase A (L$_1$)	Phase B (L$_2$)	Phase C (L$_3$)	Line Total
36 kVA, 208V, 3Ø	12 kVA	12 kVA	12 kVA	36 kVA
10 kVA, 208V, 1Ø	5 kVA	5 kVA		10 kVA
10 kVA, 208V, 1Ø		5 kVA	5 kVA	10 kVA
Three 3 kVA, 120V	6 kVA *		3 kVA *	9 kVA
	23 kVA	22 kVA	20 kVA	65 kVA

* Indicates neutral (120V) loads.

12–23 WYE PANEL BALANCING

When balancing a panelboard in VA, be sure that 3Ø loads are split one-third on each line, 208V, 1Ø loads are split one-half on each line, and 120V loads are placed on line1 and line 2.

❏ **Panel Balancing Example**

Balance and size a 208Y/120V, 3Ø phase panelboard in kVA for the following loads: 36 kVA, 208V, 3Ø heat strip; two 10 kVA, 208V, 1Ø loads; three 3 kVA, 120V loads.

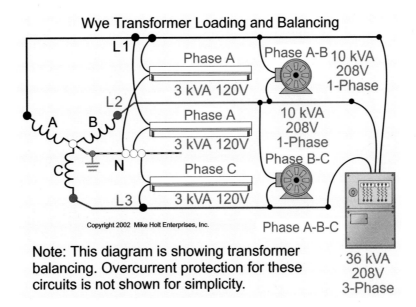

Wye Transformer Loading and Balancing

Note: This diagram is showing transformer balancing. Overcurrent protection for these circuits is not shown for simplicity.

Copyright 2002 Mike Holt Enterprises, Inc.

Figure 12-28

Wye Transformer Loading and Balancing

	Line 1	Line 2	Line 3	Line Total
36 kVA, 208V, 3Ø	12 kVA	12 kVA	12 kVA	36,000 VA
10 kVA, 208V, 1Ø	5 kVA	5 kVA		10,000 VA
10 kVA, 208V, 1Ø		5 kVA	5 kVA	10,000 VA
Two 3 kVA, 120V	6 kVA *			6,000 VA
One 3 kVA, 120V			3 kVA *	3,000 VA
	23 kVA	22 kVA	20 kVA	65,000 VA

* Indicates neutral (120V) loads.

12–24 WYE PANELBOARD AND CONDUCTOR SIZING

When selecting and sizing the panelboard and conductors, we must balance the line loads in amperes.

❏ Panelboard Conductors Sizing

Balance and size a 208Y/120V, 3Ø panelboard in amperes for the following loads: 36 kVA, 208V, 3Ø heat strip; two 10 kVA, 208V, 1Ø loads; and three 3 kVA, 120V loads.

	Line 1	Line 2	Line 3	Ampere Calculation
36 kVA, 208V, 3Ø	100A	100A	100A	$36,000/(208V \times 1.732)$
10 kVA, 208V, 1Ø	48A	48A		10,000 VA/208V
10 kVA, 208V, 1Ø		48A	48A	10,000 VA/208V
Two 3 kVA, 120V	50A*			6,000 VA/120V
One 3 kVA, 120V			25A*	3,000 VA/120V
	198A	196A	173A	

12–25 WYE NEUTRAL CURRENT

To determine the neutral current of a wye system, we use the following formula:

$$I_{Neutral} = \sqrt{(I_{Line\ 1}^2 + I_{Line\ 2}^2 + I_{Line\ 3}^2) - [(I_{Line\ 1} \times I_{Line\ 2}) + (I_{Line\ 2} \times I_{Line\ 3}) + (I_{Line\ 1} \times I_{Line\ 3})]}$$

❏ Neutral Current

Balance and size the neutral current for the following loads: 208V, 36 kVA, 3Ø heat strip; two 208V, 10 kVA, 1Ø loads; three 3 kVA, 120V loads. $L_1 = 50A$, $L_2 = 0A$, $L_3 = 25A$.

(a) 0A   (b) 25A   (c) 43A   (d) 50A

• Answer: (c) 43A

Based on the previous example.

$$I_{Neutral} = \sqrt{(I_{Line\ 1}^2 + I_{Line\ 2}^2 + I_{Line\ 3}^2) - [(I_{Line\ 1} \times I_{Line\ 2}) + (I_{Line\ 2} \times I_{Line\ 3}) + (I_{Line\ 1} \times I_{Line\ 3})]}$$

$$I_{Neutral} = \sqrt{(50^2 + 0^2 + 25^2) - [(50 \times 0) + (0 \times 25) + (50 \times 25)]} = \sqrt{(2,500 + 625) - 1,250} = \sqrt{1,875} = 43.3A$$

12–26 WYE MAXIMUM UNBALANCED LOAD

The maximum unbalanced load (neutral) is the actual current on the grounded conductor.

❏ Maximum Unbalanced Load

Balance and size the maximum unbalanced load for the following loads: 208V, 36 kVA, 3Ø heat strip; two 208V, 10 kVA, 1Ø loads; three 120V, 3 kVA loads. L1 = 50A, L2 = 0A, L3 = 25A.

(a) 0A   (b) 25A   (c) 50A   (d) 100A

• Answer: (c) 50A

The maximum unbalanced current equals the largest line neutral current.

	Line 1	Line 2	Line 3	Ampere Calculation
36 kVA, 208V, 3Ø	100A	100A	100A	36,000/(208V × 1.732)
10 kVA, 208V, 1Ø	48A	48A		10,000 VA/208V
10 kVA, 208V, 1Ø		48A	48A	10,000 VA/208V
Two 3 kVA, 120V	50A*			6,000 VA/120V
One 3 kVA, 120V			25A*	3,000 VA/120V
	198A	196A	173A	

* Indicates neutral (120V) loads.

12–27 DELTA/WYE EXAMPLE

Dishwasher (4.5 kW) 120V   Two – 3-hp motors, 208V, 1Ø
A/C motor (10-hp) 208V, 3Ø   Water Heater (10 kW) 208V, 1Ø
Heat (18 kW) 208V, 3Ø   Eight – Lighting circuits (1.5 kW), 120V
Two – Ranges (14 kW) 208V, 1Ø

	Phase A (L$_1$)	Phase B (L$_2$)	Phase C (L$_3$)	Line Total
Heat (A/C omitted *)	6,000 VA	6,000 VA	6,000 VA	18.0 kVA
Ranges 14 kW, 208V, 1Ø	7,000 VA	7,000 VA		14.0 kVA
Ranges 14 kW, 208V, 1Ø		7,000 VA	7,000 VA	14.0 kVA
Water Heater 10 kW 208V, 1Ø	5,000 VA		5,000 VA	10.0 kVA
3-hp, 208V, 1Ø Motor	1,945 VA	1,945 VA		3.89 kVA
3-hp, 208V, 1Ø Motor		1,945 VA	1,945 VA	3.89 kVA
Dishwasher 4.5 kW, 120V			4,500 VA *	4.5 kVA
Lighting (8 – 1.5 kW circuits)	6,000 VA *	3,000 VA *	3,000 VA *	12.0 kVA
	25,945 VA	26,890 VA	27,445 VA	80.28 kVA

Note: The phase totals (26 kVA, 27 kVA and 27.5 kVA) should add up to the line total (80.5 kVA). This method is used as a check to make sure all items have been accounted for and added correctly.

Motor VA

VA 1Ø = Table Volts × Table Amperes
3-hp VA = E × I = 208V × 18.7A = 3,890 VA
VA Per Phase 3,890 VA/2 = 1,945 VA per phase

VA 3Ø = Table Volts × Table Amperes × $\sqrt{3}$
10-hp, A/C, VA = Table Volts × Table Amperes × 1.732
10-hp, A/C, VA = 208V × 30.8A × 1.732 = 11,096 VA (omit, smaller than 18 kW heat)

Transformer Sizing

What size transformers are required?
(a) three 30 kVA, 1Ø transformers   (b) one 90 kVA, 3Ø transformer
(c) a or b   (d) none of these
  • Answer: (c) a or b
  Phase A = 26 kVA
  Phase B = 27 kVA
  Phase C = 28 kVA

Maximum kVA on Neutral

What is the maximum kVA on the neutral?
(a) 3 kVA   (b) 6 kVA   (c) 7.5 kVA   (d) 9 kVA
  • Answer: (c) 7.5 kVA
  120V loads only; do not count phase-to-phase (208V) loads.

Maximum Neutral Current

What is the maximum current on the neutral?

(a) 50A   (b) 63A   (c) 75A   (d) 100A

• Answer: (b) 63A

$I = P/E = 7,500 \text{ VA}/120\text{V} = 63\text{A}$.

The neutral must be sized to carry the maximum unbalanced neutral current, which in this case is 63A.

Neutral Current

What is the current on the neutral?

(a) 25A   (b) 33A   (c) 50A   (d) 63A

• Answer: (b) 33A

$$I_{Neutral} = \sqrt{(I_{Line\ 1}^2 + I_{Line\ 2}^2 + I_{Line\ 3}^2) - [(I_{Line\ 1} \times I_{Line\ 2}) + (I_{Line\ 2} \times I_{Line\ 3}) + (I_{Line\ 1} \times I_{Line\ 3})]}$$

$L_1 = 6,000 \text{ VA}/120\text{V} = 50\text{A}$

$L_2 = 3,000 \text{ VA}/120\text{V} = 25\text{A}$

$L_3 = 7,500 \text{ VA}/120\text{V} = 63\text{A}$

$$I_{Neutral} = \sqrt{(50^2 + 25^2 + 63^2) - [(50 \times 25) + (25 \times 63) + 50 \times 63)]}$$

$$I_{Neutral} = \sqrt{(2,500 + 625 + 3,969) - (1,250 + 1,575 + 3,150)} = \sqrt{7,049 - 5,975} = \sqrt{1,119} = 33\text{A}$$

Phase Load 208V, 1Ø

What is the per phase load of each 3-hp, 208V, 1Ø motor?

(a) 1,945 VA   (b) 3,910 VA   (c) 978 VA   (d) none of these

• Answer: (a) 1,945 VA

VA (1Ø) = Table Volts × Table Amperes

3-hp VA = 208V × 18.7A = 3,890 VA

208V, 1Ø load is on two transformer phases = 3,890 VA/2 phases = 1,945 VA

Phase Load 208V, 3Ø

What is the per phase load for the 10-hp, 208V, 3Ø A/C motor?

(a) 11,154 VA   (b) 3,698 VA   (c) 5,577 VA   (d) none of these

• Answer: (b) 3,698 VA

10-hp VA = Table volts × Table amperes × 1.732

VA = 208V × 30.8A × 1.732 = 11,095 VA

3Ø load is on three phases = 11,095 VA/3 phases = 3,698 VA per phase

Voltage Ratio

What is the phase voltage ratio of the transformer?

(a) 4:1   (b) 1:4   (c) 1:2   (d) 2:1

• Answer: (a) 4:1

480 primary-phase volts to 120 secondary-phase volts

❏ Balance Panel – VA

Balance the loads on the panelboard in VA.

	Line 1	Line 2	Line 3	Line Total
Heat 18 kW(A/C 11 kW/ omitted)	6,000 VA	6,000 VA	6,000 VA	18.0 kVA
Ranges 14 kW, 208V, 1Ø	7,000 VA	7,000 VA		14.0 kVA
Ranges 14 kW, 208V, 1Ø		7,000 VA	7,000 VA	14.0 kVA
Water Heater 10 kW, 208V, 1Ø	5,000 VA		5,000 VA	10.0 kVA
3-hp, 208V, 1Ø	1,945 VA	1,945 VA		3.9 kVA
3-hp, 208V, 1Ø		1,945 VA	1,945 VA	3.9 kVA
Dishwasher 4.5 kW, 120V			4,500 VA	4.5 kVA
Lighting (8 – 1.5 kW circuits)	6,000 VA	3,000 VA	3,000 VA	12.0 kVA
	25,945 VA	26,890 VA	27,445 VA	80.3 kVA

❑ **Panelboard Balancing and Sizing**

Balance the previous example loads on the panelboard in amperes.

	Line 1	Line 2	Line 3	Ampere Calculations
Heat 18 kW, Heat 3Ø	50.0A	50.0A	50.0A	18,000 VA/(208V × 1.732)
Range 14 kW, 208V, 1Ø	67.0A	67.0A		14,000 VA/208V
Range 14 kW, 208V, 1Ø		67.0A	67.0A	14,000 VA/208V
Water Heater 10 kW, 208V, 1Ø	48.0A		48.0A	10,000 VA/208V
3-hp, 208V, 1Ø Motor	18.7A	18.7A		Motor FLC
3-hp, 208V, 1Ø Motor		18.7A	18.7A	Motor FLC
Dishwasher 120V			38.0A	4,500 VA/120V
Lighting (8 – 1.5 kW circuits)	50.0A	25.0A	25.0A	1,500 VA/120V
	233.7A	246.4A	246.7A	

12–28 DELTA VERSUS WYE

Delta

Delta phase VOLTAGE is the SAME as the line voltage.

$E_{Phase} = E_{Line}$ or $E_{Line} = E_{Phase}$

Delta phase CURRENT is DIFFERENT than the line current.
Line current is greater than Phase Current by the square root of 3 (1.732).

$I_{Line} = I_{Phase} \times \sqrt{3}$ or $I_{Phase} = I_{Line}/\sqrt{3}$

Wye

Phase CURRENT is the SAME as line current.

$I_{Phase} = I_{Line}$ or $I_{Line} = I_{Phase}$

Wye Phase VOLTAGE is DIFFERENT than line voltage.
Line Voltage is greater than Phase Voltage by the square root of 3 ($\sqrt{3}$).

$E_{Line} = E_{Phas} \times \sqrt{3}$ or $E_{Phase} = E_{Line}/\sqrt{3}$

Unit 12 – Delta/Delta and Delta/Wye Transformers Summary Questions

Calculations Questions

Definitions

1. Delta-configured means that the windings of three 1Ø transformers are connected in _____ with each other. A delta-connected transformer is represented by the Greek letter delta.
    (a) series   (b) parallel   (c) series-parallel   (d) a and b

2. What 4-wire system is called the high-leg system because the voltage from one conductor to ground is between 190 and 208 volts-to-ground?
    (a) Delta   (b) Wye

3. The term _____ is used to identify the ungrounded (hot) conductors in an electrical system.
    (a) load   (b) line   (c) system   (d) grounded

4. The _____ voltage is the voltage measured between any two ungrounded (hot) conductors.
    (a) load   (b) line   (c) grounded   (d) any of these

5. • The term "phase current" is used to identify the current flowing through the transformer winding. For _____-connected systems, the phase current will be less than the line current and for _____-connected systems, the phase current will be the same as the line current.
    (a) wye, wye   (b) delta, delta   (c) wye, delta   (d) delta, wye

6. The term "phase load" is used to identify the load on the transformer winding. For delta-configured systems, the phase load will be equal to _____.
    (a) 3Ø, 240V load = line load/3   (b) 1Ø, 240V load = line load
    (c) 1Ø, 120V load = line load   (d) all of these

7. The term "phase load" is used to identify the load on the transformer winding. For wye-configured systems, the phase load will be equal to _____.
    (a) 3Ø, 208V load = line load/3   (b) 1Ø, 208V load = line load/2
    (c) 1Ø, 120V load = line load   (d) all of these

8. The term "phase voltage" is used to identify the internal voltage of a single transformer winding. For _____-connected systems, the phase voltage will be equal to the line voltage and for _____-connected systems, the phase voltage is less than the line voltage.
    (a) wye, wye   (b) delta, delta   (c) wye, delta   (d) delta, wye

9. The turns ratio is the relationship between the number of primary windings compared to the number of secondary windings. This is the same as the turns ratio between the primary phase voltage and the secondary phase voltage. For a delta/delta system, the phase voltage ratio will be _____ and for a delta/wye system, the phase voltage ratio will be _____.
    (a) 1:2, 1:4   (b) 2:1, 4:1   (c) 4:1, 2:1   (d) none of these

10. What is the turns ratio for a 480V to 208Y/120V, 3Ø delta/delta configured transformer?
    (a) 4:1   (b) 1:4   (c) 1:2   (d) 2:1

11. • The term "unbalanced load" is intended to identify the load on the secondary grounded (neutral) conductor. For _____-configured systems, the unbalanced load shall be determined by the formula:

$$I_{Neutral} = \sqrt{(I_{Line\ 1}^2 + I_{Line\ 2}^2 + I_{Line\ 3}^2) - (I_{Line\ 1} \times I_{Line\ 2} + I_{Line\ 2} \times I_{Line\ 3} + I_{Line\ 1} \times I_{Line\ 3})}$$

    (a) delta   (b) wye   (c) a or b   (d) none of these

12. The unbalanced load in amperes for a _____-configured system can be calculated by the formula:

$$I_{Neutral} = I_{Line1} - I_{Line3}$$

    (a) delta   (b) wye   (c) a or b   (d) none of these

13. _____-connected means a connection of three 1Ø transformer windings to a common point (neutral), with the other end of each winding connected to the line conductors.
    (a) Delta   (b) Wye   (c) a or b   (d) none of these

12–1 Current Flow

14. When a load is connected to the secondary winding of a transformer, current will flow through the secondary winding. The current flowing in the secondary creates an electromagnetic field that opposes the primary electromagnetic field.
    (a) True   (b) False

15. In a delta-configured transformer, the ratio of the primary-to-secondary line current will be directly proportional (the same ratio) to the voltage turns ratio.
    (a) True   (b) False

Part A – Delta/Delta Transformers

12–2 Delta Transformer Voltages

16. In a delta-configured transformer, the line voltage is equal to the phase voltage.
    (a) True   (b) False

12–3 Delta High-Leg

17. If the secondary voltage of a delta/delta transformer is 220/110V, the high-leg voltage-to-ground (or neutral) will be approximately _____.
    (a) 191V   (b) 196V   (c) 202V   (d) 208V

12–4 Delta Primary and Secondary Line Currents

18. In a delta-configured transformer, the line current equals the phase current.
    (a) True   (b) False

19. What is the primary line current for a fully loaded 45 kVA, 240/120V, 3Ø transformer?
    (a) 124A   (b) 108A   (c) 54A   (d) 43A

20. What is the secondary line current for a fully loaded 45 kVA, 480V to 240V, 3Ø transformer?
    (a) 124A   (b) 108A   (c) 54A   (d) 43A

12–5 Delta Primary or Secondary Phase Currents

21. The phase current of a transformer winding is calculated by dividing the phase load by the phase volts: $I_{Phase} = VA_{Phase}/E_{Phase}$.
    (a) True   (b) False

22. What is the primary phase current for a fully loaded 15 kVA, 480V to 240/120V, 1Ø transformer?
    (a) 124A   (b) 62A   (c) 45A   (d) 31A

23. What is the secondary phase current for a fully loaded 15 kVA, 480V to 240/120V, 1Ø transformer?
    (a) 124A   (b) 63A   (c) 45A   (d) 31A

12–6 Delta Phase Versus Line

24. A 15 kVA, 240V, 3Ø load has the following effect on a delta system:
    (a) $I_{Line} = 15,000\ VA/(240V \times \sqrt{3}) = 36A$ (b) $I_{Phase} = 5,000\ VA/240V = 21A$
    (c) $I_{Line} = I_{Phase} \times \sqrt{3} = 21A \times 1.732 = 36A$ (d) all of the above

25. • A 5 kVA, 240V, 1Ø load has the following effect on a delta/delta system:
    (a) Line Power = 5 kVA   (b) Phase Power = 5 kVA
    (c) $I_{Phase} = 5,000\ VA/240V = 21A$ (d) all of the above

26. • A 2 kVA, 120V, 1Ø load has the following effect on a delta system:
    (a) Line Power = 2 kVA   (b) $I_{Line} = 2,000\ VA/120V = 17A$
    (c) Phase Power = 3 kVA (C1 or C2 winding)   (d) a and b

12–7 Delta Current Triangle

27. The line and phase current of a delta system are not equal. The difference between line and phase current can be described by which of the following equations?
    (a) $I_{Line} = I_{Phase} \times \sqrt{3}$ (b) $I_{Phase} = I_{Line}/\sqrt{3}$
    (c) a and b   (d) none of these

12–8 and 12–9 Delta Transformer Balancing and Sizing

28. To properly size a delta /delta transformer, the transformer shall be balanced in kVA. Which of the following steps should be used to balance the transformer?
    (a) 3Ø loads, split one-third on Phase A, one-third on Phase B and one-third on Phase C.
    (b) 1Ø, 240V loads, place 100% on Phase A or Phase B.
    (c) 120V loads, place 100% on Phase C1 or C2.
    (d) all of these

29. Balance and size a 480V to 240/120V, 3Ø delta transformer for the following loads: one 18 kVA, 3Ø heat strip, two 5 kVA, 1Ø, 240V loads and three 2 kVA 120V loads. What size transformer is required?
    (a) three 1Ø, 12.5 kVA transformers
    (b) one 3Ø, 37.5 kVA transformer
    (c) a or b
    (d) none of these

12–10 Delta Panel Schedule in kVA

30. When balancing a panelboard, be sure to split the 3Ø VA load so that one-third is on each line, split the 240V, 1Ø VA load so that 50 percent is on each line and place 120V, 1Ø VA loads so that they are balanced as equally as possible on each line.
    (a) True   (b) False

31. • Balance a 120/240V delta 3Ø panelboard in kVA for the following loads: one 18 kVA, 3Ø heat strip; two 5 kVA, 1Ø, 240V loads; and three 2 kVA 120V loads.
    (a) The largest Line load is 13 kVA.   (b) The smallest Line load is 10 kVA.
    (c) Total Line load is equal to 34 kVA.   (d) all of these.

12–11 Delta Panelboard and Conductor Sizing

32. When selecting and sizing panelboards and conductors, we must balance the line loads in amperes. Balance a 120/240V, 3Ø panelboard in amperes for the following loads: one 18 kVA, 3Ø heat strip; two 5 kVA, 1Ø, 240V loads; and three 2 kVA 120V loads.
     (a) The largest line is 98A   (b) The smallest line is 81A
     (c) Each line equals to 82A   (d) a and b

12–12 Delta Neutral Current

33. What is the neutral current for: one 18 kVA, 3Ø heat strip; two 5 kVA, 1Ø, 240V loads; and three 2 kVA 120V loads?
     (a) 17A   (b) 34A   (c) 0A   (d) a and b

12–13 Delta Maximum Unbalanced Load

34. • What is the maximum unbalanced load in amperes for: one 18 kVA, 3Ø heat strip; two 5 kVA, 1Ø, 240V loads; and three 2 kVA 120V loads?
     (a) 25A   (b) 34A   (c) 0A   (d) none of these

35. What is the phase load for an 18 kVA, 3Ø heat strip?
     (a) 18 kVA   (b) 9 kVA   (c) 6 kVA   (d) 3 kVA

36. • On a delta system, what is the transformer winding phase load for a 5 kVA, 1Ø, 240V load?
     (a) 5 kVA   (b) 2.5 kVA   (c) 1.25 kVA   (d) none of these

37. What is the current on the high-leg if the only load on the high-leg is a 12 kW, 240V, 3Ø load?
     (a) 29A   (b) 43A   (c) 73A   (d) 97A

38. What is the turns ratio of a 480V to 240/120V transformer?
     (a) 4:1   (b) 1:4   (c) 1:2   (d) 2:1

Part B Delta/Wye Transformers

12–15 Wye Transformers Voltages

39. In a wye-configured transformer, the line voltage will be the same as the phase voltage.
     (a) True   (b) False

12–17 Wye Transformers Current

40. In a wye-configured transformer, the line current will be the same as the phase current.
     (a) True   (b) False

12–18 Wye Line Current

41. The line current for both delta and wye-configured 3Ø transformers can be calculated by the formula:
     $I_{Line} = VA_{Line}/(E_{Line} \times \sqrt{3})$.
     (a) True   (b) False

42. What is the primary line current for a fully loaded 22 kVA, 480V to 208Y/120V, 3Ø transformer?
     (a) 61A   (b) 54A   (c) 26A   (d) 22A

43. What is the secondary line current for a fully loaded 22 kVA, 480V to 208Y/120V, 3Ø transformer?
     (a) 61A   (b) 54A   (c) 27A   (d) 22A

12–19 Wye Phase Current

44. The phase current of a wye-configured transformer will be the same as the line current. The phase current of a delta-configured transformer is less than the line current by a factor of 1.732 ($\sqrt{3}$).
    (a) True   (b) False

45. • What is the primary phase current for a 22 kVA, 480V to 208Y/120V, 3Ø transformer?
    (a) 15A   (b) 22A   (c) 36A   (d) 61A

46. • What is the secondary phase current for a 22 kVA, 480V to 208Y/120V, 3Ø transformer?
    (a) 15A   (b) 22A   (c) 36A   (d) 61A

12–20 Wye Phase Versus Line

47. Since each line conductor from a wye transformer is connected to a different transformer winding (phase), the effects of loading on the line are the same as the phase. A 12 kVA, 208V, 3Ø load has the following effect on the system:
    (a) $I_{Line} = VA_{Line}/(E_{Line} \times \sqrt{3}) = 12,000 \text{ VA}/(208V \times \sqrt{3}) = 33A$
    (b) Phase Power = 4 kVA
    (c) $I_{Phase} = VA_{Phase}/E_{Phase} = 4,000 \text{ VA}/120V = 33A$
    (d) all of these

48. A 5 kVA, 208V, 1Ø load has the following effect on the system:
    (a) Line Power = 5 kVA
    (b) $I_{Line} = VA/E = 5,000 \text{ VA}/208V = 24A$
    (c) $I_{Phase} = VA_{Phase}/E_{Phase} = 2,500 \text{ VA}/120V = 21A$
    (d) all of these

49. A 2 kVA, 120V, 1Ø load has the following effect on the system:
    (a) Line Power = 2 kVA
    (b) $I_{Line} = VA/E = 2,000 \text{ VA}/120V = 17A$
    (c) $I_{Phase} = VA_{Phase}/E_{Phase} = 2,000 \text{ VA}/120V = 17A$
    (d) all of these

50. • What is the phase load (load per transformer winding) for a 3-hp 208V, 1Ø motor?
    (a) 1,945 VA   (b) 3,890 VA   (c) 978 VA   (d) none of these

51. • What is the phase load (load per transformer winding) for a 7.5-hp, 208V, 3Ø motor?
    (a) 2,906 VA   (b) 8,718 VA   (c) 4,359 VA   (d) none of these

12–21 and 12–22 Wye Transformer Balancing and Sizing

52. To properly size a delta/wye transformer, the secondary transformer phases (windings) or the line conductors shall be balanced. Balancing the panel (line conductors) is identical to balancing the transformer for wye configured transformers. Which of the following steps should be used to balance the transformer?
    (a) 3Ø loads, split one-third on each phase
    (b) 1Ø, 208V loads, split one-half each phase
    (c) 120V loads, place 100% on any phase
    (d) all of these

53. Balance and size a 480V to 208Y/120V, 3Ø transformer for the following loads: one 18 kVA, 3Ø heat strip; two 5 kVA, 1Ø, 208V loads; and three 2 kVA 120V loads. What size transformer is required?
    (a) three 1Ø, 12.5 kVA transformers
    (b) one 3Ø, 37.5 kVA transformer
    (c) a or b
    (d) none of these

12–23 Wye Panel Schedule in kVA

54. Balance a 208Y/120V, 3Ø panelboard in kVA for the following loads: one 18 kVA, 3Ø heat strip; two 5-kVA, 1Ø, 208V loads; and three 2 kVA, 120V loads.
    (a) The largest line is 12.5 kVA
    (b) The smallest line is 10.5 kVA
    (c) The total line load equals 34 kVA
    (d) all of these

12–24 Wye Panelboard and Conductor Sizing

55. • When selecting and sizing panelboards and conductors, we must balance the line loads in amperes. Balance a 208Y/120V, 3Ø panelboard in amperes for the following loads: one 18 kVA, 208V, 3Ø heat strip; two 5 kVA, 1Ø, 208V loads; and three 2 kVA 120V loads.
    (a) The largest line current is 108A
    (b) The smallest line is 91A
    (c) a and b
    (d) none of these

12–25 Wye Neutral Current

56. The neutral current for a 3-wire circuit from a 4-wire wye-configured system can be determined by the following formula:
$$I_{Neutral} = \sqrt{(I_{Line\ 1}^2 + I_{Line\ 2}^2 + I_{Line\ 3}^2) - [(I_{Line\ 1} \times I_{Line\ 2}) + (I_{Line\ 2} \times I_{Line\ 3}) + (I_{Line\ 1} \times I_{Line\ 3})]}$$
    (a) True   (b) False

57. • What is the neutral current for: one 18 kVA, 3Ø heat strip; two 5 kVA, 1Ø, 208V loads; and three 2 kVA 120V loads?
    (a) 17A   (b) 29A   (c) 34A   (d) A and B

12–26 Wye Maximum Unbalanced Load

58. • What is the maximum unbalanced load in amperes for: one 18 kVA, 3Ø heat strip; two 5 kVA, 1Ø, 208V loads; and three 2 kVA 120V loads.
    (a) 17A   (b) 34A   (c) 0A   (d) a and b

12–27 Delta Versus Wye

59.   Which of the following statements are true for wye systems?
      (a) Phase current is the same as line current: $I_{Phase} = I_{Line}$.
      (b) Phase voltage is the same as the line voltage.
      (c) $E_{Line} = E_{Phase} \times \sqrt{3}$
      (d) a and c

60.   Which of the following statements are true for delta systems?
      (a) Phase voltage is the same as line voltage: $E_{Phase} = E_{Line}$.
      (b) Phase current is the same as the line current.
      (c) Line current is greater than phase current: $I_{Line} = I_{Phase} \times \sqrt{3}$.
      (d) a and c

☆ Challenge Questions

Part A – Delta/Delta Transformers

The following loads are to be connected to a 3Ø, delta/delta 480V to 240V transformer:

Heat 18 kW, 230V, 3Ø heat
A/C 25-hp, 230V, 3Ø synchronous motor
Two 1-hp, 115V, 1Ø motors
Two 2-hp, 230V motors
Three 1,900W, 115V lighting loads

Balance the loads as closely as possible, then answer questions 61 through 67 based on this information.

61.   • If three 1Ø transformers are connected in series, what size transformers are required for the loads?
      (a) 10/10/10 kVA   (b) 15/15/15 kVA   (c) 10/10/20 kVA   (d) 15/15/10 kVA

62.   • The line VA load of the 25-hp, 3Ø synchronous motor would be _____ VA.
      (a) 9,700   (b) 21,100   (c) 32,000   (d) 48,000

63.   • After balancing the 120V loads, what is the maximum unbalanced neutral current in _____.
      (a) 27A   (b) 1.5A   (c) 48A   (d) 148A

64.   • The transformer phase load of one 1-hp, 230V motor would be _____.
      (a) 1,380 VA   (b) 2,760 VA   (c) 1,840 VA   (d) 1,150 VA

65.   • The line load of one of the 1-hp motors would be _____.
      (a) 1,380 VA   (b) 2,760 VA   (c) 1,840 VA   (d) 1,150 VA

66.   • The transformer phase VA load of the 25-hp, 230V, 3Ø synchronous motor is closest to _____n .
      (a) 3,520 VA   (b) 7,038 VA   (c) 21,100 VA   (d) 32,100 VA

67.   • If the only load on the high-leg is the 3Ø, 25-hp motor, the high-leg conductor would have to be sized for a minimum of _____.
      (a) 13A   (b) 66A   (c) 18A   (d) 22A

Answer the next two questions based on this information.
A 3Ø, 37.5 kVA delta/delta transformer has a primary of 480V and a secondary of 240/120V.

68.   • The primary line current is closest to _____.
      (a) 45A   (b) 90A   (c) 50A   (d) 65A

69. • The maximum size overcurrent device for the 37.5 kVA delta/delta transformer is _____.
    (a) 40A   (b) 45A   (c) 50A   (d) 60A

Answer the next question based on this information.
Seven heaters 1,500W, 120V
Five 3 kW, 240V, 1Ø heaters
Equipment circuits 30 kW, 120V
Lighting circuits 40 kW, 120V

70. • If a 3Ø, 37.5 kVA, center tapped, delta/delta transformer that has a primary voltage of 480V and secondary voltage of 240/120V has only the 120V loads listed connected to its secondary, what is the total kW load on this transformer?
    (a) 30-40 kW   (b) 50-60 kW   (c) 80-90 kW   (d) over 100 kW

71. • The total load on a center tapped 240/120V transformer (delta/delta) is 100 kW which consists of 80 kW of 120V balanced loads and 20 kW of 240V loads. What is the maximum unbalanced current on the grounded (neutral) conductor?
    (a) 167A   (b) 0A   (c) 192A   (d) 333A

72. • The phase current flowing though the secondary winding for delta/delta 3Ø transformer will be _____ if the load is 30,000 VA, 240V, 3Ø.
    (a) 125A   (b) 72A   (c) 42A   (d) none of these

73. • What is the secondary line current for a 45 kVA delta/delta transformer that has a line voltage turns ratio of 2:1, if the primary operates at 460V?
    (a) 113A   (b) 55A   (c) 36A   (d) cannot be determined

74. • A 4160/240V, 3Ø transformer is connected delta/delta. The secondary phase current is 300A and the secondary line current is 240V. How much current will be flowing in the conductors supplying the transformer (line current)?
    (a) 17A   (b) 20A   (c) 50A   (d) 30A

75. • The 3Ø transformer connection giving the highest secondary line voltage is _____. Assume the primary line voltage is 480V.
    (a) wye primary, delta secondary   (b) delta primary, delta secondary
    (c) wye primary, wye secondary   (d) delta primary, wye secondary

Part B – Delta/Wye Transformers

76. • What is the primary line current for a delta/wye 480V to 208/120V, 3Ø transformer that has a secondary line current of 200A?
    (a) 31A   (b) 72A   (c) 87A   (d) 53A

77. • The secondary line current for a wye-connected transformer will be approximately _____ if the secondary phase voltage is 277V and the load is 10 kVA, 3Ø.
    (a) 7A   (b) 12A   (c) 32A   (d) none of these

Answer the next question based on this information.
One motor, 15-hp, 208V, 3Ø
Two dryers, 1.5 kW, 208V dryers, 1Ø
Four lighting circuits, 3 kW, 120V
One cooktop, 8 kW, 208V, 1Ø
Five heaters, 2.5 kW, 208V, 1Ø each

78. • Each 2.5 kW heat strip has a line load of _____.
    (a) 1,250W   (b) 12,500W   (c) 2,500W   (d) no load

79. • The VA load per line for a 15-hp, 230V, 3Ø motor can be determined by which of these formulas?
(a) $\sqrt{3} \times$ Volts $\times$ Amperes   (b) (Volts $\times$ Amperes)/3
(c) Volts $\times$ Amperes   (d) none of these

80. • If a 3Ø motor has a total load of 12,000 VA and is connected to a delta/wye transformer, the load on one phase of the secondary will be _____.
(a) 12,000 VA   (b) 4,000 VA   (c) 3,000 VA   (d) 1,500 VA

81. • A 3Ø, 480Y to 208/120V transformer provides power to nine 2,000 VA, 120V lights and nine 2,000W, 208V heaters. The maximum neutral current is _____.
(a) 50A   (b) 75A   (c) 87A   (d) 95A

82. • There are five 2,500W, 208V heat strips on the secondary of a 3Ø, 208Y/120V transformer. What is the load per phase for each heat strip?
(a) 1,400W   (b) 1,250W   (c) 2,500W   (d) 12,500W

83. • The phase load of the 10-hp, 3Ø, 208Y/120V motor is closest to _____.
(a) 3,700 VA   (b) 5,500 VA   (c) 4,300 VA   (d) 11,100 VA

Answer the next question based on this information.
The system is a 3Ø, 208Y/120V system and the loads are as follows:
Nine 2 kVA, 3Ø, 208Y/120V lighting loads
Five 3 kVA, 120V, 1Ø loads

84. • The neutral conductor shall be sized to carry _____.
(a) 10A   (b) 25A   (c) 90A   (d) 100A

NEC Questions – Articles 725-Annex C

Article 725 Class 1, Class 2, and Class 3 Remote-Control, Signaling and Power Limited Circuits

85. Conductors of Class 2 and Class 3 circuits shall not be placed in any enclosure, raceway, cable or similar fittings with conductors of Class 1 or electric light or power conductors, except when they are _____.
(a) insulated for the maximum voltage
(b) totally comprised of aluminum conductors
(c) separated by a barrier
(d) all of these

86. • Cables and conductors of Class 2 and Class 3 circuits _____ be placed in any cable, cable tray, compartment, enclosure, manhole, outlet box, device box, raceway or similar fitting with conductors of electric light, power, Class 1, non-power-limited fire alarm circuits, and medium power network-powered broadband communications circuits.
(a) may   (b) shall not   (c) a and b   (d) none of these

87. Class 2 and 3 control conductors shall be installed in _____ when used in hoistways.
(a) rigid metal conduit   (b) electrical metallic tubing
(c) intermediate metal conduit   (d) all of these

88. Class 2 and Class 3 circuits shall be permitted within the same cable, raceway or enclosure provided that the insulation of the Class 2 circuit conductors meets the requirements for Class 3 circuits.
(a) True   (b) False

89. Raceways shall not be used as a means of support for Class 2 or Class 3 circuit conductors.
(a) True   (b) False

Article 760 Fire Alarm Systems

90. Article 760 covers the requirements for the installation of wiring and equipment of _____.
    (a) communications   (b) antennae   (c) fire alarm systems   (d) fiber optics

91. Fire alarm systems include _____.
    (a) fire detection and alarm notification   (b) guard's tour
    (c) sprinkler water flow   (d) all of these

92. Class 2, Class 3 and PLTC cables that are not terminated at equipment and not identified for future use with a tag will be considered as abandoned.
    (a) True   (b) False

93. Electrical fire alarm installations in hollow spaces, vertical shafts and ventilation or air-handling ducts shall be made so that the spread of fire or products of combustion (such as smoke) will not be _____.
    (a) possible   (b) substantially increased
    (c) permitted   (d) none of these

94. Fire alarm cables installed _____ on the surface of ceilings and sidewalls shall be supported by the structural components of the building in such a manner that the cable will not be damaged by normal building use.
    (a) exposed   (b) concealed   (c) hidden   (d) a and b

95. Exposed cables shall be secured to structural components by straps, staples, hangers or similar fittings designed and installed so as not to damage the cable.
    (a) True   (b) False

96. Fire alarm cables installed _____ to framing members shall be protected against physical damage from penetration by screws or nails by $1^1/_4$ in. separation from the framing member or by a suitable metal plate in accordance with 300.4(D).
    (a) exposed   (b) concealed   (c) parallel   (d) all of these

97. For nonpower-limited fire alarm circuits, a 18 AWG conductor is considered protected if the overcurrent device protecting the system is not over _____.
    (a) 15A   (b) 10A   (c) 20A   (d) 7A

98. TFN or other approved insulation used on a nonpower-limited fire alarm circuit is required for _____ conductors.
    (a) 16 and 18 AWG   (b) all conductors up to 600V
    (c) 14 AWG and larger   (d) all

99. Where installed exposed, fire alarm cables shall be _____.
    (a) adequately supported
    (b) installed to be protected against physical damage
    (c) a and b
    (d) none of these

100. Splices and terminations of nonpower-limited fire alarm circuits shall be made in _____ fittings, boxes, enclosures, fire alarm devices or utilization equipment.
    (a) identified   (b) listed   (c) labeled   (d) none of these

101. Exposed power-limited fire alarm circuit cables shall be installed in _____ when passing through a floor or wall to a height of 7 ft above the floor unless adequate protection can be afforded by the building construction or unless an equivalent solid guard is provided.
    (a) rigid metal conduit
    (b) rigid nonmetallic conduit
    (c) electrical metallic tubing
    (d) any of these

102. Equipment shall be durably marked where plainly visible to indicate each circuit that is _____.
(a) supplied by a power-limited fire alarm circuit
(b) a power-limited fire alarm circuit
(c) a fire alarm circuit
(d) none of these

103. Conductors of lighting or power may occupy the same enclosure or raceway with conductors of power-limited fire alarm circuits.
(a) True   (b) False

104. Cables and conductors of two or more power-limited fire alarm circuits can be installed in the same cable, enclosure or raceway.
(a) True   (b) False

105. Power-limited fire alarm circuit cables installed within buildings in an air-handling space shall use _____ conductors.
(a) FPL   (b) CL3P   (c) OFNP   (d) FPLP

106. Power-limited fire alarm cables installed as wiring within buildings shall be _____ as being resistant to the spread of fire.
(a) labeled   (b) listed   (c) identified   (d) marked

107. Coaxial cables used in power-limited fire alarm systems shall have an overall insulation rating of _____.
(a) 100V   (b) 300V   (c) 600V   (d) 1,000V

Article 770 Optical Fiber Cables and Raceways

108. Optical fiber cable that is not terminated at equipment and not identified for future use with a tag will be considered as abandoned.
(a) True   (b) False

109. All accessible portions of abandoned optical fiber cable shall be removed.
(a) True   (b) False

110. Optical fiber cable installed in a raceway shall be of a type permitted in Chapter 3 and the raceway shall be installed in accordance with Chapter 3 requirements.
(a) True   (b) False

111. Optical fiber cables installed _____ on the surface of ceilings and sidewalls shall be supported by the structural components of the building in such a manner that the cable will not be damaged by normal building use.
(a) exposed   (b) concealed   (c) hidden   (d) a and b

112. Exposed optical fiber cables shall be secured to structural components by straps, staples, hangers or similar fittings designed and installed so as not to damage the cable.
(a) True   (b) False

113. Where exposed to contact with electric light or power conductors, the noncurrent-carrying metallic members of optical fiber cables entering buildings shall be _____.
(a) grounded at the point of emergence through an exterior wall
(b) grounded at the point of emergence through a concrete floor slab
(c) interrupted as close to the point of entrance as practicable by an insulating joint
(d) any of these

114. _____ is a conductive, general-use type of optical fiber cable that may be installed as wiring within buildings.
(a) FPLP   (b) CL2P   (c) OFPN   (d) OFC

Chapter 8 Communications Systems

Article 800 Communications Circuits

115.	Communications cables that are not terminated at both ends with a connector or other equipment and not identified for future use with a tag will be considered as abandoned.
(a) True   (b) False

116.	Equipment intended to be electrically connected to a telecommunications network shall be listed for the purpose.
(a) True   (b) False

117.	Communications cables installed _____ on the surface of ceilings and sidewalls shall be supported by the structural components of the building in such a manner that the cable will not be damaged by normal building use.
(a) exposed   (b) concealed   (c) hidden   (d) a and b

118.	Exposed communications cables shall be secured to structural components by straps, staples, hangers or similar fittings designed and installed so as not to damage the cable.
(a) True   (b) False

119.	Where communications cables enter buildings, they shall _____.
(a) where practicable, be located below the electric light or power conductors
(b) not be attached to a cross-arm that carries electric light or power conductors
(c) have a vertical clearance of not less than 8 ft from all points of roofs above which they pass
(d) all of these

120.	When practical, a separation of at least _____ shall be maintained between open conductors of communication systems on buildings and lightning conductors.
(a) 6 ft   (b) 8 ft   (c) 10 ft   (d) 12 ft

121.	The metallic sheath of communications cable entering buildings shall be _____.
(a) grounded at the point of emergence through an exterior wall
(b) grounded at the point of emergence through a concrete floor slab
(c) interrupted as close to the point of entrance as practicable by an insulating joint
(d) any of these

122.	In one- and two-family dwellings, the primary protector grounding conductor for communications systems shall be as short as practicable, not to exceed _____ in length.
(a) 5 ft   (b) 8 ft   (c) 10 ft   (d) 20 ft

123.	In one- and two-family dwellings where it is not practicable to achieve an overall maximum primary protector grounding conductor length for communications systems, a separate _____ communications ground rod shall be driven and it shall be bonded to the power grounding electrode system with a 6 AWG conductor.
(a) 5 ft   (b) 8 ft   (c) 10 ft   (d) 20 ft

124.	In communication circuits, all separate electrodes can be bonded together using a minimum jumper size of _____ copper.
(a) 10 AWG   (b) 8 AWG   (c) 6 AWG   (d) 4 AWG

125.	Where communications wire and cables are installed in a raceway, the raceway shall be of a type permitted in Chapter 3 and shall be installed in accordance with Chapter 3 requirements.
(a) True   (b) False

126.	Communications plenum cable shall be _____ as being suitable for use in ducts, plenums and other spaces used for environmental air.
(a) marked   (b) identified   (c) approved   (d) listed

127. Communications _____ cable shall be listed as being suitable for use in a vertical run in a shaft, or from floor to floor, and shall also be listed as having fire-resistant characteristics capable of preventing the carrying of fire from floor to floor.
(a) plenum   (b) riser   (c) general-purpose   (d) none of these

128. Communications wires and cables shall be separated by at least 2 in. from conductors of _____ circuits.
(a) power   (b) lighting   (c) Class 1   (d) any of these

129. All accessible portions of abandoned communications cable shall be removed.
(a) True   (b) False

130. Floor penetrations requiring CMR communications cable shall contain only cables suitable for riser or plenum use except where the listed cables are _____.
(a) encased in metal raceways
(b) located in a fireproof shaft with firestops at each floor
(c) a or b
(d) none of these

Article 810 Radio and Television Equipment

131. Soft-drawn or medium-drawn copper lead-in conductors for receiving antenna systems shall be permitted where the maximum span between points of support is less than _____.
(a) 35 ft   (b) 30 ft   (c) 20 ft   (d) 10 ft

132. The receiving station outdoor antenna conductor on a span of 75 ft shall be at least a _____ if a copper-clad steel conductor is used.
(a) 10 AWG   (b) 12 AWG   (c) 14 AWG   (d) 17 AWG

133. An outdoor antenna of a receiving station with a 75 ft span using a copper-clad steel conductor shall not be less than _____.
(a) 10 AWG   (b) 12 AWG   (c) 14 AWG   (d) 17 AWG

134. The maximum open span length is 200 ft for antenna conductors of hard-drawn copper located at an amateur transmitting and receiving station, and requires a minimum conductor size of _____.
(a) 14 AWG   (b) 12 AWG   (c) 10 AWG   (d) 8 AWG

135. Unshielded lead-in antenna conductors for amateur transmitting stations attached to buildings shall be firmly mounted at least _____ clear of the surface of the building on nonabsorbent insulating supports.
(a) 1 in.   (b) 2 in.   (c) 3 in.   (d) 4 in.

136. The transmitter enclosure of a radio or TV station shall have interlocks to disconnect the power to the transmitter in the enclosure access door if the voltage between conductors is over _____.
(a) 150V   (b) 250V   (c) 350V   (d) 480V

Article 820 Community Antenna Television and Radio Distribution Systems

137. CATV cable that is not terminated at equipment and not identified for future use with a tag will be considered as abandoned.
(a) True   (b) False

138. All accessible portions of abandoned CATV cable shall be removed.
(a) True   (b) False

139. CATV cables installed _____ on the surface of ceilings and sidewalls shall be supported by the structural components of the building in such a manner that the cable will not be damaged by normal building use.
(a) exposed   (b) concealed   (c) hidden   (d) a and b

140. Exposed CATV cables shall be secured to structural components by straps, staples, hangers or similar fittings designed and installed so as not to damage the cable.
(a) True   (b) False

141. Where practical, coaxial cables for a CATV system shall be separated by at least _____ from lightning conductors.
(a) 3 in.   (b) 6 in.   (c) 6 ft   (d) 2 ft

142. The outer conductive shield of the coaxial cable shall be grounded at the building premises as close to the point of cable entrance or attachment as practical.
(a) True   (b) False

143. The conductor used to ground the outer cover of a coaxial cable shall be _____.
(a) insulated   (b) 14 AWG minimum   (c) bare   (d) a and b

144. In one- and two-family dwellings, the grounding conductor for CATV shall be as short as practicable, not to exceed _____ in length.
(a) 5 ft   (b) 8 ft   (c) 10 ft   (d) 20 ft

145. In one- and two-family dwellings where it is not practicable to achieve an overall maximum grounding conductor length for CATV, a separate _____ ground shall be used and it shall be bonded to the power grounding electrode system with a 6 AWG conductor.
(a) 5 ft   (b) 8 ft   (c) 10 ft   (d) 20 ft

146. Coaxial cable is permitted to be placed in a raceway, compartment, outlet box or junction box with the conductors of light or power circuits, or Class 1 circuits when _____.
(a) installed in rigid metal conduit   (b) separated by a permanent barrier
(c) insulated   (d) none of these

Article 830 Network-Powered Broadband Communications Systems

147. Network-powered broadband cable that is not terminated at equipment and not identified for future use with a tag will be considered as abandoned.
(a) True   (b) False

148. All accessible portions of abandoned network-powered broadband cable shall be removed.
(a) True   (b) False

149. Network-powered broadband cables installed _____ on the surface of ceilings and sidewalls shall be supported by the structural components of the building in such a manner that the cable will not be damaged by normal building use.
(a) exposed   (b) concealed   (c) hidden   (d) a and b

150. Exposed network-powered broadband cables shall be secured to structural components by straps, staples, hangers or similar fittings designed and installed so as not to damage the cable.
(a) True   (b) False

151. Where network-powered broadband communications system cables enter buildings, they shall _____.
(a) where practicable, be located below the electric light or power conductors
(b) not be attached to a cross-arm that carries electric light or power conductors
(c) have a vertical clearance of not less than 8 ft from all points of roofs above which they pass
(d) all of these

152. Where practical, a separation of at least _____ shall be maintained between any network-powered broadband communications cable on buildings and lightning conductors.
    (a) 6 ft   (b) 8 ft   (c) 10 ft   (d) 12 ft

153. In one-family and multifamily dwellings, the grounding conductor for network-powered broadband communications systems shall be as short as permissible, not to exceed _____ in length.
    (a) 5 ft   (b) 6 ft   (c) 10 ft   (d) 20 ft

154. In one- and two-family dwellings where it is not practicable to achieve an overall maximum primary protector grounding conductor length for network-powered broadband communications systems, a separate _____ communications ground rod shall be driven and it shall be bonded to the power grounding electrode system with a 6 AWG conductor.
    (a) 5 ft   (b) 8 ft   (c) 10 ft   (d) 20 ft

155. In network-powered broadband communications systems, all separate electrodes can be bonded together using a minimum size jumper of _____ copper.
    (a) 10 AWG   (b) 8 AWG   (c) 6 AWG   (d) 4 AWG

156. Network-powered broadband communications system cables shall be separated at least 2 in. from conductors of _____ circuits.
    (a) power   (b) electric light (c) Class 1   (d) any of these

Chapter 9 Tables

157. The percentage of conduit fill for two conductors is _____ percent.
    (a) 40   (b) 31   (c) 53   (d) 30

158. The percentage of conduit fill for five conductors is _____ percent.
    (a) 35   (b) 60   (c) 40   (d) 55

159. When considering the number of conductors permitted when calculating raceway conductors fill, equipment grounding or bonding conductors shall _____.
    (a) not be counted   (b) have the actual dimensions used
    (c) not be counted if in a nipple   (d) not be counted if for a wye 3Ø balanced load

160. When conduit nipples having a maximum length not exceeding 24 in. are installed between boxes _____.
    (1) the nipple can be filled to 75 percent
    (2) 310.15(B)(2) adjustment factors don't apply
    (3) 310.15(B)(2) adjustment factors apply
    (4) the nipple can be filled to 60 percent
    (a) 1 and 2   (b) 2 and 4   (c) 3 and 4   (d) 1 and 2

161. When conduit nipples that are 24 in. or less are installed, the ampacity adjustment factor for more than three current-carrying conductors _____ to this condition.
    (a) does not apply   (b) shall be applied

162. When the calculated number of conductors, all of the same size, that may be installed in a conduit or tubing includes a decimal, the next higher whole number shall be used when this decimal is _____ or larger.
    (a) 0.40   (b) 0.60   (c) 0.70   (d) 0.80

163. The inside diameter of a 1 in. IMC is _____
    (a) 1.000 in.   (b) 1.105 in.   (c) 0.826 in.   (d) 0.314 in.

164. A rigid metal conduit nipple (1½ in.) with three conductors can be filled to an area of _____
    (a) 0.882 sq in.   (b) 1.072 sq in.   (c) 1.243 sq in.   (d) 1.343 sq in.

165. The circular mil area of a 12 AWG conductor is _____.
    (a) 10,380   (b) 26,240   (c) 6,530   (d) 6,350

166. A 10 AWG single strand (solid) copper wire has a cross-sectional area of _____
      (a) 0.008 sq in.   (b) 0.101 sq in.   (c) 0.012 sq in.   (d) 0.106 sq in.

167. A 10 AWG 7 strand copper wire has a cross-sectional area of _____
      (a) 0.007 sq in.   (b) 0.011 sq in.   (c) 0.012 sq in.   (d) 0.106 sq in.

168. The area in square inches for a 1/0 AWG bare aluminum conductor is _____.
      (a) 0.087 sq in.   (b) 0.109 sq in.   (c) 0.137 sq in.   (d) 0.173 sq in.

169. A 250 kcmil bare cable has a conductor diameter of _____
      (a) 0.557 in.   (b) 0.755 in.   (c) 0.575 in.   (d) 0.690 in.

170. The ac ohms-to-neutral impedance per 1,000 ft of 4/0 AWG aluminum conductor in a steel raceway is _____.
      Note: The ohms-to-neutral impedance is listed as the bottom value in Table 9 of Chapter 9, (see the heading for more
      information). For example the ohms-to-neutral impedance per 1,000 ft for 12 AWG copper in a steel raceway is 2.0Ω
      and the ohms-to-neutral impedance per kilometer is 6.6 ohms.
      (a) 0.06Ω (b) 0.10Ω (c) 0.22Ω (d) 0.11Ω

171. The ac ohms-to-neutral impedance per 1,000 ft of 2/0 AWG copper conductor in a steel raceway is _____ .
      (a) 0.06Ω (b) 0.10Ω (c) 0.22Ω (d) 0.11Ω

172. The ac ohms-to-neutral impedance per 1,000 ft of 4/0 AWG aluminum in a steel raceway would have the same ohms-to-
      neutral impedance as 1,000 ft of _____ copper installed in a steel raceway.
      (a) 1 AWG   (b) 1/0 AWG   (c) 2/0 AWG   (d) 250 kcmil

Annex C Conduit and Tubing Fill Tables for Conductors and Fixture Wires of the Same Size

173. Annex C contains tables that list the number of conductors permitted in a raceway.
      (a) True   (b) False

Unit 12 NEC Exam – NEC Code Order 725.41 – Chapter 9, Table 8

1. When required, the maximum overcurrent protection in amperes for a not inherently limited Class 2 remote-control circuit is _____.
   (a) 1.0A   (b) 2.0A   (c) 5.0A   (d) 7.5A

2. The power source for a Class 2 circuit shall be _____.
   (a) a listed Class 2 or 3 transformer
   (b) a listed Class 2 or 3 power supply
   (c) other listed equipment marked to identify the Class 2 or Class 3 power source
   (d) any of these

3. Cables and conductors of Class 2 and Class 3 circuits _____ be placed in any cable, cable tray, compartment, enclosure, manhole, outlet box, device box, raceway or similar fitting with conductors of electric light, power, Class 1, non-power-limited fire alarm circuits and medium power network-powered broadband communications circuits.
   (a) may   (b) shall not   (c) a and b   (d) none of these

4. Article 760 covers the requirements for the installation of wiring and equipment of _____.
   (a) communications   (b) antennae
   (c) fire alarm systems   (d) fiber optics

5. Splices and terminations of nonpower-limited fire alarm circuits shall be made in _____ fittings, boxes, enclosures, fire alarm devices or utilization equipment.
   (a) identified   (b) listed   (c) labeled   (d) none of these

6. Optical fiber cable that is not terminated at equipment and not identified for future use with a tag will be considered as abandoned.
   (a) True   (b) False

7. Optical fiber cable installed in a raceway must be of a type permitted in Chapter 3 and the raceway shall be installed in accordance with Chapter 3 requirements.
   (a) True   (b) False

8. Optical fiber cables installed _____ on the surface of ceilings and sidewalls must be supported by the structural components of the building in such a manner that the cable will not be damaged by normal building use.
   (a) exposed   (b) concealed   (c) hidden   (d) a and b

9. Where exposed to contact with electric light or power conductors, the noncurrent-carrying metallic members of optical fiber cables entering buildings shall be _____.
   (a) grounded at the point of emergence through an exterior wall
   (b) grounded at the point of emergence through a concrete floor slab
   (c) interrupted as close to the point of entrance as practicable by an insulating joint
   (d) any of these

10. Where communications cables enter buildings, they shall _____.
    (a) where practicable, be located below the electric light or power conductors
    (b) not be attached to a cross-arm that carries electric light or power conductors
    (c) have a vertical clearance of not less than 8 ft from all points of roofs above that they pass
    (d) all of these

11. When practical, a separation of at least _____ shall be maintained between open conductors of communication systems on buildings and lightning conductors.
    (a) 6 ft   (b) 8 ft   (c) 10 ft   (d) 12 ft

12. Where communications wire and cables are installed in a raceway, the raceway shall be of a type permitted in Chapter 3 and shall be installed in accordance with Chapter 3 requirements.
    (a) True   (b) False

13. Floor penetrations requiring CMR communications cable shall contain only cables suitable for riser or plenum use except where the listed cables are _____.
    (a) encased in metal raceways
    (b) located in a fireproof shaft with firestops at each floor
    (c) a or b
    (d) none of these

14. The maximum open span length is 200 ft for antenna conductors of hard-drawn copper located at an amateur transmitting and receiving station, and requires a minimum conductor size of _____.
    (a) 14 AWG   (b) 12 AWG   (c) 10 AWG   (d) 8 AWG

15. The transmitter enclosure of a radio or TV station shall have interlocks to disconnect the power to the transmitter in the enclosure access door if the voltage between conductors is over _____.
    (a) 150V   (b) 250V   (c) 350V   (d) 480V

16. The outer conductive shield of the coaxial cable must be grounded at the building premises as close to the point of cable entrance or attachment as practical.
    (a) True   (b) False

17. Coaxial cable is permitted to be placed in a raceway, compartment, outlet box or junction box with the conductors of light or power circuits or Class 1 circuits when _____.
    (a) installed in rigid metal conduit   (b) separated by a permanent barrier
    (c) insulated   (d) none of these

18. Network-powered broadband cables installed _____ on the surface of ceilings and sidewalls must be supported by the structural components of the building in such a manner that the cable will not be damaged by normal building use.
    (a) exposed   (b) concealed   (c) hidden   (d) a and b

19. Where network-powered broadband communications system cables enter buildings, they shall _____.
    (a) where practicable, be located below the electric light or power conductors
    (b) not be attached to a cross-arm that carries electric light or power conductors
    (c) have a vertical clearance of not less than 8 ft from all points of roofs above which they pass
    (d) all of these

20. Where practical, a separation of at least _____ shall be maintained between any network-powered broadband communications cable on buildings and lightning conductors.
    (a) 6 ft   (b) 8 ft   (c) 10 ft   (d) 12 ft

21. The percentage of conduit fill for five conductors is _____ percent.
    (a) 35   (b) 60   (c) 40   (d) 55

22. When conduit nipples having a maximum length not exceeding 24 in. are installed between boxes _____.
    (1) The nipple can be filled to 75 percent
    (2) 310.15(B)(2) adjustment factors don't apply
    (3) 310.15(B)(2) adjustment factors apply
    (4) The nipple can be filled to 60 percent
    (a) 1 and 2   (b) 2 and 4   (c) 3 and 4   (d) 1 and 2

23. A solid bare (1 strand) 10 AWG solid copper wire has a cross-sectional area of _____
    (a) 0.008 sq in.   (b) 0.101 sq in.   (c) 0.012 sq in.   (d) 0.106 sq in.

24. A stranded bare (7 strand) 10 AWG solid copper wire has a cross-sectional area of _____
    (a) 0.007 sq in.   (b) 0.011 sq in.   (c) 0.012 sq in.   (d) 0.106 sq in.

25. The area in square inches for a 1/0 AWG bare aluminum conductor is _____
    (a) 0.087 sq in.   (b) 0.109 sq in.   (c) 0.137 sq in.   (d) 0.173 sq in.

Unit 12 NEC Exam – Random Order 110.14 – Chapter 9, Table 9

1. Raceways shall be _____ between pulling points prior to the installation of conductors.
   (a) mechanically completed
   (b) tested for ground faults
   (c) a minimum of 80 percent completed
   (d) none of these

2. Recessed incandescent luminaires shall have _____ protection and shall be identified as thermally protected.
   (a) physical   (b) corrosion   (c) thermal   (d) all of these

3. Resistors and reactors over 600V, nominal, shall be isolated by _____ to protect personnel from accidental contact with energized parts.
   (a) an enclosure   (b) elevation   (c) a or b   (d) a and b

4. RMC shall be permitted to be installed in concrete, in direct contact with the earth or in areas subject to severe corrosive influences when protected by _____ and judged suitable for the condition.
   (a) ceramic   (b) corrosion protection
   (c) backfill   (d) a natural barrier

5. RNC and fittings used above ground shall be resistant to _____.
   (a) moisture and chemical atmospheres
   (b) low temperatures and sunlight
   (c) distortion from heat
   (d) all of these

6. RNC shall be securely fastened within _____ of each box.
   (a) 6 in.   (b) 24 in.   (c) 12 in.   (d) 36 in.

7. Running threads of IMC shall not be used on conduit for connection at couplings.
   (a) True   (b) False

8. Separately installed pressure connectors shall be used with conductors at the _____ not exceeding the ampacity at the listed and identified temperature rating of the connector.
   (a) voltages   (b) temperatures   (c) listings   (d) ampacities

9. Service conductors installed as unjacketed multiconductor cable shall have a minimum clearance of _____ from windows that are designed to be opened, doors, porches, stairs, fire escapes and similar equipment.
   (a) 3 ft   (b) 4 ft   (c) 6 ft   (d) 10 ft

10. Service conductors supplying a building or other structure shall not _____ of another building or other structure.
    (a) be installed on the exterior walls
    (b) pass through the interior
    (c) a and b
    (d) none of these

11. Service drops installed over roofs shall have a vertical clearance of _____ above the roof surface.
    (a) 8 ft   (b) 12 ft   (c) 15 ft   (d) 3 ft

12. Service equipment for floating docks or marinas must be located _____ the floating structure.
    (a) next to   (b) on   (c) 100 ft from   (d) 20 ft from

13. Service heads must be located _____.
    (a) above the point of attachment   (b) below the point of attachment
    (c) even with the point of attachment   (d) none of these

14. Sheet steel boxes not over 100 cubic inches in size shall be made from steel not less than _____ thick.
    (a) 0.0625 in.   (b) 0.0757 in.   (c) 0.0750 in.   (d) 0.0250 in.

15. Single-pole breakers utilizing approved handle ties cannot be used for the circuit disconnect for gasoline dispensing equipment.
    (a) True   (b) False

16. Size 18- or 16 AWG fixture wires can be used for the control and operating circuits of X-ray and auxiliary equipment when protected by an overcurrent protection device not larger than _____.
    (a) 15A   (b) 20A   (c) 25A   (d) 30A

17. Snap switches installed in recessed boxes must have the _____ seated against the finished wall surface.
    (a) mounting yoke   (b) body   (c) toggle   (d) all of these

18. Snap switches rated _____ or less directly connected to aluminum conductors must be listed and marked CO/ALR.
    (a) 15A   (b) 20A   (c) 25A   (d) 30A

19. Snap switches shall not be grouped or ganged in enclosures unless they can be arranged so that the voltage between adjacent switches does not exceed _____, or unless they are installed in enclosures equipped with permanently installed barriers between adjacent switches.
    (a) 100V   (b) 200V   (c) 300V   (d) 400V

20. Soft-drawn or medium-drawn copper lead-in conductors for receiving antenna systems shall be permitted where the maximum span between points of support is less than _____.
    (a) 35 ft   (b) 30 ft   (c) 20 ft   (d) 10 ft

21. Soldered splices shall first be spliced or joined so as to be mechanically and electrically secure without solder and then be soldered.
    (a) True   (b) False

22. Stainless steel, copper or copper alloy clamps approved for direct burial use can be used to connect the bonding conductor to the common bonding grid of a swimming pool.
    (a) True   (b) False

23. Steel or aluminum cable tray systems can be used as equipment grounding conductors, provided the cable tray sections and fittings have been _____ marked to show the cross-sectional area of metal in channel cable trays or cable trays of one-piece construction.
    (a) legibly   (b) durably   (c) a or b   (d) a and b

24. Stock rooms and similar areas adjacent to aircraft hangars, but effectively isolated and adequately ventilated, shall be designated as _____ locations.
    (a) Class I, Division 2   (b) Class II, Division 1
    (c) Class II, Division 2   (d) nonhazardous

25. Sufficient access and _____ shall be provided and maintained about all electrical equipment to permit ready and safe operation and maintenance of such equipment.
    (a) ventilation   (b) cleanliness   (c) circulation   (d) working space

26. Supports for concealed knob-and-tube wiring shall be installed within _____ of each side of each tap or splice, and at intervals not exceeding _____.
    (a) 3 in., $2^1/_2$ ft   (b) 2 in., $3^1/_2$ ft   (c) 6 in., $4^1/_2$ ft   (d) 4 in., $6^1/_2$ ft

27. Surface metal raceways and their fittings shall be so designed that the sections can be _____.
    (a) electrically coupled together
    (b) mechanically coupled together
    (c) installed without subjecting the wires to abrasion
    (d) all of these

28. Switch, outlet and tap devices of insulating material shall be permitted to be used without boxes in exposed NMC cable.
    (a) True   (b) False

29. Switchboards shall be placed so as to reduce the probability of communicating _____ to adjacent combustible materials.
    (a) sparks   (b) backfeed   (c) fire   (d) all of these

30. Switched lampholders shall be of such construction that the switching mechanism interrupts the electrical connection to the _____.
    (a) lampholder   (b) center contact   (c) branch circuit   (d) all of these

31. Switches or circuit breakers shall not disconnect the grounded conductor of a circuit unless the switch or circuit breaker _____.
    (a) can be opened and closed by hand levers only
    (b) simultaneously disconnects all conductors of the circuit
    (c) opens the grounded conductor before it disconnects the ungrounded conductors
    (d) none of these

32. Switches, circuit breakers, motor controllers and fuses including pushbuttons, relays and similar devices that are installed in Class I, Division 1 locations shall be provided with approved explosionproof enclosures, and they shall be identified as a complete assembly for use in Class I locations.
    (a) True   (b) False

33. Table 310.70 provides the ampacities of insulated single aluminum conductor isolated in air. If the conductor size is 8 AWG, MV-90, and the voltage range is 2,001 through 5,000, then the ampacity is _____.
    (a) 64A   (b) 85A   (c) 115A   (d) 150A

34. TC cable shall be permitted to be used _____.
    (a) for power and lighting circuits
    (b) in cable trays in hazardous locations
    (c) in Class 1 control circuits
    (d) all of these

35. TC tray cable shall not be installed _____.
    (a) where it will be exposed to physical damage
    (b) as open cable on brackets or cleats
    (c) direct buried, unless identified for such use
    (d) all of these

36. Temporary wiring must be _____ immediately upon the completion of construction or purpose for which it was installed.
    (a) disconnected   (b) removed   (c) de-energized   (d) any of these

37. TFE-insulated conductors are manufactured in sizes from 14 through _____ .
    (a) 2 AWG   (b) 1 AWG   (c) 2/0 AWG   (d) 4/0 AWG

38. The _____ is defined as the area between the top of direct-burial cable and the finished grade.
    (a) notch   (b) cover   (c) gap   (d) none of these

39. The ac ohms-to-neutral resistance per 1,000 ft of 4/0 AWG aluminum in a steel raceway would have the same resistance as 1,000 ft of _____ copper installed in a steel raceway.
    (a) 1 AWG   (b) 1/0 AWG   (c) 2/0 AWG   (d) 250 kcmil

40. The alternate source for emergency systems _____ be required to have ground-fault protection of equipment.
    (a) shall   (b) shall not

41. The ampacity of branch-circuit conductors and the rating or setting of overcurrent protective devices supplying fixed electric heating equipment for pipelines and vessels shall be not less than _____ percent of the total load of the heaters.
    (a) 75   (b) 100   (c) 125   (d) 150

42. The ampacity of capacitor circuit conductors shall not be less than _____ percent of the rated current of the capacitor.
    (a) 100   (b) 115   (c) 125   (d) 135

43. The bottom of sign and outline lighting enclosures shall not be less than _____ above areas accessible to vehicles.
    (a) 12 ft   (b) 14 ft   (c) 16 ft   (d) 18 ft

44. The circular mil area of a 12 AWG conductor is _____.
    (a) 10,380   (b) 26,240   (c) 6,530   (d) 6,350

45. The conductor used to ground the outer cover of a coaxial cable shall be _____.
    (a) insulated   (b) 14 AWG minimum
    (c) bare   (d) a and b

46. The percentage of conduit fill for two conductors is _____ percent.
    (a) 40   (b) 31   (c) 53   (d) 30

47. When conduit nipples that are 24 in. or less are installed, the ampacity adjustment factor for more than three current-carrying conductors _____ to this condition.
    (a) does not apply   (b) shall be applied

48. When the calculated number of conductors, all of the same size, that may be installed in a conduit or tubing includes a decimal, the next higher whole number shall be used when this decimal is _____ or larger.
    (a) 0.40   (b) 0.60   (c) 0.70   (d) 0.80

49. Where installed exposed, fire alarm cables shall be _____.
    (a) adequately supported
    (b) installed to be protected against physical damage
    (c) a and b
    (d) none of these

50. Where practical, coaxial cables for a CATV system shall be separated by at least _____ from lightning conductors.
    (a) 3 in.   (b) 6 in.   (c) 6 ft   (d) 2 ft

Final NEC Exam – Random Order 100 – 830.58

1. Electrical systems that are grounded shall be connected to earth in a manner that will _____.
    (a) limit voltages due to lightning, line surges or unintentional contact with higher voltage lines
    (b) stabilize the voltage-to-ground during normal operation
    (c) facilitate overcurrent protection device operation in case of ground faults
    (d) a and b

2. Electrically heated smoothing irons shall be equipped with an identified _____ means.
    (a) disconnecting   (b) temperature-limiting
    (c) current limiting   (d) none of these

3. Embedded deicing and snow-melting equipment cables, units and panels shall not be installed where they bridge _____, unless provisions are made for expansion and contraction.
    (a) roads   (b) over water spans
    (c) runways   (d) expansion joints

4. ENT must be secured in place every _____
    (a) 12 in.   (b) 18 in.   (c) 24 in.   (d) 36 in.

5. Equipment in an area containing acetylene gas must be rated for a Class I, Group A location.
    (a) True   (b) False

6. Equipment intended to break current at other than fault levels shall have an interrupting rating at system voltage sufficient for the current that must be interrupted.
    (a) True   (b) False

7. Exposed conductive parts of luminaires shall be _____.
    (a) grounded   (b) painted   (c) bonded   (d) a and b

8. Exposed runs of AC cable shall closely follow the surface of the building finish or of running boards. Exposed runs shall also be permitted to be installed on the underside of joists where supported at each joist and located so as not to be subject to physical damage.
    (a) True   (b) False

9. External surfaces of pipeline and vessel heating equipment that operate at temperatures exceeding _____ shall be physically guarded, isolated or thermally insulated to protect against contact by personnel in the area.
    (a) 110 °F   (b) 120 °F   (c) 130 °F   (d) 140 °F

10. Feeders to floating buildings can be installed in _____ with approved fittings where flexible connections are required for services.
    (a) rigid nonmetallic conduit
    (b) liquidtight metallic conduit
    (c) liquidtight nonmetallic conduit
    (d) portable power cables

11. Fire alarm systems include _____.
    (a) fire detection and alarm notification
    (b) guard's tour
    (c) sprinkler water flow
    (d) all of these

12. Fittings and connectors shall be used only with the specific wiring methods for which they are designed and listed.
    (a) True   (b) False

13. Fixed electric space-heating equipment requiring supply conductors with insulation rated over _____ shall be clearly and permanently marked.
    (a) 75 °C   (b) 60 °C   (c) 90 °C   (d) all of these

14. Fixture wires shall be permitted for connecting luminaires to the _____ conductors supplying the luminaires.
    (a) service   (b) branch-circuit   (c) supply   (d) none of these

15. Fixture wires used for luminaires shall not be smaller than _____.
    (a) 22 AWG   (b) 18 AWG   (c) 16 AWG   (d) 14 AWG

16. Flat cable assemblies are suitable to supply tap devices for _____ loads. The rating of the branch circuit shall not exceed 30A.
    (a) lighting   (b) small appliance   (c) small power   (d) all of these

17. Flat cable assemblies shall have conductors of _____ special stranded copper wires.
    (a) 14 AWG   (b) 12 AWG   (c) 10 AWG   (d) all of these

18. Flat conductor cable cannot be installed in _____.
    (a) residential units   (b) schools
    (c) hospitals   (d) any of these

19. Flexible cords and cables shall be connected to devices and to fittings so that tension will not be transmitted to joints or terminal screws. This shall be accomplished by _____.
    (a) knotting the cord   (b) winding with tape
    (c) use of fittings designed for the purpose   (d) all of these

20. Flexible cords and cables used for temporary wiring shall be protected _____.
    (a) from accidental damage   (b) where passing through doorways
    (c) from sharp corners and projections   (d) all of these

21. Flexible cords shall not be used as a substitute for _____ wiring.
    (a) temporary   (b) fixed   (c) overhead   (d) none of these

22. FMC can be installed exposed or concealed where not subject to physical damage.
    (a) True   (b) False

23. FMC cannot be installed _____.
    (a) underground   (b) embedded in poured concrete
    (c) where subject to physical damage   (d) all of these

24. FMC shall be secured _____.
    (a) at intervals not exceeding $4^1/_2$ ft
    (b) within 12 in. on each side of a box where fished
    (c) where fished
    (d) at intervals not exceeding 3 ft at motor terminals

25. For cellular concrete floor raceways, junction boxes shall be _____ the floor grade and sealed against the free entrance of water or concrete.
    (a) leveled to   (b) above   (c) below   (d) perpendicular to

26. For installations consisting of not more than two 2-wire branch circuits, the building disconnecting means shall have a rating of not less than _____.
    (a) 15A   (b) 20A   (c) 25A   (d) 30A

27. For installations consisting of not more than two 2-wire branch circuits, the service disconnecting means shall have a rating of not less than _____.
    (a) 15A   (b) 20A   (c) 25A   (d) 30A

28. For installations of resistors and reactors, a thermal barrier shall be required if the space between them and any combustible material is less than _____
    (a) 2 in.   (b) 3 in.   (c) 6 in.   (d) 12 in.

29. For straight pulls, the length of a pull box shall not be less than _____ times the outside diameter, over sheath, of the largest conductor or cable entering the box on systems over 600V.
    (a) 18   (b) 16   (c) 36   (d) 48

30. Fuseholders for cartridge fuses shall be so designed that it is difficult to put a fuse of any given class into a fuseholder that is designed for a lower _____ or a higher _____ than that of the class to which the fuse belongs.
    (a) voltage, wattage   (b) wattage, voltage
    (c) voltage, current   (d) current, voltage

31. General-use branch circuits using flat conductor cable shall not exceed _____.
    (a) 15A   (b) 20A   (c) 30A   (d) 40A

32. Ground-fault circuit-interrupter (GFCI) protection in a mobile home is required for _____.
    (a) receptacle outlets installed outdoors
    (b) receptacles within 6 ft of a wet bar sink
    (c) all receptacles in bathrooms including receptacles in luminaires (light fixtures)
    (d) all of these

33. Ground-fault protection at service equipment may make it necessary to review the overall wiring system for proper selective overcurrent protection _____.
    (a) rating   (b) coordination   (c) devices   (d) none of these

34. Ground-fault protection for personnel is required for receptacles rated more than 30A or other than 125V. The ground-fault protection shall be supplied by _____.
    (a) GFCI protection device
    (b) Assured Equipment Grounding Conductor Program
    (c) AFCI protection device
    (d) a or b

35. Ground-fault protection of equipment shall be provided for solidly grounded wye electrical services of more than 150 volts-to-ground, but not exceeding 600V phase-to-phase for each service disconnecting means rated _____ or more.
    (a) 1,000A   (b) 1,500A   (c) 2,000A   (d) 2,500A

36. Ground-fault protection of equipment shall be provided in accordance with the provisions of 230.95 for solidly-grounded wye electrical systems of more than 150 volts-to-ground, but not exceeding 600V phase-to-phase for each individual device used as a building or structure main disconnecting means rated _____, or more.
    (a) 1,000A   (b) 1,500A   (c) 2,000A   (d) 2,500A

37. Grounding and bonding conductors shall not be connected by _____.
    (a) pressure connections   (b) solder
    (c) lugs   (d) approved clamps

38. Grounding-type attachment plugs shall be used only with a cord having a(n) _____ conductor.
    (a) equipment grounding   (b) isolated
    (c) computer circuit   (d) insulated

39. Hallways in dwelling units that are _____ long or longer require a receptacle outlet.
    (a) 12 ft   (b) 10 ft   (c) 8 ft   (d) 15 ft

40. Hazardous locations are classified based on the properties of the _____ that may be present and the likelihood that a flammable or combustible concentration or quantity is present.
    (a) flammable vapors   (b) flammable gases or liquids
    (c) combustible dusts or fibers   (d) all of these

41. Horizontal runs of RMC supported by openings through _____ at intervals not exceeding 10 ft and securely fastened within 3 ft of termination points shall be permitted.
    (a) walls   (b) trusses   (c) rafters   (d) framing members

42. HPD cord is permitted for _____ usage.
(a) not hard   (b) hard   (c) extra hard   (d) all of these

43. If a dwelling unit is served by 1Ø, 3-wire, 120/240V service-entrance or feeder conductors with an ampacity of _____ or greater, it shall be permissible to compute the feeder and service loads in accordance with 220.30 instead of the method specified in Part II of Article 220.
(a) 100   (b) 125   (c) 150   (d) 175

44. IMC can be installed in or under cinder fill that is subjected to permanent moisture _____.
(a) where the conduit is not less than 18 in. under the fill
(b) when protected on all sides by 2 in. of concrete
(c) where protected by corrosion protection judged suitable
(d) any of these

45. IMC shall be firmly fastened within _____ of each outlet box, junction box, device box, fitting, cabinet or other conduit termination.
(a) 12 in.   (b) 18 in.   (c) 2 ft   (d) 3 ft

46. In a commercial garage over a Class I location, equipment less than _____ above the floor level that may produce arcs, sparks or particles of hot metal, shall be of the totally enclosed type or constructed so as to prevent the escape of sparks or hot metal particles.
(a) 6 ft   (b) 10 ft   (c) 12v   (d) 18 ft

47. In a concealed FMC installation, _____ connectors shall not be used.
(a) straight   (b) angle   (c) grounding type   (d) none of these

48. In cellular concrete floor raceways, a cell is defined as a single, enclosed _____ space in a floor made of precast cellular concrete slabs, the direction of the cell being parallel to the direction of the floor member.
(a) circular   (b) oval   (c) tubular   (d) hexagonal

49. In Class _____ locations for Groups A, B, C and D, the classification involves determinations of maximum explosion pressure, maximum safe clearance between parts of a clamped joint in an enclosure and the minimum ignition temperature of the atmospheric mixture.
(a) I   (b) II   (c) III   (d) all of these

50. In communication circuits, all separate electrodes can be bonded together using a minimum jumper size of _____ copper.
(a) 10 AWG   (b) 8 AWG   (c) 6 AWG   (d) 4 AWG

51. In electrical nonmetallic tubing, the maximum number of bends between pull points cannot exceed _____ degrees including any offsets.
(a) 320   (b) 270   (c) 360   (d) unlimited

52. In general, the minimum size phase, neutral or grounded conductor permitted for use in parallel installations is _____.
(a) 10 AWG   (b) 1 AWG   (c) 1/0 AWG   (d) 4 AWG

53. In network-powered broadband communications systems, all separate electrodes can be bonded together using a minimum size jumper of _____ copper.
(a) 10 AWG   (b) 8 AWG   (c) 6 AWG   (d) 4 AWG

54. In order to _____ NUCC the conduit shall be trimmed away from the conductors or cables using an approved method that will not damage the conductor or cable insulation or jacket.
(a) facilitate installing
(b) enhance the appearance of the installation of
(c) terminate
(d) provide safety to the persons installing

55. In reference to mobile/manufactured homes, examples of portable appliances could be _____, but only if these appliances are not built-in.
    (a) refrigerators   (b) range equipment   (c) clothes washers   (d) all of these

56. In walls constructed of wood or other _____ material, electrical cabinets shall be flush with the finished surface or project therefrom.
    (a) nonconductive   (b) porous
    (c) fibrous   (d) combustible

57. Junction boxes for pool lighting shall not be located less than _____ from the inside wall of a pool unless separated by a fence or wall.
    (a) 3 ft   (b) 4 ft   (c) 6 ft   (d) 8 ft

58. Lettering on conductor insulation indicates intended condition of use. THWN is rated _____.
    (a) 75°C   (b) for wet locations
    (c) a and b   (d) not enough information

59. Lighting track is a manufactured assembly and its length may not be altered by the addition or subtraction of sections of track.
    (a) True   (b) False

60. Lighting track is a manufactured assembly designed to support and _____ luminaires that are capable of being readily repositioned on the track.
    (a) connect   (b) protect   (c) energize   (d) all of these

61. Listed boxes designed for underground installation can be directly buried when covered by _____.
    (a) concrete   (b) gravel
    (c) noncohesive granulated soil   (d) b and c

62. Luminaires designed for end-to-end connection to form a continuous assembly, or luminaires connected together by recognized wiring methods, shall be permitted to contain the conductors of a 2-wire branch circuit or one _____ branch circuit, supplying the connected luminaires and need not be listed as a raceway.
    (a) small appliance   (b) appliance
    (c) multiwire   (d) industrial

63. Luminaires shall be supported independently of the outlet box where the weight exceeds _____
    (a) 60 lbs.   (b) 50 lbs.   (c) 40 lbs.   (d) 30 lbs.

64. Luminaires shall be wired so that the _____ of lampholders will be connected to the same fixture or circuit conductor or terminal.
    (a) conductor   (b) neutral   (c) base   (d) screw-shells

65. Luminaires shall be wired with conductors having insulation suitable for the environmental conditions and _____ to which the conductors will be subjected.
    (a) temperature   (b) voltage   (c) current   (d) all of these

66. Luminaires, lampholders and receptacles shall have no live parts normally exposed to contact, but cleat-type lampholders and receptacles located at least _____ above the floor shall be permitted to have exposed contacts.
    (a) 3   (b) 6   (c) 8   (d) none of these

67. MC cable shall be supported and secured at intervals not exceeding _____.
    (a) 3 ft   (b) 6 ft   (c) 4 ft   (d) 2 ft

68. Means shall be provided to disconnect simultaneously all _____ 1Ø supply conductors to the phase converter.
    (a) ungrounded   (b) grounded   (c) grounding   (d) all of these

69. Metal conduit and metal piping within _____ of the inside walls of the pool that are not separated from the pool by a permanent barrier are required to be bonded.
    (a) 4 ft   (b) 5 ft   (c) 8 ft   (d) 10 ft

70. Metal poles that support luminaires must meet the following requirements: _____.
    (a) They must have an accessible handhole (sized 2 × 4 in.) with a raintight cover
    (b) An accessible grounding terminal must be installed accessible from the handhole
    (c) a and b
    (d) none of these

71. Metal wireways are sheet metal troughs with _____ that are used to form a raceway system.
    (a) removable covers   (b) hinged covers
    (c) a or b   (d) none of these

72. MI cable conductors shall be made of _____ with a resistance corresponding to standard AWG and kcmil sizes.
    (a) solid copper   (b) solid or stranded copper
    (c) stranded copper   (d) solid copper or aluminum

73. Motor control circuits shall be arranged so that they will be disconnected from all sources of supply when the disconnecting means is in the open position. Where separate devices are used for the motor and control circuit, they shall be located immediately adjacent to each other.
    (a) True   (b) False

74. Motor controllers and terminals of control circuit devices are required to be connected to copper conductors unless identified as approved for use with a different type of conductor.
    (a) True   (b) False

75. Motors shall be located so that adequate _____ is provided and so that maintenance, such as lubrication of bearings and replacing of brushes, can be readily accomplished.
    (a) space   (b) ventilation   (c) protection   (d) all of these

76. Network-powered broadband communications system cables shall be separated at least 2 in. from conductors of _____ circuits.
    (a) power   (b) electric light   (c) Class 1   (d) any of these

77. NM cable shall be secured in place within _____ of every cabinet, box or fitting.
    (a) 6 in.   (b) 10 in.   (c) 12 in.   (d) 18 in.

78. NM cables run horizontally through framing are considered supported and secured where such support does not exceed $4^1/_2$ ft intervals and the NM cable is securely fastened in place by an approved means within 12 in. of each box, cabinet, conduit body or other NM cable termination.
    (a) True   (b) False

79. No _____ splices or taps shall be made within or on a luminaire (fixture).
    (a) unapproved   (b) untested   (c) uninspected   (d) unnecessary

80. No seal is required if a conduit (with no unions, couplings, boxes or fittings) passes through a Class I, Division 2 location if the termination points of the unbroken conduit are in at least _____ within the unclassified location.
    (a) 6 in.   (b) 12 in.   (c) 18 in.   (d) 24 in.

81. Noninsulated busbars shall have a minimum space of _____ between the bottom of enclosure and busbar.
    (a) 6 in.   (b) 8 in.   (c) 10 in.   (d) 12 in.

82. Nonmetallic extensions shall be secured in place by approved means at intervals not exceeding _____
    (a) 6 in.   (b) 8 in.   (c) 10 in.   (d) 16 in.

83. NUCC larger than _____ shall not be used.
    (a) 1 in.   (b) 2 in.   (c) 3 in.   (d) 4 in.

84. NUCC shall be capable of being supplied on reels without damage or _____ and shall be of sufficient strength to withstand abuse, such as impact or crushing in handling and during installation, without damage to conduit or conductors.
    (a) distortion   (b) breakage   (c) shattering   (d) all of these

85. On a 4-wire, 3Ø wye circuit, where the major portion of the load consists of electrical discharge lighting, data-processing equipment or other harmonic current inducting loads, the grounded conductor shall be counted when applying 310.15(B)(2) adjustment factors.
    (a) True   (b) False

86. One conductor of flexible cords intended to be used as a(n) _____ circuit conductor shall have a continuous marker readily distinguishing it from the other conductor or conductors.
    (a) grounded   (b) equipment   (c) ungrounded   (d) all of these

87. One wiring method that is permitted in ducts or plenums used for environmental air is _____.
    (a) flexible metal conduit of any length
    (b) electrical metallic tubing
    (c) armored cable (BX)
    (d) nonmetallic-sheathed cable

88. One-inch IMC must be supported every _____.
    (a) 8 ft   (b) 10 ft   (c) 12 ft   (d) 14 ft

89. One-inch RNC must be supported every _____.
    (a) 2 ft   (b) 3 ft   (c) 4 ft   (d) 6 ft

90. Open conductors entering or leaving locations subject to dampness, wetness or corrosive vapors shall have _____ formed on them and shall then pass upward and inward from the outside of the buildings, or from the damp, wet or corrosive location, through noncombustible, nonabsorbent insulating tubes.
    (a) weather heads   (b) drip loops
    (c) identification   (d) blisters

91. Openings in ventilated dry-type _____ or similar openings in other equipment over 600V shall be designed so that foreign objects inserted through these openings will be deflected from energized parts.
    (a) lampholders   (b) motors
    (c) fuseholders   (d) transformers

92. Overcurrent protection devices are not permitted to be located _____.
    (a) where exposed to physical damage
    (b) near easily ignitable materials, such as in clothes closets
    (c) in bathrooms of dwelling units
    (d) all of these

93. Overhead conductors installed over roofs shall have a vertical clearance of _____ above the roof surface.
    (a) 8 ft   (b) 12 ft   (c) 15 ft   (d) 3 ft

94. Overhead utility service conductors that operate at not over 600V must maintain a _____ clearance in any direction to the water level, edge of water surface, base of diving platform or permanently anchored raft.
    (a) 14 ft   (b) 16 ft   (c) 20 ft   (d) 22$^1/_2$ ft

95. Overhead-service conductors from the last pole or other aerial support to and including the splices, if any, are called _____ conductors.
    (a) service-entrance   (b) service-drop
    (c) service   (d) overhead service

96. Parking garages used for parking or storage and where no repair work is done except for exchange of parts and routine maintenance requiring no use of electrical equipment, open flame, welding or the use of volatile flammable liquids, are not classified as hazardous locations.
    (a) True   (b) False

97. Plug fuses of 15A or less shall be identified by a(n) _____ configuration of the window, cap or other prominent part to distinguish them from fuses of higher ampere ratings.
    (a) octagonal   (b) rectangular   (c) hexagonal   (d) triangular

98. Plug fuses of the Edison-base type have a maximum rating of _____.
     (a) 20A   (b) 30A   (c) 40A   (d) 50A

99. Portable lamps shall be wired with _____ recognized by 400.4 and have an attachment plug of the polarized or grounding type.
     (a) flexible cable   (b) flexible cord
     (c) nonmetallic flexible cable   (d) nonmetallic flexible cord

100. Power-limited fire alarm cables installed as wiring within buildings shall be _____ as being resistant to the spread of fire.
     (a) labeled   (b) listed   (c) identified   (d) marked